Ernst-Christian Koch

High Explosives, Propellants, Pyrotechnics

Also of interest

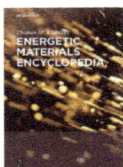

Energetic Materials Encyclopedia
Thomas M. Klapötke, 2018
ISBN 978-3-11-044139-0, e-ISBN 978-3-11-044292-2

Energetic Compounds.
Methods for Prediction of their Performance
Mohammad Hossein Keshavarz und Thomas M. Klapötke, 2020
ISBN 978-3-11-067764-5, e-ISBN 978-3-11-067765-2

Chemistry of High-Energy Materials
Thomas M. Klapötke, 2019
ISBN 978-3-11-062438-0, e-ISBN 978-3-11-062457-1

Combustible Organic Materials.
Determination and Prediction of Combustion Properties
Mohammad Hossein Keshavarz, 2018
ISBN 978-3-11-057220-9, e-ISBN 978-3-11-057222-3

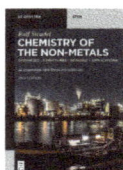

Chemistry of the Non-Metals.
Syntheses – Structures – Bonding – Applications
Ralf Steudel, 2nd Ed., 2020
ISBN 978-3-11-057805-8, e-ISBN 978-3-11-057806-5

Ernst-Christian Koch

High Explosives Propellants Pyrotechnics

—

DE GRUYTER

Author
Dr. Ernst-Christian Koch
Lutra*dyn*-Energetic Materials Science & Technology
Burgherrenstraße 132
67661 Kaiserslautern
Germany
e-c.koch@lutradyn.com

ISBN 978-3-11-066052-4
e-ISBN (PDF) 978-3-11-066056-2
e-ISBN (EPUB) 978-3-11-066059-3

Library of Congress Control Number: 2020941878

Bibliographic information published by the Deutsche Nationalbibliothek
The Deutsche Nationalbibliothek lists this publication in the Deutsche Nationalbibliografie;
detailed bibliographic data are available on the Internet at http://dnb.dnb.de.

Cover image: The cover shows a true color photograph (aperture: 16, exposure time: 1/8000 s) of
the combustion flame of ytterbium/polytetrafluorethylene. Credits: E.-C.-Koch, 1. October 2010.
Typesetting: Integra Software Services Pvt. Ltd.
Printing and binding: CPI books GmbH, Leck

www.degruyter.com

Dedicated to my family

Foreword to the English Edition

The author of this book, Dr. Ernst-Christian Koch, is a leading scientist in the area of energetic materials. This book is the translation and revised, extended, and updated version of the German edition *Sprengstoffe, Treibmittel, Pyrotechnika* (De Gruyter, 2nd edn, 2019) by the same author. With over 40 original publications, three books, two book chapters, and 15 MSIAC reports, he now shares some of his knowledge in this book. The book contains 739 entries on 708 pages in alphabetical order from AAD to Zirconium-nickel alloy. This is followed by a comprehensive "Table of Contents" and an extensive Appendix on PBX and various German and US explosive formulations. The book summarizes physical–chemical properties and sensitivities of explosives, explosive formulations, oxidizers, fuels, propellant binders, and pyrotechnics. In addition, many civil and military applications are discussed. Over 100 figures (many of them in color), 160 tables, and over 200 structural drawings elucidate the sometime complex concepts. In addition, the book also includes some novel materials such as DBX-1 and TKX-50.

The book's main emphasis is on high explosives and pyrotechnics, and within this field a very wide range of materials, formulations, and tests are summarized. Generally, all ingredients are described meticulously with the REACH classification and the CAS number. Important physical and safety data of the raw materials are given with references and, if applicable, properties such as glass transition temperatures are given. In addition, the author gives short but precise explanations for complex phenomena like erosive burning and detonation with formulae and more than 1200 citations from original papers. A few short biographies of leading scientists who have worked in the field complete this encyclopedia.

This is a must for the bookshelf of both the expert and novice entering the area of energetic materials.

West Lafayette, Purdue University, July 2020
Professor Steven F. Son, Alfred J. McAllister
Professor of Mechanical Engineering
Munich, LMU, July 2020
Professor Dr. Thomas M. Klapötke

https://doi.org/10.1515/9783110660562-202

Foreword to the First German Edition

Having worked with explosives for many years, it became obvious to me that this field has plenty of facets. There are chemical problems to solve, just like physical ones but also legal issues to consider. To meet with all of them the author, who has acquired his knowledge through working for many years in both industry and NATO panels with great expertise, has now meticulously prepared an encyclopedic overview with more than 500 entries. The individual entries are precise and rich in content as expected for an encyclopedia. Extensive references allow the reader a deeper research. The book is useful for the freshman looking to learn about the field. Also, the experienced user is able to find information on side topics of our field. This book should be in inventory of every library of the explosives industry and the corresponding institutes.

Swisttal-Heimerzheim
November 2017

Dipl.-Phys. Roland Wild,
Regierungsdirektor a. D.

https://doi.org/10.1515/9783110660562-203

Preface and Acknowledgment to First English Edition

Already when the first German edition appeared in January 2018, there was a request for an international, that is, an English version of this encyclopedia. Now I am glad to submit the first English version which is based on the second revised and enlarged German version and is again revised and expanded over the German Edition by 55 new additional entries and now over 1400 references to the primary literature. As a courtesy for the English readership, there are the German expressions after each entry header to enable and encourage access and research of the corresponding German literature.

While all statements made in previous editions are still valid, I would now also like to expand my thanks to **Mrs. Ute Skambraks** (Content Editor STEM, DG), **Mrs. Jana Habermann** (Production Manager, DG) and **Mr. David Jüngst** (Project Manager, Integra) for the excellent cooperation upon development, production, and marketing of this book.

I very much appreciate **Prof. Dr. Thomas M. Klapötke** for his revision of the second German version and for providing a list of errors as well as his and **Prof. Steve Son's** effort to write a foreword for this first English version.

Also, I very much appreciate **Dr. Werner Arnold** (Ingolstadt) for revising all entries relating to detonation and terminal ballistics.

Now I wish you an inspiring read and satisfying finds to support your work and very much looking forward to receive your comments on potential errors and necessary amendments for future editions of this encyclopedia.

Kaiserslautern, November 2020   *Ernst-Christian Koch*

https://doi.org/10.1515/9783110660562-204

Preface and Acknowledgment
to Second German Edition

The first edition of this book found a ready market. It generated encouraging feedback from the professional readership including the demand for an electronic version which led to the second edition.

I gratefully acknowledge **Dr. Karin Sora**, Vice President Science, Technology, Engineering & Mathematics (STEM), of De Gruyter (DG) for her interest to produce and publish this book.

I gratefully acknowledge **Dr. Carina Kniep**, formerly Editorial Director Physical Sciences/DG, for her valuable coordination in the very early phase of this book.

I thank very much **Mrs. Lena Stoll**, Project Editor Chemistry & Materials Science/DG and **Dr. Ria Fritz**, Project Editor Chemistry/DG, for the excellent cooperation upon development, production, and marketing of this book

I gratefully acknowledge **Mrs. Jeannette Krause**, reader and production manager/le-tex publishing services GmbH, for her always on-time and reliable communication, the careful, precise, and convincing implementation of a difficult manuscript.

This second and revised edition is expanded with 98 new and additional entries, and many new references and tables. In addition, it is now also available as an electronic version (E-Book).

Again, I am indebted to many peers for their comments and corrections.

In particular, I would like to thank

Dr. Werner Arnold, Ingolstadt, for his critical revision, correction, and expansion of many entries from the field of detonics and terminal ballistics;

Dr. Kurt Schubert, Gunzenhausen, for his painstaking revision of the first edition that led him find technical and orthographical errors;

Dr. Paul Wanninger, Adelsberg, for his scrupulous revision and recension of the first edition and many hints and comments on errors and missing information and for providing extensive information on the technology of plastic-bonded explosives developed in Germany.

Now with this book I wish you successful research and valuable momentum for your work. Furthermore, I would be very pleased to receive your feedback on errors and necessary amendments for any future edition of this book.

Kaiserslautern, April 2019   *Ernst-Christian Koch*

https://doi.org/10.1515/9783110660562-205

Preface and Acknowledgment to First German Edition

Our current world would be inconceivable without explosives. In munitions, propellants, and pyrotechnics they serve the protection and defense. However, in wrong possession ammunition and explosives can cause chaos and annihilation.

Less controversial is the use of explosives in the development of our *human habitats*. Roads, tunnels, and basically any construction operation on rocky ground are unimaginable without the use of explosives as are demolition or mining operations above, underground, and offshore. Explosives help to easily forge metal stock and weld-complicated metal combinations that would otherwise not join. Almost forgotten is the massive use of explosive rivets in the manufacture of aircrafts in the mid-twentieth century. Gas generators in nearly each automobile today protect passengers and have saved countless lives and have prevented severe injuries. Flares and smoke serve as distress signals. Fireworks crest festive and holidays. Without the plentiful energetic materials in chemical propulsion, actuators, separating bolts, and cutting cords the exploration of space would be impossible. Finally, the use of explosives facilitates the production of new materials (SHS, detonation synthesis) that are inaccessible by other means.

A German encyclopedia for high explosives, propellants, and pyrotechnics, offering direct access to the primary literature was missing so far and is presented in this book. On 460 pages, there are 590 entries with 133 tables, 82 figures and more than 200 structural formulas, and nearly 900 references.

The selection of entries naturally is subjective and is based on my current and previous field of work.

A book claiming to be an encyclopedia can and must not be the work of a single individual. It is hence that I am thankful to many current and former colleagues for hints, information, and comments.

Representative for many, I would like to thank physicist **Mr. Roland Wild**, retired deputy director of the Heimerzheim branch of the Wehrwissenschaftliches Institut für Werk- Explosiv- und Betriebstoffe, for reviewing the manuscript, for his many valuable comments, and for his efforts to write a foreword for this book.

I now wish you happy reads and lookups and would be very glad to receive your comments on errors and necessary amendments for future editions.

Kaiserslautern, November 2017 *Ernst-Christian Koch*

https://doi.org/10.1515/9783110660562-206

Abbreviations

α	Angle, °
α	Co-volume, m^3
α	Thermal expansion coefficient, K^{-1}
α_λ	Spectral mass extinction coefficient, $m^2\ g^{-1}$
γ	Isentropic exponent, –
Δ	Loading density, $g\ cm^{-3}$
$\Delta_c H$	Heat of combustion (in oxygen), $kJ\ mol^{-1}$
$\Delta_{det} H$	Heat of detonation, $kJ\ mol^{-1}$
$\Delta_{diss} H$	Heat of dissociation, $kJ\ mol^{-1}$
$\Delta_{ex} H$	Heat of explosion, $kJ\ mol^{-1}$
$\Delta_{melt} H$	Heat of fusion, $kJ\ mol^{-1}$
$\Delta_{vap} H$	Heat of vaporization, $kJ\ mol^{-1}$
$\Delta_f H$	Heat of formation, $kJ\ mol^{-1}$
ρ	Density, $g\ cm^{-3}$
ϕ	Kamlet parameter
Φ	Fluorine content, wt.-%
$\mathfrak{H}$	Hugoniot
Λ	Oxygen balance, wt.-%
μ	Fraction, –
*	Stereo center in a chemical structure
a	Temperature coefficient of Vieille's law, $mm\ s^{-1}\ MPa^{-1}$
A	Vivacity
A	Empirical constant, 3.9712 (*Kamlet*)
ADN	Ammonium dinitramide, $NH_4N_3O_4$
A_E	Activation energy, $kJ\ mol^{-1}$
A_N	Aerosol yield, $g\ g^{-1}$
AN	Ammonium nitrate, NH_4NO_3
ANFO	Ammonium nitrate fuel oil (explosive)
AP	Ammonium perchlorate, NH_4ClO_4
AZM	Ignition mixture (Ger.: *Anzündmischung*)
B	Empirical constant, 1.30 (*Kamlet*)
B_λ	Illuminating power, lx
BAM	Bundesanstalt für Materialprüfung und Forschung
BET	Specific surface area i.a.w. Brunauer–Emmet–Teller, $m^2\ g^{-1}$
Bp	Boiling point, °C
C	Mass of high explosive (*Gurney* model)
CAS	Chemical Abstracts Service registration number
cd	candela (luminous intensity), 10^3 cd = 1 kcd
c_L	Speed of sound, longitudinal, $m\ s^{-1}$
CL	China Lake
CMDB	Composite modified double base
c_p	Specific heat, $J\ K^{-1}\ mol^{-1}$
DBP	Dibutyl phthalate, $C_{16}H_{22}O_4$, [84-74-2]
DDT	Deflagration to detonation transition
dec	Decomposition
Dp	Decomposition point, °C
DM	German Ammunition Code (Ger: *Deutsches Muster*)

https://doi.org/10.1515/9783110660562-207

DMF	Dimethyl formamide, C_3H_7NO
DMSO	Dimethyl sulfoxide, C_2H_6SO
DNDA	Dinitrodiaza compounds
DOA	Dioctyl adipate, $C_{24}H_{38}O_4$
DOS	Dioctyl sebacate, $C_{26}H_{50}O_4$
DRH	Deliquescence relative humidity, % RH
DSC	Differential scanning calorimetry
d_p	Particle diameter, μm
E_a	Activation energy, kJ mol^{-1}
E_λ	Specific (light) intensity, cd s g^{-1} (VIS) or J g^{-1} sr^{-1} (IR)
EI(D)S	Extremely insensitive (detonating) substances
EILS	Energetic ionic liquids
EINECS	EG compound inventory number
ESD	Electrostatic discharge
EVA	Ethylvinyl acetate, $(C_{24}H_{44}O_4)_n$
f	Impetus (force), J g^{-1}
Friction	Friction force, N
FT	Fourier transform
FTS	Solid propellant (Ger.: *Festtreibstoff*)
G	Gibbs free enthalpy
GHS	Global harmonized system
h	Height over ground, m
H	Enthalpy
HC	Hexachlorethane, C_2Cl_6
HE	High explosive
H_i	Autoignition enthalpy, kJ
HMX	Octogen, $C_4H_8N_8O_8$
H-phrases	Hazard notes i.a.w. GHS
HOF	Hypofluoric acid, H-O-F
HPC	Hydroxypropylcellulose
HTPB	Hydroxy-terminated polybutadiene
HVD	High-velocity detonation
ICT	Fraunhofer Institut für Chemische Technologie
ICt_{50}	Mean incapacitating concentration, mg min^{-1} m^{-3}
IHE	Insensitive High Explosive
I_λ	Light intensity, cd
Impact	Impact energy, J
I_{sp}	Specific impulse, Ns kg^{-1}
IR	Infrared
κ	Heat conductivity, W K^{-1} m^{-1}
K	Empirical constant, 240.86 (*Kamlet*)
L	Latent heat, kJ mol^{-1}
LANL	Los Alamos National Laboratory
LCt_{50}	Concentration leading after exposure of 1 min to a 50% mortality, mg m^{-3}
LLNL	Lawrence Livermore National Laboratory
LOI	Limiting oxygen index, vol.-% O_2
LSGT	Large-scale gap test
LVD	Low-velocity detonation
M, m	Mass, kg
M	Mean molecular mass, g mol^{-1}
M	Mass of accelerated metal casing (*Gurney* model)

M	Mass of gaseous products per molar amount of gaseous products (*Kamlet*)
Mk	Mark
MMW	Millimetric wave
Mp	Melting point, °C
m_r	Molar mass, g mol^{-1}
m_z	Mass of igniter, kg
N	Static countermass (*Gurney* model)
N	Nitrogen content, wt.-%
N	Molar amount of gaseous products per gram explosive (*Kamlet*)
n	*Vielle's law* pressure exponent, –
NC	Nitrocellulose, variable composition
NEW	Net Explosive Weight, kg
NGl	Nitroglycerine, $C_3H_5N_3O_9$
NGu	Nitroguanidine, $CH_4N_4O_2$
NIR	Near infrared, $\lambda = 0.8-1.5$ μm
NOL	Naval Ordnance Laboratory
NTO	Nitrotriazolone, $C_2H_2N_4O_3$
p, P	Pressure, MPa
PBX	Plastic-bonded explosive
P_{CJ}	Chapman–Jouguet pressure, GPa
PETN	Nitropenta
PIB	Polyisobutylene, $(C_4H_8)_n$
P_{NoGo}	Limiting initiation pressure, GPa
P-phrases	Precautionary advice i.a.w. GHS
ppm	Parts per million
PT	Pyrotechnic composition
P_R	Red phosphorus
q	Heat of reaction, kJ g^{-1}
Q_{ex}	Heat of explosion, kJ g^{-1}
r	Radius, m
R_0	Universal gas constant
RDX	Hexogen, $C_3H_6N_6O_6$
REACH	Indicates the level of registration with ECHA: LRS, LPRS CL-SVHC, SVHC
ST	Standard temperature, 20 °C
S	Entropy
S	Surface area, m^2
SR	Superintendent research
SSGT	Small-scale gap test
t	Time, s
T	Temperature, K
T_{ad}	Adiabatic flame temperature, K/°C
TEQ	Toxicity equivalent, ng (10^{-12} g)
T_{ex5}	5 s autoignition temperature, °C
T_i	Ignition temperature, °C
TMD	Theoretical maximum density, g cm^{-3}
TNT	Trinitrotoluene, $C_7H_5N_3O_6$
T_p	Phase transfer temperature, °C
Trauzl	Lead block expansion after Trauzl, cm^3
u	Burn rate, m s^{-1}
U	Internal energy
UN	United Nations

V_D, w	Detonation velocity, $m\,s^{-1}$; $mm\,\mu s^{-1}$
v	Specific volume, $cm^3\,g^{-1}$
V_m	Free volume of rocket motor, m^3
V/V_o	Cylinder expansion, –
VIS	Visual range, $\lambda = 380$–780 nm
w	Wave velocity, $m\,s^{-1}$
x	Coordinate, m
Y_f	Yield factor, –
z	Scaled distance, $z = \frac{r}{\sqrt[3]{M}}$
$\sqrt{2E_g}$	Gurney constant, $mm\,\mu s^{-1}$
$\varnothing_{cr}$	Critical diameter, mm
$\varnothing$	Caliber, mm

Indices

ad	Adiabatic
calc	Calculated
CJ	With respect to Chapman–Jouguet condition
exp	Experimental
h	Hexagonal
k	Cubic
mnkl	Monoclin
orh	Orthorhombic
p	At constant pressure
r	Rhombohedral
s	Surface
tg	Trigonal
v	At constant volume
∞	Ambient
50	With respect to 50 % of all tests
nr	no reaction

Remarks

All titles, if they are not proper names, or foreign terms adopted in English, or read identical in German.

Title

are followed by the German translation in italic font.

Überschrift

In the text, occasionally individuals, proper names, and other entries in this encyclopedia are highlighted in *italic font*.

If protected and registered trade names are not designated as such in the text, this does not mean they can be freely used.

If there is no line with regard to "aspect" of a substance, then this substance is either colorless or the color is not known.

All heats of explosion and combustion relate to $H_2O_{(l)}$ as final product.

Values in *italic fonts* indicate calculated values.

GHS codes and description of hazards*

01 Explosives
02 Flammable
03 Oxidizing
04 Gases
05 Corrosive
06 Toxic
07 Irritant
08 Harmful
09 Environmental danger

---

*http://www.unece.org/trans/danger/publi/ghs/ghs_welcome_e.html

https://doi.org/10.1515/9783110660562-208

REACH codes*

	Substance registered in
LPRS	List of Preregistered Substances
LRS	List of Registered Substances
SVHC	Substance of Very High Concern

---

*https://echa.europa.eu/regulations/reach/understanding-reach

Legal Notice

Explosive materials (high explosives, propellants, pyrotechnics) are extremely dangerous and should only be handled and prepared by persons trained and skilled in this area and who have obtained federal licenses to do so in the appropriate facilities approved by the proper authorities (e.g. regulatory authorities) in accordance with the corresponding national legislation.

This book and the information contained therein have been carefully prepared. However, neither De Gruyter nor the author and other contributors to this book can be held liable in any way for any damage to personal or property resulting from the use or misuse of the information contained in this book.

Contents

A

AAD

AAD designates a eutectic melt-cast formulation based on ammonium nitrate (AN), 3-amino-1,2,4-triazolium nitrate, and 3,5-diamino-1,2,4-triazolium nitrate. ADD melts in the 90–99 °C range and is a shear-thinning fluid just like 2,4,6-trinitrotoluene (TNT). It has a relative deliquescence humidity (DRH) of >50%. A formulation containing 50 wt.-% HMX is called AH-55 and compares nicely with Comp B; however, it is less sensitive when compared with Comp B (Table A.1).

Table A.1: Composition and properties of AAD melt-cast high explosives.

Components	AAD	AH-55
Ammonium nitrate, NH_4NO_3 (wt.-%)	50	25
3-Amino-1,2,4-triazolium nitrate, $[C_2H_5N_4][NO_3]$(wt.-%)	25	12,5
3,5-Diamino-1,2,4-triazolium nitrate, $[C_2H_6N_5][NO_3]$ (wt.-%)	25	12,5
Octogen, HMX, $C_4H_8N_8O_8$, (wt.-%)	–	50[#]
Mp (°C)	104 (DSC)	
ρ (g cm^{-3})	1.66	1.70 (96% TMD)
V_D (km s^{-1})	8.55*	7.97 @ 24 mm
P_{CJ} (GPa)	26*	27.3 @ 24 mm
Onset (°C)	203	203
Impact (J)	21	9.16
Friction (N)	>360	>360
Spark	0.25	0.125

*Calculated with CHEETAH 2.0, [#]class II HMX.

– P. W. Leonard, D. E. Chavez, P. R. Bowden, E. G. Francois, Nitrate Salt Based Melt Cast Materials, *Propellants, Explos., Pyrotech.* **2018**, *43*, 11–14.

AASTP

The AASTP = *Allied Ammunition Storage and Transport Publication* is a series of documents (Table A.2) developed and issued by the NATO Conference of National Armament Directors (CNAD) Ammunition Safety Group (AC/326). The AASTP currently comprises five documents, dealing with the basics of storage and transportation of explosives and ammunition in the NATO military environment.

https://doi.org/10.1515/9783110660562-001

Table A.2: AASTP.

AASTP-1	Manual of NATO Safety Principles for the Storage of Military Ammunition and Explosives
AASTP-2	Manual of NATO Safety Principles for the Transport of Military Ammunition and Explosives
AASTP-3	Manual of NATO Safety Principles for the Hazard Classification of Military Ammunition and Explosives
AASTP-4	Manual of Explosives Safety Risk Analysis
AASTP-5	NATO Guidelines for the Storage, Maintenance and Transport of Ammunition on Deployed Missions or Operations

Abel test

The Abel test is the oldest method to probe the stability of nitrate ester containing energetic materials. While a sample is heated to 65.5 °C or 82.2 °C, the time taken for effective blueing of potassium iodide-starch paper via release of iodine is measured. The Abel test is among the mandatory tests for *BuNENA* and is described in STANAG 4583.

The reactions occurring in the aqueous phase of the KI-starch paper can be summarized as follows:

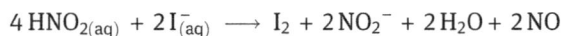

$$4\,HNO_{2(aq)} + 2\,I^-_{(aq)} \longrightarrow I_2 + 2\,NO_2^- + 2\,H_2O + 2\,NO$$

The released iodine reacts with starch to give an intense violet complex.

– *Chemical test procedures and Requirements for n-Butyl-2-nitratoethyl nitramine (n-Butyl NENA)*, STANAG 4583, Ed. 1, 18 June **2007**, NATO Standardization Agency, Brussels.

Abel, Sir Frederick Augustus (1827–1902)

Abel (Figure A.1) was born in London and worked at the Royal Military Academy starting in 1853. He was the first to succeed in preparing a stable nitrocellulose (NC) by boiling and milling the freshly nitrated cellulose fibers. He also discovered how to stabilize NC with diphenylamine, developed a gun propellant (*Cordite*), a method for testing the stability of gun propellants (*Abel test*), and introduced the ignition temperature of explosives as a general safety property.

Figure A.1: Sir Frederick Augustus Abel.

– https://en.wikipedia.org/wiki/Frederick_Abel#/media/File:Frederick_Augustus_Abel2.jpg.

Abel equation
Abel-Gleichung

The maximum gas pressure, p_{max}, reached at the end of a burn in a ballistic bomb is given by

$$p = \frac{m}{V - \alpha \cdot m}_{max}$$

$$p = \frac{\Delta}{1 - \alpha}_{max}$$

with the co-volume of the combustion gases, α; the force, f; the mass of the propellant, m; and the volume of the propellant, V (or the loading density $\Delta = m/V$). To determine f and α experimentally the powder under investigation is tested at different loading densities, Δ. Abel equation diagram with reciprocal maximum pressure versus reciprocal loading density yields a straight line with the slope correlating with the force, f via $\tan \gamma = f^{-1}$ and the intercept line giving the co-volume, α (Figure A.2).

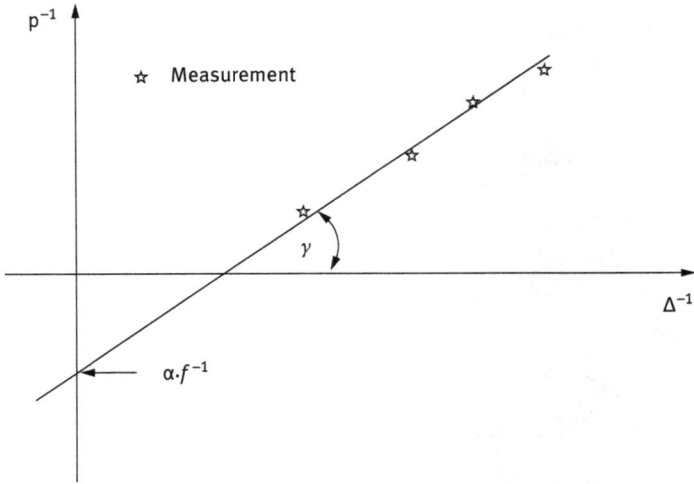

Figure A.2: Reciprocal maximum pressure versus reciprocal loading density.

– D. Grune, Studies of the Combustion of Solid Propellants at High Pressures, *Propellants Explos.* **1976**, *1*, 27–28.

Abietic acid

Abietinsäure, Sylvinsäure

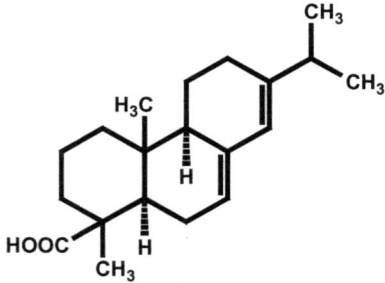

Aspect		Colorless – yellowish monoclinic plates with characteristic odor
Formula		$C_{20}H_{30}O_2$
REACH		LPRS
EINECS		208–178–3
H-phrases		315–319–335–400
P-phrases		305+351+338
CAS		[514-10-3]
m_r	$g\ mol^{-1}$	302.451
ρ	$g\ cm^{-3}$	$1.099\ g\ cm^{-3}$
Mp	°C	177.74

$\Delta_m H$	kJ mol^{-1}	19.454
Bp	°C	255 at 1.2 kPa
$\Delta_f H$	kJ mol^{-1}	−686.88
$\Delta_c H$	kJ mol^{-1}	−11.471.0
	kJ g^{-1}	−37.927
	kJ cm^{-3}	−41.681
Λ	wt.-%	−280.37

Abietic acid is the main component of colophony, a natural resin obtained by distillation of pine wood. It is soluble in acetone, alcohol, and diethyl ether but insoluble in water. Abietic acid is an important component in various binders (e.g. *B-14*) used either in pyrotechnics or as bonding agent in booster cartridges and shells. It has also been suggested as aerosol component in non-toxic smokes. Abietic acid and its oxidation products are contact allergens. Its calcium salt is known as *calcium resinate*.

- P. Groc, D. Grycza, Y. Guelou, *Smoke-Generating Composition Based on Colophony Derivatives*, US-Patent 6,436,210 B1, France, **2002**.

ABX

ABX are high-density metal containing explosive formulations based on 1,3,5-triamino-2,4,6-trinitrobenzene (TATB) (Table A.3).

Table A.3: ABX explosive compositions.

Components	B	ABX 120-1
1,3,5-Triamino-2,4,6-trinitrobenzene, TATB (wt.-%)	95	60
Silver (wt.-%)	5	3
Bismuth (wt.-%)		18
Indium (wt.-%)		19

- R. Grieggs, G. E. Turner, *ENV-RCRA-12-0083*, National Nuclear Security Administration, Los Alamos, Apr 16, **2012**.

Acardite I

Akardit I

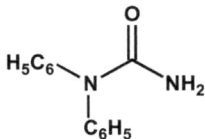

N,N-Diphenylurea		
Formula		$C_{13}H_{12}N_2O$
REACH		LPRS
EINECS		210-048-6
CAS		[603-54-3]
m_r	g mol^{-1}	212.246
ρ	g cm^{-3}	1.276
Mp	°C	189
Bp	°C	260 (dec)
$\Delta_f H$	kJ mol^{-1}	−122.7
$\Delta_c H$	kJ mol^{-1}	−6708
	kJ g^{-1}	−31.605
	kJ cm^{-3}	−40.328
Λ	wt.-%	−233.68
N	wt.-%	13.2

Acardite is used in amounts up to 0.8 wt.-% as stabilizer for NC-based gun propellants and as burn rate modifier.

Acardite II

Akardit II

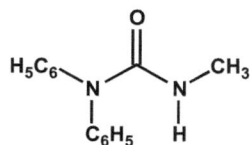

N'-Methyl-N,N-diphenylmethylurea

Formula		$C_{14}H_{14}N_2O$
REACH		LPRS
EINECS		236-039-7
CAS		[13114-72-2]
m_r	g mol^{-1}	226.278
ρ	g cm^{-3}	1.151
Mp	°C	171.2
$\Delta_f H$	kJ mol^{-1}	−106.7
$\Delta_c H$	kJ mol^{-1}	−7403.4
	kJ g^{-1}	−32.719
	kJ cm^{-3}	−37.659
Λ	wt.-%	−240.4
N	wt.-%	12.38

Acardite I serves as stabilizer and gelling agent for NC-based gun propellants.

Acardite III

Akardit III

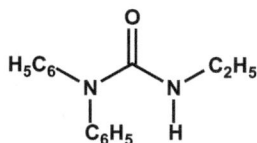

N'-Ethyl-N,N-diphenylmethylurea

Formula		$C_{15}H_{16}N_2O$
REACH		LPRS
EINECS		242-052-9
CAS		[18168-01-9]
m_r	g mol^{-1}	240.305

ρ	g cm^{-3}	1.128
Mp	°C	73.1
$\Delta_f H$	kJ mol^{-1}	−152.7
$\Delta_c H$	kJ mol^{-1}	−8036.7
	kJ g^{-1}	−33.445
	kJ cm^{-3}	−37.725
Λ	wt.-%	−246.34
N	wt.-%	11.66

Acardite III serves as a stabilizer and a gelling agent for NC-based gun propellants.

Acaroides, gum
Akaroidharz

Red gum, Yacca gum, *gummi acaroides* (Lat.), is a reddish resin which is obtained by oxidation of the sap from the Australian tree *Xanthorrhoea*. The approximate elemental composition reads $C_6H_{5.42-5.59}O_{1.15-1.66}$; CAS-No. [9000-20-8]. The heat of formation, $\Delta_f H = -450$ kJ mol^{-1}. Acaroides serves both as fuel and water-soluble binder in pyrotechnics. Between T = 70 and 110 °C, acaroides starts to release moisture. The decomposition of acaroides starts at T = 235 °C. Contaminants typically encountered in technical qualities are sand and vegetable fibers.

– W. Meyerriecks, Organic Fuels Composition and Formation Enthalpy Part II – Resins, Charcoal, Pitch, Gilsonite and Waxes, *J. Pyrotech.* **1999**, *9*, 1–19

Activation energy
Aktivierungsenergie

The activation energy, E_a (kJ), is the energy threshold (catalyzed or uncatalyzed) (Figure A.3) that has to be overcome in the course of a reaction. The sum of the bond energies of the bonds involved in a reaction limits the maximum possible activation energy (e.g. in dissociation reactions). The activation energy of explosive materials (Tables A.4–A.7) is often correlated with their sensitiveness. However, bulk properties (thermal diffusivity, crystal structure, etc.) greatly influence sensitiveness too and hence the activation energy is of only limited value to estimate the sensitivity of a material.

Energy

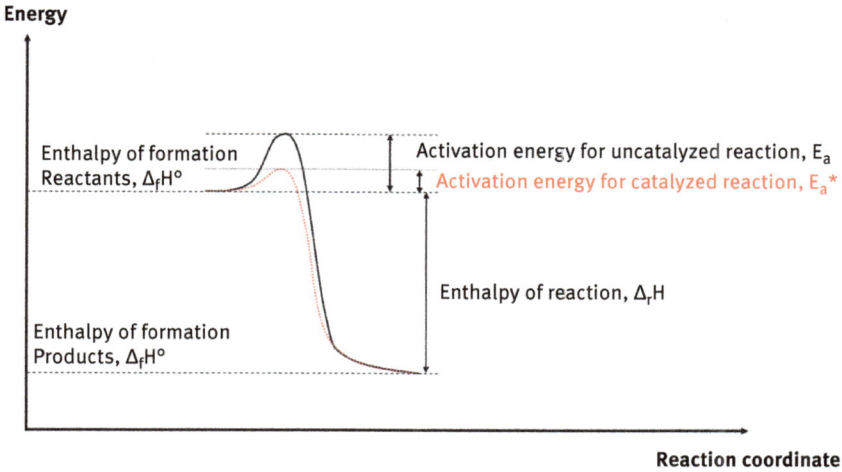

Figure A.3: Energy diagram for a chemical reaction.

Table A.4: Activation energy and BAM-impact energy for selected high explosives.

Explosive	E_a (kJ mol^{-1})	BAM-impact energy (J)
Nitroguanidine (NGu)	87.5	>50
Hexanitrostilbene (HNS)	126.8	7.5
2,4,6-Trinitrotoluene (TNT)	143.9	15
Benzotrifuroxane (BTF)	155.6	3
Dihydroxylammonium bistetrazolediolate (TKX-50)	157.9	20
Diaminotrinitrobenzene (DATB)	193.7	>50
Pentaerythritole tetranitrate (PETN)	196.6	15
Hexogen (RDX)	197.0	7.5
Octogen (HMX)	220.5	7.5
1,3,5-Triamino-2,4,6-trinitrobenzene (TATB)	250.6	>50
ε-Hexanitroisowurzitane (HNIW, CL-20)	296.6	5

Table A.5: Activation energy and BAM-impact energy for selected pyrotechnics.

Explosive	E_a (kJ mol^{-1})	BAM-impact energy (J)
TiH$_x$/KClO$_4$ (61/39)	99.8	
W-K$_2$Cr$_2$O$_7$	11	
Sb-KMnO$_4$	21	
Fe-BaO$_2$	24	
Fe-K$_2$Cr$_2$O$_7$	29–92	

Table A.5 (continued)

Explosive	E_a (kJ mol^{-1})	BAM-impact energy (J)
B/MoO$_3$	54–63	
KNO$_3$/sulfur (87.7/12.3)	59.9	
Black powder (KNO$_3$/charcoal/sulfur: 75/15/10)	61.9	
Ti/KClO$_4$ (50/50)	63	
Mg/KClO$_4$ (20/80)	66–143	
Ti/KClO$_4$/BaCrO$_4$ (38.5/38.5/23)	84	
Fe/K$_2$Cr$_2$O$_7$ (50/50)	92 ± 14	
2 K-Black powder (KNO$_3$/charcoal: 82.6/17.4)	129.4	
Mg/KNO$_3$	175 ± 5	
B-K$_2$Cr$_2$O$_7$	185	
Mg/PTFE/Viton®A(60/30/10)	188	17.8 (50%)
Zr-MoO$_2$	190–272	
Mg/NaNO$_3$/linseed oil (58/38/4) (SR-524)	256 ± 9	
Zr-MoO$_3$	272	

Table A.6: Activation energy and BAM-impact energy for selected gun propellants.

Explosive	E_a (kJ mol^{-1})	BAM-impact energy (J)
A5020 (NC/DBP/DOP/centralite: 93.5/2.5/2.5/1.5)	84.1	
JA-2 (NC/NGl/DGDN/akardite II: 59.5/14.9/24.8/0.7)	149.6	2
NK1074 (NC/NGl/centralite: 51.8/40.5/1.5)	99.5	
M30 (NC/NGl/NGu/centralite: 28/22.5/47/2.5)	251	
N11 (NC/DNDA-5,7/RDX)	100–116	

Table A.7: Activation energy and BAM-impact energy for selected primary explosives.

Explosive	E_a (kJ mol^{-1})	BAM-impact energy (J)
Mercury fulminate	84–243	
Lead azide	88–234	
Diazodinitrophenole	104	
Lead azotetrazolate, basic	130	
Lead styphnate	135–293	
Cesium picrate	180	
Silver azide	184–192	
Silver nitrotetrazole	205	
Potassium dinitrobenzofuroxane (KDNBF)	345	

- R. N. Rogers, Thermochemistry of Explosives, *Thermochim. Acta*, **1975**, *11*, 131–139.
- M. F. Foltz, C. L. Coon, F. Garcia, A. L. Nichols, III, The Thermal Stability of the Polymorphs of Hexanitrohexaazaisowurzitane, Part II, *Propellants Explos. Pyrotech.* **1994**, *19*, 131–144.
- J. Wang, S. Chen, Q. Yao, S. Jin, S. Zhao, Z. Yu, J. Li, Q. Shu, Preparation, Characterization, Thermal Evaluation and Sensitivities of TKX-50/GO Composite, *Propellants Explos. Pyrotech.* **2017**, *42*, 1104–1110.
- Z. Babar, A. Q. Malik, Investigation of the thermal decomposition of magnesium-sodium nitrate pyrotechnic composition (SR-524) and the effect of accelerated aging, *J. Saud. Chem. Soc.* **2012**, *21*, 262–269.
- T. Boddington, A. Cottrell, P. G. Laye, Activation Energies of Gasless Pyrotechnics, *11th IPS-Seminars*, 7–11 July, **1986**, Vail, 419–423.
- V. Bhingarkar, H. Gandhi, P. Phawade, H. Singh, Sensitivity and Thermal Analysis of MTV Igniter Compositions, 30th *ICT-Jata*, 29 June- 2July **1999**, Karlsruhe, Germany, 72.
- F. Zhang, Z. Liu, P. Du, Thermal Decomposition Kinetics of Nitroguanidine Propellant Under Different Pressures, *Propellants Explos. Pyrotech.* **2018**, *43*, 390–397.
- M. Heil, K. Wimmer, M. A. Bohn, Characterization of Gun Propellants by Long-Term Mass Loss Measurements, *Propellants Explos. Pyrotech.* **2017**, *42*, 706–711.
- M. A. Bohn, D. Mueller, Insensitivity aspects of NC bonded and DNDA plasticizer containing gun propellants, IMEMTS **2006**.
- G. Krien, *Thermoanalytische Ergebnisse der Untersuchung von Zünd- und Anzündstoffen*, BICT, Swisttal-Heimerzheim, **1977**, 210 pp.

Actuator, CAD/PAD
Aktuator

Actuators also known as pyromechanisms, cartridge actuated device (CAD)/propellant actuated device (PAD) are one-shot devices typically characterized by a long service life and high reliability. Actuators are very often intended for use under extreme conditions like high altitude, low temperature, under water, space-vacuum, high temperatures, or high g-loads. Actuators can be triggered mechanically, electrically, and optically. In a common actuator, the deflagration of a propellant affects pressure volume work thereby moving a piston to provide push or pull effects. The corresponding devices are called protractors and retractors. Figure A.4 shows a fired and an unfired protractor; Figure A.5 shows the drawing of a protractor. Other actuators based on the same principle are shear-bolts, cable, and reefline cutters (Figures A.6 and A.7), as well as opening or closing valves.

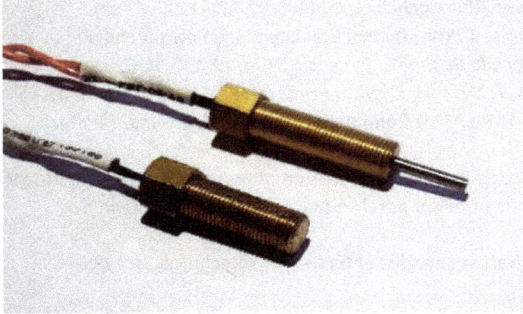

Figure A.4: Protractor after firing (above) before firing (below).

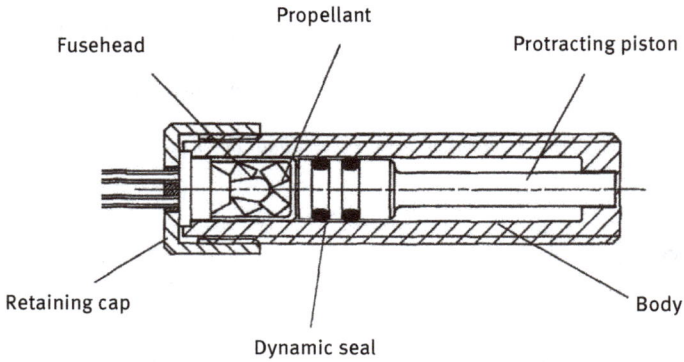

Figure A.5: Sectional view of a protractor before firing.

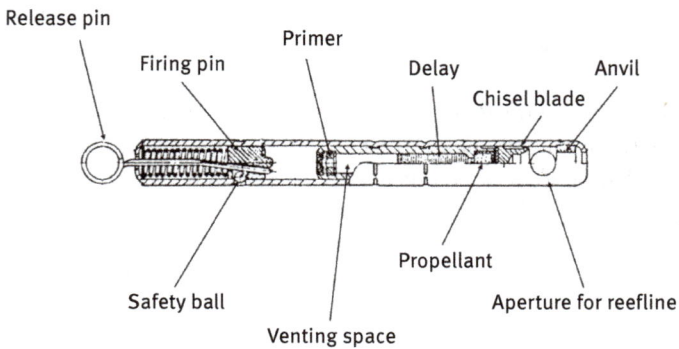

Figure A.6: Sectional view of reefline cutter, mechanical, with pyrotechnic delay.

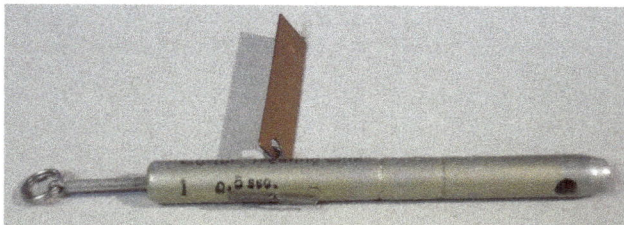

Figure A.7: Reefline cutter, as shown earlier with safety splint (not visible in Figure A.6).

In other actuators, high explosives (HEs) detonate to disintegrate large caliber joints and bolts.

- K. O. Brauer, *Handbook of Pyrotechnics*, Chemical Publishing Company, **1974**, 402 pp.
- R. T. Barbour, *Pyrotechnics in Industry*, McGraw-Hill Book, **1981**, 190 pp.

Adamsite, DM
Diphenylaminchlorarsin

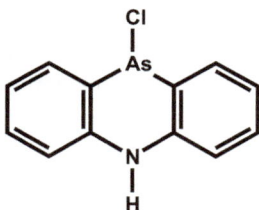

Diphenylaminchlorarsine		
Aspect		Yellow crystals, technical product: green
Formula		$AsC_{12}ClNH_9$
GHS		06, 09
H-phrases		331-301-410
P-phrases		
CAS		[578-94-9]
m_r	g mol^{-1}	277.585
ρ	g cm^{-3}	1.65
$\Delta_f H$	kJ mol^{-1}	+225 (gas phase, PM3)
$\Delta_c H$	kJ mol^{-1}	
	kJ g^{-1}	
	kJ cm^{-3}	
Mp	°C	195
Bp	°C	410(dec)

Fp	°C	113
Λ	wt.-%	−170.03
N	wt.-%	5.04
LCt_{50}	ppm	30
ICt_{50}	ppb min^{-1}	2–5

DM is one of the most effective vomiting agents. Due to its high thermal stability, it can be disseminated by pyrotechnic means or by way of detonation. DM is not prone to hydrolysis and has no significant vapor pressure at ambient temperature. Detoxification of DM can be carried out with sodium sulfide solutions which yield precipitation of insoluble As_2S_5.

- S. Franke, *Lehrbuch der Militärchemie, Band 1*, Militärverlag der Deutschen Demokratischen Republik, 2. Auflage, Leipzig, **1977**, pp. 175–180.
- N. Westphal, *Studienmaterial zur Chemie der Militärisch bedeutsamen Gifte*, Dresden, **1972**, pp. 49.

Adiabatic
Adiabatisch

The change of condition (p, V) of a thermodynamic system is called adiabatic if there is no heat transfer between the system and the environment. The corresponding state is given by the *Poisson equation:*

$$p_2 V_2^\gamma = p_1 V_1^\gamma$$

with

$$\frac{c_p}{c_v} = \gamma$$

Its representation in the p-v diagram is the *isentrope.*

- H. D. Gruschka, F. Wecken, *Gasdynamic Theory of Detonation*, Gordon & Breach, New York, **1971**, p. 17.

Aerosil®

Aerosil is a registered trade name of Evonik, Germany, for a highly disperse amorphous silicon dioxide, SiO_2, prepared by high-temperature hydrolysis with CAS-Nos. [112945-52-5] and [60842-32-2], and its enthalpy of formation $\Delta_f H = -902$ kJ mol^{-1}. Its preparation starts with the introduction of a volatile silicon compound (e.g. $SiCl_4$) into a oxyhydrogen flame (H_2/air). Under the reaction conditions, $SiCl_4$

hydrolyses to silicon dioxide and hydrochloric acid. The solid SiO_2 coagulates in the flame with other particles and yields aggregates with diameters between 10 and 100 nm equivalent to specific surface areas between 300 and 50 m^2 g^{-1}. The surface of aerosil is mainly determined by Si-OH and Si-O-Si groups. Between 80 and 150 °C the loss of physically adsorbed water can be observed in DSC experiments. Also, silanol groups may condense according to

$$Si\text{-}OH + Si\text{-}OH \longrightarrow Si\text{-}O\text{-}Si + H_2O$$

Typical bulk tapping densities range between 50 and 100 g L^{-1}. Due to its low density, it is often used as an anti-caking and free-flowing agent in inorganic salts. Though Aerosil has no effect on the chemical stability of pyrotechnic compositions, it has been observed that highly dispersed silica present at low level (>1% wt.-%) yields mechanical sensitization.

- M. Ettlinger, H. Ferch, D. Korth, *Aerosil ®, Aluminumoxid C und Titandioxid P25 für Katalysatoren, Schriftenreihe Pigmente Nummer 72*, Degussa, Frankfurt, **1988**, 24 pp.
- J. Moretti, *Sustainable Incendiary Projectiles*, SERDP & ESTCP Webinar 16 July **2015**. https://www.serdp-estcp.org/News-and-Events/Calendar/Webinar-Series-07-16-15/(language)/eng-US

AFX

The United States Air Force uses the acronym AFX followed by a three- and four-digit number to indicate HE formulations developed in their responsibility. The main ingredients of formulations in the public domain are given in Table A.8.

Table A.8: AFX-explosive formulations.

AFX -Series	Processing	Main ingredient(s)
100	Castable	RDX
200	Castable	HMX
300	–	*Not assigned or not publicly released*
400	Castable	(EAK)ethylene diammonium dinitrate, NH_4NO_3 (AN), KNO_3
500	Pressable	2,4,6-Tris-(picrylamino)-8-triazine
600	–	NTO
700	Castable	RDX and aluminum
800	Castable	HMX and aluminum
900	–	Nitroguanidine
1000	–	Foamed high explosive
1100	Castable	TNT and aluminum
1200	–	Tungsten powder (DIME)

Ageing
Alterung

Aging (A.E.)

Due to the inherent reactivity of their components and the influence of external factors such as heat, oxygen, and humidity, explosives in their life cycle undergo chemical and physical changes that affect both safety and performance. To investigate the ageing process of explosives, various analytical techniques such as microcalorimetry, heat flow calorimetry (HFC), and accelerating rate calorimetry (ARC) are used. Polymeric binders in PBX are protected from ageing by the use of antioxidants, and metal powders in pyrotechnics are coated with polymers.

- M. A. Bohn (Ed.) *Heat Flow Calorimetry on Energetic Materials*, Fraunhofer IRB Verlag, **2008**, 377 pp.

Agent
Kampfstoff, Wirk(stoff)-mittel

The term *agent* is very often used as a short form for *chemical warfare agent*. Within the field of chemical warfare agents, the prefix *agent* is an integral part of the nomenclature of certain materials such as *BZ (Agent Buzz)* (→*BZ*) as well as the so-called Rainbow Herbicides developed and massively used in the Vietnam war 1961–1971:

- Agent Blue (sodium dimethylarsenate + dimethylarsinic acid)
- Agent Green (*n*-butylester of 2,4,5-trichlorophenoxyacetic acid)
- Agent Orange (iso-octylester of 2,4,5-trichlorophenoxyacetic acid and 2,4-dichlorophenoxyacetic acid)
- Agent Pink (n-butyl- and iso-butylester of 2,4,5-trichlorophenoxyacetic acid)
- Agent Purple (n-butylester of 2,4,5-trichlorophenoxyacetic acid and 2,4-dichlorophenoxyacetic acid)
- Agent White (triisopropylammonium salt of 2,4-dichlorophenoxyacetic acid + picloram)
- F. Barnaby, *Ecological Consequences of the Second Indochina War*, SIPRI, Stockholm, **1976**, 119 pp.
- R. E. Langford, *Introduction to Weapons of Mass Destruction*, Wiley-Interscience, New York, **2004**, 232; 274–276.

Agent Defeat Weapon/Warhead

Agent Defeat Weapon/Warhead (ADW) comprise pyrotechnic compositions and/or HE charges intended to defeat and neutralize biological (bacteria, toxins, and viruses) and chemical warfare agents (nerve agents, pulmonary agents, blood agents, blister agents, lachrymatory agents, etc.). The primary effect of ADW is in the application of intense and prolonged heat to facilitate thermal decomposition of a warfare agent, this includes shock heating. In addition, secondary effects most often comprise the release of certain substances. These substances

(a) can scavenge certain elements (F, Cl, and S) to impede the *de novo* formation of toxic chemicals (basic oxides, CaO, and MgO)

(b) possess catalytic activity to affect long-term destruction and decontamination of an area (e.g. TiO_2) and

(c) possess anti-bacteriological and anti-viral effects (e.g. HF, Ag, and I_2).

– A. K. Chinnam, A. Shlomovich, O. Shamis, N. Petrutik, D. Kumar, K. Wang, E. P. Komarala, D. S. Tov, M. Sućeska, Q. L. Yan, M. Gozin, Combustion of energetic iodine-rich coordination polymer -engineering of new biocidal materials, *Chem. Eng. J.* **2018**, *350*, 1084–1091.
– *Review of Thermal Destruction Technologies for Chemical and Biological Agents Bound on Materials*, U.S. Environmental Protection Agency (EPA) Office of Research and Development (ORD) National Homeland Security Research Center (NHSRC), October **2015**, 129 pp.
– V. Weiser, J. Neutz, N. Eisenreich, E. Roth, H. Schneider, S. Kelzenberg, Development and Characterization of pyrotechnic Compositions as Counter Measures Against Toxic Clouds, *ICT-JATA*, Karlsruhe, **2005**, V-5.
– D. Windle, 'Agent defeat weapons' ready for use, *New Scientist*, 21. February **2003**.
– R. R. McGuire, D. L. Ornellas, F. H. Helm, C. L. Coon, M. Finger, Detonation Chemistry: An Investigation of Fluorine as an Oxidizing Moiety in Explosives, *Det. Symp.*, **1981**, 940–951.

Airbag

An airbag is a gas-tight textile bag that is pneumatically connected to a pyrotechnic gas generator. Figure A.8 shows a sectional view of a pyrotechnic gas generator of the first generation (until the mid-1990s) which often contained sodium azide as the main gas-generating substance (compositions 1 and 2, Table A.9). The released elemental sodium was scavenged with silica as silicate or with molybdenum sulfide and sulfur as sodium sulfide.

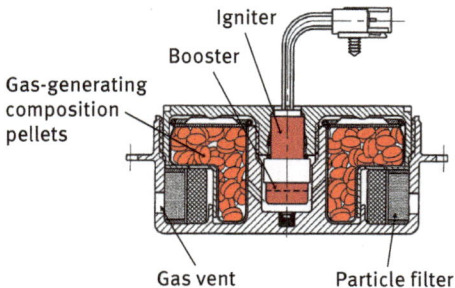

Figure A.8: Section view of a pyrotechnic gas generator.

Table A.9: First-generation (before 1995) azide-based airbag formulations.

Components	# 1	#2
Sodium azide, NaN$_3$ (wt.-%)	57	63
Potassium nitrate, KNO$_3$ (wt.-%)	13	
Silica, SiO$_2$ (wt.-%)	30	
Molybdenum sulfide, MoS$_2$ (wt.-%)		30
Sulfur, S$_8$ (wt.-%)		17

The overall simplified reaction equations read:

Composition #1

$$1\,NaN_3 + 0.15\,KNO_3 + 0.575\,SiO_2 \longrightarrow 0.5\,Na_2SiO_3 + 0.075\,K_2SiO_3 + 1.575\,N_2$$

Composition #2

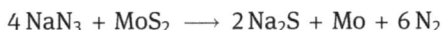

$$4\,NaN_3 + MoS_2 \longrightarrow 2\,Na_2S + Mo + 6\,N_2$$

$$2\,NaN_3 + 1/8\,S_8 \longrightarrow Na_2S + 3\,N_2$$

The high toxicity of NaN$_3$ combined with the relative low gas yield of these formulations (0.3 to 0.35 L g^{-1}) triggered the development of new azide-free compositions (Table A.10). These formulations are nearly oxygen balanced and contain high nitrogen compounds with low carbon content (e.g. guanidinium nitrate, nitroguanidine, or derivatives of tetrazole) as fuel and yield up to 0.85 L gas per g composition.

Table A.10: State-of-the-art "non-azide" airbag compositions.

Components	# 3	# 4
Ammonium nitrate (PSAN), NH$_4$NO$_3$ (wt.-%)		66.8
Potassium nitrate*, KNO$_3$ (wt.-%)		6.7
Potassium perchlorate, KClO$_4$ (wt.-%)	35	
Nitroguanidine, CH$_4$N$_4$O$_2$ (wt.-%)	15	
Guanidinium nitrate, CN$_4$H$_6$O$_3$ (wt.-%)	50	
Diammonium 5,5'-bi-1H-tetrazolate, C$_2$H$_8$N$_{16}$ (wt.-%)		21.5
Glass fibers, SiO$_2$ (wt.-%)		5
Oxygen balance, Λ (wt.-%)	−1.55	+0.55
Burn rate, u (mm s^{-1})	34.7 at 20 MPa	16.4 at 20.7 MPa
Pressure exponent, n	0.380	0.62
Gas yield	81.32 wt.-%	4 mol/100 g

*Phase stabilizer for NH$_4$NO$_3$.

There are particular requirements to be fulfilled for gas-generating compositions in use in vehicle passenger restraint systems:

- highly variable temperature of pyrolant (−40 to +70 °C);
- concentration of toxicologically relevant substances below the legal limits;
- variable ignition and combustion rate;
- high gas yield >14 mol kg^{-1};
- non-toxic ingredients and reactions products;
- sufficient chemical and thermal stability;
- low sensitiveness;
- low cost and good availability of components;
- good processability at technical and industrial scale;
- easy recycling.

There are limits for the gas compositions on either side of the airbag which are given in Table A.11.

Table A.11: Concentration limits for toxic substances according to *USCAR*.

Gas	Vehicle exterior (ppm)	Vehicle interior (ppm)
Chlorine, Cl_2	1	0.25
Carbon monoxide, CO	461	115
Carbon dioxide, CO_2	30.000	7500
Phosgene, $COCl_2$	0.33	0.08
Nitrogen monoxide, NO	75	18.75
Nitrogen dioxide, NO_2	5	1.25
Ammonia, NH_3	35	9
Hydrogen chloride, HCl	5	1.25
Sulfur dioxide, SO_2	5	1.25
Sulfane, H_2S	15	3.75
Benzene, C_6H_6	22.5	5.63
Hydrogen cyanide, HCN	4.7	1.18
Formaldehyde, H_2CO	1	0.25

Apart from their mass use in passenger restraint systems, airbags are used in other areas too such as landing cushions for planetary descent modules and emergency lifting bodies for damaged submarines. For the latter use at sea temperature, the hot reaction gasses are cooled down by contact with substances that decompose in an endothermal reaction such as ammonium oxalate:

$$(NH_4)_2C_2O_4 + \text{heat} \longrightarrow 2\,NH_3 + CO + CO_2 + H_2O$$

The ratio of gas-generating composition versus cooling agent is about 0.7/1. The standard gas volume including coolant is about 1 L/g.

- K. Menke, J. Neutz, U. Schleicher, *Pyrotechnische Kaltgasgeneratoren*, DE 102012217718A1, **2013**, Germany.
- K. Menke, J. Neutz, U. Schleicher, *Pyrotechnische Kaltgasgeneratoren als Bestandteil von Unterwasser-Rettungssystemen für U-Boote und Tauchplattformen*, in *Wehrwissenschaftliche Forschung Jahresbericht 2011 Innovative Verteidigungsforschung für ein zukunftsorientiertes Fähigkeitsprofil der deutschen Streitkräfte 11*, Bundesministerium der Verteidigung, Bonn, April **2012**, pp. 28–29.
- S. Zeuner, R. Schropp, A. Hofmann, K. H. Rödig, *Nitrocellulosefreie gaserzeugende Composition*, EP1275629B1, **2012**, Germany.
- V. Mendenhall, G. K. Lund, *Gas Generating Compositions Having Glass Fibers*, US 2010/0116384, **2010**, USA.
- USCAR Inflator Technical Requirements and Validation, *USCAR-24*, June **2004**, 93 pp.
- https://mars.nasa.gov/mer/mission/spacecraft_edl_airbags.html

Akremite

Akremite is a commercial binary explosive based on AN and carbon black that is prepared on site. It is not cap sensitive and hence requires a cap-sensitive booster charge.

Alex

Alex is the term used to describe spherical submicron sized aluminum powder. Alex is obtained from electrical explosions of **aluminum** wire in a controlled atmosphere to favor the formation of a protective oxidation layer. Due to the high surface to mass ratio, Alex reacts much faster than micron-sized Al powder. Table A.12 displays various Alex types and their typical properties.

Table A.12: Properties of various Alex types.

Type	Passivation	Coating	BET ($m^2\,g^{-1}$)	d_p (nm)	Al content (-%)	T_z^* (°C)
Alex	Air	–	11.8 ± 0.4	188	89 ± 0.2	547
C-Alex	Air	0.1 wt.-% 12DHB$	11.3 ± 0.1	192	88 ± 1.5	n.a.
CH-Alex	Air	0.1 wt.-% 12DHB + 1 wt.-% HTPB	n.a.	n.a.	89 ± 1.0	n.a.
L-Alex	Stearic acid	–	n.a.	n.a.	79	498
VF-Alex	Air	Viton ®A + §	6.9 ± 0,2	322	78 ± 1.5	440

*300 K · s^{-1} heating rate, air, 0.1 MPa; $ 1,2-dihydroxybenzene; § ester of 1 H,1 H-perfluoroundecanol with furandione.

-   Q. S. M. Kwok, R. C. Fouchard, A.-M. Turcotte, P. D. Lightfoot, R. Bowes, D. E. G. Jones, Characterization of Aluminum Nanopowder Compositions, *Propellants Explos. Pyrotech.* **2002**, *27*, 229–240.
-   A. Gromov, U. Teipel (Eds.), *Metal Nanopowders*, Wiley-VCH, Weinheim, **2014**, 417 pp.
-   E. Lafontaine, M. Comet, *Nanothermites*, Wiley-VCH, ISTE, London, **2016**, 327 pp.
-   V. E. Zarko, A. A. Gromov (Eds.), *Energetic Nanomaterials*, Elsevier, **2016**, 374 pp.

Alginates, alginic acid
Alginate, Alginsäure

Alginic acid, CAS-No.: [9005-32-7] Mp: 300 °C, is a colorless polysaccharide-bearing carboxy groups with molecular weights in the 200,000 ball park. Figure A.9 depicts the monomeric unit. Alginic acid binds 200–300 times its mass of water forming stable gels, emulsions, and thixotropic solutions. Hence, alginates (e.g. calcium alginate, CAS-No. [9005-35-0], $(C_{12}H_{14}CaO_{12})_n$) are used to stabilize W/O emulsion explosives.

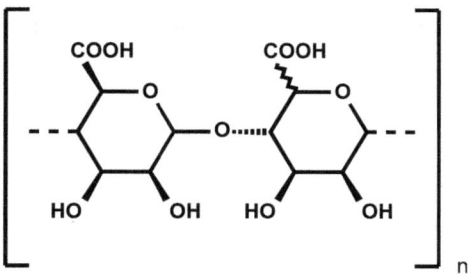

Figure A.9: Alginic acid.

Nitration of alginic acid with mixed acid yields products with a maximum of 7.69 wt.-% N.

-   H. J. Lucas and W. T. Stewart, Esters of Alginic Acid, *J. Am. Chem. Soc.* **1940**, *62*, 1070–1074.

ALICE

ALICE is a term describing the cryogenic system **aluminum/ice (Al + $H_2O_{(s)}$)**, which has been proposed as both environmentally benign rocket propellant and hydrogen generator.

-   T. L. v Pfeil, J. Tsohas, S. F. Son, Feasibility Study and Demonstration of an Aluminum and Ice Solid Propellant, *International Journal of Aerospace Engineering*, **2012**, Article ID 874076, 11 p. doi: http://dx.doi.org/10.1155/2012/874076.

- G. A. Risha, J. L. Sabourin, V. Yang, R. A. Yetter, S. F. Son, B. C. Tappan, Combustion and Conversion Efficiency of Nanoaluminum-Water Mixtures, *Combust. Sci. Tech.*, **2008**, *180*, 2127–2142. doi: http://dx.doi.org/10.1080/00102200802414873.

Alloys and alloying reactions
Legierungen und Legierungsbildung

Alloys are metallic materials based on two or more metals or metalloids (Si, B, and As). They often show strikingly different physical and chemical properties than their constituting components. Alloys are mainly produced by fusion of different metals and founding, by sintering powdered components, by reduction of mixed oxides, or by reactive milling. Alloys can be either homogeneous (solid solution of a metal in a base metal) or in case that phases are immiscible heterogeneous (adjacent crystallites of different chemical constitution). Alloys play an important role as fuels in pyrotechnics and high blast explosives ($CaSi_2$, MgB_2, Mg_2Si, Mg_3Al_4, ZrNi, $ZrAl_2$, FeSi, etc.).

In addition, the exothermic formation of an alloy can partly or fully sustain energetic applications. Examples of which are the alloying reaction between Al + Ni in the *Pyronol*® torch or *nanolaminates* such as *Nanofoil*®

$$Al + Ni \longrightarrow AlNi, \Delta_R H = -117 \, kJ \, mol^{-1}$$

and the alloying reaction between Al + Pd wires in *pyrofuze*®:

$$Al + Pd \longrightarrow AlPd, \Delta_R H = -100 \, kJ \, mol^{-1}$$

Other popular alloying reactions comprise

$$Ti + C \longrightarrow TiC, \Delta_R H = -124 \, kJ \, mol^{-1}$$

which is frequently used in igniter systems and

$$Ti + 2B \longrightarrow TiB_2, \Delta_R H = -328 \, kJ \, mol^{-1}$$

which has been suggested as a primary stage reaction in the ADW agent HTI-J-1000.

The enthalpy of formation of binary alloys can be estimated based on Miedema's work and with the Miedema calculator available as freeware on the Internet.

- O. Kubaschewski, J. A. Catterall, Metal Physics and Physical Metallurgy Volume 3, *Thermochemical Data of Alloys*, Pergamon Press, London, **1956**, 200 pp.
- A. R. Miedema, R. Boom, F. R. de Boer, On the heat of formation of solid alloys, *J. Less-Common Met.* **1975**, *41*, 283–298.
- A. R. Miedema, On the heat of formation of solid alloys II, *J. Less-Common Met.* **1976**, *46*, 67–83.
- A. R. Miedema, P. F. de Châtel, F. R. de Boer, Cohesion in Alloys – Fundamentals of a Semi-Empirical Model, *Physica* **1980** *100B*, 1–28.

- A. R. Miedema, F. R. de Boer, R. Boom, Predicting Heat Effects in Alloys, *Physica*, **1981**, *103B*, 67–81.
- "Miedema-Calculator" on the Site of the Institute of Metallurgy and Material Sciences of the Polish Academy of Sciences: http://www.entall.imim.pl/calculator/ accessed on October 15 2019.

Aluminum

Aluminium

Aluminium (B.E.)		
Aspect		Silver grey – black
Formula		Al
GSH		02
H-phrases		250-261
P-phrases		210-222-231+232-370+378a-422a-501
EINECS		231-072-3
CAS		[7429-90-5]
m_r	g mol^{-1}	26.98
ρ	g cm^{-3}	2.699
Mp	°C	660.4
Bp	°C	2330
$\Delta_m H$	kJ mol^{-1}	10.67
$\Delta_v H$	kJ mol^{-1}	290.8
c_p	J mol^{-1} K^{-1}	24.30
c_L	m s^{-1}	6390
κ	W m^{-1} K^{-1}	237
$\Delta_c H$	kJ mol^{-1}	−837.5
	kJ g^{-1}	−31.041
	kJ cm^{-3}	−83.781
Λ	wt.-%	−88.951

With decreasing particle size, Al is silver-grey to black powder. Contrary to magnesium, Al is stable in weak acidic media. However alkaline substances (hydroxides and carbonates) attack Al. Under air, Al yields a protective coating consisting mainly of the hydroxide, $Al(OH)_3$. Al reacts exothermally with a great many number of metals to give alloys, for example, Co (AlCo), Ni (AlNi)(*Pyronol*) and Pd (AlPd)(*Pyrofuze*), and Pt (AlPt). As an ingredient in HEs, Al reacts when heated to T > 2500 K behind the CJ plane only in the expanding product gases and hence does not affect the detonation velocity. However, it does enhance the overall blast pressure. Al does not affect the Gurney energy. In pyrotechnics, Al is mainly used as fuel in flash and report compositions. Commonly it is not used in signal flares or illuminants as the flame volume is very small compared to Mg. In addition, Al yields refractory complex oxides of the Spinell type with Ba and Sr compounds hence precipitating from the flame (Al_2MeO_4) early and giving rise to strong continuum radiation which adversely affects the color purity. Halogens in the flame zone preferably react with Al over Ca, Sr, Ba, and Cu further impeding the formation of the important volatile metal mono-chlorides as color emitters. Independent comprehensive studies by *Schmied* and *Kott* have addressed the properties of Al and Mg/Al-alloys in flare formulations. *Farnell* has investigated the influence of transition metal compounds on Al-based flare compositions.

Another important use of Al is as an energetic fuel in composite and composite-modified solid rocket propellants. Al serves primarily the increase of performance but also the stabilization and increase of the combustion rate over non-aluminized propellants.

- A. C. Kott, G. A. Lane, Aluminum-Fueled Flares, *IPS*, Snowmass-at-Aspen Colorado, **1970**, pp. 137–166.
- P. L. Farnell, R. P. Westerdahl, F. R. Taylor, The Influence of Transition Metal Compounds on the Aluminum-Sodium Nitrate Reaction, *IPS*, Colorado-Springs, Colorado, **1972**, pp. 271–290.
- G. R. Lakshminarayanan, G. S. Mannix, T. Carney, G. Chen, Evaluation of Alternative Fuel and Binder Materials in hand Held Signals Illuminant Compositions, *ARWEC-TR-99008*, Picatinny Arsenal, November **2000**.
- A. Hahma, Ignition and Combustion of Aluminum in High Explosives, *J. Pyrotech.* **2007**, *26*, 24–46.
- H. Ritter, S. Braun, High Explosive Containing Ultrafine Aluminum ALEX, *Propellants Explos. Pyrotech.* **2001**, *26*, 311–314.

Aluminum hydride
Aluminiumhydrid, α-Alan

α-Alane		
Formula		AlH_3
UN-HD		4.3
REACH		LPRS
EINECS		232-053-2
CAS		[7784-21-6]
m_r	g mol^{-1}	30.005
ρ	g cm^{-3}	1.477
dc	°C	157 (H_2-release)
$\Delta_f H$	kJ mol^{-1}	−45
$\Delta_c H$	kJ mol^{-1}	−1221
	kJ g^{-1}	−40.693
	kJ cm^{-3}	−60.104
Λ	wt.-%	−159.97

AlH_3 forms six different polymorphs. The stabilized α-AlH_3 which results from doping the crystal lattice is the only polymorph of technical relevance. The polycrystalline material is under investigation as an energetic ingredient in gelled propellants, pyrotechnic formulations, and HEs.

- H. Fong, E. McLaughlin, P. E. Penwell, M. A. Petrie MARK A, D. Stout, US201615184962, *Crystallization and Stabilization in the Synthesis of Microcrystalline Alpha Alane*, USA, **2016**.
- T. Bazyn, R. Eyer, H. Krier, N. Glumac, Combustion Characteristics of Aluminum Hydride at Elevated Pressure and Temperature, *J. Propul. Power* **2004**, *20*, 427–431.

Amatex

Amatex are melt-castable HE formulations based on TNT, AN, and hexogen (RDX) (Tables A.13 and A.14). Amatex are prepared by melting corresponding proportions of amatol and Comp B. The non-ideal behavior of Amatex correlates with the AN content. Amatex is nicely stable in the VST.

Table A.13: Composition of Amatex high explosives.

	Amatex 5	Amatex 9	Amatex 20	Amatex 20 K	Amatex 30	Amatex 40
TNT (wt.-%)	50	41	40	40	40	40
NH_4NO_3[*](wt.-%)	25	50	40	36	30	20
RDX (wt.-%)	25	9	20	20	30	40
KNO_3 (wt.-%)	–	–	–	4		
ρ (g cm^{-3})			1.615	1.60	1.625	1.650

*Median 500 µm diameter.

Investigations with both Amatex 20 and Amatex 40 have demonstrated that disregard the particle size of the AN (whether 15 or 500 µm) only about 50 wt.-% of it undergo reaction in the chemical reaction zone (CRZ).

Table A.14: Performance of amatex high explosives.

Parameter	Amatex 20	Amatex 20 K	Amatex 40
V_D (m s^{-1})	7009	6790	7545
P_{CJ} (GPa)	20.5		
$\Delta_f H$ (kJ kg^{-1})	–	–1954	
ρ (g cm^{-3})	1.61		1.66
$\varnothing_{cr}$ (mm)	18	19 >> 25	

Table A.14 (continued)

Parameter	Amatex 20	Amatex 20 K	Amatex 40
$\Delta_{melt}H$ (J g^{-1})	38.6	38.6	
c_p (J g^{-1} K^{-1})	–	1.67	
Dp (°C)	–	156	
A_E (kJ mol^{-1})	137	138	
T_{5ex} (°C)	240		
Impact (J)	7.6		

Amatex 20 shows a greater shock sensitivity in the NOL-SSGT compared to TNT, but is less sensitive than Comp B. Amatex 9 was used extensively in World War II-UK aerial ordnance.

Amatol

Amatol are melt-castable HEs based on TNT and AN (Tables A.15 and A.16). Amatols were used for the first time in World War I by the UK to stretch the TNT stocks (Amatol -80/20 and Amatol 50/50). In World War II, Germany fielded Amatol-39, -40, and -41 which were melt-castable compositions of a different kind based on 2,4-dinitroanisole (DNAN), 1,3-dinitrobenzene, ethylenediammonium dinitrate, hexogen (RDX), and AN.

Table A.15: Composition of amatol explosives.

	Amatol 80/20	Amatol 60/40	Amatol 50/50	Amatol 45/55	Amatol 40/60	Amatol 30/70
TNT (wt.-%)	20	40	50	55	60	70
NH_4NO_3 (wt.-%)	80	60	50	45	40	30
ρ (g cm^{-3})	1.46	1.61	1.59		1.54	1.63

The depth of the dent (d_d) obtained in the Plate-dent test with amatol 60/40 increases with decreasing particle size of the AN (d_p = 275 > 68 > 15 µm) (d_d = 1.88 > 2.20 > 2.54 mm) and indicates an increase of the CJ pressure. The length of the CRZ in Amatol 80/20 is 4 mm at ρ = 1.67 g cm^{-3}.

Table A.16: Performance of amatol explosives.

Parameter	Amatol 80/20	Amatol 60/40	Amatol 50/50	Amatol 45/55	Amatol 40/60	Amatol 30/70
V_D (m s^{-1})	5200	5760	5975	6470	6500	6370
P_{CJ} (GPa)			14.67			
$\Delta_f H$ (kJ kg^{-1})		-2840				
$\Delta_{det} H$ (kJ kg^{-1})	4270					
ρ (g cm^{-3})	1.46	1.60	1.58		1.54	1.63
TMD (g cm^{-3})	1.71					1.675
$\varnothing_{cr}$(mm)	80					
$\Delta_{melt} H$ (J g^{-1})	20.4	38.6	48.2			
c_p (J g^{-1} K^{-1})			1.602			
Dp (°C)		218				
$T_{ex-5\ s}$ (°C)	280–300	270	254–265			
Trauzl (cm^3)	385	360	350	353	335	
Impact (5 kg cm)	25	15	19	–	21	–

The 50% initiation pressure in the NOL-LSGT for amatol −60/40 = 3.30 GPa.

Amid powder

Amidpulver

Millon and *Reset* proposed a sulfur-free black powder in 1843 with additional AN designated *Amid powder*

- 40–45 wt.-% potassium nitrate
- 35–38 wt.-% AN
- 14–22 wt.-% charcoal

3-Amino-1,2,4-triazolium nitrate, ATrzN

3-Amino-1,2,4-triazoliumnitrat

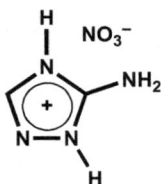

Formula		$C_2H_5N_5O_3$
CAS		[13040-74-9]
m_r	g mol^{-1}	147.093
ρ	g cm^{-3}	1.5
Δ_fH	kJ mol^{-1}	−171
$\Delta_{ex}H$	kJ mol^{-1}	−525.8
	kJ g^{-1}	−3.575
	kJ cm^{-3}	−5.362
Δ_cH	kJ mol^{-1}	−1330.6
	kJ g^{-1}	−9.046
	kJ cm^{-3}	−13.569
Λ	wt.-%	−38.07
N	wt.-%	47.613
Mp	°C	120
Dp	°C	195
Friction	N	>355
Impact	J	38.9

3-Amino-1,2,4-triazolium nitrate is used as in ingredient in gas generators and melt-cast explosives (*AAD*).

- L. I. Bagal, M. S. Pevzner, V. A. Lopyrev, Basicity and Structure of 1,2,4-triazole Derivatives, *Khim. Geterot. Soed.* **1966**, *3*, 440–442.
- M. M. Williams, W. S. McEwan, R. A. Henry, The heats of combustion of substituted triazoles, tetrazoles and related high nitrogen compounds, *J. Phys. Chem.* **1957**, *61*, 261–267

5-Amino-1 *H*-tetrazole, 5-AT

5-Amino-1 H-tetrazol

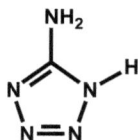

Formula		$C_1H_3N_5$
GHS		02, 07
REACH		LPRS
EINECS		224-581-7
CAS		[4418-61-5]
m_r	g mol^{-1}	85.068
ρ	g cm^{-3}	1.502
$\Delta_f H$	kJ mol^{-1}	207.78
$\Delta_c H$	kJ mol^{-1}	1030.3
	kJ g^{-1}	−12.108
	kJ cm^{-3}	−18.187
Λ	wt.-%	−65.83
N	wt.-%	82.33
Mp	°C	203 (dec)

5-Amino-1*H*-tetrazole (5-AT) is hygroscopic and forms a monohydrate. It is frequently used as a high nitrogen fuel in gas generators. Certain heavy metal salts of 5-AT are currently considered as primary explosives.

- H. Stadler, K. Ballreich, H. Gawlick, *Pyrotechnisches Gemisch*, DE1446918A, **1966**, Germany.

4-Amino-1-methyl-1,2,4-triazolium nitrate
4-Amino-1-methyl-1,2,4-triazoliumnitrat

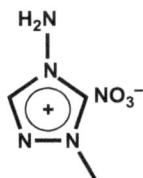

C1 N		
Formula		$C_3H_7N_5O_3$
CAS		[13040-75-0]
m_r	g mol^{-1}	161.12
ρ	g cm^{-3}	1.4
Δ_fH	kJ mol^{-1}	−228.40
$\Delta_{ex}H$	kJ mol^{-1}	−610.30
	kJ g^{-1}	−3.789
	kJ cm^{-3}	−5.303
Δ_cH	kJ mol^{-1}	−1952.6
	kJ g^{-1}	−12.119
	kJ cm^{-3}	−16.966
Λ	wt.-%	−64.55
N	wt.-%	43.47
Mp	°C	−54 (g)
Dp	°C	249
Friction	N	>355
Impact	J	20

4-AMNT is currently considered as a non-volatile plasticizer having a greater stability than NGl or DNDA. 4-AMNT is also investigated as monergol; however, stable combustion only occurs at pressure ≥12 MPa with u ~ 10 mm s^{-1}. Modified with both powdered Al and methylcellulose 4-AMNT shows stable combustion already at pressure ≥2 MPa.

– U. Schaller, T. Keicher, H. Krause, S. Schlechtriem, EILs – suitable substances for future energetic applications? *ICT -Jata*, Karlsruhe, **2016**, V9.

4-Amino-3,7-dinitrotriazolo-[5,1 c][1,2,4]triazine, DPX-26

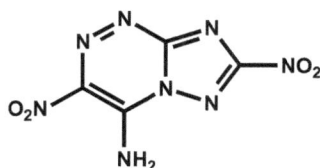

4-Amino-3,7-dinitrotriazolo-[5,1 c][1,2,4]triazin

Aspect		Creme
Formula		$C_4H_2N_8O_4$
CAS		[1941251-35-9]
m_r	g mol^{-1}	226.11
ρ	g cm^{-3}	1.86
$\Delta_f H$	kJ mol^{-1}	387
$\Delta_{ex}H$	kJ mol^{-1}	−1140.3
	kJ g^{-1}	−5.043
	kJ cm^{-3}	−9.380
$\Delta_c H$	kJ mol^{-1}	−2246.9
	kJ g^{-1}	−9.937
	kJ cm^{-3}	−18.483
Λ	wt.-%	−35.38
N	wt.-%	49.56
Mp	°C	232 (*dec*)
Friction	N	>360 ABL
Impact	J	29 LANL (type 12 tool)
V_D	m s^{-1}	*8700 at 1.86 g cm^{-3}*
P_{CJ}	GPa	*32 at 1.86 g cm^{-3}*

DPX-26 is an experimental HE prepared only recently with good thermal stability, low sensitivity, and a performance comparable to RDX. The oxidation of DPX-26 with hypofluoric acid (HOF) yields the 2-*N*-oxide (DPX-27) which due to a higher oxygen balance is predicted to have a higher P_{CJ} but also a higher sensitiveness.

– D. G. Piercey, D. E. Chavez, B. L. Scott, G. H. Imler, D. A. Parrish, An Energetic Triazolo-1,2,4-Triazine and Its Oxide, *Angew. Chem.* **2016**, *128*, 15541–15544.

1-Amino-3-methyl-1,2,3-triazolium nitrate

1-Amino-3-methyl-1,2,3-triazoliumnitrat (1-AMTN)

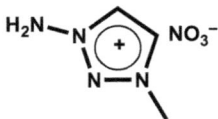

$H_2N-N\overset{+}{\underset{N=N}{\bigcirc}}$ NO_3^-

Aspect		Pale yellow crystals
Formula		$C_3H_7N_5O_3$
CAS		[944132-38-1]
m_r	g mol^{-1}	161.12
ρ	g cm^{-3}	1.615
$\Delta_f H$	kJ mol^{-1}	71
$\Delta_{ex}H$	kJ mol^{-1}	−799.51
	kJ g^{-1}	−4.962
	kJ cm^{-3}	−8.014
$\Delta_c H$	kJ mol^{-1}	−2252.1
	kJ g^{-1}	−13.978
	kJ cm^{-3}	−22.574
Λ	wt.-%	−64.55
N	wt.-%	43.47
Mp	°C	88
Dp	°C	185
Friction	N	>355
Impact	J	>20

1-AMTN is considered an insensitive alternative to TNT. The critical diameter at $\rho = 1.63$ g cm^{-3} is $\varnothing_{cr} > 18$ mm.

- A. Brand, T Hawkins, G. Drake, I. M. K. Ismail, G. Warmoth, L. Hudgens, Energetic Ionic Liquids as TNT Replacements, *JANNAF 33rd Propellant & Explosives Development & Characterization Subcommittee (PEDCS) / 22nd Safety & Environmental Protection Subcommittee (SEPS) Joint Meeting*, Sandestin Beach, **2006**.
- G. Drake, G. Kaplan, L. Hall, T. Hawkins, J. Larue, A new family of energetic ionic liquids 1-amino-3-alkyl-1,2,3-triazolium nitrates, *J. Chem. Cryst.* **2007**, *37*, 15–23.
- Z. Wang, J. Zhang, J. Wu, X. Yin, T. Zhang, Replacement of 2,4,6-trinitrotoluene by two eutectics formed between 4-amino-1,2,4-triazolium nitrate and 4-amino-1,2,4-triazolium perchlorate, *RSC Adv.* **2016**, *6*, 44742–44748.

7-Amino-4,6-dinitrobenzofuroxane, ADNBF

7-Amino-4,6-dinitrobenzofuroxan

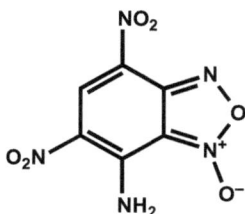

Formula		$C_6H_3N_5O_6$
CAS		[97096-78-1]
m_r	g mol^{-1}	241.12
ρ	g cm^{-3}	1.902
Δ_fH	kJ mol^{-1}	154
$\Delta_{ex}H$	kJ mol^{-1}	−1260.5
	kJ g^{-1}	−5.228
	kJ cm^{-3}	−9.943
Δ_cH	kJ mol^{-1}	−2943.9
	kJ g^{-1}	−12.209
	kJ cm^{-3}	−23.222
Λ	wt.-%	−49.77
N	wt.-%	29.05
Mp	°C	270 (dec)
Impact	cm	28–41; 2.5 kg
V_D	m s^{-1}	*8220 at 1.901 g cm^{-3}.*
P_{CJ}	GPa	*32.9*

ADNBF is an experimental insensitive HE that was prepared at LLNL in the 1980s for the first time. In comparison to the structurally related 4,6-dinitrobenzofuroxane, ADNBF is less sensitive to shock and impact due to the additional amino group allowing for a better stabilization. ADNBF is the energetic filler in PBXC-18.

– M. L. Chan, C. D. Lind, P. Politzer, Shock Sensitivity of Energetically Substituted Benzofuroxans, *9th Det. Symp.*, Portland, USA, **1989**, 566–572.

3-Amino-5-nitro-1,2,4-triazole, ANTA

3-Amino-5-nitro-1,2,4-triazol

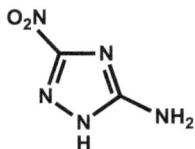

Aspect		Yellow needles
Formula		$C_2H_3N_5O_2$
CAS		[58794-77-7]
m_r	g mol^{-1}	129.078
ρ	g cm^{-3}	1.819 (α-polymorph)
$\Delta_f H$	kJ mol^{-1}	87.86
$\Delta_{ex}H$	kJ mol^{-1}	−548.2
	kJ g^{-1}	−4.247
	kJ cm^{-3}	−7.725
$\Delta_c H$	kJ mol^{-1}	−1303
	kJ g^{-1}	−10.101
	kJ cm^{-3}	−18.373
Λ	wt.-%	−43.38
N	wt.-%	54.26
Mp	°C	227 (*dec*)
Friction	N	168
Impact	cm	>177 with 2.5 kg (type 12 tool)
V_D	m s^{-1}	7710 @ 1.752 g cm^{-3} with 5 wt.-% Kel-F 800

3-Amino-5-nitro-1,2,4-triazole is an insensitive HE with moderate energy content.

−   R. L. Simpson, P. F. Pagoria, A. R. Mitchell, C. L. Coon, Synthesis, Properties
    and Performance of the High Explosive ANTA, *Propellants Explos. Pyrotech.*
    1994, *19*, 174–179.

Aminoguanidine-5,5′-azotetrazolate, AGTZ

Aminoguanidin-5,5′-azotetrazolat

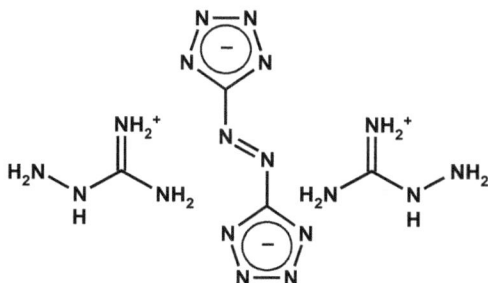

Aspect		Yellow crystals
Formula		$C_4H_{14}N_{18}$
CAS		[862107-15-1]
m_r	$g.mol^{-1}$	314.276
ρ	$g\ cm^{-3}$	1.54
Δ_fH	$kJ\ mol^{-1}$	434.30
$\Delta_{ex}H$	$kJ\ mol^{-1}$	−613.9
	$kJ\ g^{-1}$	−1.953
	$kJ\ cm^{-3}$	−3.008
Δ_cH	$kJ\ mol^{-1}$	−4009.2
	$kJ\ g^{-1}$	−12.757
	$kJ\ cm^{-3}$	−19.646
Λ	wt.-%	−76.36
N	wt.-%	80.22
Mp	°C	218 (*dec*)
Friction	N	>355
Impact	J	15

AGTZ is used as a high nitrogen fuel in gas-generating compositions.

– U. Bley, R. Hagel, J. Havlik, A. Hoschenko, P. S. Lechner, *Pyrotechnisches Mittel*, EP1890986B1, RUAG, **2013**.

Aminonitroguanidine, ANQ

Aminonitroguanidin

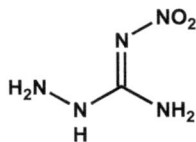

ANQ		
Formula		$C_1H_5N_5O_2$
CAS		[27256-18-4]
m_r	$g\ mol^{-1}$	119.083
ρ	$g\ cm^{-3}$	1.767
$\Delta_f H$	$kJ\ mol^{-1}$	25
$\Delta_{ex}H$	$kJ\ mol^{-1}$	−537.5
	$kJ\ g^{-1}$	−4.513
	$kJ\ cm^{-3}$	−7.975
$\Delta_c H$	$kJ\ mol^{-1}$	−1133.1
	$kJ\ g^{-1}$	−9.515
	$kJ\ cm^{-3}$	−16.813
Λ	wt.-%	−33.59
N	wt.-%	58.81
Mp	°C	190 (*dec*)
Friction	N	190
Impact	J	>20
V_D	$m\ s^{-1}$	8365 @ 1.61 $g\ cm^{-3}$
P_{CJ}	GPa	28.1 @ 1.61 $g\ cm^{-3}$
$\sqrt{2E_g}$	$mm\ \mu s^{-1}$	2.55 @ 1.61 $g\ cm^{-3}$
P_{NoGo}	GPa	39 (BICT-SSWGT)

ANQ is obtained in moderate yield by the reaction of nitroguanidine with hydrazine. ANQ is a HE with a performance similar to RDX; however, it has greater sensitivity to shock than latter. A more recent synthetic pathway starts with *N*-nitro-*S*-methylisothiourea.

- H. H. Licht, B. Wanders, L'explosif aminonitroguanidine (ANQ), *Rapport CO 206/91*, ISL -Saint Louis, France, **1991**.
- B. Fuqiang, H. Huan, Jia Siyuan, L. Xiangzhi, W. Bozhou, *Method for synthesizing 1-amino-3-nitroguanidine*, CN-Patent 105503661A, **2015**, China.

5-Aminotetrazolium nitrate, ATN

5-Aminotetrazoliumnitrat

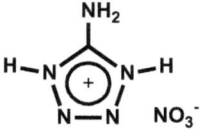

Formula		$C_1H_4N_6O_3$
CAS		[43146-62-9]
m_r	g mol^{-1}	148.081
ρ	g cm^{-3}	1.807 @ 296 K, 1.847 @ 200 K
$\Delta_f H$	kJ mol^{-1}	−21
$\Delta_c H$	kJ mol^{-1}	−944.2
	kJ g^{-1}	−6.376
	kJ cm^{-3}	−11.522
Λ	wt.-%	−10.80
N	wt.-%	56.75
Dp	°C	167 under air, 193 under N_2
Friction	N	324
Impact	J	5
Koenen	mm	>10 mm (type: ?)

- G. Ma, T. Zhang, J. Zhang, K. Yu, Thermal decomposition and molecular structure of 5-aminotetrazolium nitrate, *Thermochim. Acta* **2004**, *423*, 137–141.
- M. von Denffer, T. M. Klapötke, G. Kramer, G. Spieß, J. M. Welch, Improved Synthesis and X-Ray Structure of 5-Aminotetrazolium Nitrate, *Propellants Explos. Pyrotech.* **2005**, *30*, 191–195.

Ammon powder

Ammonpulver

Ammon powder was the designation for a sulfur-free low signature black powder:
- 80–90 wt.-% AN
- 10–20 wt.-% charcoal

Similarly, as the amid powder, it was used until World War I as an artillery propellant with only minimal muzzle flash and low smoke formation.

Ammonal

Ammonal are melt-castable HEs based on the ternary system *TNT*, *AN*, and *aluminum* (Tables A.17, A.18, and Figure A.10).

Table A.17: Composition of ammonal explosives.

wt.-%	DE-Ammonal-I	DE-Ammonal-II	US-Ammonal-I	US-Ammonal-II	Fp	Fp-19	Fp-13-113	A-8	A-45
TNT	30	31	67	12	40	55	20	12	45
NH_4NO_3[*)	54	44.9	22	72	40	35	70	80	40
aluminum	16	24.1	11	16	20	10	10	8	15
ρ (g cm^{-3})					1.65	1.62			

Table A.18: Properties of ammonal explosives.

Parameter	US-Ammonal-I	Fp	A-8	A-45
V_D (m s^{-1})			3900	6050
ρ (g cm^{-3})	1.65	1.74		
$\Delta_{melt}H$ (J g^{-1})	64.6	38.6		
c_p (J g^{-1} K^{-1})		1.268		
Dp (°C)		250		
$\varnothing_{cr}$ (mm)			35	70
Trauzl (cm^3)			430	450

These formulations were codenamed *Füllpulver* (Fp) 19 and 13-113 in wartime Germany or Grammonal A-8 and A-45 in the former USSR. However, there are records of other compositions bearing the same designation "ammonal" containing additional ingredients. The detonation velocity of ammonal is lower than the corresponding Amatols. However, the highly exothermal afterburn of aluminum yields a significant extension of the positive pressure phase and hence increases the overall mechanical energy as is evident from the greater volume obtained in the Trauzl test. Ammonal were used as underwater explosives in depth charges and torpedo warheads. In contrary to the Amatols, the Ammonals are more sensitive to friction and impact due to their aluminum content and react upon bullet impact. It is hence that the Amatols were replaced in German torpedo warheads with formulations based on TNT/hexanitrodiphenyl-amine/aluminum (*Schießwolle)* which do not detonate upon bullet impact.

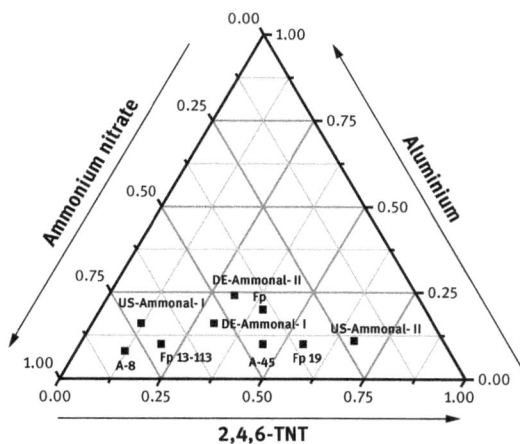

Figure A.10: Explosives based on the ternary system Al/AN/TNT.

Ammonium 2,4,5-trinitroimidazolate, ATNI

Ammonium-2,4,5-trinitroimidazolat

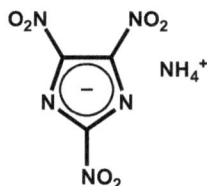

Formula		$C_3H_4N_6O_6$
CAS		[63839-60-1]
m_r	g mol^{-1}	220.101
ρ	g cm^{-3}	1.835
$\Delta_f H$	kJ mol^{-1}	−87
$\Delta_{ex}H$	kJ mol^{-1}	−1107.6
	kJ g^{-1}	−5.032
	kJ cm^{-3}	−9.234
$\Delta_c H$	kJ mol^{-1}	−1665.2
	kJ g^{-1}	−7.566
	kJ cm^{-3}	−13.883
Λ	wt.-%	−14.54
N	wt.-%	38.18
Mp	°C	248
DSC-onset	°C	320
Impact	cm	50.3 (type 12)
V_D	m s^{-1}	8560 @ 1.835 g cm^{-3}
P_{CJ}	GPa	33

– M. D. Coburn, *Ammonium 2,4,6-trinitroimidazole*, US 4028154, **1977**, USA.

Ammonium 3,5-diaminopicrate, ADAP

Ammonium-3,5-diamino-2,6-dinitrophenolat

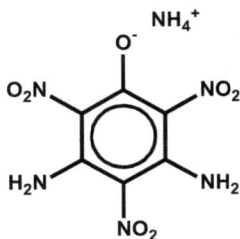

Aspect		Yellowish solid
Formula		$C_6H_8N_6O_7$
CAS		[352543-47-6]
m_r	g mol^{-1}	276.1637
ρ	g cm^{-3}	>1.884
Δ_fH	kJ mol^{-1}	−287.76±52.47
$\Delta_{ex}H$	kJ mol^{-1}	−1247.16
	kJ g^{-1}	−4.516
	kJ cm^{-3}	−8.508
Δ_cH	kJ mol^{-1}	−3216.7
	kJ g^{-1}	−11.648
	kJ cm^{-3}	−21.944
Λ	wt.-%	−52.14
N	wt.-%	30.43
Mp	°C	267(dp)
DSC-onset	°C	238
Impact$_{Rotter}$	FOI	90
V_D	m s^{-1}	*7764* (EXPLO 5.0)
P_{CJ}	GPa	*26.8* (EXPLO 5.0)

ADAP is an experimental insensitive high explosive.

- A. J. Bellamy, A. E. Contini, J. Padfield, P. Golding, M. F. Mahon, Synthesis of the Ammonium Salt of 6-Amino-2-Hydroxy-3,5-Dinitropyrazine and a Comparison of its Properties with Those of Ammonium 3,5 Diaminopicrate (ADAP), *Propellants Explos. Pyrotech.* **2009**, *34*, 377–384.
- A. J. Bellamy, S. Ward, P. Golding, Synthesis of Ammonium Diaminopicrate (ADAP), a New Secoondary Explosive, *Propellants Explos. Pyrotech.* **2002**, *27*, 59–61.

Ammonium azide
Ammoniumazid

Aspect		Colorless platelets
Formula		NH_4N_3
REACH		LPRS
EINECS		240-827-6
CAS		[12164-94-2]
m_r	g mol^{-1}	60.059
ρ	g cm^{-3}	1.3459
Δ_fH	kJ mol^{-1}	114
Dp	°C	160
Λ	wt.-%	−53.28
N	wt.-%	93.29
S	g L^{-1}	201.6 at 20 °C

Though an ionic substance ammonium azide has very little cohesive energy and hence is very volatile and already at ambient temperature completely dissociates into HN_3 and NH_3. It has no practical use other than serving as a model substance with high nitrogen content.

- H. D. Fair, R. F. Walker, *Energetic Materials 1- Physics and Chemistry of the Inorganic Azides*, Plenum Press, New York, **1977**, p.72–73.

Ammonium ceric(IV) nitrate
Cerammoniumnitrat

Ammonium hexanitrato cerate(IV)		
Aspect		Orange red
Formula		$(NH_4)_2[Ce(NO_3)_6]$
H-phrases		272-315-319-355
P-phrases		221-210-305+351+338-302+352-321-501a
REACH		LPRS
EINECS		240-827-6
CAS		[16774-21-3] bzw. [10139-51-2]
m_r	g mol^{-1}	548.222
ρ	g cm^{-3}	2.49
Δ_fH	kJ mol^{-1}	−2370 (est.)
Λ	wt.-%	+25.79
N	wt.-%	20.44

Ammonium ceric(IV) nitrate is soluble in both water and alcohol and bears a slightly stinging odor. *Jennings-White* and *Barr* independently investigated the potential of A. in pyrotechnics and propellants. It is used as a phase stabilizer and burn rate modifier for AN.

- C. Jennings-White, Some Esoteric Firework Materials, *Pyrotechnica*, **1990**, *XIII*, 26–32.
- G. M. Barr, Alternative Oxidation Systems for Propellants and Pyrotechnics, *Transfer Report MPhil/PhD*, Cranfield University, June, **2010**.

Ammonium chlorate
Ammoniumchlorat

Formula		NH_4ClO_3
CAS		[10192-29-7]
REACH		LPRS
EINECS		233-468-1
m_r	g mol^{-1}	101.49
ρ	g cm^{-3}	1.93 (XRD), 1.91 (flotation)
$\Delta_f H$	kJ mol^{-1}	−231.8
$\Delta_{ex}H$	kJ mol^{-1}	−194.96
	kJ g^{-1}	−1.921
	kJ cm^{-3}	−3.706
Λ	wt.-%	+23.65
N	wt.-%	13.801
Dp	°C	>60
Friction	N	>360
Impact	J	4
V_D	m s^{-1}	3300 @ 0.9 g cm^{-3} (iron tube $\varnothing$ 26 mm, 4 mm wall)
Trauzl	cm^3	245–254
Kast	mm	1.52

Ammonium chlorate is a very unstable compound that decomposes even in very pure state at ambient temperature and pressure according to

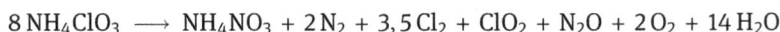

$$8\,NH_4ClO_3 \longrightarrow NH_4NO_3 + 2\,N_2 + 3,5\,Cl_2 + ClO_2 + N_2O + 2\,O_2 + 14\,H_2O$$

When heated to T > 75 °C explosion occurs after a delay between 70 and 10 min depending on the age of the sample. Thus, aged material (44 days) explodes after 10 min whereas pristine AC may take up to 70 min to explode. The activation energy for the decomposition is ~100 kJ mol^{-1}. AC may form accidentally from the double decomposition of mixtures from chlorates and ammonium salts according to

$$NH_4X + MClO_3 \longrightarrow NH_4ClO_3 + MX$$

explaining the great explosion hazard associated with these mixtures.

- H. Kast, Über explosible Ammonsalze, *Z. Schieß. Sprengw.*, **1926**, *21*, 204–209.
- R.B. Gillespie, P. K. Gantzel, K. N. Trueblood, The Crystal Structure of Ammonium Chlorate, *Acta Cryst.* **1962**, *15*, 1271–1272.
- F. Solymosi, Kinetik und Mechanismus der thermischen Zersetzung von Ammoniumhalogenaten im festen Zustand, in V. Boldyrev, K. Meyer, *Festkörperchemie*, VEB Deutscher Verlag für Grundstoffindustrie, Leipzig, **1973**, 424–441.

Ammonium chloride

Ammoniumchlorid, Salmiak

Formula		NH_4Cl
H-phrases		302-319
P-phrases		264-280-301+312-305+351+338-337+313-501a
REACH		LRS
EINECS		235-186-4
CAS		[12125-02-9]
m_r	g mol^{-1}	53.491
ρ	g cm^{-3}	1.527
Sbl	°C	340
$\Delta_f H$	kJ mol^{-1}	−314
Λ	wt.-%	−44.87
N	wt.-%	26.19

Ammonium chloride is a slightly hygroscopic (DRH 80%), colorless salt. It is used in pyrotechnics in smoke compositions together with $KClO_3$. Even though these compositions are prone to undergo metathesis to yield KCl and the highly unstable (see *ammonium chlorate*)

$$KClO_3 + NH_4Cl \longrightarrow KCl + NH_4ClO_3$$

this is not commonly observed. *Shimizu* assumed that the lower solubility product of $KClO_3$ ($K_{L,\ 20\,°C} = 0.355$) compared to NH_4ClO_3, ($K_{L,\ 20\,°C} = 28.787$) impedes the reaction and thereby shifts the equilibrium to the left.

In addition to its use in smoke compositions NH_4Cl is also used as chlorine source in colored flares and strobe formulations (see Table S.9, p. 619).

- T. Shimizu, *Fireworks*, Pyrotechnica Publications, **1981**, pp. 342.

Ammonium dichromate(VI)

Ammoniumdichromat

Aspect		Orange red
Formula		$(NH_4)_2Cr_2O_7$
GHS		03,05,06,08,09
H-phrases		200-301-330-334-350-360-372-314-272-312-317-400-410
P-phrases		221-301+310-303+361+353-305+351+338-320-373-401a-405-501a
REACH		LRS, CL-SVHC
EINECS		232-143-1
CAS		[7789-09-5]
m_r	g mol^{-1}	252.065
ρ	g cm^{-3}	2.155
$\Delta_f H$	kJ mol^{-1}	−1806.65
Dp	°C	180
Λ	wt.-%	0
N	wt.-%	11.11

Ammonium dichromate (VI) is nicely soluble in water and is used as a ballistic modifier in solid rocket propellants.

Ammonium dinitramide, ADN

Ammoniumdinitramid

ADN		
Formula		$NH_4[N(NO_2)_2]$
REACH		LPRS
EINECS		604-184-9
CAS		[140456-78-6]
m_r	g mol^{-1}	124.056
ρ	g cm^{-3}	1.831
$\rho_{100\ °C}$	g cm^{-3}	1.560
c_p	J g^{-1} K^{-1}	1.8
$\Delta_{melt}H$	J g^{-1}	130
$\Delta_f H$	kJ mol^{-1}	−150.6
$\Delta_{ex}H$	kJ mol^{-1}	−420
	kJ g^{-1}	−3.386
	kJ cm^{-3}	−6.199
Λ	wt.-%	+25.79
N	wt.-%	45.16
Mp	°C	92.9
Dp	°C	120 (onset)

E_A	kJ mol^{-1}	134
S	g kg^{-1}	3570 (at 20 °C)
DRH	% RH	55.1 (at 20 °C) non-prilled material
Friction	N	>360
Impact	J	12 (prills), 8 (powder)
V_D	m s^{-1}	5990 @ 1.76 g cm^{-3} (cast with 1 wt.-% MgO) @ 100 mm $\varnothing$
P_{CJ}	GPa	14 @ 43 mm $\varnothing$ and 1.66 g cm^{-3}
$\varnothing_{cr}$	mm	25–40 (depending on density and confinement)

ADN is a high-energy oxidizer. It has been reported first in 1991 by *Bottaro* in the US but soon after was claimed by *Pak* as having been found and used in the USSR long before. ADN nowadays (2020) is still an experimental ingredient in solid rocket propellants as a replacement for ammonium perchlorate (AP) and is investigated as solution in either water or liquid ammonia (Diver's liquid) as liquid monopropellant as replacement for carcinogenic hydrazine. As ADN is chlorine free it allows for both a low tactical signature and a reduced environmental impact as the main combustion products are water and nitrogen. Table A.19 displays the burn rate regime for pure ADN.

Table A.19: Vieille's law for pure ADN.

Pressure interval	$r = a\,p^n$	
	a	n
0.2–6.08	16.8	0.54
6.08–9.12	75.9	−0.774
10.13–35.46	21.11	0.316

The combustion properties of ADN are much differently than AP which impede using ADN as a simple drop-in replacement in standard AP/Al/HTPB-based propellants as is depicted in Figure A.11. Most prominent for ADN is the huge pressure exponent when formulated with inert binders of the HTPB type (solid red diamonds in Figure A.11).

However, with energetic GAP binder ADN yields relatively small pressure exponents with and without additional aluminum (open red squares and solid red circles) in Figure A.11.

Pristine ADN crystals have irregular shape, are prone to mechanical damage, and hygroscopic. Hence, ADN is typically processed to prills with and without reactive coatings which increases mechanical stability, reduces viscosity in casting operations, and also reduces tendency to take up water.

Due to its non-planar dinitramide anion ($\angle$O-N-N-N: −28.3°) and correspondingly reduced charge delocalization in the anion ADN, it is prone to undergo

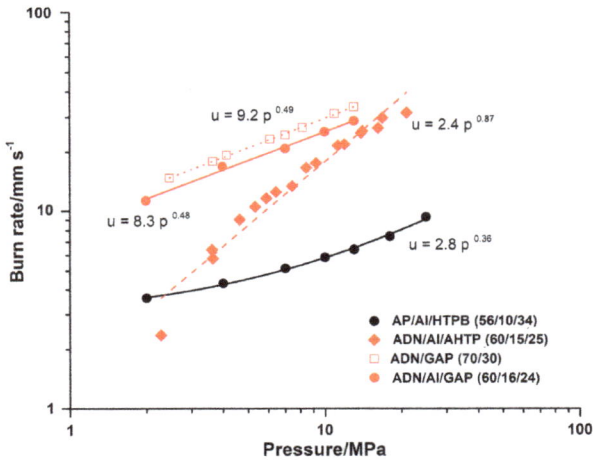

Figure A.11: Burn rate of AP/Al/HTPB compared to different ADN-based propellants.

thermal decomposition through internal proton transfer on the surface of ADN crystals yielding the highly unstable dinitramic acid as follows:

$$[NH_4]^+\,[N(NO_2)_2]^- \longrightarrow NH_3 + HN(NO_2)_2$$

Hence, stabilizers for ADN are weak bases such as hexamine, akardite, or zeolites. Figure A.12 depicts the mass loss for ADN under hot storage with and without stabilization after *Bohn*. Molten ADN is an extremely sensitive material capable to detonate even at charge diameters of 4 mm.

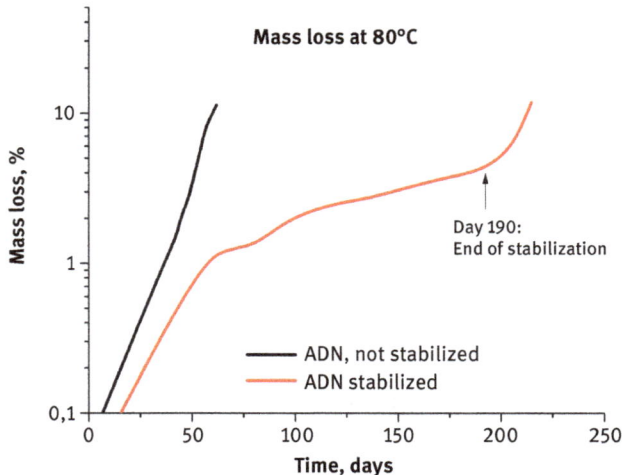

Figure A.12: Mass loss of ammonium dinitramide with and without stabilizer.

– M. Comet, C. Schwartz, F. Oudot, F. Schnell, D. Spitzer, Hazardous Properties of Molten Ammonium Dinitramide, Propellants Explos. *Pyrotech.* **2020**. doi: https://doi.org/10.1002/prep.202000037.

– U. Schaller, V. Weiser, J. Hürttlen, T. Keicher, H. Ciezki, S. Schlechtriem, Investigations on liquid carbon-free gas generating Compositions based on ADN, *IMEMTS*, 22 October **2019**, Sevilla, Spain.

– N. Wingborg, M. Skarstind, M. Sjöblom, A. Lindborg, M. Brantlind, J. Johansson, S. Ek, M. Liljedahl, J. Kjellberg, Ammonium Dinitramide, *50th ICT-JATA*, 25–28. June **2019**, Karlsruhe, Germany, V-29.

– V. Gettwert, M. A. Bohn, S. Fischer, V. Weiser, Performance of ADN/GAP Propellants Compared to Al/AP/HTPB, *IMEMTS*, 18–21 May **2015**, Rome, Italy.

– M. Bohn, Review of some peculiarities of the stability and decomposition of HNF and ADN, *18th NTREM*, Pardubice, Czech Republic, **2015**, 4–25.

– P. Thakre, Y. Duan, V. Yang, Modeling of ammonium dinitramide (ADN) monopropellant combustion with coupled condensed and gas phase kinetics, *Combust. Flame* **2014**, *161*, 347–362.

– N. Wingborg, J. Johansson, M. Johansson, M. Liljedahl, Solid ADN Propellant Development, AIAA 2013-3723, 49th AIAA/ASME7SAE/ASEE Joint Propul. Conf. 14–17 July, **2013**, San Jose, CA, USA.

– J. de Flon, S. Andreasson, M. Liljedahl, C. Oscarsson, M. Wanhatalo, N. Wingborg, Solid Propellants based on ADN and HTPB, AIAA 2011-6136, 47th AIAA/ASME7SAE/ASEE Joint Propul. Conf. 31 July 1 August, **2011**, San Diego, CA, USA.

– A. Hahma, H. Edvinsson, H. Östmark, The properties of Ammonium Dinitramine (ADN): Part 2: Melt Casting, *J. Energ. Mater.* **2010**, *28*, 114–138.

– M. Rahm, T. Brinck, The anomalous solid state decomposition of ammonium dinitramide: a matter of surface polarization, *Chem. Commun.* **2009**, 2869–2898.

– N. Wingborg, Ammonium Dinitramide-Water: Interaction and Properties, *J. Chem. Eng. Data* **2006**, *51*, 1582–1586.

– S. Löbbecke, H. H. Krause, A. Pfeil, Thermal Analysis of Ammonium Dinitramide Decomposition, *Propellants Explos. Pyrotech.* **1997**, *22*, 184–188.

– G. B. Manelis, Thermal decomposition of dinitramide ammonium salt, *26th ICT Jata*, **1995**, Karlsruhe, Germany, V 15.

– Z. Pak, Some Ways to Higher Environmental Safety of Solid Rocket Propellant Application, *AIAA-93-1755, AIAA/SAE/ASME/ASEE 29th Joint Propulsion Conf.*, June 28–30, **1993**, Monterey, CA.

– J. C. Bottaro, R. J. Schmidt, P. E. Penwell, D. S. Ross, *Manufacture of dinitramide salts for rocket propellants*, WO9119670A1, USA, **1991**.

– J. Johansson, N. Latypov, S. Ek, M. Skarstind, H. Skifs, *Synthesis of Ammonium Dinitramide, ADN* US10112834, Sweden, **2018**.

Ammonium nitrate, AN

Ammoniumnitrat, Ammonsalpeter

Formula		NH_4NO_3
GHS		03, 07
H-phrases		H271-H315-H319-H335
P-phrases		P210-P221-P283-P305+P351+P338-P405-P501a
UN		1942
REACH		LRS
EINECS		229-347-8
CAS		[6484-52-2]
m_r	g mol^{-1}	80.043
ρ	g cm^{-3}	1.725
$\Delta_f H$	kJ mol^{-1}	−365.56
$\Delta_{ex}H$	kJ mol^{-1}	−127.2
	kJ g^{-1}	−1.590
	kJ cm^{-3}	−2.742
Λ	wt.-%	+19.99
N	wt.-%	35.00
Mp	°C	169.9
Bp	°C	210
Dp	°C	249

AN is a hygroscopic (DRH = 65%) powder or prilled material. It is the main ingredient in the so-called **ammonium nitrate–fuel oil** (ANFO) explosives. It is not a common oxidizer in hot burning pyrotechnics but frequently used in gas-generating compositions As AN exhibits a series of phase transitions (Table A.20); it is commonly stabilized with $CsNO_3$ or KNO_3 to suppress the phase transitions. Thus, treated AN is designated phase stabilized AN. According to *Price* AN is a group I explosive. AN has an TNT equivalent of about 50 %.

Table A.20: Phase transitions of non-stabilized AN:

Transition	I – II	II – III	III – IV	IV –V	V –VI
T [°C]	125.2	84.2	32.1	−16	−170

- E. G. Mahadevan, *Ammonium Nitrate Explosives for Civil Applications*, Wiley-VCH, Weinheim, **2013**, 214 pp.
- D. Price, Critical Parameters for Detonation Propagation and Initiation of Solid Explosives, *NSWC-TR-80-339*, NSWC, Dahlgren, September **1981**, 94 pp.

Ammonium nitratocuprate(II)

Ammoniumnitratocuprat(II)

Aspect		Dark blue green crystals
Formula		$(NH_4)_3[Cu(NO_3)_4](NO_3)$
CAS		[215184-39-7]
m_r	g mol^{-1}	427.6862
ρ	g cm^{-3}	1.936
$\Delta_f H$	kJ mol^{-1}	−2500
Λ	wt.-%	+29.93
N	wt.-%	26.20
Mp	°C	151
$T_{p(or→c)}$	°C	144

Ammonium nitratocuprate(II) decomposes in moist atmosphere slowly to give copper(II) nitrate and AN.

- V. Morozov, A. A. Fedorova, S. I. Troyanov, Synthesis and Crystal Structure of Alkali Metal and Ammonium Nitratocuprates(II) $M_3[Cu(NO_3)_4](NO_3)$ (M = K, NH$_4$, Rb) and $Cs_2[Cu(NO_3)_4]$, *Z. Anorg. Allg. Chem.* **1998**, *624*, 1543–1547.

Ammonium oxalate

Ammoniumoxalat

Formula		$(NH_4)_2C_2O_4$
H-phrases		H302-H312
P-phrases		P280-P301+P312-P302+P352-P312-P322-P501a
REACH		LPRS
EINECS		238-135-4
CAS		[1113-38-8]
m_r	g mol^{-1}	124.097
ρ	g cm^{-3}	1.48
$\Delta_f H$	kJ mol^{-1}	−1123
Λ	wt.-%	−51.57
N	wt.-%	22.57
Mp	°C	212 (dec)

In moist atmosphere ammonium oxalate forms the monohydrate CAS-No. [6009-70-7], ρ = 1.582 g cm^{-3}, $\Delta_f H$ = −1425 kJ mol^{-1}, which releases the crystal water at T = 78 °C. Ammonium oxalate serves as endothermal additive in cold gas generators and as a component in ablating coolers.

— H. Ratz, H. Gawlick, W. Spranger, G. Marondel, W. Siegelin, *Druckgaserzeugender kühle Gase liefernder Treibsatz*, DE-Patent 1806550B2, Dynamit Nobel **1976**.

Ammonium perchlorate, AP

Ammoniumperchlorat

Formula		NH_4ClO_4
GHS		01
H-phrases		H200-H271
P-phrases		P210-P221-P283-P373-P401a-P501a
UN		1442
REACH		LRS
EINECS		232-235-1
CAS		[7790-98-9]
m_r	g mol^{-1}	117.489
ρ	g cm^{-3}	1.95
$\Delta_f H$	kJ mol^{-1}	−295.77
$\Delta_{ex}H$	kJ mol^{-1}	−131.4
	kJ g^{-1}	−1.118
	kJ cm^{-3}	−2.181
Λ	wt.-%	+34.04
N	wt.-%	11.92
Mp	°C	130
$T_{p(or\rightarrow c)}$	°C	240
Dp	°C	274

AP is non-hygroscopic. It is among the most important pyrotechnic oxidizers and is a main ingredient in many civilian and military composite propellants, e.g. the Ariane 5 rocket uses 162 tons AP. It is also an important ingredient in insensitive HE formulations such as PAX-21 (DNAN/RDX/AP/MNA: 34/35/30/0.7). AP is used in colored signal compositions of 75 wt.-% AP–25 wt.-% M(acac)$_x$ (M = Ca, Sr, Ba, Mo, Zr, Cu, La)(*Dumont*) and propellants burning with a colored trail of 29 wt.-% AP – 30 wt.-% M(NO$_3$)$_2$ (M = Sr, Ba) – 40 wt.-% Mg$_3$Al$_4$ – 10 wt.-% polyester (*Shimizu*), strobe flares, and spectrally adapted IR decoys, as well as IR decoy compositions for kinematic decoy flares.

AP is incompatible with magnesium:

$$Mg + 2\,NH_4ClO_4 \longrightarrow 2\,NH_3 \uparrow + H_2 \uparrow + Mg(ClO_4)_2$$

Hence, only passivated magnesium should be used with AP or preferably magnalium alloy. Mixtures of water-soluble chlorates with AP are extremely dangerous through their propensity to yield sensitive and hazardous *ammonium chlorate*. AP is also incompatible with alkali and alkaline earth carbonates:

$$2\,NH_4ClO_4 + MCO_3 \longrightarrow 2\,NH_3 + M(ClO_4)_2 + CO_2 + H_2O, \quad M = Mg, Ca, Sr, Ba$$

$$2\,NH_4ClO_4 + M_2CO_3 \longrightarrow 2\,NH_3 + 2\,MClO_4 + CO_2 + H_2O, \quad M = Li, Na, K$$

Thermal decomposition of AP at T < 180 °C starts with minute sublimation of AP and subsequent dissociation:

$$NH_4ClO_{4(s)} \longrightarrow NH_4ClO_{4(g)} \longrightarrow NH_3 + HClO_4$$

with ammonium and perchloric acid being adsorbed at the crystal surface. At 240 °C a phase transition occurs from the orthorhombic into the cubic version. At higher temperature between 300 and 350 °C, the adsorbed NH_3 and $HClO_4$ react to give H_2O, O_2, and NO. At T > 430 °C, finally the decomposition of AP is complete. The exact decomposition mechanism is strongly dependent on the reaction conditions.

– J. L. Dumont, Pyrotechnische Sätze zur Herstellung von farbigem Feuerwerk, EP252803B1, France, **1992**.
– T. Shimizu, Compositions containing luminance agents and propellants for signal rockets, JP52120113A, Japan, **1977**.

Ammonium picrate, Explosive D, and Dunnite

Ammoniumpikrat

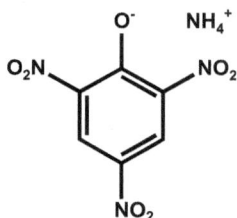

Aspect		Yellow to red crystals
Formula		$C_6H_6N_4O_7$
REACH		LPRS
EINECS		205-038-3
CAS		[131-74-8]
m_r	g mol^{-1}	246.136
ρ	g cm^{-3}	1.717
$\Delta_f H$	kJ mol^{-1}	−386
$\Delta_{ex}H$	kJ mol^{-1}	−1035.1
	kJ g^{-1}	−4.208
	kJ cm^{-3}	−7.225

$\Delta_c H$	kJ mol^{-1}	−2832.6
	kJ g^{-1}	−11.508
	kJ cm^{-3}	−19.760
Λ	wt.-%	−52
N	wt.-%	22.76
Mp	°C	280
Dp	°C	320
Impact	J	>20
Trauzl	cm^3	280
V_D	m s^{-1}	7338 at 1.717 g cm^{-3}, 6850 at 1.55 g cm^{-3}, 4990 at 1.0 g cm^{-3}

Ammonium picrate was used extensively due to its low mechanical and thermal sensitivity in armor piercing naval artillery ammunition. The large stocks of it are considered as starting material for the synthesis of TATB.

- M. Coburn, P. Hsu, G. S. Lee, A. R. Mitchell, P. F. Pagoria. R. Schmidt, *Synthesis and Purification of 1,3,5-triamino-2,4,6-trinitrobenzene (TATB)*, US7057072B, **2006**, USA.

Amorces
Zündplättchen

Amorces, also called toy caps, are small plastic cups or paper tapes containing small amounts of → *Armstrong's mixture*. They are used in toy guns and other toys that are, for example, dropped and initiated upon impact on a hard surface to yield a loud noise. In the EU, the amount of Armstrong's mixture in those amorces has been reduced by legislation in 2001 to prevent sustained hearing disorder with children and teenagers. According to the relevant DIN EN 71 for "toys used in close vicinity to the ear" the emission-sound pressure level, Lp_A, has been lowered from 92 dB (A) to 80 dB (A) explaining the lowering of the charge mass in amorces.

- *Bundestags-Drucksache 15/1159*, Bonn, 10.06.**2003**; http://dipbt.bundestag.de/doc/btd/15/011/1501159.pdf

ANFO
Ammonium nitrate – Fuel Oil Explosives

ANFO are pourable compositions based on prilled AN fuel oil with an approximate mass ratio 94.3 wt.-% AN, 5.5 wt.-% fuel oil, and additives. This composition is just oxygen balanced (Λ = 0 wt.-% O_2), and Table A.21 shows the properties of ANFO with AN prills having high density and low density.

Table A.21: Properties of ANFO.

Parameter	ANFO–HDAN	ANFO–LDAN
V_D (m s^{-1})	1500–1800	4000
ρ (g cm^{-3})	0.90–0.98	0.70–0.82
$\varnothing_{cr(open)}$ (cm)	22.8	6.3
$\varnothing_{cr(confined)}$ (cm)	10.1	4.4

- E. G. Mahadevan, *Ammonium Nitrate Explosives for Civil Applications*, Wiley-VCH, Weinheim, **2013**, 214 pp.

Anthracene

Anthracen

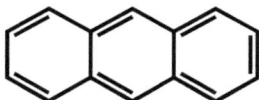

Aspect		Pearlescent crystals
Formula		$C_{14}H_{10}$
GHS		07,09
H-phrases		400-410-317
P-phrases		261-280-302+352-321-363-501a
UN		3077
REACH		LRS, SVHC
EINECS		204-371-1
CAS		[120-12-7]
m_r	g mol^{-1}	178.24
ρ	g cm^{-3}	1.252
$\Delta_f H$	kJ mol^{-1}	129,16
$\Delta_c H$	kJ mol^{-1}	−7067.6
	kJ g^{-1}	−39.655
	kJ cm^{-3}	−49,647
Λ	wt.-%	−296.23
Mp	°C	218
Bp	°C	340
$\Delta_m H$	kJ mol^{-1}	29.3 (215 °C)
$\Delta_{sub} H$	kJ mol^{-1}	98 ± 10 (47 °C)
$\Delta_{vap} H$	kJ mol^{-1}	78.5 (125 °C)

Technical-grade anthracene appears creme to beige. It was used in wartime Germany to replace shellac which had to be imported as binder in pyrotechnic illuminating compositions. It is used in smoke compositions (A/HC/Mg: 20/60/20) where it absorbs heat thereby slowing down the combustion process and yielding soot via cyclodehydrogenation. *Nourdin* proposes anthracene as a component in incendiary mixtures. *Nielson* has developed black-body-type propellant compositions (Mg/AP/A/HTPB: 15/33/35/17) for use in kinematic propelled decoy flares. Dermatological conditions following handling of anthracene are due to contamination of it with higher benzenoid compounds.

- E. Nourdin, *Composition incendiaire et projectile incendiaire dispersant une telle composition*, EP0663376 A1, **1994**, France.
- D. B. Nielson, D. M. Lester, *Black body decoy flare compositions for thrusted applications and methods of use*, US5834680, **1998**, USA.

Anthraquinone
9,10-Anthrachinon

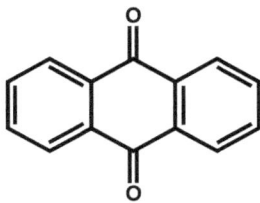

9,10-Anthraquinone, 9,10-anthracenedione,		
Aspect		Pearlescent crystals
Formula		$C_{14}H_8O_2$
GHS		08, 07
H-phrases		351-317
P-phrases		261-280-302+352-321-405-501a
UN		
REACH		
EINECS		201-549-0
CAS		[84-65-1]
m_r	g mol^{-1}	208,216
ρ	g cm^{-3}	1.438
Δ_fH	kJ mol^{-1}	−188,49
Δ_cH	kJ mol^{-1}	−6464.1
	kJ g^{-1}	−31.046
	kJ cm^{-3}	−44.644
Λ	wt.-%	−230.53
Mp	°C	286
Bp	°C	380
Δ_mH	kJ mol^{-1}	32.6 (285 °C)
$\Delta_{sub}H$	kJ mol^{-1}	114 ± 20 (197 °C)
$\Delta_{vap}H$	kJ mol^{-1}	64.3 (301 °C)

Anthraquinone can be used as carbon source in IR-opaque obscurant formulations and has been investigated as flare fuel in spectral aerial countermeasures.

- E.-C. Koch, Pyrotechnic Countermeasures. III: The Influence of Oxygen Balance of an Aromatic Fuel on the Color Ratio of Spectral Flare Compositions, *Propellants Explos. Pyrotech.* **2007**, *32*, 365–370
- J. Schneider, H. Büsel, *Composition generating an IR-opaque smoke*, US5,389,308, Germany, **1995**.

Antimony
Antimon

Aspect		Silver gray to black
Formula		Sb
GHS		06, 09
H-phrases		301-332-411
P-phrases		261-301+310-304+340-321-405-501a
UN		2871
REACH		LRS
EINECS		231-146-5
CAS		[7440-36-0]
m_r	g mol^{-1}	121.757
ρ	g cm^{-3}	6.691
Mp	°C	631
Bp	°C	1587
$\Delta_m H$	kJ mol^{-1}	21.985
$\Delta_b H$	kJ mol^{-1}	165.8
c_p	J K^{-1} mol^{-1}	25.23
c_L	m s^{-1}	3140
κ	W m^{-1} K^{-1}	25.9
$\Delta_c H$	kJ mol^{-1}	−485.5
	kJ g^{-1}	−3.987
	kJ cm^{-3}	−26.680
Λ	wt.-%	−19.71

Sb is used in fireworks to produce white flames together with KNO_3 and sulfur. Sb was used in wartime Germany as fuel in gasless delay compositions with $KMnO_4$ as oxidizer ($KMnO_4$/Sb: 67/33). *Moghaddam* has investigated the combustion behavior of Sb/$KMnO_4$ mixtures. *Brown et al (1998)* studied the influence of various Sb-particle geometries in the system Sb/$KMnO_4$ on the burn rate. *Held* studied Sb/element compositions as *coruscative* liners in shaped charges.

– A. Z. Moghaddam, Combustion reactions of antimony + potassium permanganate mixtures. Rate of propagation and influence of loading pressure, *Thermochimica Acta* **1993**, *223*, 193–200.
– M. E. Brown, S. J. Taylor, M. J. Tribelhorn, Fuel – Oxidant Particle Contact in Binary Pyrotechnic Reactions, *Propellants Explos. Pyrotech.* **1998**, *23*, 320–327.

Antimony(III) sulfide, Stibnit

Antimonsulfid, Grauspießglanz

Aspect		Black or orange red
Formula		Sb_2S_3
GHS		07, 09
H-phrases		302-332-411
P-phrases		261-273-301+312-304+340-312-501a
UN		1549
REACH		LRS
EINECS		215-713-4
CAS		[1345-04-6] or [1317-86-8]
m_r	g mol^{-1}	339.68
ρ	g cm^{-3}	4.64
Mp	°C	565 (under inert atmosphere)
Dp	°C	250
$\Delta_f H$	kJ mol^{-1}	−147.28
$\Delta_c H$	kJ mol^{-1}	−1715
	kJ g^{-1}	−5.049
	kJ cm^{-3}	−23.427
Λ	wt.-%	−42.39
c_p	J K^{-1} mol^{-1}	117.87
κ	W m^{-1} K^{-1}	25.9

Similar to sulfur antimony(III) sulfide is applied to lower the ignition temperature of many pyrotechnic compositions. Combined with its high hardness it is advantageously used in friction formulations as both fuel and sensitizer. It is used in sensitive report charges with potassium chlorate or perchlorate and aluminum. It is also applied as fuel in glitter formulations and white fire mixtures. The freshly precipitated material appears orange red and is 28 kJ mol^{-1} higher in energy than the aged black modification and is not to be confused with the likewise colored antimony(V) pentasulfide, Sb_2S_5.

- C. J. White, Glitter Chemistry, *J. Pyrotech.* **1998**, *8*, 53–70.
- C. K. Kelley, Thermal Analysis Study of Antimony Sulfides, *WRDC-TR-89-2099*, Wright-Patterson AFB OH, July **1989**, 84 pp.

Antimony(V) sulfide
Diantimonpentasulfid, Goldschwefel

Aspect		Orange red
Formula		Sb_2S_5
GHS		02, 07
H-phrases		228-315-319-335
P-phrases		210-241-305+351-338-302+352-405-501a
UN		3178
REACH		LPRS
EINECS		215-255-5
CAS		[1315-04-4]
m_r	g mol^{-1}	403.80
ρ	g cm^{-3}	4.12
Dp	°C	74-76
Λ	wt.-%	−51.51

Antimony(V) sulfide is occasionally used as an ingredient in slow burning delay compositions.

- B. Sebela, P. Rychetsky, *Slowly Burning Pyrotechnical Composition*, CZ-Patent 184172B1, Czech Republic, **1978**.
- M. Miszczak, M. Nita, R. Warchoł, X-ray Investigations of Combustion Phenomena Occurring in Confined Pyrotechnic Charges, *Centr. Eur. J. Energ. Mater.* **2015**, *12*, 553–561.

AOP

The AOP = *Allied Ordnance Publication* is a series of publications distributed by the NATO Standardization Office (NSO) dealing with ammunition and the energetic materials contained therein. As short notice changes and amendments of AOPs do not require ratification in the CNAD but STANAGs (Standardization Agreements) do, it has been agreed in 2014 that technical details are to be documented in AOPs only from this date on. Selected important AOPs are given in Table A.22.

Table A.22: Selected energetic material and ammunition related AOPs.

AOP-2	Identification of Ammunition
AOP-6	Catalogue of Interchangeable Ammunition
AOP-7	Manual of Data Requirements and Tests for the Qualification of Explosive Materials for Military Use
AOP-8	NATO Fuze Characteristics Data
AOP-15	Guidance on the Assessment of the Safety and Suitability for Service of Non-Nuclear Munitions for NATO Armed Forces
AOP-16	Fuzing Systems – Guidelines for STANAG 4187
AOP-20	Manual of Tests for the Safety Qualification of Fuzing Systems
AOP-21	Initiation Systems Characterization and Safety Test Methods and Procedures for Detonating Explosive Components
AOP-22	Design Criteria and Test Methods for Inductive Setting of Large caliber Projectile Fuzes
AOP-26	NATO Catalogue of Qualified Explosives
AOP-29	NAAG – Surface-to-Surface Artillery Panel
AOP-31	Demolition Materiel Design Principles
AOP-32	Demolition Materiel Assessment and Testing of Safety and Suitability for Service
AOP-34	Vibration Test Method and Severities for Munitions Carried in Tracked Vehicles
AOP-36	NATO Hand Book of Standard Smoke Munitions
AOP-38	Specialist Glossary of Terms and Definitions on Ammunition Safety
AOP-39	Guidance on the Assessment and Development of Insensitive Munitions
AOP-40	Ammunition Data Sheets
AOP-42	Integrated Design Analysis for Munition Initiation and Other Safety Critical Systems
AOP-43	Electro-Explosive Devices Assessment and Test Methods for Characterization
AOP-46	The Scientific Basis for the Whole Life Assessment of Munitions
AOP-48	Explosives, Nitrocellulose-Based Propellants, Stability Test Procedures and Requirements Using Stabilizer Depletion
AOP-56	Compendium of Chemical and Physical Tests for Analysis of Energetic Materials Against their Applicable NATO STANAG
AOP-57	Test for Measuring the Burning Rate of Solid Rocket Propellants with Subscale Motors
AOP-58	Methods for Analyzing Data from Tests designed to Measure the Burning Rate of Solid Rocket Propellants with Subscale Motors
AOP-59	Non-Intrusive Methods for Measuring the Burning Rate of Solid Rocket Propellants
AOP-60	In-Service Surveillance of Munitions Conditions Monitoring of Energetic Materials
AOP-62	In-Service Surveillance of Munitions – General Guidance
AOP-64	In-Service Surveillance of Munitions – Condition Monitoring of Energetic Materials

As long as these documents are not restricted, they are directly available from the NSO website:

Weblink: https://nso.nato.int/nso/

Aquarium test

The aquarium test serves the determination of the detonation velocity, V_D (m s^{-1}), and the detonation pressure, P_{CJ} (GPa). Water is optically transparent and inert and hence allows the observation of the shock wave emanating from a detonating charge through the change of the index of refraction (Schlieren method). Alternatively, other transparent media, for example, PMMA can be used too.

- J. K. Rigdon, I. B. Akst, An Analysis of the „Aquarium Technique" as a Precision Detonation Pressure Measurement Gage, *Det. Symp.*, **1970**, 59–66.

ARC – accelerated rate calorimetry

ARC can be used to determine the thermal stability of explosives. Therefore, the sample is heated stepwise in a very sensitive adiabatic calorimeter (Figure A.13). The sample is left for a certain period to determine any change of sample temperature. The result of the measurement is the time to reach the maximum reaction rate (*time to maximum rate*) as depicted in Figure A.13.

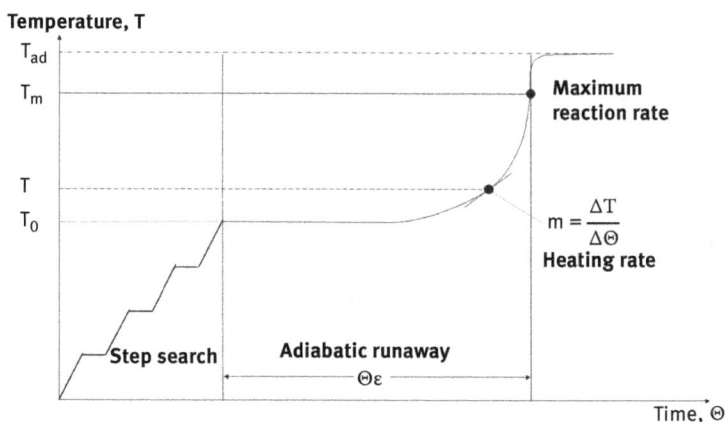

Figure A.13: ARC diagram (after *Townsend*).

- D. I. Townsend, Accelerating Rate Calorimetry, *I. Chem. E. Symposium Series No. 68*, 3/Q1-3/Q-14, **1981**.

Argon flash and argon bomb

Argon-Blitz, Argon-Bombe

Upon detonation of HEs in an atmosphere the incident shock waves heat up the gas by adiabatic compression to high temperatures 8,000–10,000 K causing a flash of bright light. This light can be used in turn as an illumination source in high-speed photography of detonation processes. Molecular gases upon shock dissociate ($\Delta_{diss}H(O_2) = 498$ kJ mol^{-1}, $\Delta_{diss}H(N_2) = 945$ kJ mol^{-1}) and hence absorb considerable energy, thereby limiting the accessible temperatures to ~8,000 K and consequently limiting the radiation intensity. In contrary, monoatomic gases such as argon can be heated up to temperatures in excess of 10,000 K. This is also favored by a low heat capacity of Ar. An argon bomb is a container filled with argon having an optically transparent window on one side and a small HE charge on the opposite site as is indicated in Figure A.14. In a very simple implementation, a balloon is filled with argon and a piece of detonation cord stripped to its back is used to shock heat the argon.

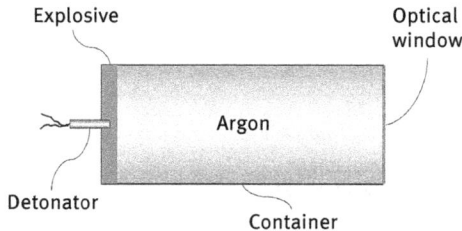

Figure A.14: Argon bomb after *Davis*.

- W. C. Davis, T. R. Salyer, S. I. Jackson, T. D. Aslam, Explosive-Driven Shockwaves in Argon, *13th International Detonation Symposium*, 1035–1044, **2006**.
- M. Held, High-Speed Photography, J. Carleone (Eds.) *Tactical Missile Warheads*, Vol 155 *Progress in Astronautics and Aeronautics*, A.R. Seebass (Eds.), American Institute of Astronautics and Aeronautics, Washington, **1993**, pp. 609–673.

Armor or armour

Panzerung

An armor mainly protects against blast waves and ballistic threats like projectiles (bullets, fragments), kinetic energy rods, shaped charges, and explosively formed penetrators. In general, there are three types of armor such as passive, reactive, and active armor.

Passive armor

Passive Panzerung

Projectile

For a passive armor struck by a cylindrical projectile, many empirical correlations have been developed to describe the perforation such as the *Wollmann* formula:

$$\left(\frac{P}{L}\right) = \left(1 - \frac{D}{L}\right)\mu\left(1 - e^{\frac{v}{0.6}}\right)^8 + 2.64 \cdot \frac{D}{L}\left(\frac{v}{4}\right)^{\frac{2}{3}}$$

with

D = Rod diameter (m)

$\mu = \sqrt{\frac{\rho_p}{\rho_T}}$

L = Rod length (m)

P = Penetration depth (m)

ρ_p = Density of penetrator (kg m^{-3})

v = Rod velocity (m s^{-1})

ρ_T = Density of target (kg m^{-3})

If an armor plate is inclined at a yaw angle β relative to the incoming projectile, the penetration can be described accordingly:

$$P = (P_0 - P_1)e^{-a\left(\frac{\beta}{b}\right)^2} + P_1$$

with

$$b = \sin^{-1}\left(\frac{H}{D} - \frac{1}{2}\left(\frac{L}{D}\right)\right)$$

$$P_0 = L\left(\frac{P}{L}\right)$$

$$P_1 = D\left(\frac{P}{L}\right)$$

$$a = 0.2\left(\frac{L}{D}\right)^{-0.8}$$

Figure A.15 depicts how inclination of armor can also increases the effective thickness of armor explaining the sharp front contours of modern armored vehicles.

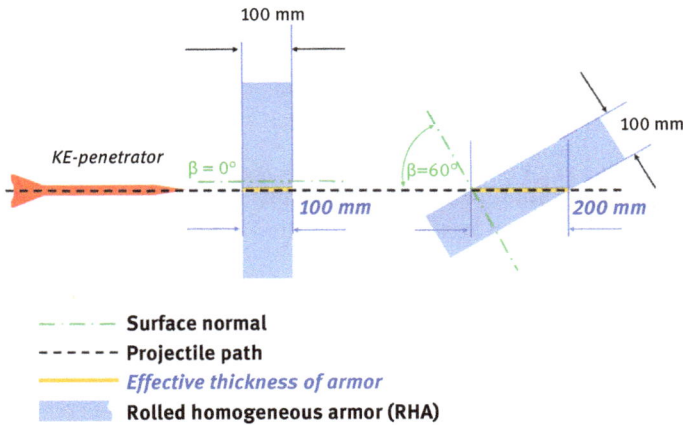

Figure A.15: Inclination of armor increases effective thickness of armor.

Shaped charge jet

The crater depth formed by a shaped charge jet in a semi-infinite armor plate is determined by the density of the jet, ρ_j, its velocity, v_j and the yield strength, σ, and density of the armor plate, ρ_T. It is known that at $v_j \le v_j^*$ the jet starts to stick in the target.

$$v_j^* = 1.25 \cdot \sqrt{2\sigma_T} \cdot \left(\rho_j \cdot \rho_T\right)^{-\frac{1}{4}}$$

The sticking of the jet in an armor plate is due to the tapering of the crater increasing with yield strength of the armor plate. However, also other materials and target configurations can cause a tapering of the crater and resulting in an eventual closure of the crater causing the jet to stick. Glass armor upon impact spalls massively, thereby increasing its volume and consequently affecting an almost instantaneous closure of the crater leading to rapid sticking of a jet.

Reactive armor
Reaktive Panzerung

With reactive armor, the counteraction is always triggered by the projectile or jet itself. It can be distinguished between

    explosive reactive armor
    self-limiting explosive reactive armor
    non-explosive reactive armor
    non-energetic reactive armor (NERA)

In general, reactive armor consists of armor plates driven by a chemical or physical mechanism:

- Fast lateral movement of the armor plates results in increased erosion and consumption of the jet and reducing its residual penetration depth. In this context reactive armor plates have a "dynamic thickness" which is indeed greater than their actual physical thickness.
- The shockwaves from detonating fills (explosive reactive armor) deflect and influence the jet causing reduced penetration.
- The armor plates collide with the jet and cause the jet particles to impinge one next to the other but not in the same crater thereby reducing the penetration depth.
- The jet particles burst upon collision with the crater walls mainly orthogonal to the direction of flight. Hence, multiplate spaced armor with air interspace allows for expansion of particle spray. By this, large parts of the jet can be consumed reducing the mass impinging on the target.
- Oxidative erosion of jets by armor elements filled with dense oxidizers ($2PbO \cdot PbO_2$, WO_2, etc.).
- In certain geometrical configurations (high-density polyethylene packed in steel) can lead to refocusing of the shockwave back onto the crater and therefore closing the crater with the jet sticking in it. Armor working in accordance with this concept are termed NERA.

Active armor
Aktive Panzerung (ADS)

Active armor starts with sensors that detect and track the threat by means of radar and laser. Then tailored blast and fragment charges are projected against the incoming threat in an effort to achieve deflection and destruction of the warhead or deflection and vibrational load on a kinetic energy penetrator with the aim to reduce its effect on the target.

- U. Deisenroth, *Aktive Panzerung gegen Hohlladungen*, DE3132008C1, **1999**, Germany.
- E. Wollmann, K. Hoog, G. Koerber, B. Wellige, *Performance of Ballistic Terminal Performance at Incidence*, RI 10/96, Institut Franco-Allemand De Recherches De Saint-Louis, ISL, **1996**.
- C. Fauquignon, P.-Y. Chanteret, Entwicklung der Panzerungen gegen Hohlladungen in G. Weihrauch (Eds.), *Ballistische Forschung im ISL 1945–1994*, ISL, Saint Louis, France, **1994**, 75–84.
- W. Trinks, Hohlladungen und Panzerschutz, *Jahrbuch der Wehrtechnik, 8*, **1974**, pp. 155–163.

Armstrong's mixture

Armstrongsche Mischung

This is a composition made up of

- 75 wt.-% potassium chlorate
- 25 wt.-% red phosphorus

with a small amount of calcium carbonate to buffer acid traces.

Armstrong's mixture is known since the nineteenth century and due to its extreme sensitiveness and high brisance, it was initially considered as a primary explosive. Although it is advised to prepare small amounts of Armstrong's mixture under an aqueous solvent only to reduce the immense friction sensitiveness, terrible accidents regularly occur!

- R. R. Rollins, Potassium Chlorate/Red Phosphorus Mixtures, *7th Symposium on Explosives and Pyrotechnics*, Philadelphia, **1981**, III-11.
- Press report from an accident in **2016** http//www.rsc.org/eic/2016/01/technician-injury-explosion-bristol

Arrested reactive milling, ARM

ARM is a tribochemical activation of condensed reactive binary or multicomponent mixtures constituted from particulate ingredients. Therefore, the mixtures are subjected to an intense milling process at ambient or cryogenic (cryo-milling) temperatures that eventually stops right before the onset of a reaction. The high mechanical energy input yields a comminution of the ingredients and an increase in specific surface area. With metallic particles an ablation of inhibiting oxide layers occurs and in general the number of particle-contact points between the ingredients is vastly increased which affects reduced diffusion pathways, consequently causes reduced activation energy (lowered ignition temperature), but also increases sensitiveness of the materials prepared in this manner.

- T. R. Sippel, S. F. Son, L. J. Groven, Altering Reactivity of Aluminum with Selective Inclusion of Polytetrafluoroethylene through Mechanical Activation, *Propellants Explos. Pyrotech.* **2013**, *38*, 286–295.
- E. L. Dreizin, Metal-based reactive nanomaterials, *Prog. Energ. Combust. Sci.* **2009**, *35*, 141–167.
- M. Schoenitz, T. S. Ward, E. L. Dreizin, Arrested Reactive Milling for In-Situ Production of Energetic Nanocomposites for Propulsion and Energy-Intensive Technologies in Exploration Missions, AIAA2005-717, *43rd AIAA Aerospace Sciences Meeting and Exhibition*, 10–13 January, **2005**, Reno, USA.

Arrhenius equation

Arrhenius Gleichung

Arrhenius (1859–1927) found that for a reaction

$$A \xrightarrow{k} B$$

the rate k is proportional to an exponential function bearing the inverse temperature in the exponent

$$k \sim e^{1/T}$$

Expanding the exponent with both the Boltzmann and the Avogadro constant yields the expression

$$k = A \cdot e^{-\frac{E_a}{RT}}$$

where A = pre-exponential factor, E_a = activation energy (kJ mol^{-1}), and R = gas constant.

To determine the activation energy, first, the order of the reaction is ascertained. Then the rate constant is plotted as ln k against 1/T which according to the earlier equation should yield a straight line. The slope of which yields the activation energy. Typical activation energies for selected energetic materials are depicted in Tables A.4–A.7 for different groups of energetic materials in ascending order.

- M. A. Libermann, Introduction to Physics and Chemistry of Combustion, Springer, Berlin, **2008**, 54.

ARX

ARX is the TLA for the **A**ustralian **r**esearch e**x**plosive nomenclature system. It is used for experimental PBX and melt-cast compositions. The explosives are categorized into the following classes:

ARX 1000–1999 Pressed Compositions
ARX 2000–2999 Inert Binder Compositions
ARX 3000–3999 Energetic Binder Compositions
ARX 4000–4999 TNT-Based and Other Melt-Castables
ARX-5000 → yet to be assigned

- M. D. Cliff, R. M. Dexter, Nomenclature and Cataloguing of Experimental Explosive Compositions, DSTO-TN-0284, DSTO Aeronautical and Maritime Research Laboratory, May **2000**, 17 pp.

Asphaltum

Asphalt

Asphaltum is a naturally occurring mixture of high boiling hydrocarbons with the approximate elemental composition $C_6H_{8.42}O_{0.07}N_{0.2}S_{0.008}$, CAS-Nr. [12002-13-6], $\rho = 1.05$ g cm^{-3}, and $\Delta_f H = -27.6$ kJ mol^{-1}. Asphaltum starts to melt at T = 260 °C. In the past, it has been used extensively as fuel and binder in pyrotechnic flares as well as bonding agent between HE and casings in warheads.

ASTM – American Association for Testing Materials

The ASTM International (initially American Society for Testing and Materials) is an international organization concerned with standardization. It is located in West Conshohocken, PA, USA, and publishes technical standards for goods and services. Until 2019 more than 13,000 ASTM Standards have been published with the main area of work devoted to the development of standardized test and analytical procedures.

– Website https://www.astm.org/ accessed on October 15 2019

ATEC, acetyltriethyl citrate

Acetyltriethylcitrat

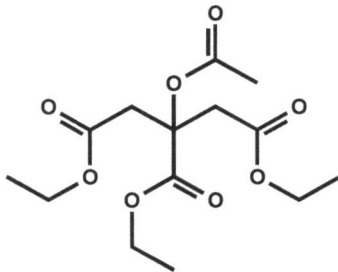

Citroflex ®		
Formula		$C_{14}H_{22}O_8$
REACH		LPRS
EINECS		201-066-5
CAS		[77-89-4]
m_r	g mol^{-1}	318.324
ρ	g cm^{-3}	1.135
$\Delta_f H$	kJ mol^{-1}	−1738.87

$\Delta_c H$	kJ mol^{-1}	−6914.5
	kJ g^{-1}	−21.722
	kJ cm^{-3}	−24.654
Λ	wt.-%	−155.81
Mp	°C	−43
Bp	°C	132

ATEC is an odorless clear high boiling liquid. It is a non-energetic plasticizer for PVC, PVA, CA, CAB, and PU. It is used frequently in pharmaceutical tablets and is considered non-toxic.

- R. L. Simmons, Effect of plasticizer on performance of XM-39 LOVA, *27th ICT-JATA*, 25–28 June **1996**, Karlsruhe, P-53.

Augmentation flare
Leuchtkörper für Luftziel

Augmentation flares provide a towed or self-propelled aerial target with an appropriate visible and/or infrared signature to facilitate detection and electrooptical tracking at tactical distances. For this purpose, augmentation flares have long burn times $t_b > 40$ s and must provide a constant level infrared radiant intensity, in the > 600 W sr^{-1} range in the PbSe band.

- E.-C. Koch, *Metal-Fluorocarbon Based Energetic Materials*, Wiley-VCH, Weinheim, **2012**, p. 153–155.

Autoignition temperature
Zündtemperatur

The autoignition temperature is the temperature at which a combustible or explosive material when brought to spontaneously ignites. The underlying mechanism leading to autoignition is called *thermal explosion*. *McIntyre* lists a great variety of autoignition temperatures for pyrotechnic compositions

- J. Harris, Autoignition Temperature of Military High Explosives by Differential Thermal Analysis, *Thermochim. Acta* **1976**, 14 183–199.
- F. L. McIntyre, A Compilation of Hazard and test Data for Pyrotechnic Compositions, *ARLCD-CR -80047*, Dover, NJ, **1980**.

Azides

Azide

Azides designate the salts of the hydrazoic acid, HN_3. *Sodium azide*, NaN_3, was the main ingredient in the first-generation gas-generating composition for automobile airbags. However, due to the high toxicity of NaN_3, it has been replaced in the meantime with less toxic and more efficient high nitrogen compounds. *Lead azide* at the moment is still the most important primary explosive in civilian and military detonators followed by *silver azide*. The term azide also designates organic compounds bearing the $-N_3$ group. For example, *GAP,* glycidyl-azide polymer an important energetic binder or trimethylsilyl azide, $(H_3 C)_3 Si-N_3$, an important agent to introduce the azide groups into organic substrates.

- H. D. Fair, R. F. Walker, *Energetic Materials – Physics and Chemistry of the Inorganic Azides* Volume 1, **1977**, 503 S.; Ibid *Energetic Materials – Technology of the Inorganic Azides* Volume 2, Plenum Press, **1977**, 296 pp.
- Trimethylsilylazid, in G. Brauer, *Handbuch der Präparativen Anorganischen Chemie*, Band II, **1978** pp. 710.

AZM

AZM is an acronym preceding a three- or four-digit number designating an ignition mixture (Ger.: **Anzündm**ischung) in the authority of the German Armed Forces.

For example,

AZM 09530 boron/potassium nitrate/binder, or

AZM 09670 barium chromate/zirconium/titanium/potassium perchlorate/NC.

Azo-bis(2,2′,4,4′,6,6′-hexanitrodiphenyl), ABH

1,2-Bis(2,2′,4,4′,6,6′-hexanitro[1,1′-biphenyl]-3-yl)diazen

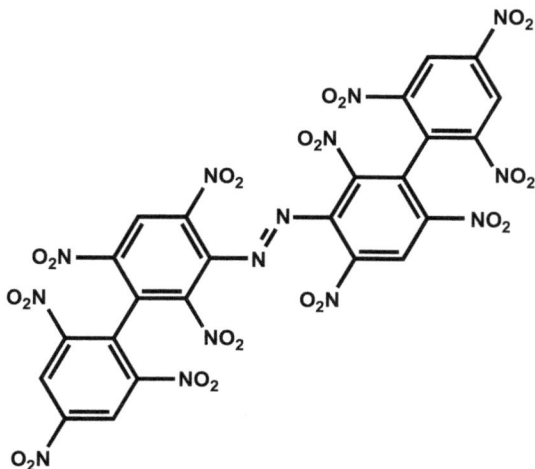

Aspect		Adobe-red brown crystals
Formula		$C_{24}H_6N_{14}O_{24}$
CAS		[23987-32-8]
m_r	g mol^{-1}	874.391
ρ	g cm^{-3}	1.78
$\Delta_f H$	kJ mol^{-1}	478.23
Ω	wt.-%	−49.4
N	wt.-%	22.42
$\Delta_{ex}H$	kJ mol^{-1}	−4472.2
	kJ g^{-1}	−5.114
	kJ cm^{-3}	−9.104
$\Delta_c H$	kJ mol^{-1}	−10,780.2
	kJ g^{-1}	−12.329
	kJ cm^{-3}	−21.945
DSC onset	°C	275
Friction	N	>360
Impact	cm	40 with type 12 tool
V_D	m s^{-1}	7600 at 1.78 g cm^{-3}
P_{CJ}	GPa	26.9

ABH has been developed for use with thermally challenging applications and exploding bridgewire detonators.

– D. V. Sickman, M. J. Kamlet, *Azobis(2,2′,4,4′,6,6′-hexanitrobiphenyl)*, US3461112, USA, **1969**.

3,3′-Azo-bis(6-amino-1,2,4,5-tetrazine), DAAT

3,3′-Azo-bis(6-amino-1,2,4,5-tetrazin)

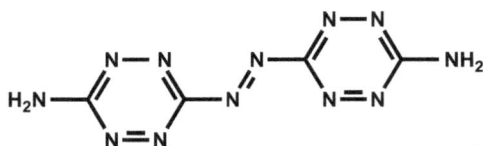

Formula		$C_4H_4N_{12}$
CAS		[303749-95-3]
m_r	g mol^{-1}	220.155
ρ	g cm^{-3}	1.84
$\Delta_f H$	kJ mol^{-1}	1035
$\Delta_c H$	kJ mol^{-1}	−3180.7
	kJ g^{-1}	−14.448
	kJ cm^{-3}	−26.584
Λ	wt.-%	−72.67
N	wt.-%	76.35
DSC-onset	°C	288
Friction	N	324
Impact	J	5
V_D	m s^{-1}	7400 at 1.64 g cm^{-3}
P_{CJ}	GPa	24.0 at 1.64 g cm^{-3}

The oxidation of 3′-azo-bis(6-amino-1,2,4,5-tetrazine) (DAAT) (e.g. with HOF) yields DAATO$_n$ (n being the index of oxidation). DAATO$_n$ are typically oxygen overbalanced and hence considerably sensitive. They possess very short ignition delay and yield high burn rates.

- A. N. Ali, M. M. Sandstrom, D. M. Oschwald, K. M. Moore, S. F. Son, Laser Ignition of DAAF, DHT and DAATO$_{3.5}$. *Propellants, Explos., Pyrotech.* **2005**, *30*, 351–355.

Azobis(isobutyronitrile), AIBN

Azobisisobutyronitril

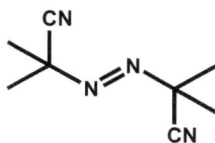

Azoisobutyric acid nitrile, AZDN		
Formula		$C_8H_{12}N_4$
GHS		01, 02, 07
H-phrases		242-302-332-412
P-phrases		210-280-273
UN		2952
REACH		LRS
EINECS		201-132-3
CAS		[78-67-1]
m_r	$g\ mol^{-1}$	164.21
ρ	$g \cdot cm^{-3}$	1.10
$\Delta_f H$	$kJ\ mol^{-1}$	246.02
$\Delta_c H$	$kJ\ mol^{-1}$	−5109
	$kJ\ g^{-1}$	−31.114
	$kJ\ cm^{-3}$	−34.225
$\Delta_{dec} H$	$kJ\ g^{-1}$	−1.3 (*Roberts et al.*)
Λ	wt.-%	−214.35
N	wt.-%	34.12
Impact	J	3 (smoke)
Friction	N	363 (smoke)
Mp	°C	101(dec)
Koenen	mm, type	1,5, A

AIBN serves as radical starter in organic chemistry. Though AIBN does not propagate a detonation shock in the (UN Test Series 1a) (only deflagration) and yields a negative result in the Koenen test too it does yield a positive result in the time/pressure test (UN Test Series 1c); hence, it is an explosive substance. In UN Test Series 3c (adiabatic self-heating), AIBN yields a positive result!

–   T. A. Roberts, M. Royle, Classification of Energetic Industrial Chemicals for Transport, *ICHEME Symp. Ser.* **1991**, *124*, 191–208.

Azodicarbonamide, ADCA

Azobisformamid

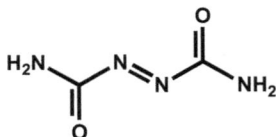

Celogen ®		
Formula		$C_2H_4N_4O_2$
GHS		08
H-phrases		334
P-phrases		261-284-304+340-342+311
UN		3242
WGK		1
REACH		LRS
EINECS		204-650-8
CAS		[123-77-3]
m_r	$g\,mol^{-1}$	116.079
ρ	$g \cdot cm^{-3}$	1.66
$\Delta_f H$	$kJ\,mol^{-1}$	−292
$\Delta_c H$	$kJ\,mol^{-1}$	−1066.7
	$kJ\,g^{-1}$	−9.189
	$kJ\,cm^{-3}$	−15.254
Λ	wt.-%	−55.13
N	wt.-%	48.26
Mp	°C	225

In smoke formulations, azodicarbonamide prevents clogging of vent holes by condensed reaction products. Hence, it is used in (→ *KM smoke*). Azodicarbonamide is also used as flame expander in illuminating compositions.

- C. Bernardy, *Pyrotechnische Zusammensetzung für Leucht und Antriebszwecke sowie ihre Verwendung*, DE2629949B2, France, **1977**.

Azoxytriazolone, AZTO

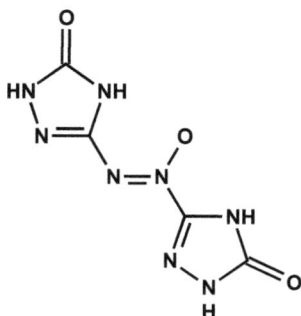

Azoxytriazolon

(Z)-1,2-bis(5-oxo-4,5-dihydro-1H-1,2,4-triazol-3-yl)diazene 1-oxide		
Aspect		Yellow-green crystals
Formula		$C_4H_4N_8O_3$
CAS		[960607-22-1]
m_r	g mol^{-1}	212.15
ρ	g cm^{-3}	1.905
Δ_fH	kJ mol^{-1}	11.0
$\Delta_{ex}H$	kJ mol^{-1}	−703.2
	kJ g^{-1}	−3.314
	kJ cm^{-3}	−6.314
Δ_cH	kJ mol^{-1}	−2156.7
	kJ g^{-1}	−10.167
	kJ cm^{-3}	−19.369
Ω	wt.-%	−52.8
N	wt.-%	52.8
DSC onset	°C	267
Friction	N	>360
Impact	J	>50
V_D	m s^{-1}	8062
P_{CJ}	GPa	27.99

AZTO is isomeric with *3,3′-diamino-4,4′-azoxyfurazan, DAAF)*. It was discovered by Lynn Wallace upon the attempted electrochemical oxidation of acidic aqueous solutions of (→ *NTO*) by comproportionation according to the following reaction:

$$2\,C_2H_2N_4O_3 + 6\,H^+ + 6\,e \longrightarrow C_4H_4N_8O_3 + 3\,H_2O$$

Further reduction of AZTO also yields between 3 and 10 wt.-% of azotriazolone, $C_4H_4N_8O_2$, CAS-Nr.: [1352233-46-5], which inherently forms co-crystals with AZTO.

Hence, a large-scale pure isolation in substance is impeded. However, the simple access to AZTO, its high density, low sensitiveness, and low solubility make this an interesting sustainable insensitive HE for large charges.

– M. P. Cronin, A. I. Day, L. Wallace, Electrochemical remediation produces a new high-nitrogen compound from NTO wastewaters, *J. Hazmat.* **2007**, *149*, 527–531.

– C. Underwood, C. Wall, A. Provatas, L. Wallace, New high nitrogen compounds Azoxytriazolone (AZTO) and azotriazolone as insensitive energetic materials, *New J. Chem.* **2012**, *36* 2613–2617.

B

B/KNO$_3$

Boron/potassium nitrate/binder compositions (18–24/70–77/5–6) designated: AZM O 953X, SR43, are used as highly energetic igniter mixtures. The combustion yields a high amount of condensed reaction products (e.g. boron nitride, BN and potassium metaborate, KBO$_2$). Figure B.1 shows the burn rate, heat of explosion, and ignition temperature for it. B/KNO$_3$ is insensitive to friction and requires an impact energy of E$_s$ > 20 J.

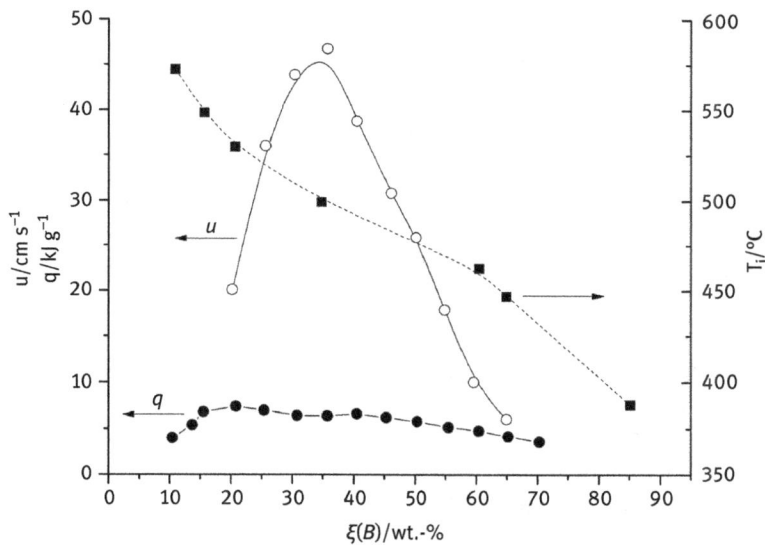

Figure B.1: Heat of explosion, q, ignition temperature, T$_i$, and burn rate, u, of B/KNO$_3$ as function of stoichiometry after *Berger*.

- B. Berger, Bestimmung der Abbrandcharakteristika sowie der Sicherheitskenndaten verschiedener binärer pyrotechnischer Anzünd- und Verzögerungsysteme, *ICT Jahrestagung*, **1986**, Karlsruhe, P-72.
- J. Stupp, Über die chemische Stabilität von Bor -Kaliumnitrat -Anzündmischungen, *ICT Jata*, **1985**, Karlsruhe, V-33.
- D. Barišin, I. Batinić-Haberle, The Influence of the Various Types of Binder on the Burning Characteristics of the Magnesium-, Boron-, and Aluminum-Based Igniters, *Propellants Explos. Pyrotech.* **1994**, *19*, 127–132.

https://doi.org/10.1515/9783110660562-002

B-14 binder

Binder B-14

B-14 is a carbon-black filled binder produced by *Nicotech GmbH* residing in Aetingen, Switzerland. B-14 is based on abietic acid–styrene copolymer. B-14 is available with either butanol or xylene as solvent base. B-14 is applied as a bonding agent in pyrotechnics (e.g. tracer) and in the field of high explosives (HEs) (booster).

Bacon, Roger (1214–1292)

Roger Bacon was born in Ilchester and died in Oxford. He was a Franciscan monk and was possibly the first in Europe to secretly document the composition of black powder in an anagram verse. He also developed a purification method for potassium nitrate based on solubility.

– G. I. Brown, *The Big Bang – A History of Explosives*, Sutton Publishing, **1998**, 5–8.

Ballistic bomb

Ballistische Bombe

A ballistic bomb serves the investigation of the ballistic performance of propellants and ignition elements. A ballistic bomb consists of a high pressure resistant metal vessel having screwable closing plugs, a pressure sensor, and electric terminals to attach an igniter (Figure B.2). With a ballistic bomb, the time-resolved pressure evolution can be determined as p_{max} that varies with loading density, Δ (g cm^{-3}); the force of the powder f can be determined ($\rightarrow$ *Abel equation*).

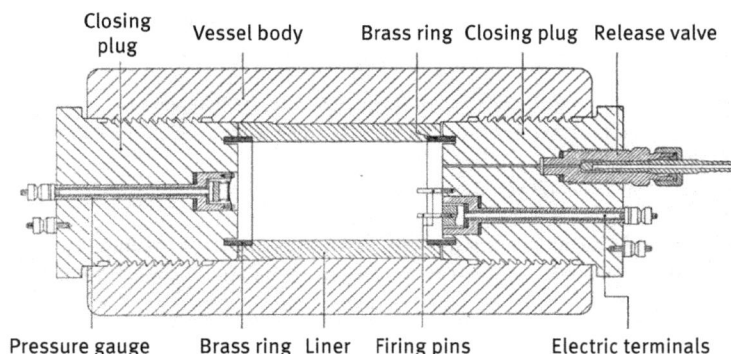

Figure B.2: Cross-sectional view of a ballistic bomb withstanding pressures of up to 1 GPa.

Ballistic mortar

Ballistischer Mörser

A ballistic mortar serves the determination of the working capacity of an HE. Therefore a 10 g charge of the explosive under investigation is initiated in the mortar. The expanding detonation gases propel a steel projectile and the recoil forces move the mortar that is suspended as a pendulum. The degree of deflection is a measure for the working capacity and is compared to standard reference HEs such as blasting gelatin or TNT.

– H. Ahrens, International Study Group for the Standardization of the Methods of Testing Explosives (formerly European Commission for the standardization of the tests of explosives – until 1976)– recommendations prepared up to now by the study group, *Propellants Explos.* **1977**, 2, 7–20.

Ballistics

Ballistik

Ballistics is the science of gun shot. As part of the large field of applied mechanics, ballistics also includes the movement of bodies against aerodynamic drag and under the influence of gravitational forces. The corresponding subsections of ballistics are as follows:
– Internal ballistics (interior ballistics, *A.E.*) deals with the thermochemical properties of propellants, the conversion of stored chemical energy into kinetic energy, and basically all processes starting with the ignition of the propellant and ending with the projectile leaving the gun barrel.
– Transient ballistics deals with the processes occurring upon emerging of a projectile from a muzzle.
– External ballistics (exterior ballistics, *A.E.*) deals with the flight of the projectile
– Terminal ballistics deals with the interaction between projectile and a target structure.

In the field of rocket propulsion, the term *internal ballistics* covers all chemical and physical processes occurring up to the expansion of the combustion gases to ambient pressure which is more or less the *exit plane* of the rocket engine.

- B. Kneubuehl, Ballistik, Springer, Berlin, **2019**, 437 p.
- Z. Rosenberg, E. Dekel, *Terminal Ballistics*, Springer, **2012**, 323 pp.
- Thermodynamic Interior Ballistic Model with Global Parameters", *NATO STANAG* 4367 Edition 3, **2000**.
- Definition and Determination of Ballistic Properties of Gun Propellant", *NATO STANAG 4115*, Edition 2, **1997**.
- G. Weihrauch, *Ballistische Forschung im ISL*, ISL, **1994**, 393 p.
- H. Krier, M. Summerfeld, (Eds.). *Interior Ballistics of Guns*, Volume 66 Progress in Astronautics and Aeronautics, AIAA, **1979**, 384 p.
- W. Wolff, *Raketen und Raketenballistik*, Deutscher Militärverlag, **1964**, 342 p.
- H. Athen, *Ballistik*, 2. Aufl. Quelle & Meyer, **1958**, 258 p.
- R. E. Kutterer, *Ballistik*, 3. Aufl., Vieweg, **1959**, 304 p.
- N. N. *Internal Ballistics*, His Majesty's Stationary Office, London, **1951**, 311 p.
- T. Vahlen, *Ballistik*, 2. Aufl. deGruyter, **1942**, 267 p.
- C. Cranz, K. Becker, Ballistik Vols. 1–3+Supplementary Volume **1926–1935**.

Ballistite

Ballistit

Ballistite, invented in 1888 by Alfred Nobel, was the first smoke-free double-base propellant that was manufactured on an industrial scale. It was based on nitrocellulose and nitroglycerine and was stabilized with diphenylamine.

- 49.5 wt.-% nitrocellulose
- 49.5 wt.-% nitroglycerine
- 1.0 wt.-% diphenylamine

BAM, Bundesanstalt für Materialprüfung und Forschung

The BAM, which translates to Federal Institute for Materials Research and Testing, is a federal agency in the responsibility of the German Ministry of Economic Affairs and Energy with its headquarters located in Berlin. A part of BAM's activities such as the examination and approval of explosives and items containing explosives is defined by the German Explosive law. The corresponding department within BAM is the department II.3 Explosives. The BAM maintains extensive testing facilities allowing for all of those tests on a 12 km^2 testing ground at Horstwalde south of Berlin.

- www.bam.de
- https//www.tes.bam.de/de/mitteilungen/tts/

BAM friction apparatus
BAM Reibapparat

The BAM friction apparatus serves the determination of the friction sensitiveness of solid, pasty, and liquid substances in accordance with UN Test Series 3(b) (i). The machine consists of an electric motor which drives an eccentric to move a pusher rod.

A load arm can hold different weights at different positions to allow for friction forces between 50 and 360 N applied to a porcelain peg fixed to it. This peg is placed on the edge of a conical sample heap (V = 10 mm^3) on a porcelain plate (Figure B.3) the latter which is driven by the pusher rod in a single reciprocating motion. Similar as with the BAM impact machine, the threshold friction forces not producing any reaction or the 50% energies are reported.

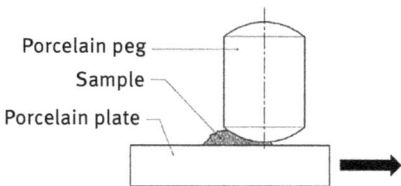

Porcelain peg
Sample
Porcelain plate

Figure B.3: Sample holder in BAM friction apparatus and direction of movement of porcelain plate.

– *Recommendations on the Transport of Dangerous Goods – Manual of Tests and Criteria* – 5th rev. Ed. United Nations, **2009**, pp. 104–107.
– H. Koenen, K. H. Die, K.-H. Swart, Sicherheitstechnische Kenndaten explosionsfähiger Stoffe, I. Mitteilung, *Explosivstoffe*, **1961**, *9*, 30–42.

BAM impact machine
BAM Fallhammer

The BAM impact machine serves the determination of the threshold energy for impact sensitiveness of an explosive material in accordance with UN Test Series 3 (a) (ii). The impact machine consists of a frame with a variable weight (m = 1, 2, and 5 kg) that is guided on vertical tracks. The weight can be released from heights between h = 0.15 and 0.75 m and impacts on the sample holder (Figure B.4). The sample (V = 40 mm^3) is placed between two steel cylinders along the radius of the lower cylinder. If liquids are to be tested then the upper cylinder is restrained 2 mm above the lower cylinder by

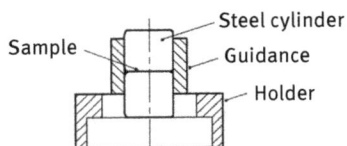

Figure B.4: Sample holder of BAM impact machine.

an outer rubber ring. The impact energy, E impact, is determined by $m \cdot h \sim E$ [N · m]. A 1 kg weight is assumed to exert a weight force about equivalent to 10 N. Impact energies reported are either threshold energies that do not affect a reaction or 50% energies in which 50% of the trials lead to a reaction (explosion, burning, flashes, smoke, smell, and discoloration of the sample).

According to *Licht*, explosives that are liquids at ambient temperature exhibit an increased sensitiveness. This is due to capillary forces which attract the liquid between the lateral surfaces of the cylinders and the outer guidance and subject the sample to shear forces. In the same manner, low melting explosives show an increased sensitiveness towards impact when the samples are placed along the outer radius.

- *Recommendations on the Transport of Dangerous Goods – Manual of Tests and Criteria* – 5th rev. Ed. United Nations, **2009**, pp. 75–82.
- H. H. Licht, Die mechanische Empfindlichkeit von niedrig schmelzenden Explosivstoffen, *ISL -R 107/96*, ISL, Saint-Louis, 28. February **1996**, 11 pp.
- H. Koenen, K. H. Die, K.-H. Swart, Sicherheitstechnische Kenndaten explosionsfähiger Stoffe, I. Mitteilung, *Explosivstoffe*, **1961**, *9*, 30–42.

Baratol

Replacing ammonium nitrate with barium nitrate, $Ba(NO_3)_2$, in Amatol yields B. (Tables B.1 and B.2). As $Ba(NO_3)_2$ is not hygroscopic, Baratol has been used as a replacement for amatols and ammonals.

Table B.1: Baratol – explosive mixtures – composition.

Baratol	24/76	28/72	30/70	33/67	35/65	20/80	10/90
TNT (wt.-%)	24	28	30	33	35	80	90
$Ba(NO_3)_2$ (wt.-%)	76	72	70	67	65	20	10

Table B.2: Baratol – explosive mixtures – performance.

Baratol	24/76	28/72	30/70	35/65	33/67	10/90
V_D (m s^{-1})	4925	5000	5120	5150		5900
P_{CJ} (GPa)	14	15.4		13.5		
$\Delta_{det}H$ (kJ kg^{-1})	3100					
ρ (g cm^{-3})	2.619	2.45	2.452	2.35	2.55	1.65
$\varnothing_{cr}$ (mm)	43.2					
$\Delta_m H$ (J g^{-1})	23.1				31.8	
c_p (J g^{-1} K^{-1})	0.657				0.841	
Dp (°C)	240					
T_{5ex} (°C)	385					
c_L (m s^{-1})	2900					
κ (W cm^{-1} K^{-1})	49.57					
Impact (J)	47.5					

The high density and low detonation velocity have favored the use of B (24/76) together with composition B in explosive lenses for plane wave generators. The formulation B. (24/76) very often in the literature is simply referred to as Baratol. Only about 20–30% of the nitrate-based oxygen is reacting before reaching the CJ plane. The 50% initiation pressure of Baratol (24/76) in the LANL-LSGT equals 17.57 GPa.

Barbara, Saint
Heilige Barbara

Saint Barbara is a Christian martyr who lived in the fourth century and was sentenced to death for her belief by emperor Maximinus Daja in Nikomedia (today Izmit Turkey). Because of the lacking reference in authentic Christian writings her observance on 4th December (Ger.: *Barbarafeier, St Barbara Fest*) was removed from the General Roman Calendar in the 2nd Vatican Council in 1969, though she remains in the Catholic Church's list of saints (Figure B.5). Saint Barbara is the patron of miners, tunnellers, armorers, military engineers, gunsmiths, and all those dealing with explosives and weapons.

In Germany, the so-called *Barbarameldung (Babara note)* of the artillerymen summarizes the relevant data of the military meteorological data such as direction of wind, wind speed, and air pressure.

Figure B.5: Saint Barbara as woodwork by Tilman Riemenschneider (around 1510) Bayerisches Nationalmuseum, Munich (Source Wikipedia: https://commons.wikimedia.org/wiki/File:Tilman_Riemenschneider_Barbara-1.jpg).

- H.-H. Kritzinger, F. Stuhlmann, *Artillerie und Ballistik in Stichworten*, Springer Verlag, Berlin, **1939**, pp. 33.

Barium – toxicity and radiotoxicity

Barium Toxizität und Radiotoxizität

The Ba-ion is a muscle cytotoxin. Water-soluble Ba salts after oral uptake ($\geq$2 mg kg^{-1} body weight) hence affect the gastrointestinal tract (e.g. nausea and diarrhea) and the heart (e.g. arrhythmia and bradycardia). The inhalation of Ba-containing dust yields breathing disorders (bronchoconstriction) and at higher levels will cause the same toxic effects as reported earlier.

Steinhauser has dealt with the possible contamination of barium and strontium salts with radium which is due to the chemical similarity. In natural ores such as baryte, part of Ba may be replaced by radium (^{226}Ra and ^{228}Ra) as radiobaryte Ba$_{x-y}$ (Ra)$_y$SO$_4$ (with y = ppm). Therefore, he investigated the activity of fireworks. However, the main activity of the samples is due to ubiquitous ^{40}K (A(^{40}K) = 10–20 Bq g^{-1}), while any activity associated with both ^{226}Ra and ^{228}Ra was determined to be two orders of magnitude smaller A = 16–260 mBq g^{-1} and can be considered negligible.

- S. Moeschlin, *Klinik und Therapie der Vergiftungen*, Thieme, **1986**, pp. 207–209.
- G. Steinhauser, A. Musilek, Do pyrotechnics contain radium? *Enviro. Res. Lett.* **2009**, 4, 034006 (6. S.)

Barium carbonate

Bariumcarbonat

Formula		BaCO$_3$
GHS		07
H-phrases		302
P-phrases		264-270-301+312-330-501a
EINECS		208-167-3
CAS		[513-77-9]
m_r	g mol^{-1}	197.339
ρ	g cm^{-3}	4.43
Mp	°C	1360 (dec)
c_p	J mol^{-1} K^{-1}	85.35
$\Delta_f H$	kJ mol^{-1}	−1216.3
Λ	wt.-%	0

BaCO$_3$ serves as a color agent in fireworks.

Barium chlorate monohydrate

Bariumchloratmonohydrat

Formula		$Ba(ClO_3)_2 \cdot H_2O$
GHS		03, 07, 09
H-phrases		271-302-332-411
P-phrases		210-221-283-306+360-371+380+375-501a
UN		1445
EINECS		236-760-7
CAS		[10294-38-9]
m_r	$g\ mol^{-1}$	322.248
ρ	$g\ cm^{-3}$	3.180
Mp	°C	406 (dec)
c_p	$J\ g^{-1}\ K^{-1}$	0.658
$\Delta_f H$	$kJ\ mol^{-1}$	−1066
Λ	wt.-%	+29.79
Koenen	mm, type	<1, 0

For a long time, barium chlorate monohydrate was the only color agent and oxidizer to produce green flames of high saturation. Table B.3 shows metal-free formulations of high saturation. Composition I requires alcohol for moistening; compositions II and III require water.

Table B.3: Metal-free barium chlorate containing illuminants (amounts in wt.-%).

	I	II	III
$Ba(ClO_3)_2 \cdot (H_2O)$	64	79	53
$Ba(NO_3)_2$		6.5	
$KClO_3$	18		28
Lactose		1.7	
Shellac	18		
Anthracene		10.5	
Acaroides, gum			10
Charcoal			5
Dextrin			4
Methylcellulose		2.3	

However, the low thermal stability of it and the high sensitiveness of the formulations containing it have led to a replacement of "barium chlorate monohydrate" by barium nitrate. Barium chlorate monohydrate loses its crystal water around 130 °C. Its exothermal decomposition occurs right at the melting point according to

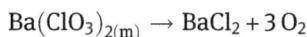

$$Ba(ClO_3)_{2(m)} \rightarrow BaCl_2 + 3\,O_2$$

– K. H. Stern, *High Temperature Properties and Thermal Decomposition of Inorganic Salts with Oxyanions*, CRC Press, **2001**, pp. 193.

Barium chromate
Bariumchromat

Aspect		Yellow crystals
Formula		$BaCrO_4$
GHS		03, 07, 08, 09
H-phrases		272-350-400-410-302-332-317
P-phrases		210-221-302+352-321-405-501a
UN		1479
REACH		LPRS
EINECS		233-660-5
CAS		[10294-40-3]
m_r	g mol^{-1}	253.324
ρ	g cm^{-3}	4.498
Mp	°C	1400 (dec)
$\Delta_f H$	kJ mol^{-1}	−1445.99
Λ	wt.-%	+9.47

Barium chromate is the most common oxidizer in gasless delay and ignition compositions. Very often it is used together with potassium perchlorate as auxiliary oxidizer and metallic fuels from Group IV. Figure B.6 depicts a ternary diagram for the burn rate and oxygen balance of delay compositions containing $BaCrO_4/KClO_4/W$ after *Weingarten*.

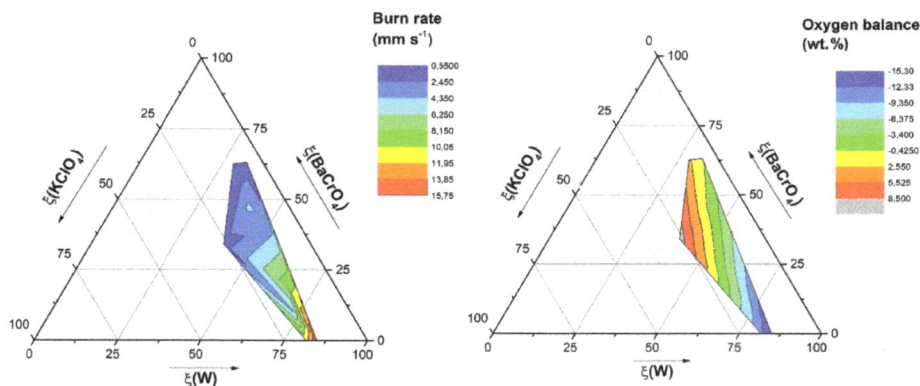

Figure B.6: Burn rate (top) and oxygen balance (bottom) of $BaCrO_4/KClO_4/W$.

– G. Weingarten, *Pyrotechnic Delay Systems*, in F. B. Pollard, J. H. Arnold (Eds.), *Aerospace Ordnance Handbook*, Prentice-Hall, **1966**, pp. 254–317.

Barium hexafluorosilicate

Bariumhexafluorosilikat

Formula		BaSiF$_6$
REACH		LPRS
EINECS		241-189-1
CAS		[17125-80-3]
m$_r$	g mol^{-1}	279.416
ρ	g cm^{-3}	4.279
Mp	°C	1580
Δ$_f$H	kJ mol^{-1}	−2952.19
Λ	wt.-%	0

Barium hexafluorosilicate serves as a flux material in delay and tracer compositions.

- G. Faber, H. Florin, P.-J. Grommes, P. Röh, *Verzögerungssätze mit langen Verzögerungszeiten*, EP 0 332 986 B1, **1993**, Germany.
- U. Ticmanis, M. Kaiser, M. Künstlinger, K. Redecker, Stabilität von Nitrocellulose in Anzündmischungen, *ICT -Jata*, **2000**, Karlsruhe, P-65.

Barium nitrate

Bariumnitrat

Formula		Ba(NO$_3$)$_2$
GHS		07
H-phrases		302-332
P-phrases		261-264-301+312-304+340-312-501a
UN		1446
EINECS		233-020-5
CAS		[10022-31-8]
m$_r$	g mol^{-1}	261.34
ρ	g cm^{-3}	3.24
Mp	°C	590 (dec)
c$_p$	J g^{-1} K^{-1}	0.578
Δ$_f$H	kJ mol^{-1}	−992.07
Λ	wt.-%	+30.61
N	wt.-%	10.72

Barium nitrate is a non-hygroscopic salt. Its thermal decomposition occurs in accordance with

$$Ba(NO_3)_2 \rightarrow BaO_{(s)} + 2\,NO_2 + 0.5\,O_2$$

In the absence of halogens barium nitrate yields a yellowish white flame, which is due to the continuum emitter $BaO_{(s)}$. In the presence of chlorine compounds however (e.g. PVC or $KClO_3$) transient $BaCl_{(g)}$ forms due to thermal excitation yields intensive green light. Barium nitrate is used together with fine aluminum powder in report formulations. These have a high ignition temperature and are safer to handle then $Al/KClO_4$.

- N. M. Varyonykh, N. V. Obeziyaev, Y. E. Sheludyak, Peculiarities of Combustion of $Mg/Sr(NO_3)_2$ and $Mg/Ba(NO_3)_2$ Mixtures, *IPS*, **2002**, 391-396.
- N. M. Varyonykh, N. V. Obeziyaev, Y. E. Sheludyak, Thermal parameters of the burning wave for stoichiometric mixtures of different oxidizers with metals, *IPS*, **2005**, P-144.
- N. M. Varyonykh, N. V. Obeziyaev, Y. E. Sheludyak, Combustion peculiarities of mixtures of barium nitrate with different metals, *IPS*, **2006**, pp. 711-725.
- N. M. Varyonykh, N. V. Obeziyaev, Y. E. Sheludyak, Thermal parameters of the burning wave for barium nitrate/magnesium/ organic additive pyrotechnic mixtures, *IPS*, 2008, pp. 585–590.

Barium oxalate

Bariumoxalat

Formula		BaC_2O_4
GHS		6.1
H-phrases		302-312-332
P-phrases		261-280-302+352-304-340-322-501
UN		1564
EINECS		208-216-9
CAS		[516-02-9]
m_r	g mol^{-1}	225.38
ρ	g cm^{-3}	2.658
Dp	°C	346 (-CO)
c_p	J g^{-1} K^{-1}	0.578
$\Delta_f H$	kJ mol^{-1}	−1371
Λ	wt.-%	−7.1
S	g L^{-1}	9.3 at 18 °C
	g L^{-1}	22.8 at 100 °C
Spec		JAN-B-660

Barium oxalate forms several hydrates

– $BaC_2O_4 \cdot \frac{1}{2} H_2O$		$\Delta_f H = -1522$ kJ mol^{-1}
– $BaC_2O_4 \cdot H_2O$,	CAS-No. [13463-22-4],	$\Delta_f H = -1663$ kJ mol^{-1}
– $BaC_2O_4 \cdot 2 H_2O$		$\Delta_f H = -1967$ kJ mol^{-1}
– $BaC_2O_4 \cdot 3,5 H_2O$		$\Delta_f H = -2409$ kJ mol^{-1}

Barium oxalate is used as a burn rate retardant and cooling agent in green burning tracers and NIR tracers.

Barium perchlorate

Bariumperchlorat

Formula		$Ba(ClO_4)_2$
GHS		03, 07
H-phrases		271-302-332
P-phrases		210-221-283-306+360-371+380+375-501a
UN		1447
REACH		LPRS
EINECS		236-710-4
CAS		[13465-95-7]
m_r	$g\ mol^{-1}$	336.231
ρ	$g\ cm^{-3}$	3.20
Mp	°C	473 (dec)
T_p	°C	283
T_p	°C	355
$\Delta_f H$	$kJ\ mol^{-1}$	−799.98
Λ	wt.-%	+38.07

Barium perchlorate is a highly hygroscopic crystalline material which upon exposure to moist air forms the trihydrate ($(Ba(ClO_4)_2 \cdot 3\ H_2O$, [10294-39-0], ρ = 2.740 g·cm^{-3}, $\Delta_f H$ = −1996 kJ·mol^{-1}). It is freely soluble in water and alcohol. *Wasmann* reported about the use of barium perchlorate in strobe formulations (also *strontium perchlorate*).

- F. W. Wasmann, Pulsierend abbrennende pyrotechnische Systeme, *ICT -Jata*, **1975**, Karlsruhe, 239–250.

Barium peroxide

Bariumperoxid

Formula		BaO_2
GHS		03, 07
H-phrases		272-302-332
P-phrases		210-220-221-261-280-501a
UN		1449
REACH		LPRS
EINECS		215-128-4
CAS		[1304-29-6]
m_r	$g\,mol^{-1}$	169.329
ρ	$g\,cm^{-3}$	4.96
Mp	°C	>700 (dec)
$\Delta_f H$	$kJ\,mol^{-1}$	−634.29
Λ	wt.-%	+9.45

BaO_2 mainly serves as oxidizer in VIS- and NIR-tracer formulations (*Henry*). In addition, BaO_2 is also an important oxidizer in fast burning delay compositions together with iron, manganese, and molybdenum. *Ellern* described the use of BaO_2 as oxidizer in delay formulations containing metallic selenium and tellurium. BaO_2/aluminum with $\xi(BaO_2) > 0.7$ are used as fast first fires. Due to significant slag formation, mixtures of it with Mg (Ger.: *Zündkirsche*) are used as igniters for thermites. BaO_2/aluminum shows adiabatic self-heating in the presence of moisture.

- G. H. Henry III., M. A. Owens, L. T. Jarret, M. A. Tucker, F. N. Bone, *Infrared Tracer for Ammunition*, US 5.811.724, **1998**, USA.
- J. Stupp, Untersuchungen über Selbstentzündungsreaktionen eines bariumperoxidhaltigen Glimmspursatzes, *CTI-Einführungssymposium*, 13–15 June **1973**, 14 pp. + Anhang.

Barium picrate

Bariumpikrat

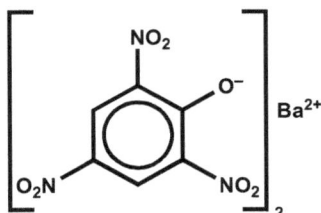

Aspect		Yellow crystal powder
Formula		$BaC_{12}H_4N_6O_{14}$
CAS		[25733-98-6]
m_r	g mol^{-1}	593.532
Dp	°C	403 (water free)
	°C	333 (pentahydrate)
$\Delta_f H$	kJ mol^{-1}	−1032
Λ	wt.-%	−35.04
N	wt.-%	14.16

Barium picrate is relatively insensitive towards friction and impact.

- S. Kaye, *Encyclopedia of Explosives and Related Items, Volume 8*, US Army Armament Research and Development Command, Dover, NJ, **1978**, pp. P-279.

Barium sulfate

Barium sulphate (B.E.)
Bariumsulfat (D)

Baryte		
Formula		$BaSO_4$
REACH		LRS
EINECS		231-784-4
CAS		[7727-43-7]
m_r	g mol^{-1}	233.40
ρ	g cm^{-3}	4.50
Mp	°C	1580
T_p	°C	1149 (rhombic → monoclinic)
$\Delta_f H$	kJ mol^{-1}	−1473.19
Λ	wt.-%	+6.86

$BaSO_4$ has been proposed as an auxiliary oxidizer for strobe compositions and dark tracers. It has also been suggested to use it together with high energetic fuels such as B, Si, Ti, and Zr in delay formulations.

- H. Weber, *Pyrotechnischer Leuchtsatz mit intermittierender Strahlungsemission*, DE3313521A1, **1984**, Germany.
- D. Funke, H. Zöllner, *Pyrotechnischer Verzögerungssatz militärischer Verzögerungselemente*, DE102014018792A1, **2015**, Germany.
- S. M. Tichapondwa, W. W. Focke, O. del Fabbro, J. Gisby, C. Kelly, A Comparative Study of Si-$BaSO_4$ and Si-$CaSO_4$ Pyrotechnic Time-Delay Compositions, *J. Energ. Mater.* **2016**, *34*, 342–356.

Baronal

Baronal is a composition based on barium nitrate, aluminum, and TNT (Tables B.4 and B.5). It was used in World War II in the USA as an underwater explosive.

Table B.4: Baronal composition.

	Baronal
TNT (wt.-%)	35
$Ba(NO_3)_2$ (wt.-%)	50
Aluminum (wt.-%)	15
ρ (g cm^{-3})	2.32

Table B.5: Baronal properties.

Parameter	Baronal
V_D (m s^{-1})	5450
$\Delta_f H$ (kJ kg^{-1})	−1715
$\Delta_{det} H$ (kJ kg^{-1})	−4748
ρ (g cm^{-3})	2.32
$\Delta_{melt} H$ (J g^{-1})	33.7
$T_{ex-5\,s}$ (°C)	345
Impact (J)	6

Base bleed

Increasing the range of a gun projectile can be achieved by increasing the muzzle velocity, v_0, providing additional thrust (e.g. **R**ocket-**A**ssisted-**P**rojectile = RAP) and via an improvement of the ballistic coefficient of the projectile itself, c_w. The ballistic coefficient itself is the sum of its contributions

$$c_w = c_{Ww} + c_{WR} + c_{WB}$$

with c_{Ww} = profile (shape) drag, c_{WR} = skin friction drag, and c_{WB} = base drag. Up to supersonic velocities the base drag (c_{WB}) accounts for about 50% of the overall drag with narrow pointy projectiles. The drag of the base itself can be reduced by giving it a conical shape (*bottail*). *Schroeder* in 1951 found that allowing air to stream from the projectile base increases the pressure in the wake of the base and therefore reduces the drag. In a practical embodiment of *Schroeders* findings, a pyrotechnic composition is burned at the base of a projectile – similarly as a tracer at the bottom of a bullet. The jet parameter I has been found to be inversely proportional to the base drag:

$$I = \frac{m'}{p \cdot v \cdot A} \ (m^{-1})$$

With m' = mass consumption rate (kg s^{-1}), p = static air pressure (kg m^{-2}), v = free stream velocity (m s^{-1}), and A = exit surface area (m^2).

Projectiles bearing base bleed by average have a range 15% higher than common projectiles. In addition, the terminal velocity is 20% higher which reduces influence by shear wind and hence causes greater accuracy. Typical base bleed formulations resemble propellant or illuminant formulations (AP/HTPB or Mg/nitrate/binder). Figure B.7 shows reduction of base drag with increase of jet parameter, I, for a general formulation of the type:

- magnesium, Mg
- strontium nitrate, $Sr(NO_3)_2$
- ammonium perchlorate, NH_4ClO_4
- calcium resinate, $C_{40}CaH_{58}O_4$

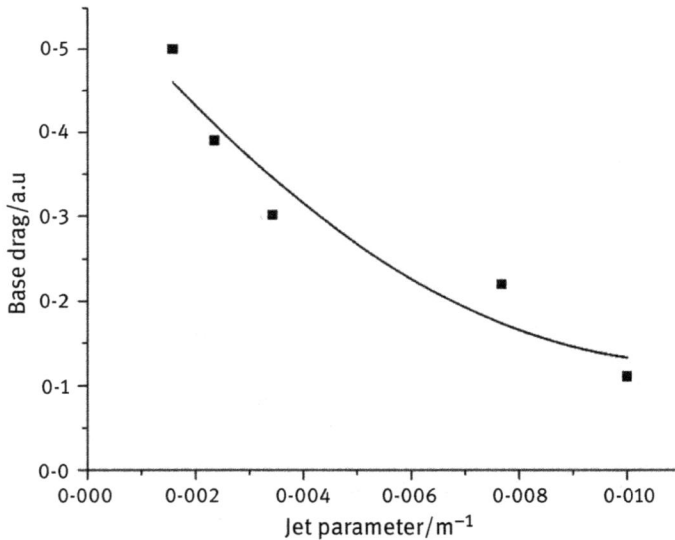

Figure B.7: Base drag as a function of jet parameter of a base bleed charge.

- K. K. Kuo, J. N. Fleming (Eds.), *Base Bleed*, Hemisphere Publishing, New York, **1991**, 314 pp.
- S. N. B. Murthy (Eds.), *Aerodynamics of base combustion, Volume 40, Progress in Astronautics and Aeronautics*, **1976**, 522 pp.
- E. M. Cortright, A. H. Schroeder, *Preliminary investigation of Effectiveness of Base Bleed in Reducing Drag of Blunt Base Bodies in Supersonic Stream*, NACA, **1951**, 24 pp.

Basic copper nitrate, BCN

Basisches Kupfernitrat

Aspect		Light green crystals
Formula		$Cu_2(OH)_3(NO_3)$
REACH		LRS
EINECS		439-590-3; 601-793-1
CAS		[12158-75-7]
m_r	$g\ mol^{-1}$	240.12
ρ	$g\ cm^{-3}$	3.389
$\Delta_f H$	$kJ\ mol^{-1}$	−864.56
Λ	wt.-%	+29.98
N	wt.-%	5.83

BCN is slightly soluble in water and consequently not hygroscopic. It can be used as an oxidizer in gas-generating compositions and has been tested as an oxidizer for green burning fireworks formulations.

- G. Steinhauser, K. Tarantik, T. M. Klapötke, Copper in Pyrotechnics, *J. Pyrotech.* **2008**, *27*, 3–13.
- J. P. Auffredic, D. Louer, M. Louer, Décomposition thermique topotactique de l'hydroxynitrate de cuivre $Cu_2(OH)_3NO_3$ étude thermodynamique et structurale, *Journees de Calorimetrie et d'Analyse Thermique*, **1978**, *Vol. 9A, B13*, 97–107.

BATEG and TEGDA

1,2-Bis(2-azidoethoxy)ethan

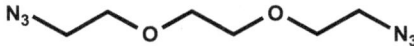

Bisazidotriethylenglykole, 1,2-bis (2-azidoethoxy)ethane		
Formula		$C_6H_{12}N_6O_2$
CAS		[59559-06-7]
m_r	g mol^{-1}	200.198
ρ	g cm^{-3}	1.15
$\Delta_f H$	kJ mol^{-1}	+215; *+240*
$\Delta_c H$	kJ mol^{-1}	−4295.1
	kJ g^{-1}	−21.454
	kJ cm^{-3}	−24.672
Λ	wt.-%	−127.87
N	wt.-%	41.98
P_{vap}	Pa	0.176
T_g	°C	−110.1
Friction	N	144
Impact	J	3

BATEG is a plasticizer for gun propellants that has been synthesized for the first time by Alvarez et al.

- F. Alvarez, N. V. Latypov, E. Holmgren, M. Wanhatalo, New Ingredients for CMDB Propellants, *NTREM 2008*, Pardubice, 9.-11. April **2008**, pp. 442–447.
- J. Böhnlein Mauß, T. Keicher, U. Schaller, M. Helfrich, BATEG – Ein Neuer Weichmacher für NC-Basierte TLP? *Workshop Treib- und Explosivstoffe*, 16–17. November **2016**, FhG-ICT, Pfinztal.

BCHMX, *cis-octahydro-1,3,4,6-tetranitroimidazo[4,5-d]imidazole*

Bicyclo-HMX, K-55

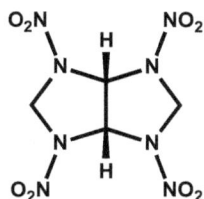

Formula		$C_4H_6N_8O_8$
CAS		[473796-67-7]
m_r	g mol^{-1}	294.139
ρ	g cm^{-3}	1.861
$\Delta_f H$	kJ mol^{-1}	236.5
$\Delta_{ex}H$	kJ mol^{-1}	-1782.8
	kJ g^{-1}	-6.061
	kJ cm^{-3}	-11.280
$\Delta_c H$	kJ mol^{-1}	-2668.1
	kJ g^{-1}	-9.071
	kJ cm^{-3}	-16.881
Λ	wt.-%	-16.32
N	wt.-%	38.0962
Mp	°C	268 (dec)
DTA onset	°C	227
Friction	N	88–250
Impact	J	2.5–3.5
V_D	m s^{-1}	8650 at 1.80 g cm^{-3}.
		1400 at 0.71 g cm^{-3}
P_{CJ}	GPa	33.9 at 1.80 g cm^{-3}.

Similar as the carbonyloge sorguyl, BCHMX is pretty sensitive to hydrolysis.

- L. Lewzuk, P. Koślik, J. Rećko, Performance of BCHMX in Small Charges, *Propellants Explos. Pyrotech.* **2020**, *45*, 581–586.
- A. K. Hussein, A. Elbeih, M. Jungová, S. Zeman, Explosive Properties of a High Explosive Composition Based on cis-1,3,4,6-Tetranitrooctahydroimidazo-[4,5-d]imidazole and 1,1-Diamino-2,2-dinitroethene (BCHMS/FOX-7), *Propellants Explos. Pyrotech.* **2018**, *43*, 472–478.
- D. Klasovitý, S. Zeman, A. Ružička, M. Jungová, M. Roháč, cis-1,3,4,6-tetranitrooctahydroimidazo- [4,5-d]imidazole (BCHMX), its properties and initiation reactivity, *J. Haz. Mat.* **2009**, *164*, 954–961.
- R. Gilardi, J. L. Flippen-Anderson, R. Evans, cis-2,4,6,8-Tetranitro-*1 H,5 H*-2,4,6,8-tetraazabicyclo [3.3.0]octane, the energetic compound 'bicyclo-HMX', *Acta Cryst. E* **2002**, *58*, o972-o974.

Becker, Richard (1887–1955)

Richard Becker was a German physicist. Starting in 1913, he worked in the field of shock waves and detonation. Together with *Chapman* and *Jouguet* and with his academic offspring (*Werner Döring*) he contributed significantly to the development and understanding of detonations. Becker was the first to calculate the detonation velocity of nitroglycerine entirely based on theoretical considerations. He independently developed an equation of state to describe the detonation products that today is known as Becker–Kistiakowski–Wilson equation of state.

- R. Becker, Zur Theorie der Detonation, *Z. Elektrochem.* **1917**, *23*, 40–49;
- R. Becker, Eine Zustandsgleichung für Stickstoff bei großen Dichten, *Z. Physik*, **1921**, *4* 393–409,
- R. Becker, Stoßwelle und Detonation, *Z. Physik*, **1922**, *8*, 321- 362.
- R. Becker, Physikalisches über feste und gasförmige Sprengstoffe, *Z. Tech. Phys.* **1922**, *3*, 249.
- Eintrag in der Deutschen Nationalbibliothek: http://d-nb.info/gnd/118079433

Bengal lights
Bengalische Flammen

Bengal lights describe mostly slow burning bare pyrotechnic formulations and those items containing consolidated formulations in mild consumable tubes (zinc or cardboard). Typical formulations are given in Table B.6.

Table B.6: Composition of Bengal lights.

Component (wt.-%)	Blue	Green	Yellow	Flesh	Red	Lilac	White
Acaroid resin	–	13	–	–	–	10	–
Ammonium perchlorate	–	–	–	–	–	50	–
Antimony sulfide, Sb_2S_3	–	–	–	–	–	–	22
Barium carbonate	–	–	–	9	–	–	–
Barium chlorate		45					
Barium nitrate	–	26	–	–	–	–	–
Benzoic acid	–	–	–	–	–	10	–
Chloroprene	10	–	–	–	–	–	–
Charcoal of lime tree	–	–	–	–	3	–	–
Gum Arabic		6					
Potassium chlorate	–	10	–	51	9	–	–
Potassium nitrate	–	–	–	–	–	–	67
Potassium perchlorate	60	–	75	–	–	20	–
Copper(II) oxide	10	–	–	–	–	5	–
Lactose monohydrate	20	–	–	15	–	–	–
Sodium oxalate	–	–	10	–	–	–	–

Table B.6 (continued)

Component (wt.-%)	Blue	Green	Yellow	Flesh	Red	Lilac	White
Shellac	–	–	15	–	–	–	–
Sulfur	–	–	–	–	21	–	11
Strontium carbonate	–	–	–	–	2	5	–
Strontium nitrate	–	–	–	–	65	–	–
Strontium sulfate	–	–	–	25	–	–	–

Benite

Benite designates extruded igniter mixtures from black powder (either 2 or 3 C) and nitrocellulose (see also *Eimite*, *oxite*, and *Ex-98*) (Tables B.7 and B.8).

Table B.7: Composition of benite.

Components (wt.-%)	NKP-S 526	NKP-S-536	A
Nitrocellulose	30	38	39.5 (13.15 wt.-% N
Potassium nitrate	51	48	44,3
Charcoal (lime tree)	10.2	12	9.4
Sulfur	6.8	–	6.3
Akardite I	1	1	
Dibutyl phthalate	1	1	
Ethylcentralite			0.5

Table B.8: Properties of benite.

Parameter	NKP-S 526	NKP-S-536	A
Density (g cm^{-3})	0.7–0.75	0.7	
Ignition temperature (°C)	–	281	
Activation energy (kJ mol^{-1})	59	77	
Specific heat (J g^{-1} K^{-1})	0.989	1.008	
Heat of explosion, calc. (J g^{-1})	3033	3298	
Combustion temperature, adiabatic (°C)	2127	–	
Force constant (J g^{-1})			553.5

– H. Hassmann, *Igniter Assembly Containing Strands of Benite*, US Patent 3.182.595, USA, **1965**.
– F. Volk, M. Hund, D. Müller, Determination of the performance of ignition powders, *3e Congres International de Pyrotechnie du Groupe de Travail de Pyrotechnie Spatiale et 12e International Pyrotechnics Seminar*, 8–12 June **1987**, Juan-les-Pins, France, 97.

Benzotriazole

1 H-Benzotriazol

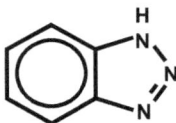

Though benzotriazole has a high heat of decomposition, it is not an explosive substance when subjected to UN Test Series 1 and, for example, it does not propagate the detonation in the 1 inch steel tube. It is used in spectrally adapted flare formulations and gas-generating compositions.

Formula		$C_6H_5N_3$
GHS		07
H-phrases		302-332-319-419
P-phrases		273-280 h-305+351+338
REACH		LRS
EINECS		202-394-1
CAS		[95-14-7]
m_r	g mol^{-1}	119.126
ρ	g cm^{-3}	1.34
Δ_fH	kJ mol^{-1}	236.48
$\Delta_{ex}H$	kJ mol^{-1}	−309
	kJ g^{-1}	−2.594
	kJ cm^{-3}	−3.476
Δ_cH	kJ mol^{-1}	−3312.2
	kJ g^{-1}	−27.804
	kJ cm^{-3}	−37.258
Λ	wt.-%	−194.74
N	wt.-%	35.27
Mp	°C	100
Bp	°C	350
Friction	N	>355
Impact	J	>40
Koenen	mm, type	1 mm (O)
Trauzl	ml	9

- J. S. Brusnahan, R. Pietrobon, M. Morgan, L. V. Krishnamoorthy, Organic Fuels in Spectral Flare Compositions, *PARARI 2011*, **2011**, Brisbane, Australien.
- M. Malow, K. D. Wehrstedt, S. Neuenfeld, On the explosive properties of 1 H-benzotriazole and 1 H-1,2,3-triazole, *Tetrahedron Lett.* **2007**, *48*, 1233–1235.

Benzotrifuroxane, BTF

Benzotrifuroxan, Benzo[1,2-c:3,4-c':5,6-c'']tris[1,2,5]oxadiazol-1,4,7-trioxid

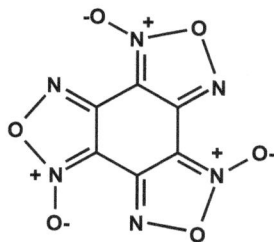

Formula		$C_6N_6O_6$
CAS		[3470-17-5]
m_r	g mol^{-1}	252.103
ρ	g cm^{-3}	1.901
$\Delta_f H$	kJ mol^{-1}	−602
$\Delta_{ex}H$	kJ mol^{-1}	−1522.2
	kJ g^{-1}	−6.038
	kJ cm^{-3}	−11.478
$\Delta_c H$	kJ mol^{-1}	−2966.1
	kJ g^{-1}	−11,766
	kJ cm^{-3}	−22.366
Λ	wt.-%	−38.08
N	wt.-%	33.34
Mp	°C	199
T_{crit}	°C	248−250
Impact	cm	22.7 mit Tool 12
V_D	m s^{-1}	8410 at 1.82 g cm^{-3}.
P_{CJ}	GPa	33.8 at 1.82 g cm^{-3}.
T_{CJ}	K	3990
$\varnothing_{cr}$	mm	0.5−1

BTF is a hydrogen-free HE of interest for the synthesis of nanodiamonds.

– Dolgoborodov, M. Brazhnikov, M. Makhov, S. Gubin, I. Maklashova, Detonation performance of high-dense BTF charges, *Journal of Physics Conference Series 500*, **2014**, 052010.

Berger, Beat (1949–2014)

Beat Berger was a Swiss chemist and an internationally renowned pyrotechnic specialist. Starting in 1979, he worked all of his professional life for the explosive laboratory of the Swiss Armed Forces in Thun (*"Gruppe Rüstungsdienste"* later renamed *"armasuisse"*). Berger did research into the basics of sensitiveness and performance of pyrotechnic compositions and explored innovative fuels, oxidizers, and binders.

–   E.-C. Koch, Beat Berger 1949–2014, *Propellants Explos. Pyrotech.* **2014**, *39*, 634–635.

Berger, Ernest Edouard Frédéric (1876–1934)

Ernest E. B. Berger was a French chemist. He served with the chemical laboratory of the French Artillery in Versailles from 1908 until 1923. During World War I in 1916, he developed a new class of pyrotechnic obscurants based on organic halogen compounds and halophilic metals known as *Berger mixture*. It is less known that he also developed obscurants based on RP/$CaSO_4$ and he invented formulations to fight germs and pests which can be considered progenitors for today's agent defeat weapons.

–   E.-C. Koch, 1916–2016, The Berger smoke mixture turns 100, *Propellants Explos. Pyrotech.* **2016**, *41*, 779–781.

Berger mixture
Berger-Mischung

Named after the French chemist *E. E. F. Berger*, the Berger mixture is based on chlorocarbon compounds and metals or alloys forming hydrolysable chlorides. For many decades hexachloroethane (HC) has been the chlorocarbon in smoke mixtures. Hence, for simplicity, those mixtures have been also termed HC, HC smoke or simply HC. A typical first-generation HC mix was constituted from
–   45 wt.-% zinc
–   55 wt.-% hexachloroethane

The mix reacts according to the following equation:

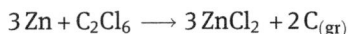

$$3\,Zn + C_2Cl_6 \longrightarrow 3\,ZnCl_2 + 2\,C_{(gr)}$$

However, as metallic zinc reacts readily with moisture to give hydrogen those formulations are prone to severe degradation. Hence as early as in the 1940s, zinc was consecutively replaced with zinc oxide which is reacted in situ with aluminum to form Zn in accordance with the below chain of reactions.

HC type C
- 5.0 wt.-% aluminum
- 48.5 wt.-% zinc oxide
- 46.5 wt.-% hexachloroethane

$$2\,Al + 3\,ZnO \longrightarrow Al_2O_3 + 3\,Zn + 625\,kJ$$

$$3\,Zn + C_2Cl_6 \longrightarrow 3\,ZnCl_2 + 2\,C_{gr} + 1042\,kJ$$

$$2\,C + 2\,ZnO + 20\,kJ \longrightarrow 2\,Zn + 2\,CO, \quad \Delta H > 0$$

Zinc chloride, $ZnCl_2$, formed upon combustion is a very effective aerosol. As it possesses a very low deliquescent relative humidity of ~10% it attracts water even under arid conditions.

Reaction of $ZnCl_2$ with moisture yields the coordination compound $[Zn(H_2O)_2]Cl_2$, which is a strong acid (pH = 1). Consecutive water uptake yields the weaker acid $[Zn(H_2O)_6]Cl_2$. Those acids are very corrosive towards metals and dissolve starch, cellulose, and proteins and hence also cause severe lesions to living tissue such as skin or the mucous membranes. Accidental inhalation of HC smoke hence has led to injuries and even fatalities in the past. It is because of these issues that HC today with many armed forces is no longer admitted as both training and tactical obscurant (*obscurants*).

- E.-C. Koch, *Metal-Halocarbon Pyrolant Combustion*, Handbook of Combustion, Vol. 5 New Technologies, Eds. M. Lackner, F. Winter, A. K. Agrawal, Wiley-VCH, **2010**, 355–402.
- *Hexachlorethane Smoke* in *Toxicity of Military Smokes and Obscurants, Volume 1*, National Academy Press, Washington, **1997**, pp. 127–160.R

Bergmann–Junk–Siebert test

The Bergmann–Junk–Siebert test is a stability test for nitrocellulose and gun propellants. Therefore, the material to be tested is heated for a certain time (see Table B.9) while nitrous gases form. The resulting nitric oxides are eluted with water and determined by titration. The test is not suitable to test triple base gun propellants containing nitroguanidine (NGu). The amine groups in NGu react with NO_2 thereby scavenging it and preventing any autocatalytic decomposition.

Table B.9: Bergmann–Junk–Siebert test.

Substance	Amount (g)	Test temperature (°C)	Duration (h)	Consumption of 0.01 N NaOH to neutralize the eluted nitric acids
Nitrocellulose	2	132	2	≤12.5 mL/g
Single base (no NGl)	5	132	5	≤10 to 12 mL/g
Double base (≤10 wt.-% NGl)	5	115	16	≤10 to 12 mL/g
Double base >10 wt.-% NGl)	5	115	8	≤10 to 12 mL/g
Double-base rocket propellant	5	115	16	≤10 to 12 mL/g

- Committee of Experts on the Transport of Dangerous Goods and on the Globally Harmonized System of Classification and Labelling of Chemicals, *Stability tests for nitrocellulose*, 12 October **2017**.
- *Bundeswehrprüfvorschrift* TL-1376-0600 M 2.22.1

Berthelot, Marcellin (1827–1907)

Marcellin Berthelot was a French chemist and also a minister for education and foreign affairs in the III. French Republic. He founded the thermochemistry of explosives and in 1883 wrote a monography in two volumes on this topic. Together with *Paul Vieille* he developed the ballistic bomb and also recognized the shock wave as a necessary characteristic of the detonation of an explosive.

- M. Berthelot, *Sur la Force des Matières Explosives d'après la Thermochimie*, Gauthier-Villars, **1883**, 2 Volumes,. 408 & 445 pp.

Berthollet, Claude Louis (1748–1822)

Claude Louis Berthollet was a French chemist. He discovered the chloric acid, $HClO_3$, their salts, the chlorates, and their oxidizing capabilities. In 1786, he developed the so-called *white gun powder* which, however, was too hazardous for use in guns but can be considered a predecessor of the chlorate blasting explosives introduced later in the late nineteenth century.

White gun powder
- 60 wt.-% potassium chlorate, $KClO_3$
- 20 wt.-% potassium hexacyanoferrate(II), $K_4[Fe(CN)_6]$
- 20 wt.-% sucrose, $C_{12}H_{22}O_{11}$

- K. A. Hofmann, U. R. Hofmann, *Anorganische Chemie*, 9. Auflage, Friedrich Vieweg, Braunschweig, **1941**, p. 662.

Beryllium

Aspect		Silver grey
Formula		Be
GSH		06, 08
H-phrases		301-330-350i-372-315-319-317-335
P-phrases		260-301+310–304+340–305+351+338-320405-501a
EINECS		231-150-7
REACH		LRS
UN		1567
CAS		[7440-41-7]
m_r	g mol^{-1}	9.102
ρ	g cm^{-3}	1.85
Mp	°C	1287
Bp	°C	2468
$\Delta_m H$	kJ mol^{-1}	7.9
$\Delta_v H$	kJ mol^{-1}	292
c_p	J mol^{-1} K^{-1}	20.79
c_L	m s^{-1}	12720
κ	W m^{-1} K^{-1}	200
$\Delta_c H$	kJ mol^{-1}	−609.40
	kJ g^{-1}	−66.952
	kJ cm^{-3}	−123.86
Λ	wt.-%	−177.53

Beryllium is an important alloy component for non-sparking tools. Its gravimetric heat of combustion is the highest among the elements, whereas its volumetric heat of combustion is second after boron. Beryllium is of interest as a high-energy fuel, however, both its high price and its complicated toxicology have impeded further consideration. Although formally satisfying *Glassman's criterion* for vapor-phase combustion it predominantly shows surface combustion.

- C. Strupp, Beryllium metal I. experimental results on acute oral toxicity, local skin and eye effects, and genotoxicity, *Ann. Occup. Hyg.* **2011**, *55*, 30–42.
- C. Strupp, Beryllium metal II. a review of the available toxicity data. *Ann. Occup. Hyg.* **2011**, *55*, 43–56.
- C. K. Law, A Simplified Theoretical Model for the Combustion for the Vapor Phase Combustion of Metal Particles, *Combust. Sci. Technol.* **1973**, *7*, 197–212.
- A. Macek, R. Friedman, J. M. Semple, Techniques for the study of combustion of beryllium and aluminum particles, in H. G. Wolfhard, I. Glassman, L. Green Jr. (Eds.) *Heterogeneous Combustion*, Vol. 15 *Progress in Astronautics and Aeronautics*, Acedemic Press, New York, **1964**, 3–16.

Bickford, William (1774–1834)

William Bickford was an English engineer. In 1831, he invented the safety fuze "*Bickford fuze*," and started a business producing it.

– N. N., *Davey Bickford Smith 1839–1989*, Elyps Editions, France, **1989**, 67 + IV.

BICT – WIWEB

The BICT, Bundesinstitut für Chemisch Technische Untersuchungen which translates into *Federal Research Institute for Chemical Technical Investigations* was a former institute in the responsibility of the German ministry of defence. Since 1972, it was located on an extensive 14 hectare-sized premises near Swisttal-Heimerzheim, North Rhine-Westphalia. With a staff of 150 technicians, engineers, and scientists it had a unique role defined by German explosive law to test and approve explosives, propellants, and pyrotechnics for use in ammunition. At BICT, its progenital organization CTI and under the late years umbrella WIWEB, many well-known test procedures had been developed and established (e.g. *BICT gap test* and *Koenen test*). In the early 2000, it was the leading institute in Germany in the area of explosives detection. In 2007, after a 10-year budgetary struggle German Ministry of Defence took the decision to shut down the institute effective 30 June 2009. Internationally renowned scientist active at the BICT were *Dr. Friedrich Trimborn* (1933–2009) and Dr. *Carl-Otto Leiber*.

– C.-O. Leiber, Friedrich Trimborn (1933–2009), *Propellants Explos. Pyrotech.* **2010**, *35*, 5–6.
– N.N., *Stellungnahme zum Wehrwissenschaftlichen Institut für Werk-, Explosiv- und Betriebsstoffe (WIWEB), Erding*, Wissenschaftrat, **2008**, 79 pp, accessed at https://www.wissenschaftsrat.de/download/archiv/8784-08.pdf

BICT gap test

The BICT gap test serves the determination of the shock sensitivity of HEs with unconfined critical diameters well below 20 mm. For this test, the cylinder of the explosive under investigation (acceptor) is subjected to the detonation shock transmitted from a HWC charge (donor) and separated by a layer of de-aerated water in a PMMA tube (Fig. B.8). To witness a successful transmission of detonation, the acceptor charge is connected to a detonation cord. Variations of this test use a mild steel plate on top of the acceptor charge to check for dents and punch holes. The shock pressure of the donor charge incident on the acceptor charge can be adjusted by variation of the height of the water layer. Table B.10 shows typical No-go pressures (nR = no reaction) and the corresponding height of the water column (mm) for configurations where no transmission of detonation occurs.

Table B.10: No-Go gap test values for different high explosives.

	ρ (g cm^{-3})	h_{nR} (mm)	P_{nR} (GPa)
Comp B	1.68	18	2.09
HWC	1.63	22	1.61
Hexyl	1.50	20	1.86
PETN/wax (93/7)	1.60	29	0.98
Pentolite	1.65	23	1.50
Picric acid	1.58	17	2.24
Tetryl	1.53	24	1.40
TNT	1.53	22	1.61
TNT	1.58	7	4.95
TNT	1.61	6	5.31
PBXW-11	1.79		1.96
	1.81		2.60
	1.82		3.04
PBXN-9	1.73		3.88
	1.75		4.22
	1.78		4.58
KS32	1.63	10	3.88
PBXN-5		20	1.86
PBXN-109		12	3.29

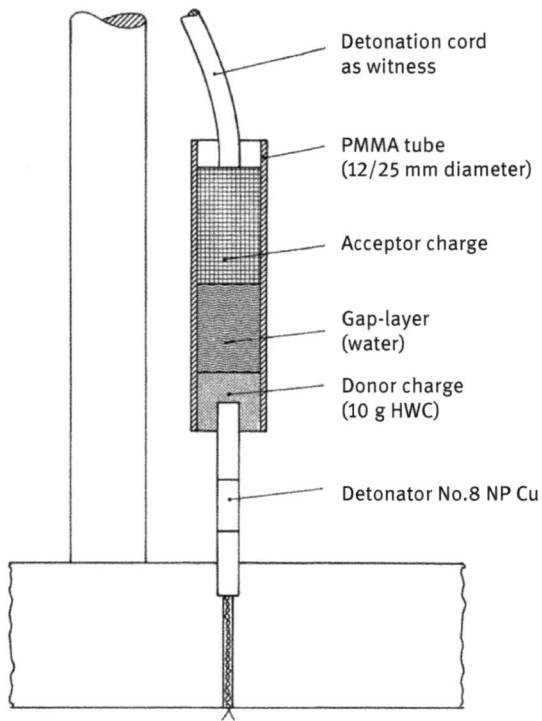

Detonation cord
as witness

PMMA tube
(12/25 mm diameter)

Acceptor charge

Gap-layer
(water)

Donor charge
(10 g HWC)

Detonator No.8 NP Cu

Figure B.8: Setup of BICT gap test.

- F. Trimborn, Eine einfache Versuchsanordnung zum „Gap-Test ", *Explosivstoffe*, **1967**, *15*, 169–175.
- F. Trimborn, R. Wild, Shock-Wave Measurements in Water for calibrating the BICT -Gap-Test, *Propellants Explos. Pyrotech.* **1982**, *7*, 87–90.

Bis(2-chloro-2,2-dinitroethyl)formal, CEFO

Bis(2-chlor-2,2-dinitroethyl)formal

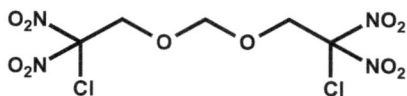

Formula		$C_5Cl_2H_6N_4O_{10}$
CAS		[5946-59-8]
m_r	g mol^{-1}	353.028
ρ	g cm^{-3}	1.631 (at 25 °C)
$\Delta_f H$	kJ mol^{-1}	−406
$\Delta_{ex}H$	kJ mol^{-1}	−1628.6
	kJ g^{-1}	−4.613
	kJ cm^{-3}	−7.524
$\Delta_c H$	kJ mol^{-1}	−2317.9
	kJ g^{-1}	−6.566
	kJ cm^{-3}	−10.709
Λ	wt.-%	−9.06
N	wt.-%	15.87
Mp	°C	
Bp	°C	110-115 at 133 Pa
E_a	kJ mol^{-1}	160
Dp	°C	156
V_D	m s^{-1}	*7066 at TMD*
P_{CJ}	GPa	*22.7 at TMD*

- H. E. Ungnade, L. W. Kissinger, Esters and Ethers of 2-Substituted 2,2-Dinitro-1-alkanols, *J. Org. Chem.* **1966**, *31*, 369–371.
- M. E. Hill, K. G. Shipp, *Process for Acetal Preparation*, US Patent 3,526,667, USA, **1970**.

Bis(2,2,2-trinitroethyl)nitramine, BTNNA, and HOX

Bis(2,2,2-trinitroethyl)nitramin

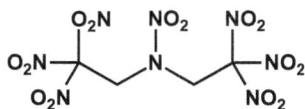

Formula		$C_4H_4N_8O_{14}$
CAS		[19836-28-3]
m_r	g mol^{-1}	388.121
ρ	g cm^{-3}	1.92
$\Delta_f H$	kJ mol^{-1}	−28

$\Delta_{ex}H$	kJ mol^{-1}	−2054.3
	kJ g^{-1}	−5.293
	kJ cm^{-3}	−10.163
$\Delta_c H$	kJ mol^{-1}	−2117.7
	kJ g^{-1}	−5.456
	kJ cm^{-3}	−10.476
Λ	wt.-%	+16.49
N	wt.-%	28.87
Mp	°C	94
Bp	°C	175 (dec)
$\Delta_{vap}H$	kJ mol^{-1}	84.77
Friction	N	120
Impact	J	2.5
V_D	m s^{-1}	8420 @ 1.92 g cm^{-3}
$\sqrt{2E_g}$	mm µs	2.81

BTTNA due to its positive oxygen balance is also termed **high oxygen explosive** (HOX). It has been investigated as an oxidizer in aluminized formulations. With bases BTNNA readily forms salts.

– H. Ritter, S. Braun, High Explosives Containing Ultrafine Aluminum ALEX, *Propellants Explos. Pyrotech.* **2001**, *26*, 311–314.

Bis(2,2,2-trinitroethyl)formal, TEFO

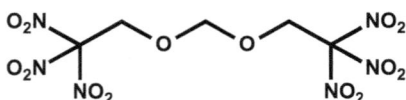

Formula		$C_5H_6N_6O_{14}$
CAS		[6263-74-7]
m_r	g mol^{-1}	374.133
ρ	g cm^{-3}	1.75 (monoclinic); 1.817(orthorhombic)
$\Delta_f H$	kJ mol^{-1}	−403.3 ± 2.67
$\Delta_{ex}H$	kJ mol^{-1}	−2315.3
	kJ g^{-1}	−6.188
	kJ cm^{-3}	−10.644
$\Delta_c H$	kJ mol^{-1}	−2422.1
	kJ g^{-1}	−6.474
	kJ cm^{-3}	−11.329
Λ	wt.-%	+4.28
N	wt.-%	22.46
Mp	°C	65.3

E_a	kJ mol^{-1}	109.44
Dp	°C	204.9
Impact	J	1.5
Friction	N	100
V_D	m s^{-1}	8624 at 1.817
P_{CJ}	GPa	30.45 at 1.817

TEFO is unusually thermally stable in the molten state and is an alleged ingredient in the PFM-1 antipersonal mine explosive charge.

- M. E. Hill, K. G. Shipp, *Process for Acetal Preparation*, US Patent 3,526,667, USA, **1970**.
- E. E. Baroody, G. A. Carpenter, Enthalpies of Formation of Bitetrazole and Bis(2,2,2-trinitroethyl)formal, *J. Chem. Eng. Data* **1979** *24*, 3–6.
- T. M. Klapötke, B. Krumm, R. Moll, S. F. Rest, CHNO Based Molecules Containing 2,2,2-Trinitroethoxy Moieties as Possible High Energy Dense Oxidizers, *Z. Anorg. Allg. Chem.* **2011**, *637*, 2103–2110.

Bis(2,2-dinitropropyl)acetal, BDNPA

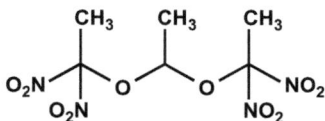

Formula		$C_8H_{14}N_4O_{10}$
REACH		LPRS
EINECS		610-610-4
CAS		[5108-69-0]
m_r	g mol^{-1}	326.22
ρ	g cm^{-3}	1.366
$\Delta_f H$	kJ mol^{-1}	−642
$\Delta_{ex}H$	kJ mol^{-1}	−1429.5
	kJ g^{-1}	−4.382
	kJ cm^{-3}	−5.986
$\Delta_c H$	kJ mol^{-1}	−4507.0
	kJ g^{-1}	−13.816
	kJ cm^{-3}	−18.872
Λ	wt.-%	−63.76
N	wt.-%	17.18

Mp	°C	33-35
Bp	°C	150 bei 1.33 Pa
$\Delta_{vap}H$	kJ mol^{-1}	93.01
Dp	°C	220 for the eutectic with BDNPF

BNDPA is used in blends with BDNPF as an energetic plasticizer in HEs (e.g. PBX9501, PAX-3) and gun propellants. The eutectic with BDNPF melts at −15 °C and has a glass transition temperature of −65.2 °C.

Bis(2,2-dinitropropyl)formal, BDNPF

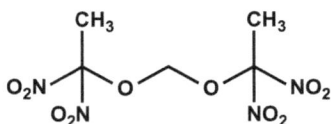

Formula		$C_7H_{12}N_4O_{10}$
REACH		LPRS
EINECS		611-807-8
CAS		[5917-61-3]
m_r	g mol^{-1}	312.193
ρ	g cm^{-3}	1.411
$\Delta_f H$	kJ mol^{-1}	−597
$\Delta_{ex}H$	kJ mol^{-1}	−1417.8
	kJ g^{-1}	−4.541
	kJ cm^{-3}	−6.408
$\Delta_c H$	kJ mol^{-1}	−3872.6
	kJ g^{-1}	−12.405
	kJ cm^{-3}	−17.503
Λ	wt.-%	−51.25
N	wt.-%	17.94
Mp	°C	31
Bp	°C	152 at 1.33 Pa
$\Delta_{vap}H$	kJ mol^{-1}	84.77

This is used in combination with BDNPF as an energetic plasticizer in HEs and propellants.

Bis(2-chloroethyl) sulfide, HD

Senfgas, S-Lost

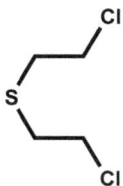

mustard		
Aspect		Oily yellowish liquid with garlic-type smell
Formula		$C_4 Cl_2 H_8 S$
CAS		[505-60-2]
m_r	g mol^{-1}	159.074
ρ	g cm^{-3}	1.2741
$\Delta_f H$	kJ mol^{-1}	−200
$\Delta_c H$	kJ mol^{-1}	−2715.6
	kJ g^{-1}	−17.071
	kJ cm^{-3}	−21.751
Λ	wt.-%	−104.00
Mp	°C	14.4
Bp	°C	217 (dec)
LCt_{50}	ppm min^{-1}	400
ICt_{50}	ppm min^{-1}	150

HD is a persistent and extremely stable blister agent, decomposing only at T ~ 500 °C. HD containing ammunition dumped into the Baltic Sea after World War II continues to causes countless pieces of encrusted HD to be washed ashore. Those pieces often lead to severe chemical burns to beachcombers who erroneously take those pieces for amber.

- J. F. Bunnett, M. Mikolajczyk, *Arsenic and Old Mustard: Chemical Problems in the Destruction of Old Arsenical and "Mustard" Munitions*, Kluwer, Publisher, Dordrecht, **1998**, 200 pp.
- S. Sass and P. M. Davis, *Laboratory Research on the Incineration of Mustard*, Edgewood Arsenal, Maryland, **1972**, 16 pp.

Bis(3-nitro-1,2,4-triazol-1-yl)-5-nitropyrimidine, BNTNP

5-Nitro-4,6-bis(3-nitro-1 H-1,2,4-triazol-1-yl)-pyrimidin

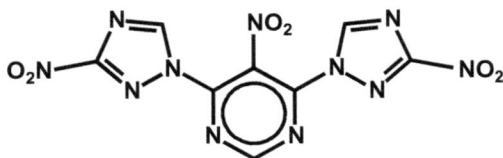

Formula		$C_8H_3N_{11}O_6$
CAS		[124777-88-4]
m_r	$g\ mol^{-1}$	349.183
ρ	$g\ cm^{-3}$	1.82
$\Delta_f H$	$kJ\ mol^{-1}$	638.1
$\Delta_{ex}H$	$kJ\ mol^{-1}$	−1724.4
	$kJ\ g^{-1}$	−4.938
	$kJ\ cm^{-3}$	−8.988
$\Delta_c H$	$kJ\ mol^{-1}$	−4215.0
	$kJ\ g^{-1}$	−12.071
	$kJ\ cm^{-3}$	−21.970
Λ	wt.-%	−52.69
N	wt.-%	44.13
DSC onset	°C	261
Dp	°C	356
Friction	N	>355
Impact	J	>50
V_D	$m\ s^{-1}$	8420 at 1.82 $g\ cm^{-3}$.

BNTNP is an explosive that is relatively insensitive towards thermal and mechanical stimuli.

– J.-P. Freche, F. Laval, C. Wartenberg, *Preparation of 5-nitro-4,6-bis-(3-nitro-1 H-1,2,4-triazol-1-yl) pyrimidine and explosive material containing it*, EP320369, **1987**, France.
– H. H. Licht, S. Braun, M. Schäfer, B. Wanders, H. Ritter, Nitrotriazole Chemische Struktur und explosive Eigenschaften, *ICT -Jata*, **1998**, V-47.

Bis(nitrimino)triazinone, DNAM

4,6-Bis(nitroamino)-1,3,5-triazin-2(1 H)-on

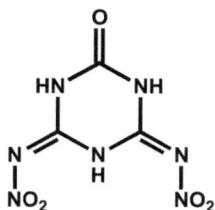

Formula		$C_3H_3N_7O_5$
CAS		[19899-80-0]
m_r	g mol^{-1}	217.101
ρ	g cm^{-3}	1.998 (at 150 K)
Δ_fH	kJ mol^{-1}	−114
$\Delta_{ex}H$	kJ mol^{-1}	−932.5
	kJ g^{-1}	−4.295
	kJ cm^{-3}	−8.582
Δ_cH	kJ mol^{-1}	−1495.3
	kJ g^{-1}	−6.888
	kJ cm^{-3}	−13.762
Λ	wt.-%	−18.42
N	wt.-%	45.16
Mp	°C	228
Friction	N	216
Impact	J	>50
V_D	m s^{-1}	9162 at 1.998 g cm^{-3}.
P_{CJ}	GPa	*36.94* at 1.998 g cm^{-3}.

DNAM was once suggested as an energetic filler in solid rocket propellants. However, despite its low sensitiveness and good performance it is prone to rapid hydrolysis.

- P. N. Simoes, L. M. Pedroso, A. M. Matos Beja, M. Ramos Silva, E MacLean, A. A. Portugal, Crystal and Molecular Structure of 4,6-Bis(nitroimino)-1,3,5-triazinan-2-one Theoretical and X-ray Studies, *J. Phys. Chem. A* **2007**, *111*, 150–158.

Bis(trinitromethyl)benzene, HNX

1,4-Bis(trinitromethyl)benzol

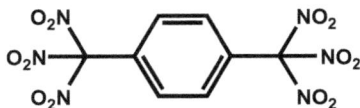

HNX = hexanitro-*p*-xylene		
Formula		$C_8H_4N_6O_{12}$
CAS		[200443-67-0]
m_r	g mol^{-1}	376.153
ρ	g cm^{-3}	1.805
$\Delta_f H$	kJ mol^{-1}	72.8
$\Delta_{ex}H$	kJ mol^{-1}	−2143.1
	kJ g^{-1}	−5.698
	kJ cm^{-3}	−10.284
$\Delta_c H$	kJ mol^{-1}	−3792.6
	kJ g^{-1}	−10.083
	kJ cm^{-3}	−18.199
Λ	w.%	−25.52
N	wt.-%	22.34
Dp	°C	129.7
DSC onset	°C	103
E_a	kJ mol^{-1}	109.4
V_D	m s^{-1}	8438 at 1.805 g cm^{-3}
P_{CJ}	GPa	32.8 at 1.805 g cm^{-3}

HNX is a sensitive HE that upon slow pyrolysis (Fig B.9) affords the tetranitroquino-dimethane, CAS-No. [130651-73-9], which is a more powerful electron acceptor than tetracyanodiquinomethane.

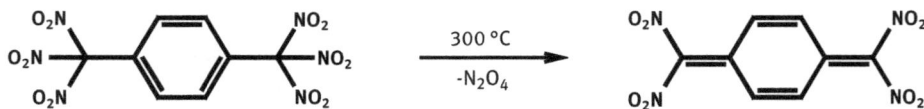

Figure B.9: Pyrolysis of HNX i.a.w. *Nielsen*.

- A. T. Nielsen, *Nitrocarbons*, Wiley-VCH, Weinheim, **1995**, 112–113.
- H. Yan, X.-P. Guan, Thermal studies of 2-Azido-1,1-dinitromethyl compounds and trinitromethyl compounds, *J. Energ. Mat.* **1997**, *15*, 283–288.
- I. Giles, *personal communication*, **2020**.

Bis-isoxazol-5,5′-bismethylene dinitrate, BIDN

3,3′-Bis-isoxazol-5,5′-bismethylendinitrat

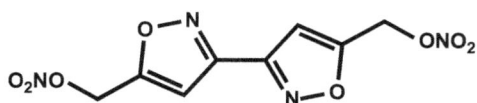

Formula		$C_8H_6N_4O_8$
CAS		[2289600-01-5]
m_r	g mol^{-1}	286.159
ρ	g cm^{-3}	1.585
Δ_fH	kJ mol^{-1}	−139
$\Delta_{ex}H$	kJ mol^{-1}	−1327.3
	kJ g^{-1}	−4.638
	kJ cm^{-3}	−7.352
Δ_cH	kJ mol^{-1}	−3866.7
	kJ g^{-1}	−13.512
	kJ cm^{-3}	−21.417
Λ	wt.-%	−61.5
N	wt.-%	19.58
Mp	°C	92
Dp	°C	189
Friction	N	>355
Impact	J	11
V_D	m s^{-1}	7060 at 1.585 g cm^{-3}.
P_{CJ}	GPa	19.3 at 1.585 g cm^{-3}.

BIDN has been proposed as an energetic plasticizer for propellants.

– L. A. Wingard, P. E. Guzmán, E. C. Johnson, J. J. Sabatini, G. W. Drake, E. F. C. Byrd, Synthesis of Bis-Isoxazole-Bis-Methylene Dinitrate a Potential Nitrate Plasticizer and Melt-CasTable Energetic Material, *ChemPlusChem* **2017**, *82*, 195–198.

Bis-isoxazol-5,5′-tetrakis(methylene nitrate), BITN

3,3′-Bis-isoxazol-5,5′-tetrakis(methylennitrat)

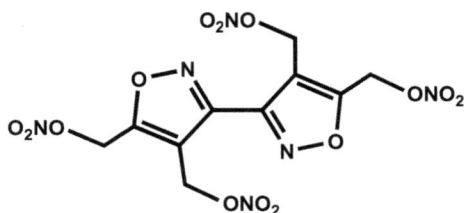

Formula		$C_{10}H_8N_6O_{14}$
CAS		[2095832-55-4]
m_r	g mol^{-1}	436.20256
ρ	g cm^{-3}	1.786
$\Delta_f H$	kJ mol^{-1}	−395
$\Delta_{ex}H$	kJ mol^{-1}	−2223.8
	kJ g^{-1}	−5.098
	kJ cm^{-3}	−9.105
$\Delta_c H$	kJ mol^{-1}	−4683.5
	kJ g^{-1}	−10.737
	kJ cm^{-3}	−19.176
Λ	wt.-%	−36.67
N	wt.-%	19.27
Mp	°C	121.9
Dp	°C	193.7
Friction	N	60
Impact	J	3
V_D	m s^{-1}	7837 at 1.786 g cm^{-3}.
P_{CJ}	GPa	27.1 at 1.786 g cm^{-3}.

BITN is a candidate replacement for lead azide.

– L. A. Wingard, E. C. Johnson, P. E. Guzmán, J. J. Sabatini, G. W. Drake, E. F. C. Byrd, R. C. Sausa, Synthesis of Biisoxazoletetrakis(methyl nitrate) a Potential Nitrate Plasticizer and Highly Explosive Material., *Eur. J. Org. Chem.* **2017**, 1765–1768.

Bismuth citrate

Bismuthcitrat

Formula		$C_6H_5BiO_7$
REACH		LRS
EINECS		212-390-1
CAS		[813-93-4]
m_r	g mol^{-1}	398.08
ρ	g cm^{-3}	3.46

Bismuth citrate has been proposed as a ballistic modifier in rocket propellants.

- L. Warren, *Minimum Signature Isocyanate Cured Propellants Containing Bismuth Compounds as Ballistic Modifiers*, US-Patent 6168677, **2001**, USA.

Bismuth oxide

Bismuthoxid

Aspect		Yellow-beige crystalline powder
Formula		Bi_2O_3
GHS		07
H-phrases		315-319-335-
P-phrases		261-302+352-305+351+338-321-405-501a
REACH		LRS
EINECS		215-134-7
CAS		[1304-76-3]
m_r	g mol^{-1}	465.958
ρ	g cm^{-3}	8.90
c_p	J g^{-1} K^{-1}	0.244
T_p	°C	730
Mp	°C	825
Bp	°C	860
$\Delta_f H$	kJ mol^{-1}	−573.9
Λ	wt.-%	+6.86

Thermochromic Bi_2O_3 (cold = yellow, hot = redbrown) is a versatile oxidizer in nanothermites (*Comet*) and delay compositions together with tin oxide and group 4 metals (*Boberg*). A formulation containing Bi_2O_3, boron, and Viton®A is known as SR57 and is used for take over cup pellets in decoy flare sequencers (*Davies*). Bi_2O_3 has replaced lead(II,IV) oxide both as a combustion modifier in double-base propellants and as an oxidizer in crackling stars *(Jennings-White)*. It was also proposed as an oxidizer in inorganic smoke formulations *(Graff)*.

- E. Lafontaine, M. Comet, *Nanothermites*, ISTE, Wiley, **2016**.
- N. Davies, T. T. Griffiths, E. L. Charsley, J. A. Rumsey, Studies on gasless delay compositions containing boron and bismuth trioxide, *ICT -Jahrestagung*, Karlsruhe, **1985**, V-15.
- T. Boberg, *Pyrotechnischer Verzögerungssatz*, DE3218997A1, **1982**, Sweden.
- J. C. Cackett, Monograph on Pyrotechnic Compositions, Ministry Of Defence, Royal Armament Research and Development Establishment, **1965**.
- G. U. Graff, *Pyrotechnic Composition for Producing Yellow Smoke*, US Patent 1920254, **1933**, USA.

Bismuth salts with energetic anions

Bismuthsalze mit energetischen Anionen

Recently, *Nesveda* has disclosed a series of bismuth oxide and hydroxide salts with energetic anions that could serve as less toxic replacements for common lead and mercury-based primary materials. Table B.11 shows the individual anions and the assumed sum formula of the corresponding salts, their observed explosion temperature and sensitiveness to friction in comparison to tetrazene.

Table B.11: Sensitiveness of different bismuth salts.

Anion	Sum formula	T_{ex} (°C)	R*	0%$^\$$	100%
Azide	$BiON_3$	320	No		
Picrate	$C_6H_2(NO_2)_3OBiO$	240	No		
Styphnate	$C_6H(NO_2)_3(OBiO)_2$	220	No		
Trinitrophloroglucine	$C_6(NO_3)_3(OBiO)_3$	175	No		
Dinitroazidophenolate	$C_6H_2N_3(NO_2)_2(OBiO)\ Bi_2O_3$	200	No		
Dinitrobenzofuroxanate	$C_6H_2(NO_2)_2N_2O_2BiO$	220	No		
Azotetrazolate	$N_{10}C_2Bi_2(OH)_4$	180	Yes	500	2800
Diazoaminotetrazolate	$N_{11}C_2Bi_3(OH)_6$	240	Yes	700	2000
Bistetrazolate	$N_8C_2Bi_2(OH)_4$	280	Yes	5000	8000
Bistetrazolylhydrazinate	$N_{20}C_4H_4(BiO)_5OH$	150	Yes	6000	12,000
Tetrazene	$C_2H_8N_{10}O$	110	Yes	150	500

*Sensitiveness to friction in (mortar + pestel)
$^\$$Friction force in g

- J. Nesveda, *Bismuth-Based Energetic Materials*, US 2016/0280614, **2016**, Czech Republic.

Bismuth subnitrate
Bismuthsubnitrat

Basic bismuth nitrate, C.I. Pigment White 17,		
Formula		$4\,BiNO_3(OH)_2 \cdot BiO(OH)$
REACH		LRS
EINECS		215-136-8
CAS		[1304-85-4]
m_r	g mol^{-1}	1461.99
ρ	g cm^{-3}	4.928
Mp	°C	260 (dec)
$\Delta_f H$	kJ mol^{-1}	−2500 (estimated)
Λ	wt.-%	+10.94
N	wt.-%	3.83

Bismuth subnitrate serves as oxidizer in nanothermites and crackling compositions. The thermal decomposition up to 350 °C can be summarized as follows:

$$4\,BiNO_3(OH)_2 \cdot BiO(OH) \;\longrightarrow\; Bi_5O_7NO_3 + 9/2\,H_2O + 3\,NO + 9/4\,O_2.$$

At T = 630 °C the decomposition can be described as follows:

$$Bi_5O_7NO_3 \;\longrightarrow\; 5/2\,Bi_2O_3 + NO + 0.75\,O_2.$$

- C. J. White, Lead-Free Crackling Microstars, *Pyrotechnica*, **1992**, *XIV*, 30–32.
- H. Kodama, Synthesis of a New Compound, $Bi_5O_7NO_3$, by Thermal Decomposition, *J. Solid State Chem.* **1994**, *112*, 27–30.

Black match
Stoppine

A black match is an open burning ignition cord based on slow burning black powder. The black match consists of a cotton thread of the thickness ~ 2 mm impregnated with a saturated solution of potassium nitrate. This thread is passed through a thick aqueous paste of meal black powder and dextrin and forced through a funnel to scrape of any supernatant material. The wet threads prepared this way are mounted for drying and cut when dry. The burn rate of black match is in the 6 cm s^{-1} ballpark. As open burning fuses black match was used intensively in fireworks, with model rocket engines and for testing energetic materials. However, their extreme ignitability by sparks and the scaling off of brittle pieces of black powder has led to a major replacement by covered ignition fuses such as Visco® fuse or Bickford fuse (Figure B.10).

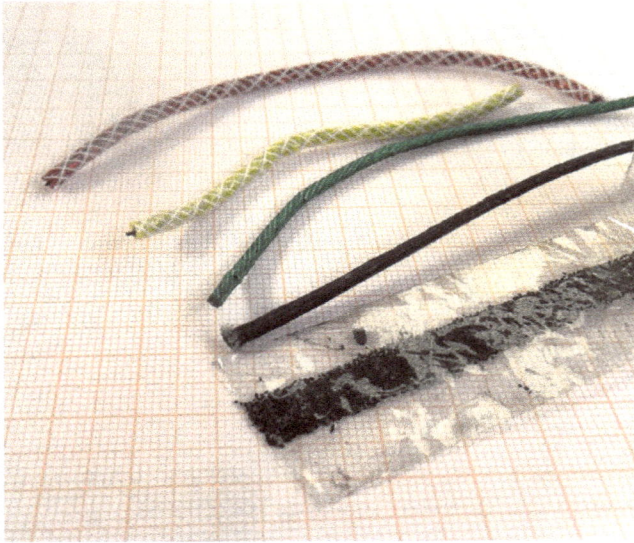

Figure B.10: Black powder-based fuzes (quickmatch red, yellow, Viscofuze®, black match and tapematch® from top to bottom) on millimeter paper.

Black powder

Schwarzpulver, Schießpulver

Black powder (also called gun powder) is the progenitor of all explosives and pyro-technic compositions. Though black powder in use since about thousand years, it is still a popular and sometimes even unmatched ingredient for first fires, igniters, safety fuses, expulsion charges, and rocket propellant for both civilian and military applications (Figure B.11). It is reproducibly manufactured from sustainable compo-nents and – provided it is kept away from water or organic solvents – black powder is virtually not ageing!

The production of black powder starts with the milling of the binary premix from charcoal (starting wood types for charcoal production greatly varies with area: *alnus, rhamnus* (GBR), *fagus* (DEU), *acer* (USA), and sulfur in a ball mill. The mill charge is passed by a strong magnet to catch steel particles that could have come off from the mill. In parallel, potassium nitrate is milled in a pin mill. Both the bi-nary premix and the ground potassium nitrate together with some water are fed into the brass pan mill. There it is thoroughly mixed and the high pressure exerted by the heavy wheels yields plastic deformation of both sulfur and potassium nitrate forcing these substances into the open porosities of the charcoal. The moist material is then hydraulically consolidated to slabs. Those slabs are granulated and the cor-responding grains are separated by size and finally coated with a little amount of

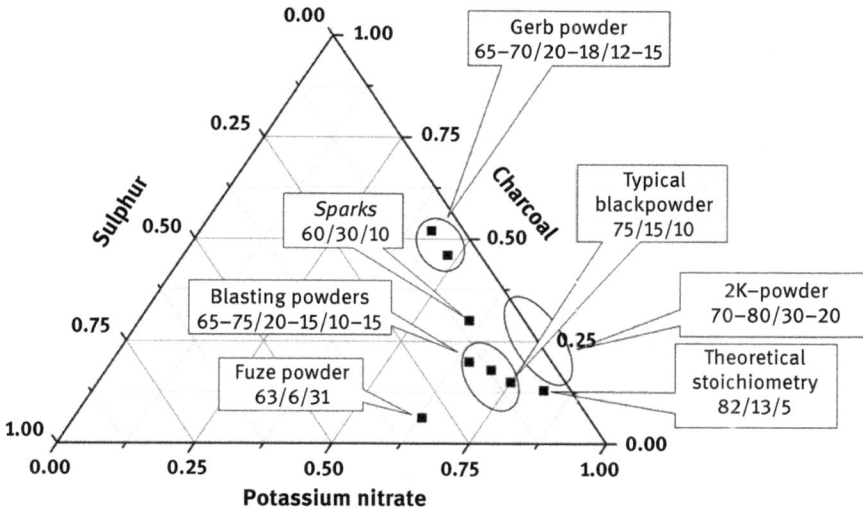

Figure B.11: Black powder types in ternary diagram, in wt.-%.

graphite to account for reduced electrostatic sensitivity and provide protection against water. While further drying the grains to a residual moisture content between 0.6 and 0.9 wt.-%, the dust fraction is removed and fed back into the process.

The formal explosion reaction of 10 kg 3 K-black powder can be most accurately described according to *Seel*.

With the 3 K-powder composition reading
- 75 wt.-% potassium nitrate
- 15 wt.-% charcoal
- 10 wt.-% sulfur

and the general formula $(C_6H_2O)_n$ assumed for charcoal we can write:

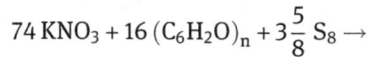

$$74\,KNO_3 + 16\,(C_6H_2O)_n + 3\frac{5}{8}\,S_8 \longrightarrow$$

- Gaseous products $\quad 35\,N_2 + 56\,CO_2 + 14\,CO + 3\,CH_4 + 2\,H_2S + 4\,H_2$
- Condensed products $19\,K_2CO_3 + 7\,K_2SO_4 + 8\,K_2S_2O_3 + 2\,K_2S_2 + 2\,KSCN$

That is, explosion of 10 kg BP yields 24.5 m³ gas at a temperature of 2900 K. At ambient temperature (T = 298 K), the permanent gas volume is 2.55 m³. The mean molecular mass of combustion products is ~ 36 g mol⁻¹. In addition, at ambient temperature 5.85 kg of the reaction products are condensed.

The burn rate, u (mm s⁻¹), of black powder (Fig. B.12) is determined by a series of factors listed in Table B.12.

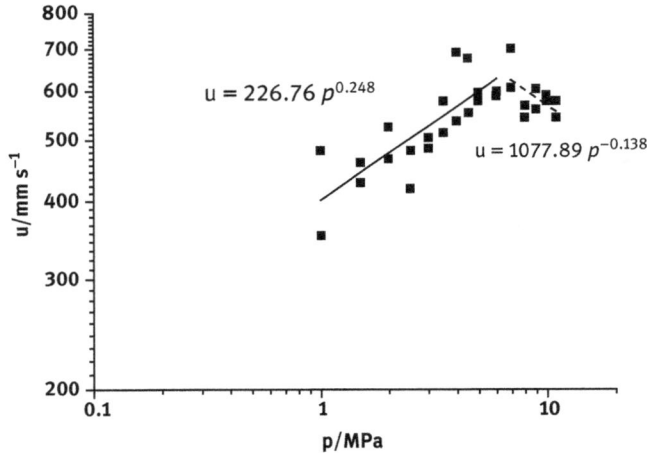

Figure B.12: Burn rate and Vieille's law for grained black powder.

Table B.12: Factors influencing the burn rate of black powder.

Factor	Influence
Composition	u increases with increasing $\xi(KNO_3)$ up to 0.75
Grain size	Increasing grain size yields decreasing u
Grain density	Increasing density (decreasing porosity) yields decreasing u
Charcoal type	Increasing free surface area yields increasing u; decreasing hydrogen content yields a decrease in u
Water content	u decreases with increasing water content
Type of mixing	Powder mixed in a pan mill burns faster than other powders

Figure B.12 shows the pressure sensitivity of the combustion of grained black powder (*Berger*). At pressures P = 7–8 MPa, a negative pressure exponent is indicative of a fundamental change in combustion mechanism from convective to laminar.

Black powder is thermally sensitive resulting in its supreme ignitability. However, it also effects a high sensitivity towards electrostatic discharges E > 16 mJ and ignition by adiabatic compression of dust!

The ignition temperature depending on the composition and grain size of 3-K powders is between T = 260 and 320 °C. The impact sensitivity for 3-K Pulver is 10 J (BAM machine).

Typical black powder-based ignition means are displayed in Figure B.10 on p. 120. Qualified military-grade black powders are *Y-593-2* (Germany) or *G21* (UK) (Table B.13).

Table B.13: Some properties of 3-K black powder.

Density (g cm^{-3})	1.6–2.1
Gas volume (cm^3)	280–360
Heat of explosion (kJ g^{-1})	2.386–3.379
Activation energy (kJ mol^{-1})	56–62 (ignition delay measurements)
	250–400 (from DSC and DTA measurements)
Specific heat (J g^{-1} K^{-1})	0.928
Ignition (°C)	240 (DSC)
Force (J g^{-1})	255–295 (at loading density, $\Delta = 0.2$ g cm^{-3})

Charges of 3-K powder at 4 mm diameter filling density of $\rho = 0.5$ g cm^{-3} yield burn rates up to $V_D = 1500$ m s^{-1} which is clearly in the low-velocity detonation regime (LVD).

Warning:

Safety fuzes (bickford fuzes) according to Leiber upon clamping, cladding, or other means of mechanical damage after a regular (laminar) ignition have been observed to transition to LVD and thereby have caused severe accidents by failure to delay and premature ignition!

- F. Seel, Geschichte und Chemie des Schwarzpulvers – Le charbon fait la poudre, *Chemie in unserer Zeit*, **1988**, *22*, 9–16.
- F. Seel, Sulfur in History: The Role of Sulfur in „Black Powder", *Studies in Inorganic Chemistry*, **1984**, *5*, 55–66.
- V. Weiser, S. Kelzenberg, E. Roth, N. Eisenreich B. Berger, B. Haas, Influence of Particle Size on the Pressure Combustion of Igniters – Experiments and Theoretical Considerations, *IPS-Seminar*, Fort Collins, **2006**, 21–30.
- C. O. Leiber, *Assessment of Safety and Risk with a Microscopic Model of Detonation*, Elsevier, **2003**, pp. 232.
- E. Gock, J. Knop, U. Waldek, Hochenergetische Schwarzpulver, *Sprenginfo* **2013**, *35*, (3) 20–30.
- MIL-P-00223C accessed on October **2019** at http://everyspec.com/MIL-SPECS/MIL-SPECS-MIL-P/MIL-P-00223C_27980/

Blast

The term "blast" describes the shock wave travelling through the air and its effects on target structures. The term is also used expanded to "*air blast*". Explosive charges that yield a strong air blast are designated "blast explosives" and the shock waves in air are termed "blast waves". Figure B.13 shows the idealized pressure–time profile of a blast wave typically observed in a distance r to the detonating charge of mass M.

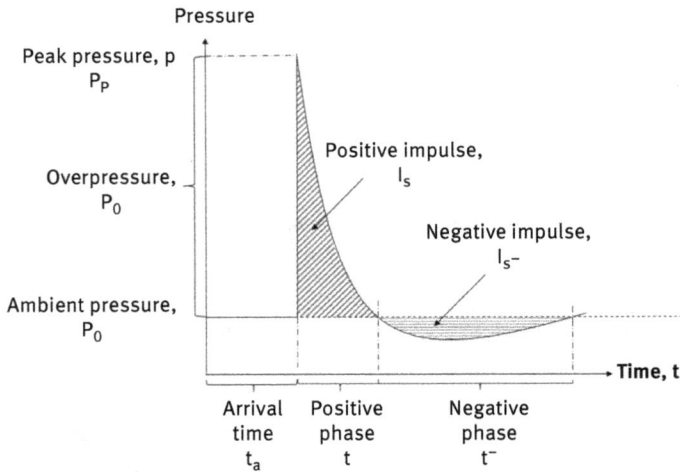

Figure B.13: Idealized pressure–time curve of a blast wave.

In accordance with *Hopkinson* and *Cranz* at equal values for z the ratio from distance (r), and cubic root of the mass of the explosive, $\sqrt[3]{M}$

$$z = \frac{r}{\sqrt[3]{M}}$$

the same peak pressure (P_p) is obtained. z is called scaled distance. Figure B.14 depicts the peak pressure for varying z for both TNT and octol (90/10). The impulse of the positive phase of a blast wave can be calculated with

$$I_s = \int_{t_a}^{t_a + t} P_u(t)dt$$

The scaled impulse $\frac{I_s}{\sqrt[3]{M}}$ is a function of z too and is shown in Figure B.15.

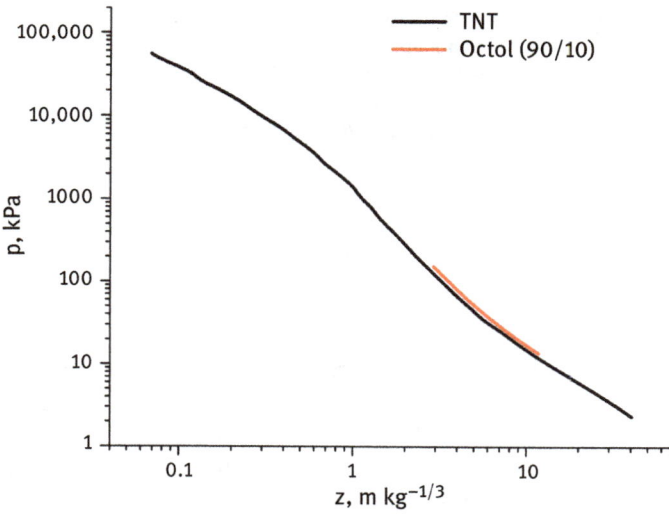

Figure B.14: Static peak pressure against scaled distance z for TNT and octol.

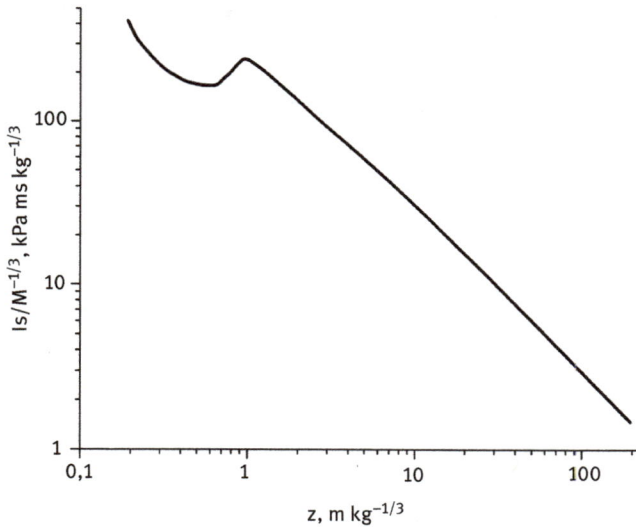

Figure B.15: Scaled impulse against scaled distance for TNT.

Based on the concept of scaled distance, large-scale effects can be studied with model setups (benefit of using small amounts of explosives).

- C. N. Kingery, Air Blast Parameters versus Distance for Hemispherical TNT Surface Bursts, *Report 1344*, US-ARMY Materiel Command, September **1966**, 77 pp.
- M. Held, Blast Waves in Free Air, *Propellants Explos. Pyrotech.* **1983**, *8*, 1–7.
- V. Karlos, G. Solomos, Calculation of Blast Loads for Application to Structural Components, *Administrative Arrangement No JRC 32253-2011 with DG-HOME Activity A5 – Blast Simulation Technology Development*, Ispra, **2013**, 58 pp.

Blasting cap
Sprengkapsel

A blasting cap is an initiating device that is triggered by an electrical impulse or thermal input (safety fuze and shock tube).

Blasting caps at minimum contain three different charges:
(1) A thermally sensitive primary explosive such as lead styphnate to facilitate ignition by a hot bridgewire or take over fire from a *bickford fuze* or *shock tube*
(2) A charge of powerful primary explosive (e.g. lead azide or silver azide)
(3) An output charge of a secondary HE (e.g. PETN)

Blasting caps may also include a precise pyrotechnic short-term delay (second–millisecond range) that is indicated by a tag attached to it.

- P.W. Cooper, S. R. Kurowski, *Introduction to the Technology of Explosives*, Wiley-VCH, Weinheim, **1996**, pp. 102–111.

BNCP, tetrammin-*cis*-bis(5-nitro-2H2*H*-tetrazolato-N²) cobalt perchlorate

Tetraammin-cis-bis(5-nitro-2H-tetrazolato-N²) cobalt(III)perchlorat

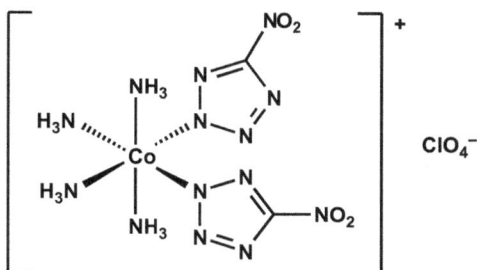

Aspect		Yellow-gold crystals
Formula		$C_2ClCoH_{12}N_{14}O_8$
CAS		[117412-28-9]
m_r	g mol^{-1}	454.592
ρ	g cm^{-3}	2.05
$\Delta_f H$	kJ mol^{-1}	± 0
$\Delta_{ex} H$	kJ mol^{-1}	−1509.2
	kJ g^{-1}	−3.320
	kJ cm^{-3}	−6.806
Λ	wt.-%	−8.8
N	wt.-%	43.14
DSC onset	°C	269
Impact	J	4.25–6.4
Friction	N	6
V_D	m s^{-1}	2700 at 1.54 g cm^{-3}.

BNCP is a coordination compound with a sensitiveness ranging between secondary and primary explosives. BNCP can be initiated to detonation by laser illumination with delay times as short as 50 µs. Hence, it is of considerable interest for instant response protection systems such as emergency escape systems and reactive defense measures.

- H. Scholles, R. Schirra, H. Zöllner, A Fast Low-Energy Optical detonator, *IPS*, **2016**, Grand Junction, USA, 422–428.
- R. Matyáš, J. Pachman, *Primary Explosives*, Springer, **2013**, 241–244.
- J. W. Fronabarger, W. B. Sanborn, T. Massis, Recent Activities in the Development of the Explosive BNCP, *22nd IPS Seminar*, Fort Collins, **1996**, 645–652.

Bonding agent
Haftvermittler

Bonding agents are substances that improve the bonding between a polymeric matrix (e.g. HTPB and GAP) and a crystalline (AP, HMX, etc.) or metallic filler (Al, Si, etc.) on a microscopic scale. Bonding agents improve the tensile strength of a HE or propellant and thereby reduce failure upon mechanical load (high acceleration or deceleration). In addition, bonding agents reduce the vulnerability of energetic materials against unwanted accidental initiation or ignition by counteracting delamination and increase of free surface area.

Typical bonding agents belong to
- alcohols:
  - triethanolamine
- aziridines:
  - tris[l-(2- methyl)-aziridinyl] phosphine oxide)
  - 1,1'-isophthaloylbis(2-methylaziridine)
- nitriles:
  - tetraethylene-pentaaminacrylonitrile
- oxazolines
  - 1,3-bis(4-methyl-4,5-dihydrooxazol-2-yl)benzene

Bonding agents that act on a macroscopic scale are tacky materials (asphaltum, B-14 binder, etc.) painted on the inside of shells and warhead casings.

- P. F. Aiello, R. W. Hunter, A. P. Manzara, Room-temperature storable bonding agent, *27th ICT-Jata*, **1996**, P-153.
- K. Hori, A. Iwama, F. Fukuda, On the Adhesion Between Hydroxyl-Terminated Polybutadiene Fuel-Binder and Ammonium Perchlorate. Performance of bonding agents, *Propellants Explos. Pyrotech.* **1985**, *10*, 176–180.

Booster
Zündverstärker

To successfully transfer the detonation from a detonator onto the main charge many explosives require a booster charge. Therefore, the booster explosive requires a low critical initiation energy to allow direct initiation by a detonator. The booster itself must be capable to deliver a stimulus $E_{cr} = \frac{P^2 \tau}{\rho_0 w}$ that is sufficient to initiate the main charge, where P represents the shock pressure (GPa), τ the duration (μs), ρ the density of the unshocked explosive (g cm^{-3}), and w the shock velocity of the explosive (m s^{-1}).

For many decades, tetryl was the booster explosive of choice, but its toxicity led to its replacement by both RDX-(CH-6) and HMX-based (PBXN-5) formulations. Table B.14 depicts the critical initiation energy, E_{cr}, for selected HEs.

Table B.14: Critical initiation energy for selected explosives after *Walker and Wasley*.

Explosive	Density (g cm^{-3})	Critical energy (J cm^{-2})
Lead azide	4.93	0.15
PETN	1.0	12
Tetryl	1.66	42
Comp A-5	1.71	52
PBX-9404	1.84	63
H$_2$+O$_2$	8.10^{-4}	~ 100
LX-04	1.865	109
Comp B3	1.727	~125
TNT (pressed)	1.645	142
Comp B	1.715	~150
TNT (cast)	1.6	~420
X-0219 (TATB/Kel-F: 90/10)	1.93	~950
Nitromethane	1.13	~1700

– F. E. Walker, R. J. Wasley, A General Model for the Shock Initiation of Explosives, *Propellants Explos.* **1976**, *1*, 73–80.

Boranes

Borane

Boranes also termed borohydrides are compounds of the general composition B_nH_{n+4}, B_nH_{n+6}, B_nH_{n+8} or B_nH_{n+10}. Many boranes are endothermic compounds. They are also highly reactive (pyrophoric, prone to hydrolysis) and very often highly toxic. The oxidation with oxygen or halogens is a highly exothermic process:

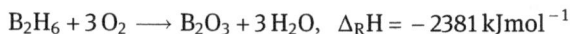

$$B_2H_6 + 3O_2 \longrightarrow B_2O_3 + 3H_2O, \quad \Delta_RH = -2381 \text{ kJmol}^{-1}$$

$$B_2H_6 + 6F_2 \longrightarrow 2BF_3 + 6HF, \quad \Delta_RH = -3948 \text{ kJmol}^{-1}$$

The alkyl-substituted boranes, the carbaboranes (e.g. ethyldecaborane, $B_{10}H_{13}$-C_2H_5 [26747-87-5], $\Delta_fH = -110$ kJ mol^{-1}), very often are not pyrophoric and hence allow easier handling. The carbaboranes have been studied intensively in the USA between 1950 and 1960 as potential high energy fuels for rockets and scramjets (U.S. Air Force Project: HEF, U.S. Navy Project: ZIP). However, it was overlooked that the initial combustion product, B_2O_3, has a higher molecular mass than CO_2, that B_2O_3 precipitates at pretty high temperatures already (bp 2065 °C) and therefore lowers the performance. In addition, B_2O_3, B_4C, BN precipitate on turbine parts and hence cause damages. The immense toxicity and hydrolytic sensitiveness of those compounds did not help either and with the intercontinental ballistic missiles (ICBMs) taking the lead in strategic nuclear deterrence both programs were brought to an end by around 1960. Since then boranes and their derivatives have been playing a very minor role in energetic materials only. Boranes have been suggested as pyrophoric starters for fuel air explosives. The alkali salts of some high-molecular boranes of the type, $M_2[B_{10}H_{10}]$, the hydridoborates, are used in very fast burning igniter mixtures ($\rightarrow$ Hivelite).

- R. T. Holzmann (Ed.), *The production of Boranes and related Research*, Academic Press, New York, **1968**.
- J. D. Clark, *Ignition! An Informal History of Liquid Rocket Propellants*, Rutgers University Press, **1972**, 195 pp.
- A. Dequasie, *The Green Flame*, Wiley, New York, **1991**, 220 pp.

Borazines

Borazine

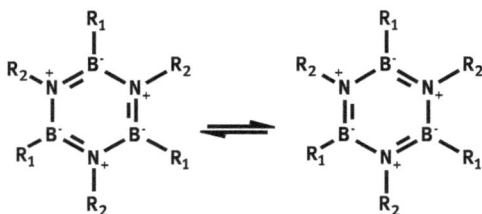

Figure B.16: Valence bond formulas for borazines.

Borazines are boron nitrogen compounds that are structurally similar with benzene and its derivatives. In contrary to benzene which has non-polar C–C bond, the B–N bond is strongly polar (see the earlier resonance structures) and hence also much more reactive than the former. Since some years theoretical and experimental work addresses the properties and synthesis of those borazines substituted with explosophores. For the yet hypothetical *N′,N″,N‴*-triamino-*B′,B″,B‴*-trinitroborazine (in Figure B.16: $R_1 = NO_2$, $R_2 = NH_2$) at a prognosticated density of the structurally related 1,3,5-triamino-2,4,6-trinitrobenzol ($\rho = 1.937$ g cm^{-3}) ($\Delta_f H = -357$ kJ mol^{-1}) a detonation velocity of $V_D = 13.3$ km s^{-1} and a detonation pressure of $P_{CJ} = 50.9$ GPa has been predicted (Table B.15).

Table B.15: Calculated performance of HMX, CL-20, and TATB compared with BN-analogue TATB derivatives.

	HMX	CL-20	TATB	NNN-BBB-TATB
Formula	$C_4H_8N_8O_8$	$C_6H_6N_{12}O_{12}$	$C_6H_6N_6O_6$	$B_3H_6N_9O_6$
CAS No.	[2691-41-0]	[135285-90-4]	[3058-38-6]	[1384979-28-5]
Density (g cm^{-3})	1.905	2.044	1.937	1.937
$\Delta_f H$ (kJ mol^{-1})	84	431	−154	−357
V_D (m s^{-1})	9.310	10.065	8.108	13.290
P_{CJ} (GPa)	39.49	48.23	31.07	50.92
$V/V_{0 =2.20}$ (kJ cm^{-3})	−7.61	−9.11	−5.48	−8.21

- E.-C. Koch, T. M. Klapötke, Boron-Based High Explosives, *Propellants Explos. Pyrotech.* **2012**, *27*, 335–344.
- M. A. Rodriguez, T. T. Borek, 2,4-Bis(dimethylamino)-1,3,5-trimethyl-6-(nitrooxy)borazine, *Acta Cryst.* **2013**, *E69*, o634.

Boric acid

Borsäure

Formula		B(OH)$_3$
GHS		08
H-phrases		H360FD
P-phrases		P201-P308+P313
REACH		SVHC, LRS
EINECS		233-139-2
CAS		[10043-35-3]
m_r	g mol^{-1}	61.832
ρ	g cm^{-3}	1.435
$\Delta_f H$	kJ mol^{-1}	−1094
Λ	wt.-%	0
Mp	°C	169 (−H$_2$O)

B(OH)$_3$ is an ingredient in insensitive HEs (Boracitol) for use in plane wave generators and explosive lenses for the implosion initiation of fission bombs. In this regard, the high mass amount of boron assists also in absorption of thermal neutrons. Other boric acid-based insensitive HEs (LBR-6) are used in explosive reactive armor (ERA) modules (Table B.16). In LBR-6, boric acid serves mainly as a flame retardant to avoid combustion and deflagration of ERA modules hit by small caliber projectiles.

Table B.16: B(OH)$_3$ containing explosive formulations.

	Boracitol	LBR-6*
Boric acid, B(OH)$_3$ (wt.-%)	60	22
TNT (wt.-%)	40	–
RDX (wt.-%)	–	48
HMX (wt.-%)	–	6
PDMS (wt.-%)	–	24
ρ (g cm^{-3})	1.550	1.560
V_D (m s^{-1})	4860	5500
$\Delta_{det}H$ (kJ g^{-1})	1.67	
Friction (N)	>360	288
Impact (J)	>50	20

Low burn rate explosive.

− E. Sokol, S. Friling, I. Shaked, N. Aviv, Y. Cohen-Arazi, LBR6-A Novel Class 1.5D Composition for Insensitive Explosive Reactive Armor (I-ERA), *IMEMTS*, **2007**, Miami, 7 pp.

Boron carbide
Borcarbid

Aspect		Black crystal powder
Formula		B$_4$C
GHS		07
H-phrases		332-315-319-335
P-phrases		261-302-352-305+351+338-321-405-501a
REACH		LRS
EINECS		235-111-5
CAS		[12069-32-8]
m_r	g mol^{-1}	55.251
ρ	g cm^{-3}	2.52
$\Delta_f H$	kJ mol^{-1}	−62.68
$\Delta_c H$	kJ mol^{-1}	−2877.3
	kJ g^{-1}	−52.077
	kJ cm^{-3}	−132.235
Λ	wt.-%	−231.66
Mp	°C	2350
$\Delta_m H$	kJ mol^{-1}	104.6

Due to its high mechanical strength and its low density, B$_4$C is used as a structural material in advanced armor. Its nature as a refractory material led to overlook its easy oxidation in oxygen-containing atmospheres. B$_4$C serves as a fuel in air-breathing propulsion systems and has been investigated for the first time as a fuel and color agent in slow burning green flare formulations by *Jennings-White* in 1997.

- J. Sabatini, J. C. Poret, R. N. Broad, Boron Carbide as a Barium-Free Green Light Emitter and Burn-Rate Modifier in Pyrotechnics, *Anwt. Chem.* **2011**, *123*, 4720–4722.
- M. Steinbrück, A. Meier, U. Stegmaier, L. Steinbock, *Experiments on the Oxidation of Boron Carbide at High Temperatures*, FZKA 6979, Forschungszentrum Karlsruhe, May **2004**, 117 pp.
- C. Jennings-White, S. Wilson, Lithium, Boron, Calcium, *Pyrotechnica* **1997**, *XVII*, 24–29.
- B. Natan, D. W. Netzer, Boron carbide combustion in solid-fuel ramjets using bypass air. Part I Experimental investigation, *Propellants Explos. Pyrotech.* **1996**, *21*, 289–294.

Boron nitride

Bornitrid

Aspect		Beige powder
Formula		BN
GHS		07
H-phrases		319-335
P-phrases		261-280-305+351+338-304+340-405-501a
REACH		LRS
EINECS		233-136-6
CAS		[10043-11-5]
m_r	g mol^{-1}	24.82
ρ	g cm^{-3}	3.48
$\Delta_f H$	kJ mol^{-1}	−250.9
$\Delta_c H$	kJ mol^{-1}	−385.05
	kJ g^{-1}	−15.51
	kJ cm^{-3}	−54.106
c_p	J K^{-1} mol^{-1}	19.7
Λ	wt.-%	−96.71
Mp	°C	2230 (Sbl.)
$\Delta_{subl} H$	kJ mol^{-1}	902

Boron nitride (BN) exists in three different polymorphs (α,β,γ). The graphite like α-phase also termed hexagonal BN due to its lubricating properties is frequently used as a phlegmatizer in pressable formulations. It has also been suggested as a wear reducing additive in gun propellants and as an additional fuel in gas-generating compositions but burns very poorly in binary systems with common oxidizers.

- T. Manning et al., Innovative boron nitride-doped propellants, *Def. Tech.* **2016**, 12, 69–80.
- R. Hagel, U. Bley, *Thermische Frühzündmittel*, EP1697277B1, **2004**, Germany.
- R. J. Blau, D. A. Flanigan, *Non-Azide Gas Generant Compositions Containing Dicyanamide Salts*, WO 95/18780, **1995**, USA.

Boron, amorphous

Bor amorph

Aspect		Dark brown fine powder
Formula		B
GHS		06,02
H-phrases		301-228-332-335
P-phrases		210-241-301+310-321-405-501a
UN		3178
REACH		LRS
EINECS		231-151-2
CAS		[7440-42-8]
m_r	g mol^{-1}	10.811
ρ	g cm^{-3}	2.34
Mp	°C	2250
Bp	°C	~3660
$\Delta_{subl}H$	kJ mol-1	504.5
κ	W cm^{-1} K^{-1}	1.16
Δ_cH	kJ mol^{-1}	−635.95
	kJ g^{-1}	−58.82
	kJ cm^{-3}	−137.65
Λ	wt.-%	−221.99

While there are five different crystalline modifications of boron known, only the vitreous-amorphous boron is used in energetic materials. Micrometric crystalline boron of unknown modification does not burn/react in pyrotechnic compositions *(Sabatini)*. Amorphous boron typically has particle sizes in the low figure micrometric range. The oxidation of amorphous boron starts at T = 465 °C and goes up to T = 1200 °C and a mass increase of about 87 wt.-% is observed. Amorphous boron at ambient temperature slowly reacts with moisture and oxygen to yield mixed oxide and acid layers. As a highly energetic fuel, boron finds use in air breathing propulsion fuel gas generators, in ignition mixtures together with all kinds of oxidizers, for example, KNO_3, Fe_2O_3, $BaCrO_4$, $KClO_4$, Bi_2O_3, and other transition metal oxides *(B/KNO$_3$)* (AZM O 953X, SR43). As boron combustion yields the green emitting species $BO_{2(g)}$ *(Roth, 2001)*, it has been repeatedly studied since World War II as a potential fuel in green signaling compositions. However, in the presence of carbon and oxygen, boron forms refractory continuum emitting species such as *B_4C* and *BN* which cause a washout of the green hue and hence very poor saturation results. Several volatile boron compounds are very strong emitters in the infrared range and are depicted in Table B.17. It is hence that boron has occasionally been proposed as fuel in spectrally matched flare compositions.

Table B.17: Infrared emission properties of volatile boron compounds in comparison to CO_2.

	Band head (μm)	Band strength at 298 K (cm^{-2} atm^{-1})
$H^{11}BO_2$	2.72	650
	4.94	1375
$O^{11}BF$	5.00	1760
^{11}BF	7.44	630 ± 200
$^{11}BF_3$	6.89	3290
$^{11}B_2H_6$	3.90	1078
CO_2	4.30	2700

- J. J. Sabatini, J. C. Poret, R. N. Broad, Use of Crystalline Boron as Burn Rate Retardant Toward the Development of Green-Colored Handheld Signal Formulations, *J. Energ. Mater.* **2011**, *29*, 360–368.
- E. Roth, Y. Piltzko, V. Weiser, W. Eckl, H. Poth, M. Klemenz, Emissionsspektren brennender Metalle, *ICT-Jata*, Karlsruhe, **2001**, P-163.
- J. R. Dawe, M. D. Cliff, Metal Dinitramides New Novel Oxidants for the Preparation of Boron Based Flare Compositions, *IPS-Seminar*, Monterey, **1998**, 789–810.
- D. W. Herbage, S. L. Salvesen, *Spectrally balanced infrared flare pyrotechnic composition*, US Patent 5472533, **1995**, USA.
- R. Strecker, A. Harrer, *Composit-Festtreibstoff mit stabilem Abbrand*, DE 28 20 969 C1, **1991**, Germany.
- N. Eisenreich, W. Liehmann, Emission spectroscopy of Boron Ignition and Combustion in the Range of 0.2 to 5.5 μm, *Propellants Explos. Pyrotech.* **1987**, *12*, 88–91.

Boronite

Boronites were explosive mixtures investigated in the 1940s. Boronites were based on TNT ammonium nitrate and amorphous boron (Table B.18). Boronites, however, did not meet the expectations with regard to working capacity and were inferior to comparable formulations containing aluminum.

Table B.18: Boronite.

Boronite	A	B	C
TNT (wt.-%)	10	20	36
NH_4NO_3 (wt.-%)	83	75	62
Boron (wt.-%)	7	5	2

Brisance
Brisanz

The brisance – derived from the French verb *briser* = *to shatter* – is used synonymously with the term *"working capacity"* of HEs.

Bruceton method
Bruceton-Methode

The Bruceton method is named after the Explosives Research Laboratory (ERL), Bruceton, PA, USA. It is a statistical method to determine the 50% impact energy e (J), by 25 experiments. However, any other sensitiveness test can be conducted in accordance with this method.

The test starts at a randomly selected stimulus level e_1. Depending on the outcome of the test, X|e

$$X|e = \begin{cases} 1 \text{ explosion} \\ 0 \text{ no explosion} \end{cases}$$

the stimulus e_2 is lowered or raised by a unit value. The 50% impact energy, M, is derived as follows:
(a) Determination of the number of positive (+) and negative (–) outcomes. The lower number is used in the further determination.
(b) The level i with the lowest height of fall for the observed reaction (+ or –) is assigned the value i = 0. For all higher levels, the values I = 1, 2, . . . are assigned.
(c) $N = \Sigma n_i$. The total number of reactions (+ or –) at the ith level:

$$A = \Sigma i \cdot n_i$$

$$B = \Sigma i^2 \cdot n_i$$

(d) $M = C + D \left[\frac{A}{N} \pm \frac{1}{2} \right]$ with C represents $\log_{10}$ the step height for i = 0 and D the $\log_{10}$ interval. The plus sign is used with negative reactions; the minus is used with positive reactions.
(e) The standard deviation (S) of M is determined by the following equation:

$$S = 1.620\,D \left[\left((NB - A^2)/N^2 \right) + 0.029 \right]$$

The validity condition for the Bruceton method, $0.5 \leq S/D \leq 2.0$, must be satisfied and documented.

For the use of the BAM impact machine in accordance with STANAG 4489 30 experiments are required. *Wild et al.* have criticized the Bruceton method and declared it not suitable to determine the sensitiveness of explosives.

−   R. Wild, E. Von Collani, Modelling of Explosives Sensitivity Part 1: The Bruceton Method, *Economic Quality Control* **2002**, *17*, 113–122.
−   Explosives, Impact Sensitivity Test, *NATO STANAG 4489*, Ed. 1 Military Agency for Standardization (MAS), Brussels, **1999**.

BTNEU, *N,N*-bis-(2,2,2,-trinitroethyl)urea
Di-(2,2,2-trinitroethyl)-harnstoff

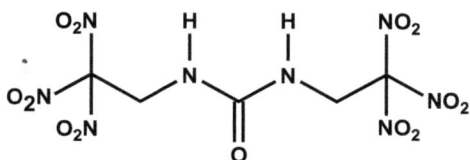

Formula		$C_5H_6N_8O_{13}$
CAS		[918-99-0]
m_r	g mol^{-1}	386.15
ρ	g cm^{-3}	1.906
Δ_fH	kJ mol^{-1}	−342.29
$\Delta_{ex}H$	kJ mol^{-1}	−2377.9
	kJ g^{-1}	−6.158
	kJ cm^{-3}	−11.460
Δ_cH	kJ mol^{-1}	−2482.6
	kJ g^{-1}	−6.430
	kJ cm^{-3}	−12.255
Λ	wt.-%	0
N	wt.-%	29.02
Mp	°C	185 (dec)

−   O. H. Johnson, *Plasticized High Explosive and Solid Propellant Composition*, US3389026, **1968**, USA.

Bullet impact test
Beschusstest

The bullet impact test is an insensitive munitions test to probe the vulnerability of a munition or a generic test vehicle filled with an explosive towards impact by small arms projectiles. In a common procedure defined in AOP4241, a rapid burst of three 12.7 × 99 mm armor-piercing bullets (e.g. M2 AP) is fired at $v_0 = 850 \pm 20$ m s^{-1} from a distance between 20 and 30 m within a 10 cm circle of the test object. While certain ammunitions may tolerate single impacts and just fracture with no indication for an energetic reaction, secondary and tertiary hits in rapid succession into the fractured and hence pre-stressed area of the energetic material can result in reactions ranging from ignition over to violent shock-to-detonation-transition (SDT).

- *NATO Standard AOP-4241, Bullet Impact Munition Test Procedures*, Ed. A. Ver. 1 November **2018**, NATO Standardization Office, Brussels, 24 pp.

Butacen®-7

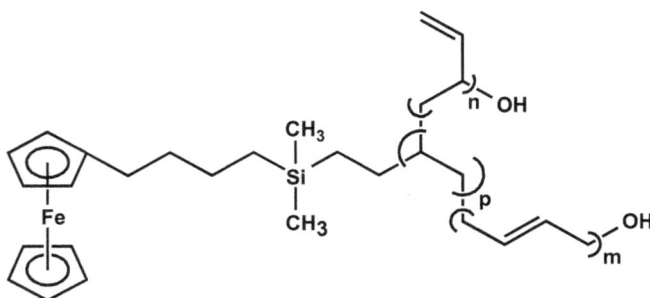

Aspect		Light yellow liquid
Formula		$(C_{10}Fe_{0.206}H_{14.985}O_{0.093}Si_{0.18})_x$
CAS		[125856-62-4]
m_r	g mol^{-1}	153.262
ρ	g cm^{-3}	1.015
$\Delta_f H$	kJ mol^{-1}	−64
$\Delta_c H$	kJ mol^{-1}	−6261.6
	kJ g^{-1}	−40.856
	kJ cm^{-3}	−41.469
Λ	wt.-%	−291.94
Fe content	wt.-%	7.05
T_g	°C	−65

Butacen®-7 (the number relating to the iron content) is a ferrocene-based modifier not prone to migrate in the cured propellant. It is obtained by coupling of ferrocene with the HTPB prepolymer. Due to a cumbersome multistep synthesis, Butacen is however very expensive which reduces widespread use of it.

– B. Finck, J. C. Mondet, Combustion de Propergols Composites Solides Utilisant un Derive Ferrocenique Greffe dans le Liant, *ICT-Jata*, **1988**, Karlsruhe, P-72.

Butantriol trinitrate, BTTN

1,2,4-Butantrioltrinitrat

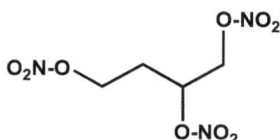

Aspect		Pale yellowish liquid
Formula		$C_4H_7N_3O_9$
REACH		LPRS
EINECS		229-697-1
CAS		[6659-60-5]
m_r	g mol^{-1}	241.114
ρ	g cm^{-3}	1.52
$\Delta_f H$	kJ mol^{-1}	−406
$\Delta_{ex}H$	kJ mol^{-1}	−1345.2
	kJ g^{-1}	−5.579
	kJ cm^{-3}	−8.480
$\Delta_c H$	kJ mol^{-1}	−2168.5
	kJ g^{-1}	−8.994
	kJ cm^{-3}	−13.670
Λ	wt.-%	−16.59
N	wt.-%	17.43
Mp	°C	−5.8
Bp	°C	n.a.
Friction	N	<353
Impact	J	1

BTTN is used as an energetic plasticizer in rocket solid propellants. During World War II, BTTN was used in Germany as plasticizer in so-called "tropenfesten" (Engl. *tropicalized*) gun propellants for use in the North African theater.

BZ – 3-chinuclidinyl benzilate

3-Chinuclidinylbenzilat

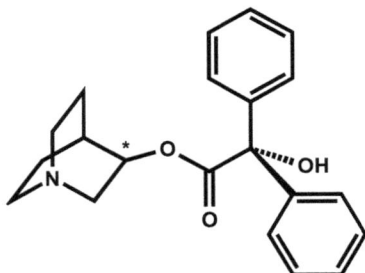

1-Azabicyclo[2.2.2]oct-8*-yl 2-hydroxy-2,2-diphenyl-acetate

Formula		$C_{21}H_{23}NO_3$
CAS		[6581-06-2]
m_r	g mol^{-1}	337.412
ρ	g cm^{-3}	1.33
$\Delta_f H$	kJ mol^{-1}	−400 (estimated)
$\Delta_c H$	kJ mol^{-1}	−11151.0
	kJ g^{-1}	−33.048
	kJ cm^{-3}	−43.954
Λ	wt.-%	−239.46
N	wt.-%	4.15
Mp	°C	189–190 (racemic mixtures 166–168)
Bp	°C	322
LCt$_{50}$	ppm min^{-1}	200.000
ICt$_{50}$	ppm min^{-1}	110

BZ is a colorless and odorless psychotoxic chemical agent. It is sufficiently stable to be disseminated by pyrotechnic smoldering compositions.

–   S. Franke, *Lehrbuch der Militärchemie Band 1*, Militärverlag der Deutschen Demokratischen Republik, 2. Überarbeitete Auflage, **1977**, pp. 214.

C

C4 → Composition C4

C6

C6 is a former German World War II *"Substitute explosive"* with the following composition:

- 50 wt.-% methylammonium nitrate
- 35 wt.-% sodium nitrate
- 15 wt.-% hexogen
- TMD 1.698 g cm^{-3}

Calcium

Aspect		Silver
Formula		Ca
GHS		02
H-phrases		261
P-phrases		231+232-233-280-370+378a-402+404-501a
UN		1401
EINECS		231-179-5
CAS		[7440-70-2]
m_r	g mol^{-1}	40.078
ρ	g cm^{-3}	1.55
Mp	°C	839
Bp	°C	1482
c_p	J g^{-1} K^{-1}	0.632
$\Delta_m H$	kJ mol^{-1}	8.54
$\Delta_v H$	kJ mol^{-1}	153.6
κ	W m^{-1} K^{-1}	200
c_L	m s^{-1}	4180
Λ	wt.-%	−39.92
$\Delta_c H$	kJ mol^{-1}	−635.1
	kJ g^{-1}	−15.847
	kJ cm^{-3}	−24.562

https://doi.org/10.1515/9783110660562-003

Calcium is a base metal whose reactivity with oxygen and water is about between that of lithium and magnesium. Due to its comparatively low Mohs hardness (Ca: 1.75 and Mg: 2.5) Ca cannot be milled to powders $d < 500$ μm. In flash compositions, it outperforms magnesium yielding the highest radiant intensities and the highest specific energies.

- S. Lopatin, D. Hart, *Calcium Containing Pyrotechnic Compositions for High Altitudes*, US 3261731, USA, **1966**.

Calcium chromate
Calciumchromat

Aspect		Yellow crystalline powder
Formula		Ca[CrO$_4$]
GHS		03, 09, 07
H-phrases		272-400-410-302
P-phrases		210-220-221-280-301+312-501a
UN		3087
REACH		LPRS
EINECS		237-366-8
CAS		[10060-08-9]
m_r	g mol^{-1}	156.074
ρ	g cm^{-3}	3.120
Mp	°C	1020 (dec)
c_p	J g^{-1} K^{-1}	0.859
Δ$_f$H	kJ mol^{-1}	−1379
Λ	wt.-%	+15.38

Rogers reports about delay compositions based on boron/CaCrO$_4$. An according formulation is used as igniter composition in the impulse cartridge M 796 in use with MJU 8A/B-style flare cartridges. *Ellern* gives caloric data on various mixture with both boron and zirconium.

- J. W. Rogers Jr., The Characterization and performance of thirteen boron/calcium chromate pyrotechnic blends, *IPS*, **1982**, Steamboat Springs, 556–573.

Calcium dipotassium styphnate

Calciumdikaliumstyphnat

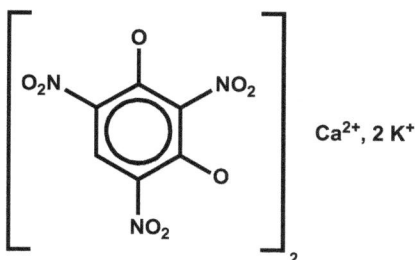

Ca^{2+}, 2 K^+

Calcium dipotassium tricinate		
Formula		$C_{12}H_2N_6O_{16}CaK_2$
CAS		[942278-90-2]
m_r	g mol^{-1}	604.45
Δ_fH	kJ mol^{-1}	−2300 (estimated)
Δ_cH	kJ mol^{-1}	−3704.5
	kJ g^{-1}	−6.129
Λ	wt.-%	−29.12
N	wt.-%	13.90
Dp	°C	345
Friction	N	9
Impact	J	3

Calcium dipotassium styphnate is used as a non-toxic primary explosive in ignition compositions.

– U. Bley, R. Hagel, A. Hoschenko, P. S: Lechner, *Salze der Styphninsäure*, EP1966120B1, **2012**, Germany.

Calcium iodate monohydrate

Calciumiodatmonohydrat

Aspect	Crème powder
Formula	$Ca(IO_3)_2 \cdot H_2O$
GHS	03, 07
H-phrases	272-315-319-335
P-phrases	221-210- 305+351+338-302+352-405-501a

		Monohydrate	Anhydrous
UN		1479	
REACH		LRS	
EINECS		232-191-3	
		Monohydrate	Anhydrous
CAS		[10031-32-0]	[7789-80-2]
m_r	g mol^{-1}	407.902	389.89
ρ	g cm^{-3}	4.25	4.59 (anhydrous)
$\Delta_f H$	kJ mol^{-1}	−1293.32	1002.49
Λ	wt.-%	+23.53	+24.62
Dp	°C	195 (-H$_2$O)	550

Calcium iodate has been used in pink inorganic colored smoke (free iodine). While for this kind of application the material is obsolete, today it has been revisited recently for the area of agent defeat applications to fight bacteria, germs, and toxins. The thermal decomposition of calcium iodate at 550 °C proceeds with disproportionation to Ca$_5$(IO$_6$)$_2$ and I$_2$.

- G. U. Graff, *Pyrotechnic composition for producing pink smoke*, US-Patent 2091977, **1970**, USA.
- S. Wang X. Liu, M. Schoenitz, E. L. Dreizin, Nanocomposite Thermites with Calcium Iodate Oxidizer, *Propellants Explos. Pyrotech.* **2017**, *42*, 284–292.

Calcium nitrate

Calciumnitrat

Nitrocalcit		
Aspect		Felt-like needles
Formula		Ca(NO$_3$)$_2$
GHS		03, 07
H-phrases		272-315-319-335
P-phrases		210-221-302+352-305+351+338-405-501a
UN		1454
REACH		LRS
EINECS		233-332-1
CAS		[10124-37-5]
m_r	g mol^{-1}	164.10
ρ	g cm^{-3}	2.504
Mp	°C	563 (dec)
$\Delta_m H$	kJ mol^{-1}	21
c_p	J g^{-1} K^{-1}	0.911

$\Delta_f H$	kJ mol^{-1}	−938.22
Λ	wt.-%	+48.75
N	wt.-%	17.07

Ca(NO$_3$)$_2$ in moist air quickly forms a series of hydrates such as Ca(NO$_3$)$_2 \cdot 4$ H$_2$O, [13477-34-4] (Mp 39.7 °C, Z 132 °C, ρ 1.890 g cm^{-3}) and Ca(NO$_3$)$_2 \cdot$ x H$_2$O, [35054-52-5]. It has been proposed as component in blasting caps. While its high hygroscopicity (low relative deliquescent humidity = 54%) hampers use in common pyrotechnic production, the high oxygen content, the low molecular weight, and the uncomplicated environmental and occupational health properties of Ca(NO$_3$)$_2$ merit further consideration. *Ettarh* and *Galwey* have investigated the thermochemical decomposition of Ca(NO$_3$)$_2$ which is found to be determined by two competing reactions:

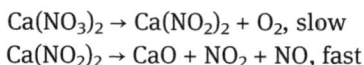

$$Ca(NO_3)_2 \rightarrow Ca(NO_2)_2 + O_2, \text{ slow}$$
$$Ca(NO_2)_2 \rightarrow CaO + NO_2 + NO, \text{ fast}$$

– C. Ettarh, A. K. Galwey, A kinetic and mechanistic study of the thermal decomposition of calcium nitrate, *Thermochimica Acta* **1996**, *288*, 203–219.

Calcium oxalate hydrate
Calciumoxalathydrat

Aspect		White to crème fine powder
Formula		CaC$_2$O$_4 \cdot$ H$_2$O
GHS		07
H-phrases		302-312
P-phrases		280-301+312-302+352-312-322-501a
UN		3288
EINECS		209-260-1
CAS		[14488-96-1]
m_r	g mol^{-1}	146.12
ρ	g cm^{-3}	2.2
$\Delta_f H$	kJ mol^{-1}	−1675
c_p	J K^{-1} mol^{-1}	153
Λ	wt.-%	−10.95
Dp	°C	125 (-H$_2$O)
Dp	°C	420 (-CO)
Dp	°C	660 (-CO$_2$)

Calcium oxalate is used as a burn rate modifier and a cooling agent in tracer formulations.

Calcium perchlorate

Calciumperchlorat

Formula		$Ca(ClO_4)_2$
GHS		03, 07
H-phrases		272-315-319-335
P-phrases		221-210-305+351+338-302+352-405-501a
UN		1455
REACH		LPRS
EINECS		236-768-0
CAS		[13477-36-6]
m_r	g mol^{-1}	238.98
ρ	g·cm^{-3}	2.651
Mp	°C	270 (dec)
$\Delta_f H$	kJ mol^{-1}	−736.76
Λ	wt.-%	+53.56
c_p	J K^{-1} mol^{-1}	233

Successive water uptake yields the tetrahydrate, $Ca(ClO_4)_2 \cdot 4\ H_2O$ [15627-86-8], $\Delta_f H =$ −1949 kJ mol^{-1}, ρ = 2.10 g cm^{-3} and the hexahydrate, $Ca(ClO_4)_2 \cdot 6\ H_2O$ [20624-28-6], $\Delta_f H \sim$ −2600 kJ mol^{-1}, ρ = 1.97 g cm^{-3}. Flash compositions based on calcium perchlorate and aluminum outperform any other system with regard to specific energy (cd s g^{-1}). *Wasmann* used calcium perchlorate as an oxidizer in plastic-bonded strobe flares.

- E. Hennings, H. Schmidt, W. Voigt, Crystal structures of $Ca(ClO_4)_2$ 4H$_2$O and $Ca(ClO_4)_2$ 6H$_2$O, *Acta Cryst.* **2014**, *E70*, 489–493.
- F. W. Wasmann, Pulsierend abbrennende pyrotechnische Systeme, *ICT Jahrestagung*, **1975**, Karlsruhe, 239–250.
- S. Lopatin, Sea-Level and High Altitude Performance of Experimental Photoflash Compositions, *Technical Report FRL-TR-29*, Picatinny Arsenal, October **1961**, USA.

Calcium peroxide
Calciumperoxid

Formula		CaO_2
GHS		03, 05
H-phrases		314-272
P-phrases		301+330+331-280-305+351+338-310-210
UN		1457
REACH		LPRS
EINECS		215-139-4
CAS		[1305-79-9]
m_r	g mol^{-1}	72.08
ρ	g·cm^{-3}	3.23
Mp	°C	275 (dec)
$\Delta_f H$	kJ mol^{-1}	−652.70
Λ	wt.-%	+22.2
c_p	J K^{-1} mol^{-1}	82.8

CaO_2 is a slightly hygroscopic substance slowly forming the octahydrate, CAS-No. [60762-59-6], in moist air. CaO_2 is occasionally used in tracer formulations.

Calcium phosphate
Calciumorthophosphat

Aspect		White to crème fine powder
Formula		$Ca_3(PO_4)_2$
GHS		07
H-phrases		315-319-335
P-phrases		261-305+351+338-302+352-321-405-501a
REACH		LRS
EINECS		235-330-6
CAS		[7758-87-4]
m_r	g mol^{-1}	310.18
ρ	g cm^{-3}	3.14
$\Delta_f H$	kJ mol^{-1}	−4120
c_p	J K^{-1} mol^{-1}	230
Λ	wt.-%	0
Mp	°C	1800

Calcium phosphate is sometimes used in small amounts as an anti-caking agent with various nitrates and perchlorates of high density. Calcium phosphate can serve as an oxidizer with strong reducing agents such as Mg or Al. Large matches (length ~ 10 cm and 1 cm diameter; Fig. C.1) comprising thick dippings of $Al/Ca_3(PO_4)_2$/binder serve as a deterrent against root vole (Herz cartridge)s (Ger.: *Wühlmauspatrone*). Those matches are ignited and placed in the mouse holes. Upon combustion Ca_3P_2 forms, which immediately with the moist soil yields highly toxic PH_3.

Figure C.1: Commercial Herz cartridge.

Calcium resinate

Calciumresinat

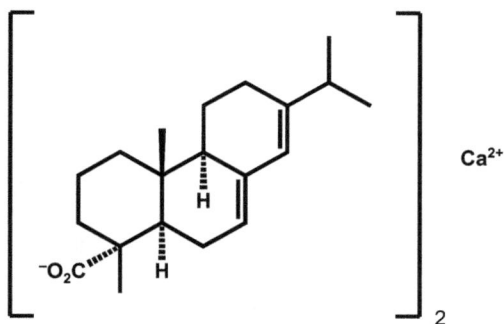

Abietic acid calcium salt		
Aspect		Orange crystals with conchoidal fracture
Formula		$C_{40}CaH_{58}O_4$
REACH		LPRS
EINECS		236-677-6
CAS		[13463-98-4]
m_r	g mol^{-1}	642.986
ρ	g cm^{-3}	0.962
$\Delta_f H$	kJ mol^{-1}	−1800 (estimated)
$\Delta_c H$	kJ mol^{-1}	−22864.8
		−35.561
	kJ cm^{-3}	−34.210
Λ	wt.-%	−263.76
Dp	°C	220
E_a	kJ mol^{-1}	50

Calcium resinate is used as a binder in pyrotechnic tracers and has been used in the past as a binder in booster explosives.

- A. Korczyński, H. Proga, Parametry kinetyczne termolizy abietynianońow metali grupy II A, *Prezm. Chem.* **1996**, *75*, 141–142.

Calcium silicide
Calciumsilicid

Aspect		Grey to black powder with metallic luster
Formula		$CaSi_2$
GHS		02
H-phrases		261
P-phrases		231+232-280-233-370+378a-402+404-501a
UN		1405
REACH		LPRS
EINECS		234-588-7
CAS		[12013-56-8]
m_r	g mol^{-1}	96.249
ρ	g cm^{-3}	2.50
Mp	°C	990
$\Delta_f H$	kJ mol^{-1}	−151.04
$\Delta_c H$	kJ mol^{-1}	−2305.4
	kJ g^{-1}	−23.952
	kJ cm^{-3}	−59.880
Λ	wt.-%	−83.11
Dp	°C	720

$CaSi_2$ reacts with hot water and more rapidly with dilute acids to yield low molecular and potentially self-inflammable silanes. The low volatility and high exothermicity of the main oxidation products of $CaSi_2$:x

- Calcium silicate, $CaSiO_3$ (Mp = 1544 °C, $\Delta_f H = -1634$ kJmol^{-1})
- Silicon dioxide, SiO_2 (Mp = 1705 °C, $\Delta_f H = -908$ kJmol^{-1})

make it the first choice in low gas forming heat compositions. In addition, $CaSi_2$ due to the high heat of formation and the high hygroscopicity of its chlorides ($CaCl_2$, $SiCl_4$) is used in obscurant formulations together with chlorocarbon compounds. $CaSi_2$ has also been used as additional fuel in blast-enhanced explosive formulations.

- T. Urbanski, *Chemie und Technologie der Explosivstoffe, Band III*, VEB Deutscher Verlag für Grundstoffindustrie, Leipzig, **1964**, pp. 231.

Calcium stearate

Calciumstearat

Steatit		
Formula		$C_{36}CaH_{70}O_4$
REACH		LPRS
EINECS		216-472-8
CAS		[1592-23-0]
m_r	g mol^{-1}	607.029
ρ	g cm^{-3}	1.03
$\Delta_f H$	kJ mol^{-1}	−2772
$\Delta_c H$	kJ mol^{-1}	−22,033.7
	kJ g^{-1}	−36.298
	kJ cm^{-3}	−37.387
Λ	wt.-%	−274.11
Mp	°C	160

Calcium stearate is used as a phlegmatizer and as a molding agent in booster explosives and pyrotechnic compositions.

Calcium sulfate hemihydrate

Alabastergips

Plaster of Paris		
Formula		$CaSO_4 \cdot 0,5\ H_2O$
REACH		LPRS
EINECS		231-900-3
CAS		[26499-65-0]
m_r	g mol^{-1}	145.15
ρ	g cm^{-3}	2.70
$\Delta_f H$	kJ mol^{-1}	−1576
c_p	J K^{-1} mol^{-1}	120
Λ	wt.-%	+11.02
Dp	°C	200 (−H_2O)

Berger was the first to propose the use of calcium sulfate as a mild oxidizer for red phosphorus (P_R). $P_R/CaSO_4$ compositions, hence, were standard payloads for many decades in the so-called *marine location marker* (or *float smoke and flame signals*) in the UK and the USA. $CaSO_4$ was used as a nitrate replacement in German World War II illuminating compositions together with MgAl alloy as fuel. Today $CaSO_4$ is used occasionally as an oxidizer in nanothermites and more frequently in mine-clearing torches together with aluminum (*Dragon-Pyrotorch*). Upon contact with water, it forms the dihydrate in an exothermic reaction:

$$2\,CaSO_4 \cdot 0,5\,H_2O + 3\,H_2O \longrightarrow 2\,CaSO_4 \cdot 2\,H_2O + 322\,kJ$$

CAS		[10101-41-4]
m_r	g mol^{-1}	172.17
ρ	g cm^{-3}	2.32
$\Delta_f H$	kJ mol^{-1}	−2022
c_p	J K^{-1} mol^{-1}	186
Λ	wt.-%	+9.29
Dp	°C	128 (−1.5 H$_2$O)

- N. Mosses, Smoke Compositions Based on Phosphorus, *Technical Note ARM .617*, Royal Aircraft Establishment, Farnborough, UK, March **1958**, 33 pp.
- A. Craib, *The Development of a Pyrotechnic Torch for Mine Destruction and Capable of Local Manufacture*, Disarmco Ltd & Cranfield University, accessible at http//www.gichd.org/filead min/pdf/LIMA/PyrotorchDRAGON_public.pdf
- M. Comet, G.Vidick, F. Schnell, Y. Suma, B. Baps, D. Spitzer, Sulfates-Based Nanothermites: An Expanding Horizon for Metastable Interstitial Composites, *Angew. Chem. Int. Ed.* **2015**, *54*, 4458–4462.

Calcium tetrachlorophthalate

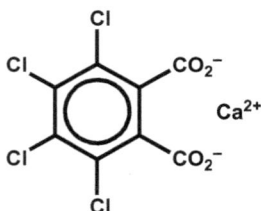

Calciumtetrachlorophthalat

Formula		$C_8Cl_4O_4Ca$
CAS		[97508-22-0]
m_r	g mol^{-1}	341.99
$\Delta_f H$	kJ mol^{-1}	−1400 (estimated)
$\Delta_c H$	kJ mol^{-1}	−2543
	kJ g^{-1}	−7.436
Λ	wt.-%	−56.14

The alkaline earth metal salts (Ca, Sr, and Ba) of tetrachlorophthalic acid were suggested as early as 1945 by *Kränzlein et al.* as additives for signaling compositions of high intensity and saturation. The underlying concept was the observation that the alkaline earth metal monochlorides – responsible for the coloration of the flame – would form in situ upon decomposition of those salts. The calcium salt was again suggested for illuminating compositions (1978) and tracer formulations (1985) together with either strontium or barium nitrate.

- G. Kränzlein, H. Rathsburg, E. Diefenbach, *Farbige Leuchtsätze hoher Intensität*, DE 750642, **1945**, Germany.
- I. Schmied, G. Marondel, *Leuchtsatz hoher spezifischer Leistung*, DE 2550114, **1978**, Germany.
- I. Schmied, G. Marondel, H. Gawlick, *Farbiger Leuchtspursatz für kleine Kalibermunition*, DE 2415847, **1985**, Germany.

Camphor
Campher

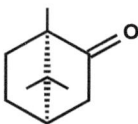

1,7,7-Trimethylbicyclo[2.2.1]heptan-2-one		
Formula		$C_{10}H_{16}O$
GHS		02, 07
H-phrases		228-315-319
P-phrases		210-280 g-305+351+338
UN		2717
REACH		LRS
EINECS		200-945-0
CAS		[76-22-2]
m_r	g mol^{-1}	152.236
ρ	g cm^{-3}	0.962
$\Delta_f H$	kJ mol^{-1}	−319
$\Delta_c H$	kJ mol^{-1}	−5902.8
	kJ g^{-1}	−38.775
	kJ cm^{-3}	−37.301
Λ	wt.-%	−283.76
Mp	°C	179.5
Bp	°C	205
$\Delta_m H$	kJ mol^{-1}	6.8
$\Delta_v H$	kJ mol^{-1}	54.4

Camphor is used as a gelling agent for nitrocellulose.

Cap sensitivity test

The cap sensitivity test is part of *UN Test Series 5*. It serves the determination of the sensitivity of an explosive towards the stimulus of a blasting cap. In accordance with the test procedure a cylindrical charge of the explosive under investigation ($\varnothing$ = 86 mm and l = 162 mm) is placed in a cardboard cylinder and is subjected to the detonation of a standard detonator inserted concentrically on top.

– *Recommendations on the Transport of Dangerous Goods – Manual of Tests and Criteria* – 5th rev. Ed. United Nations, **2009**, pp. 130–133.

Capsaicin, OC

(E)-N-(4-Hydroxy-3-methoxybenzyl)-8-methyl-6-nonensäureamid

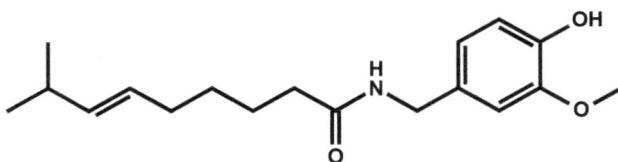

8-Methyl-N-vanillyl-6-noneneamide

Formula		$C_{18}H_{27}NO_3$
GHS		06, 08, 05
H-phrases		301-315-317-318-334-335
P-phrases		261-280-280-284-301+330+331+310-305+351+338+310
UN-Number		1544
EINECS		206-969-8
CAS		[404-86-4]
m_r	g mol^{-1}	305.412
ρ	g cm^{-3}	1.188
Mp	°C	62-65
Bp	°C	210-220
Fp	°C	113
P_{vap}	Pa	2×10^{-6}
Λ	wt.-%	−243.60
N	wt.-%	4.59

OC (Lat.: *oleresin capsicum*) is a chemical irritant. It effects extreme irritation of the respiratory system, the mucous membranes, and the skin. It is used in dispensers mostly as diluted solution with substances improving the transcutaneous mobility (Ger.: *Schlittensubstanz*) such as dimethyl sulfoxide.

Caseless ammunition

Hülsenlose Munition

Caseless ammunition lacks the typical metal (e.g. brass) casing and only consists of a propellant mono-grain, an igniter, and the projectile.

It is hence that caseless ammunition has been considered both from the standpoint of saving critical components (wartime Germany) and more generally to reduce the weight and volume of individual rounds, thereby allowing a greater amount of ammunition to be carried. Figure C.1 depicts the caseless 4.73 × 33 mm ammunition (DM18) that has been designed specifically for the German G11 assault rifle. Figure C.2 depicts the propellant mono-grain, the igniter cup, projectile, and plastic plug.

Figure C.2: Components of 4.7 x 33 mm caseless ammunition (DM 18) after *Drake00*.

The brass casing of a common cartridge apart from holding the grain also withstands certain mechanical impact; it protects the actual grain from dirt and moisture and most importantly its heat capacity (~0.377 J g^{-1} K^{-1}) serves in cooling the chamber to avoid overheating.

Hence the challenges in the design of caseless ammunition very much focus on the propellant that has to possess a

- mechanical stability to meet with tactical requirements
- thermal stability to avoid cook-off in the hot chamber.

While nitrocellulose can be tailored to meet with the mechanical demands it is inherently prone to thermal decomposition and cook-off. Hence the depicted DM 18 was based on an nitrocellulose-free propellant containing *polynitropolyphenylene* (→).

- Drake00 taken from the English Wikipedia, CC BY-SA 3.0, https://commons.wikimedia.org/w/index.php?curid=2132367

Catocen®

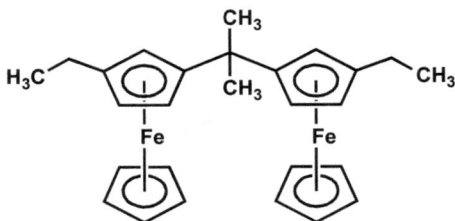

2,2-Bis(ethylferrocenyl)propane		
PLUTORAC EFP		
Aspect		Orange yellow liquid
Formula		$C_{27}H_{32}Fe_2$
REACH		LPRS
EINECS		310-202-3
CAS		[37206-42-1]
m_r	g mol^{-1}	468.245
ρ	g cm^{-3}	1.27
Δ_fH	kJ mol^{-1}	225
Δ_cH	kJ mol^{-1}	−16,247.6
	kJ g^{-1}	−34.699
	kJ cm^{-3}	−44.068
Λ	wt.-%	−246.02
Fe content	wt.-%	23.88

Catocen is a combustion modifier for ammonium perchlorate containing composite propellants. Even though catocen carries bulky alkyl chains it still shows a tendency to migrate upon heat treatment.

- M. Talbot, T. T. Foster, *Dicyclopentadienyl iron compounds*, US3673232, **1972**, USA.
- K. Menke, P. Gerber, E. Geissler, G. Bunte, H. Kentgens, R. Schoffl, Ferrocene migration and mechanical stresses for an end burning propellant grain, *ICT -Jata*, Karlsruhe, **1999**, V-28.

Cellulose acetate butyrate, CAB
Celluloseacetatbutyrat

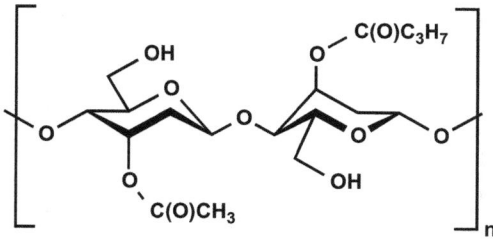

Formula		$C_{15}H_{22}O_8$
REACH		LPRS
EINECS		618-381-2
CAS		[9004-36-8]
m_r	g mol^{-1}	330.35
ρ	g cm^{-3}	1.22
$\Delta_f H$	kJ mol^{-1}	−1629
$\Delta_c H$	kJ mol^{-1}	−7417.9
	kJ g^{-1}	−22.456
	kJ cm^{-3}	27.396
Λ	wt.-%	−159.83

CAB is a non-energetic plasticizer for insensitive gun propellants and high explosives.

Cellulose acetate nitrate, CAN

Celluloseacetatnitrat

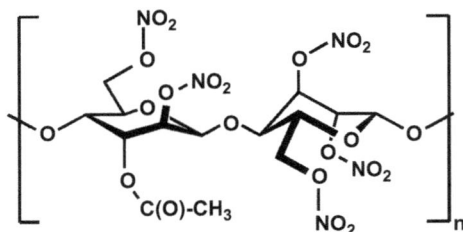

Aspect		Colourless flocs
CAS		[9032-48-8]
ρ	g cm^{-3}	1.27
$\Delta_f H$	kJ mol^{-1}	−762
Λ	wt.-%	−246.02
Mp	°C	230 (deflagration)
Impact	J	3.4

CAN is an energetic plasticizer for gun propellants.

- T. G. Manning, J. Wyckoff, E. Rozumov, J. Laquidara, V. Panchal, C. Knott, C. Michienzi, G. Johnston, B. Vaughan, S. Velarde, Scale −Up of the Insensitive Energetic Binder, Cellulose Acetate Nitrate (CAN), *2012 Insensitive Munitions & Energetic Materials Technology Symposium*, Las Vegas, Nevada, **2012**.
- J. Kimura, T. Shimidzu, J. Maruyama, H. Hayashi, New LO VA Propellants Based on a Desensitized Nitrocellulose Cellulose Acetate Nitrate (CAN), *ADPA Meeting 655*, San Diego, **1996**.

Centralite I, carbamite and ethylcentralite

Centralit I, Ethylcentralit

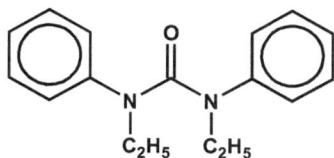

1,3-Diethyl-1,3-diphenylurea		
EC		
Formula		$C_{17}H_{20}N_2O$
REACH		LRS
EINECS		201-654-2
CAS		[85-98-3]
m_r	g mol^{-1}	268.359
ρ	g cm^{-3}	1.14
$\Delta_f H$	kJ mol^{-1}	−105
$\Delta_c H$	kJ mol^{-1}	−9443.1
	kJ g^{-1}	−35.189
	kJ cm^{-3}	−40.116
Λ	wt.-%	−256.36
N	wt.-%	10.44
Mp	°C	79
Bp	°C	326.5

Centralite I is a stabilizer for nitrocellulose.

Centralite II

Centralit II

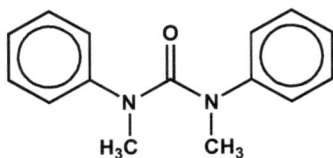

1,3-Dimethyl-1,3-diphenylurea

Formula		$C_{15}H_{16}N_2O$
REACH		LRS
EINECS		210-283-4
CAS		[611-92-7]
m_r	g mol^{-1}	240.305
Δ_fH	kJ mol^{-1}	−73.22
Δ_cH	kJ mol^{-1}	−8116.2
	kJ g^{-1}	−33.775
Mp	°C	122
Λ	wt.-%	−246.34
N	wt.-%	11.66
Dp	°C	350

Centralite II is a stabilizer for nitrocellulose.

Centralite III

Centralit III

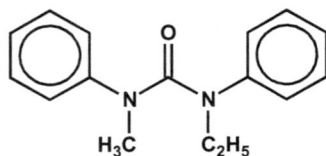

1-Ethyl-3-methyl-1,3-diphenylurea

Formula		$C_{16}H_{18}N_2O$
REACH		LPRS
EINECS		224-747-9
CAS		[4474-03-7]
m_r	g mol^{-1}	254.332
Δ_fH	kJ mol^{-1}	−127
Δ_cH	kJ mol^{-1}	−8741.8

	kJ g^{-1}	−34.372
Λ	wt.-%	−251.63
N	wt.-%	11.01
Mp	°C	74

Centralite III is a stabilizer for nitrocellulose.

Centralite IV

Centralit IV

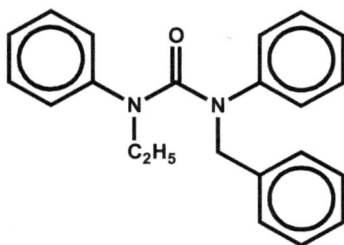

1-Benzyl-3-ethyl-1,3-diphenylurea		
Formula		C$_{22}$H$_{22}$N$_2$O
CAS		[1905331-46-5]
m$_r$	g mol^{-1}	330.41
ρ	g cm^{-3}	1.143
Δ$_f$H	kJ mol^{-1}	−154
Δ$_c$H	kJ mol^{-1}	−11647.6
	kJ g^{-1}	−35.250
	kJ cm^{-3}	−40.291
Λ	wt.-%	−261.48
N	wt.-%	8.48

Centralite IV is a stabilizer for nitrocellulose.

CERV

CERV is a thermochemical code, developed by *J. J. Gottlieb* of the University of Toronto. CERV code is different from other thermochemical codes in that it determines the Gibbs-free energy via the reaction variables instead of via the compositional variables (e.g. mole numbers) as is done by codes like NASA-CEA or ICT. In contrary to NASA-CEA or ICT code, CERV code is also more robust in dealing with systems containing considerable amounts of condensed products (such as pyrotechnics) and hence yields more appropriate and more realistic results than the former.

- E.-C. Koch, R. Webb, V. Weiser, Review of Thermochemical Codes, O-138, NATO -Munitions Safety Information Analysis Center, (MSIAC), Brussels, September **2010**, 35 pp.
- J. J. Gottlieb, *Study of Internal Ballistics of a Closed Vessel*, Institute for Aerospace Studies, University of Toronto, June **2000**, 726 pp.

Cesium dinitramide

Caesiumdinitramid, CsDN

Aspect		Yellow crystalline powder
Formula		$CsN(NO_2)_2$
CAS		[140456-77-5]
m_r	g mol^{-1}	238.923
ρ	g cm^{-3}	3.05
Mp	°C	87
Bp	°C	175 (dec)
$\Delta_f H$	kJ mol^{-1}	−273.55
Λ	wt.-%	+23.44
N	wt.-%	17.59
Friction	N	>360
Impact	J	>20

Though CsDN possesses a low melting point, T = 86 °C, it shows the highest decomposition temperature of all alkali dinitramides T > 175 °C. Correspondingly it has been investigated by *Berger* for use in replacement formulations for lead-based primaries.

- B. Berger, H. Bircher, M. Studer, M. Wälchli, Alkali Dinitramide Salts. Part 1 Synthesis and Characterization, *Propellants Explos. Pyrotech.* **2005** *30*, 184–190.
- B. Berger, J. Mathieu, P. Folly, Alkali Dinitramide Salts. Part 2: Oxidizers for Special Pyrotechnic Applications, *Propellants Explos. Pyrotech.* **2006**, *31*, 269–277.
- M. D. Cliff, M. W. Smith, Thermal characteristics of alkali metal dinitramide salts, *J. Energ. Mater.* **1999**, *17*, 69–86.

Cesium nitrate

Caesiumnitrat

Formula		$CsNO_3$
GHS		03
H-phrases		272
P-phrases		210-220-221—280-370+378a-501a
UN		1451
REACH		LRS
EINECS		232-146-8
CAS		[7789-18-6]
m_r	g mol^{-1}	194.91
ρ	g·cm^{-3}	3.685
Mp	°C	141
Bp	°C	584
$T_{p(h\leftrightarrow k)}$	°C	160
Dp	°C	584
$\Delta_f H$	kJ mol^{-1}	−505.97
Λ	wt.-%	+20.52
N	wt.-%	7.19

$CsNO_3$ is non-hygroscopic. The thermal decompcsition of $CsNO_3$ at 548 °C proceeds is similar to that of KNO_3. $CsNO_3$ is used as an oxidizer in infrared screening smokes. Its use is based on the observation that Cs compounds most easily undergo ionization in flames (IE = 3.89 eV). Those ions are good condensation nuclei for atmospheric moisture and hence increase the aerosol yield. The easy thermal excitation of Cs compounds in the near IR at λ = 852 and 894 nm is exploited in NIR-illuminating compositions. Finally, the easy ionization of cesium is used in the exoatmospheric generation of electron clouds (CsHex) and for use in pyrotechnic plasma in MHD (Magnetohydrodynamic) applications and rocket motor diagnostics.

CsHEX
- 34 wt.-% cesium nitrate
- 21 wt.-% aluminum
- 19 wt.-% RDX
- 26 wt.-% TNT
  $\Delta_{ex}H$ = 5000 kJ kg^{-1}

Similar to $RbNO_3$, $CsNO_3$ does not give any characteristic flame coloration, despite showing two weak blue lines at λ = 455 and 459 nm in its UV-VIS spectrum.

$CsNO_3$ stabilizes the II and V phases of NH_4NO_3. Co-precipitates of $CsNO_3$ and cesium dodecahydroborate are used as fillers for high speed igniters ($\rightarrow$ *Hivelite*).

- E.-C. Koch, Special Materials in Pyrotechnics II. Caesium and Rubidium Compounds in Pyrotechnics, *J. Pyrotech.*, **2002**, *15*, 9–24.
- B. Berger, S. D. Brown, E. L. Charsley, J. J. Rooney, R. P. Claridge, T. T. Griffiths, A study of the pyrotechnic performance of the silicon-caesium nitrate pyrotechnic system, *IPS-Seminar*, Fort Collin, **2002**, 743–753.

Cesium perchlorate

Caesiumperchlorat

Formula		$CsClO_4$
GHS		03, 07
H-phrases		272-315-319-335
P-phrases		210-221-302+352-305+351+338-405-501a
UN		1481
REACH		LPRS
EINECS		236-643-0
CAS		[13454-84-7]
m_r	g mol^{-1}	232.356
ρ	g·cm^{-3}	3.327
$\Delta_f H$	kJ mol^{-1}	−434.72
Λ	wt.-%	+27.54
Mp	°C	577
$T_{p(rh\leftrightarrow cub)}$	°C	222

$CsClO_4$ is non-hygroscopic. Its mixtures with aluminum have been used to prepare pyrotechnic plasma. At the optimum stoichiometry ($CsClO_4$/Al: 75/25) T_f = 3549 °C, more than 99% of the Cs is ionized. $CsClO_4$ has been suggested as an oxidizer delivering electrons for use in a pyrotechnic strobe flare recognizable by radar. *Berger* has investigated $CsClO_4$/Ti. Upon decomposition of $CsClO_4$, a eutectic with CsCl forms yielding a melting range between 400 and 500 °C in which the oxidation of Ti particles can occur.

- E.-C. Koch, Special Materials in Pyrotechnics II. Caesium and Rubidium Compounds in Pyrotechnics, *J. Pyrotech.*, **2002**, *15*, 9–24.
- B. Berger, B. Haas, V. Weiser, Y. Plitzko, Temperaturmessungen an Titan/Caesiumperchlorat - Mischungen, *ICT-Jata*, **2002**, Karlsruhe, P-135.

Cesium picrate

Caesiumpikrat

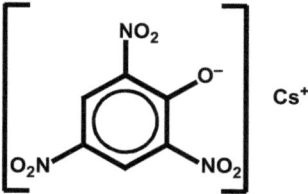

Formula		$CsC_6H_2N_3O_7$
CAS		[3638-61-7]
m_r	g mol^{-1}	361.003
ρ	g·cm^{-3}	2.481
$\Delta_f H$	kJ mol^{-1}	−560
Λ	wt.-%	−28.81
N	wt.-%	11.64
Mp	°C	320 (dec.)
$T_{p(???\leftrightarrow???)}$	°C	280
Impact	J	7.25
E_a	kJ mol^{-1}	180

Cesium picrate has been used as a seeding agent for HE charges in detonative driven MHD generators. The pressure dependent combustion of cesium picrate has been studied by *Fogel'zang et al.*

- E. H. Jager, F. R. Thomanek, Untersuchungen über sprengstoffbetriebene MHD-Generatoren, *J. Appl. Math. Phys.* **1974**, *25*, 47–54.
- B. S. Svetlov, A. E. Fogel'zang, Combustion of Fast-Burning Explosives, *Combustion Explos. Shock.* **1969**, *5*, 46–51.
- M. Stammler, Thermal properties and spectra of the picrates of the group IA elements and of the Tl$^+$, NH$_4^+$, [(CH$_3$)$_4$ N]$^+$, and [(C$_2$H$_5$)$_4$N]$^+$ Ions, *Explosivstoffe*, **1968**, *16*, 154–163.

CH-6

CH-6 is a booster explosive formulation (Table C.1) based on phlegmatized RDX:

97.5 wt.-%	Hexogen
1.5 wt.-%	Calcium stearate
0.5 wt.-%	Graphite
0.5 wt.-%	Polyisobutylene

Table C.1: Properties of CH-6.

Parameter	CH-6
V_D (m s^{-1})	8091
P_{CJ} (GPa)	27.8
ρ (g cm^{-3})	1.67
c_p (J g^{-1} K^{-1})	1.182
Dp (°C)	190
E_a (kJ mol^{-1})	172
No-Go-BICT-SSWGT (GPa)	1.25
E-Impact NOL (J)	6,6

- MIL-C-21723 B, Military Specification Composition CH-6.

Chapman, David-Leonard (1869–1958)

Chapman was a British physicist who together with Jouguet developed a theory to describe gas detonations.

- E. J. Bowden, Obituary: David Leonard Chapman, 1869–1958, *Bibliogr. Mem. Fellows Roy. Soc.* **1958**, 4, 34–44.

Charcoal
Holzkohle

Charcoal is a highly porous (70–85 vol.-% pore volume), carbon-rich, non-volatile solid (apparent density ~ 0.4 g cm^{-3} and true density ~ 1.4 g cm^{-3}) that is obtained by carbonization of wood or other vegetable organic matter (rice straw, bamboo, etc.). Therefore, the starting material is heated under anaerobic conditions to temperatures up to 275 °C which triggers an exothermic decomposition causing a further temperature rise up to about 400 °C. With increasing carbonization temperature, the hydrogen content of the charcoal decreases.

The typical composition of the dry wood is given in Table C.2. The average sum formula for wood reads $C_{42}H_{60}O_{28}$.

Table C.2: Composition of dry wood after *Flügge*.

	C (wt.-%)	H (wt.-%)	O (wt.-%)	N (wt.-%)	Ash (wt.-%)
Dry wood	49	6	44	0,1	0,9

The pyrolysis of wood can be written in a simplified manner:

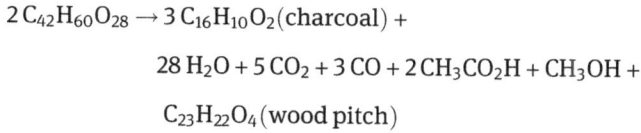

$$2\,C_{42}H_{60}O_{28} \rightarrow 3\,C_{16}H_{10}O_2(\text{charcoal}) +$$

$$28\,H_2O + 5\,CO_2 + 3\,CO + 2\,CH_3CO_2H + CH_3OH +$$

$$C_{23}H_{22}O_4\,(\text{wood pitch})$$

The typical yield of charcoal from carbonization is about 35 wt.-% based on the mass of the dry wood. The composition of different charcoals, their density, and enthalpy of formation is given in Table C.3.

Table C.3: Chemical composition of different charcoals and enthalpy of formation.

Charcoal	C	H	O	N	Ca	Density (g cm^{-3})	$\Delta_f H$ (kJ mol^{-1})	m_r (g mol^{-1})
Maple	10	5.085	1.340	0.035	0.028		−140.6	151.196
Beech	10	3.857	0.873	–	0.022	1.362	+2.5	138.847
Oak	10	4.925	1.408	0.074	0.195		−282.26	170.011
Alder	10	4.774	1.234	0.039	0.026	1.342	−36.40	146.254
Buckthorn	10	6.696	2.117	–	0.009		−206.69	161.091
Pine	10	4.600	1.056	0.021	0.036		+61.32	149.577

Charcoal possesses a high specific surface area (S_A = 50–80 m^2 g^{-1}) and is easily ignited. The ignition temperature decreases with increasing content of volatiles (hydrogen) (Figure C.3). The absorption of moisture is depicted in Figure C.4. The specific heat, c_p, at ambient temperature = 1 J g^{-1} K^{-1}.

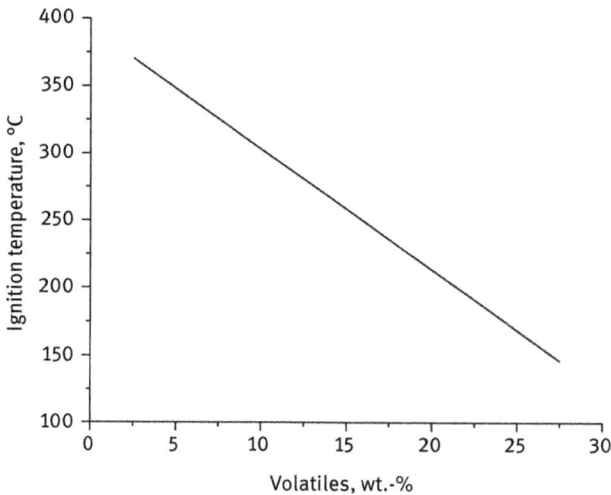

Figure C.3: Ignition temperature as a function of content of volatiles (hydrogen content).

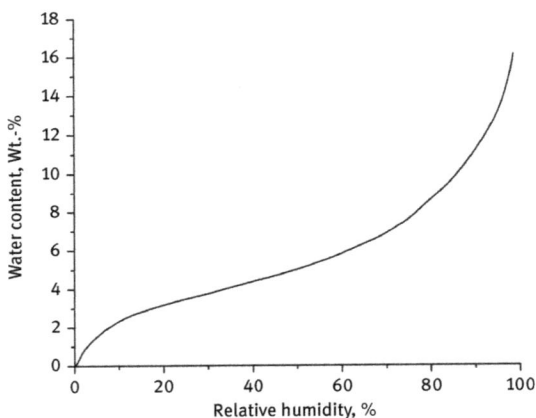

Figure C.4: Water content of charcoal as a function of relative humidity at 20 °C.

- W. Meyerriecks, Organic Fuels: Composition and Formation Enthalpy – Part II – Resins, Charcoal, Pitch, Gilsonite, and Waxes, *J. Pyrotech.* **1999**, *9*, 1–19.
- F. Flügge, Holzverkohlung, in K. Winnacker, E Winnacker (Eds.) *Chemische Technologie – Organische Technologie I*, Carl Hanser, Munich, **1952**, pp. 497–530.

CHEETAH

CHEETAH is a combined thermochemical-kinetic code to predict the performance of explosives and propellants. Releases after version 2.0 are available only for users authorized by the US Department of Defense. The latter have to agree on a *quid pro quo* with the DOD and report back all calculations done with and any modifications carried out on CHEETAH code. At the time of writing this book (fall 2020), version 11.0 is in use by the community. A comparison between EXPLO 5.0 and on publicly reported data of CHEETAH 7.0 has been published by *Sućeska*. The POC for CHEETAH at LLNL is Sorin Bastea: Email: sbastea@llnl.gov.

- M. Sućeska, EXPLO 5.0, Theoretical Background and Capabilities, *14th Workshop on Pyrotechnic Combustion Mechanism*, June 25, **2018**, Kaiserslautern Germany.
- L. E. Fried, W. M. Howard, P. C. Souers, *Cheetah 2.0 User's Manual*, UCRL-MA-11541 Rev. 5, Lawrence Livermore National Laboratory, August 20, **1998**, USA.

Chemical warfare agents
Chemische Kampfstoffe

Chemical warfare agents (CWA) are pure substances and mixtures that serve the lethal, damaging, and incapacitating effects on human, animals, plants, and the environment including the modification of weather.

Of all 197 countries on earth, 194 countries have joined in the international *Chemical Weapons Convention* since 1993 and have signed and ratified the treaty so far (status 2018). The treaty countries commit themselves not to use any CWA in military conflicts, not to produce or to store CWA and to dispose and to destroy any CWA present in the responsibility of the corresponding country (including legacy munitions from previous wars such as World War I and World War II).

Egypt and North Korea have not joined in the treaty. Israel has signed the treaty in 1993 but has not ratified it so far.

CWAs can be separated into three groups:
- **lethal agents**
- **incapacitating agents**
- **environmental agents**

The lethal agents include blood agents, blister agents, pulmonary agents, and nerve agents. The incapacitating agents include lacrymogens, psycho-chemical agents, and nettle agents. The environmental agents finally include herbicides and weather modifying substances.

- https://www.opcw.org/news/article/south-sudan-to-join-chemical-weapons-convention/
- S. Franke, *Lehrbuch der Militärchemie, Band I + II*, 2. Auflage, Militärverlag der Deutschen Demokratischen Republik, Leipzig **1976**, 512 + 615 pp.

Chlorate explosives
Chloratsprengstoffe

The first ever prepared chlorate explosive was the so-called *white gun powder* invented by *Berthollet* in 1786. However, this composition had a much greater brisance than common black powder and hence would shatter and destroy weapons. While this brought any further improvements of gun propellants to a temporary halt, it paved the way for chlorate-based explosives as more powerful replacements for black powder in mining and landscaping. Thus, in the late nineteenth and the early twentieth centuries, a huge number of formulations containing either sodium or potassium chlorate and various liquid or pasty organics including paraffins and nitroaromatic compounds were prepared and extensively used. Chlorate-based explosives began to vanish starting in the 1930s due to their extraordinary high mechanical sensitivity. However, in wartime Germany, they saw a brief renaissance and were used to overcome the shortage in nitro compounds. Chlorate-based explosives were used wherever severe set-back forces upon regular handling of munitions were absent that is in anti-personal mines, depth charges, and similar applications. Nowadays, chlorate-based explosives are no more.

Chlorinated paraffins

Chlorparaffine, CP

Chlorinated paraffins (CS) are liquid, pasty, and solid chloroalkyl compounds with chlorine contents between 15 and 70 wt.-%. They find use as binder and chlorine source in both pyrotechnics and HE formulations (e.g. PAX-46).

PAX-46

- 85.0 wt.-% hexogen
- 10.5 wt.-% Chlorez wax
- 4.5 wt.-% Paroil

$\rho = 1.76$ g cm^{-3}

DXP-1#94

- 66.2 wt.-% octogen, NSO 137 < 637 μm
- 27.8 wt.-% octogen, NSO 152
- 4.5 wt.-% chloroparaffin, Leuna CP52, liquid
- 1.5 wt.-% PVC, Solvin 266 SF

$\rho = 1.801$ g cm^{-3} (87% TMD)

Typical chlorinated paraffins are, for example, *Chlorez* [8029-39-8] (EINCES:264-150-0, REACH: LRS) and *Paroil* [63449-39-8], both manufactured by *Dover Chemical*. The high chlorine content of those binders leads to advantageous cook-off properties. Cl radicals released by pyrolysis of CPs scavenge hydrocarbon radicals and thereby act as chain stoppers. While short-chain CPs (C$_{10-13}$) bear a toxicological risk, the long-chain CPs (C$_{18-30}$) are currently considered non-toxic.

- A. Hahma, J. Licha, O. Pham-Schönwetter, *Insensitive Sprengstoffwirkmasse*, EP2872464B1, **2017**, Germany.
- N. N., *Kurzkettige chlorierte Paraffine – Stoffflußanalyse, Schriftenreihe Umwelt, Nr. 354*, Bundesamt für Umwelt, Wald und Landschaft, Bern, **2003**, 97 pp, accessible at http//www. pops.int/documents/meetings/poprc/prepdocs/annexesubmissions/Short-chained%20chlori nated%20paraffins%20Switzerland%20info2.pdf
- http//www.doverchem.com/Products/Chlorez%C2%AEResinousChlorinatedAlkanes.aspx
- http//www.doverchem.com/Products/Paroil%C2%AELiquidChlorinatedAlkanes.aspx
- E. Leibnitz, H.-G. Könnecke, E. Schmidt, C. Schuhler, Zur Chlorierung der Alkane. IV, *J. Prakt. Chem.* **1962**, *15*, 155–174.

Chlorine trifluoride

Chlortrifluorid, N-Stoff, CTF

Aspect		Colorless gas, slightly yellowish green liquid
Formula		ClF_3
GHS		03, 06, 05, 08, 09
H-phrases		270-280-330-314-370-372-400
REACH		LPRS
EINECS		232-230-4
CAS		[7790-91-2]
m_r	g mol^{-1}	92.448
ρ	g cm^{-3}	1.8094 (@ 25 °C)
$\Delta_f H$	kJ mol^{-1}	−163
$\Delta_v H$	kJ mol^{-1}	24.7
Mp	°C	−76.3
Bp	°C	11.75
P_{vap}	Pa	1.42×10^3 @ 20 °C

The heavily corrosive ClF_3 is a strong breathing poison which even at low concentration can be detected by its characteristic sweet smell. ClF_3 is also a potent oxidizer and was eventually considered after its first preparation by *Ruff* to become a mask breaking agent (*N-Stoff*) as it reacts hypergolically with active charcoal. Its hypergolicity with many fuels made it also interesting as a potential oxidizer for rocket propulsion and was correspondingly investigated in the early 1950s.

- J. D. Clark, *Ignition*, 4th Reprint Rutger University Press, **2010**, 195 pp.
- A. Dadieu, R. Damm. E. W. Schmidt, *Raketentreibstoffe*, Springer Verlag, Vienna, **1968**, p. 633ff.
- O. Ruff, H. Krug, Über ein neues Chlorfluorid-ClF_3, *Z. Anorg. Allg. Chem.* **1930**, *190*, 270–276.

Chloroacetophenone, CN

Chloracetophenon

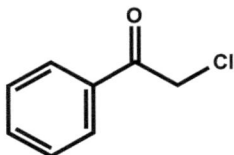

Aspect		Colorless, technical grade is about gray
Formula		C_8H_7ClO
CAS		[532-27-4]
m_r	g mol^{-1}	154.59
ρ	g cm^{-3}	1.324
Mp	°C	58
Bp	°C	245
Fp	°C	88
P_{vap}	Pa	1.733 @ 20 °C
$\Delta_{sub}H$	kJ mol^{-1}	90.7
Λ	wt.-%	−186.29
LCt_{50}	ppm	>10,000
ICt_{50}	ppm min^{-1}	80
$\Delta_f H$	kJ mol^{-1}	−175 (estimated)
$\Delta_c H$	kJ mol^{-1}	−3923.0
	kJ g^{-1}	−25.576
	kJ cm^{-3}	−33.598

CN is sufficiently thermally stable to allow both slow dissemination by burning pyrotechnics or fast expulsion by bursting charges. It is stable towards hydrolysis and hence decomposes only slowly in the environment.

- S. Franke, *Lehrbuch der Militärchemie, Band 1*, Militärverlag der Deutschen Demokratischen Republik, 2. Auflage, Leipzig, **1977**, pp. 148–151.

Chlorobenzylidenemalon dinitrile, CS

2-Chlorbenzylidenmalonsäuredinitril

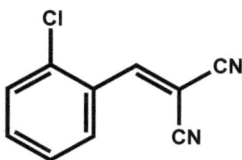

o-Chlorobenzylidene malononitrile, tear gas

Formula		$C_{10}H_6ClN_2$
CAS		[2698-41-1]
m_r	g mol^{-1}	188.61
ρ	g cm^{-3}	1.41
$\Delta_f H$	kJ mol^{-1}	350
$\Delta_c H$	kJ mol^{-1}	−4949.2
	kJ g^{-1}	−26.240
	kJ cm^{-3}	−36.998
Mp	°C	96
Bp	°C	315
P_{vap}	Pa	0.45 @ 20 °C
Λ	wt.-%	−186.62
N	wt.-%	14.85
LCt_{50}	ppm min^{-1}	10,000
ICt_{50}	ppm min^{-1}	20

CS has a weak pepper-like smell. Based on its thermal stability it can be disseminated from burning pyrotechnics.

Chlorotrinitrobenzene, CTNB

Pikrylchlorid

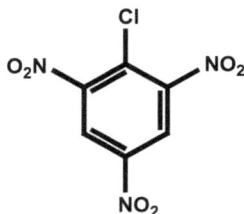

1-Chloro-2,4,6-trinitrobenzenepicryl chloride		
Aspect		Light yellowish needles
Formula		$C_6ClH_2N_3O_6$
GHS		01, 06, 09
H-phrases		201-300-310-330-410
P-phrases		260-264-273-280-289-301+310
REACH		LPRS
EINECS		201-864-3
CAS		[88-88-0]
m_r	g mol^{-1}	247.551
ρ	g cm^{-3}	1.797
$\Delta_f H$	kJ mol^{-1}	26.82
$\Delta_{ex} H$	kJ mol^{-1}	−1118.5
	kJ g^{-1}	−4.518
	kJ cm^{-3}	−8.119
$\Delta_c H$	kJ mol^{-1}	−2673.8
	kJ g^{-1}	−10.801
	kJ cm^{-3}	−19.409
Λ	wt.-%	−42.01
N	wt.-%	16.97
Mp	°C	83
$\Delta_m H$	kJ mol^{-1}	18.15
Dp	°C	395
P_{vap}	Pa	5 @ 83 °C
Friction	N	>355
Impact	J	16
V_D	m s^{-1}	7200 @ 1.4 g cm^{-3}
Trauzl	cm^3	315

Ullmann reaction of picryl chloride affords hexanitrodiphenyl.

– T. Urbanski, *Chemie und Technologie der Explosivstoffe, Band I*, VEB Verlag für Grundstoffindustrie, **1961**, pp. 207.

Chlorovinylarsine dichloride and L-1, lewisite

Lewisit

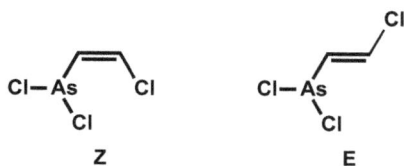

		Z		E

2-Chlorethenyldichlorarsine

		Z		E
Aspect		Dark brown oily liquid		
Formula		$AsC_2Cl_3H_2$		
		Z		E
CAS		[541-25-3]		[541-25-3]
m_r	$g\ mol^{-1}$		207.318	
ρ	$g\ cm^{-3}$	1.8598		1.8793
Mp	°C	−44.7		−2.4
Bp	°C	169.8		196.6
P_{vap}	kPa	208.25		53.33
$\Delta_{vap}H$	$kJ\ mol^{-1}$			53.4
Λ	wt.-%	−38.59		
LCt_{50}	$ppm\ min^{-1}$	1.3		
ICt_{50}	$ppm\ min^{-1}$	0.3		
Δ_fH	$kJ\ mol^{-1}$	+140 (estimated)		

The geranium-type smell of technical-grade lewisite is due to AsH_3 which forms upon thermolysis.

Chloropicrin, PS

Chlorpikrin

Trichloronitromethane, vomiting gas

Aspect		Colorless viscous liquid, technical grades appear yellowish green
Formula		CCl_3NO_2
CAS		[76-06-2]
m_r	$g\ mol^{-1}$	164.375
ρ	$g\ cm^{-3}$	1.657
Mp	°C	−69.2
Bp	°C	113
P_{vap}	kPa	2.254 at 20 °C
$\Delta_{sub}H$	$kJ\ mol^{-1}$	39.3 at 15 °C
Λ	wt.-%	0
LCt_{50}	$ppm\ min^{-1}$	20,000
ICt_{50}	$ppm\ min^{-1}$	200

Chloropicrin is a colorless oily strongly refractive liquid with a pungent odor. It decomposes beyond 400 °C completely to NOCl and the highly toxic *phosgen*.

Christe, Karl Otto (1936*)

Karl Otto Christe is a native German chemist and expert for fluorine and high nitrogen compounds. He studied chemistry starting in 1957 at the TU Stuttgart, Germany, and obtained his PhD with Josef Goubeau (1901–1990). Since his immigration to the USA in 1962, he has been doing research initially in industry and as an advisor to the Air Force Research Laboratory and since 1994 as a professor at the University of Southern California in the field of fluorine, high nitrogen, and oxygen-rich materials. Many of the compounds made by Christe for the first time are prospect oxidizers and HEs. Christe was the first to recognize the potential of energetic ionic liquids as versatile explosive and propellant materials. Christe has been awarded many times for his pioneering work in the field of fluorine chemistry. Inter alia he received the prestigious Henri Moissan Medal (2000), the Alfred Stock-Gedächtnispreis (2006), the Tolman Award (2011) and the M. Frederick Hawthorne Award in Main Group Inorganic Chemistry (2021). He is an elected member of the European Academy of Science and Arts (2010) and a Fellow of the American Association for the Advancement of Science (2017).

- K. O. Christe, Recent Advances in the Chemistry of N_5^+, N_5^- and High-Oxygen Compounds, *Propellants Explos. Pyrotech.* **2007**, *32*, 194–204.
- K. O. Christe, Polynitrogen chemistry enters the ring A cyclo-N_5^- anion has been synthesized as a stable salt and characterized, *Science*, **2017**, *355*, 351.
- K. O. Christe, D. A. Dixon, M. Vasiliu, R. Haiges, B. Hu, How Energetic Are Cyclo-Pentazolates? *Propellants Explos. Pyrotech.* **2019**, *44*, 263–266.
- Entry in the catalogue of the German National Library: http://d-nb.info/gnd/106710699

Chromium(III) oxide
Chrom(III)oxid, Chromoxydgrün

Aspect		Dark green powder
Formula		Cr_2O_3
GHS		07
H-phrases		302-332-317
P-phrases		261-280-302+352-304+340-321-501a
REACH		LRS
EINECS		215-160-9
CAS		[1308-38-9]
m_r	g mol^{-1}	151.99
ρ	g cm^{-3}	5.21

$\Delta_f H$	kJ mol^{-1}	1140
Λ	wt.-%	0
Mp	°C	2275
Sblp	°C	4000
c_p	J K^{-1} mol^{-1}	120

Cr_2O_3 is used as an oxidizer in delay compositions and has been used in legacy infrared flares for great heights (e.g. SI-119) and most recently in nanothermites.

SI-119
- 49 wt.-% zirconium
- 31 wt.-% molybdenum trioxide, MoO_3
- 20 wt.-% chromium oxide, Cr_2O_3
- +2 wt.-% nitrocellulose

- M. Comet, V. Pichot, B. Siegert, E. Fousson, J. Mory, F. Moitrier, D. Spitzer, Preparation of Cr_2O_3 nanoparticles for superthermites by the detonation of an explosive nanocomposite material, *J. Nanopart. Res.* **2011**, *13*, 1961–1969.
- C. A. Knapp, *New Infrared Flare and High-Altitude Igniter Compositions*, Feltman Research and Engineering Laboratories, Picatinny Arsenal, Dover, N. J., July **1959**.

E-Cinnamic acid, CA
trans-Zimtsäure

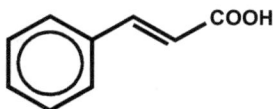

Formula		$C_9H_8O_2$
GHS		07
H-phrases		319
P-phrases		305+351-338
EINECS		205-398-1
CAS		[140-10-3]
m_r	g mol^{-1}	148.16
$\rho_{25\ °C}$	g·cm^{-3}	1.25
Mp	°C	133–136
Sblp	°C	300
$\Delta_f H$	kJ mol^{-1}	−315
$\Delta_c H$	kJ mol^{-1}	−4370.0
	kJ g^{-1}	−29.495
	kJ cm^{-3}	−36.869
Λ	wt.-%	−215.97
c_p	J K^{-1} mol^{-1}	197.5

Cinnamic acid serves as aerosol in training smokes. These formulations originated in the UK. Table C.4 shows the composition of *PN 868* and *PN 907* and published performance data.

Table C.4: Composition and performance of training smokes based on cinnamic acid.

	PN 868	PN 907
Lactose monohydrate (wt.-%)	32.5	26 ± 1.0
Potassium chlorate (wt.-%)	32.5	26 ± 1.0
E-Cinnamic acid (wt.-%)	20.0	33 ± 1.0
Kaolin (wt.-%)	15.0	15 ± 0.5
α_{VIS} at 22 °C, 50%RH	–	3.0
$\alpha_{3-5\ \mu m}$ at 20 °C, 50%RH	–	0.080
u (mm s^{-1})	–	0.39
Y_F at 20 °C, 50% RH	–	0.82

Cinnamic acid aerosols do not show any indication for adverse health effects with test animals and not even with voluntary male test persons exposed to the aerosol.

- K.J. Smit, P. Berry, A. Lee, N. Potticaryand L. Redman, Pyrotechnic composition effects on cinnamic acid smoke obscuration in the infrared and visible regions, *IPS*, **2001**, Adelaide, Australia, pp. 631–636.
- T. C. Marrs, H. F. Colgrave, J. A. G. Edginton, N. L. Cross, Repeated Dose Inhalation Toxicity of Cinnamic Acid Smoke, *J. Hazard. Mater.* **1989**, *21*, 1–13.

CL-14, 5,7-diamino-4,6-dinitrobenzofuroxane
5,7-Diamino-4,6-dinitrobenzofuroxan

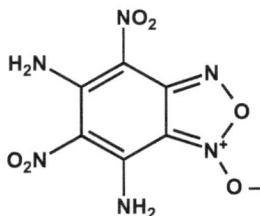

Formula		$C_6H_4N_6O_6$
CAS		[117907-74-1]
m_r	g mol^{-1}	256.135
ρ	g cm^{-3}	1.942
$\Delta_f H$	kJ mol^{-1}	86

$\Delta_{ex}H$	kJ mol^{-1}	−1272.2
	kJ g^{-1}	−4.967
	kJ cm^{-3}	−9.645
Δ_cH	kJ mol^{-1}	−3018.8
	kJ g^{-1}	−11.786
	kJ cm^{-3}	−22.888
Λ	wt.-%	−49.97
N	wt.-%	32.81
Dp	°C	287
Friction	N	Insensitive
Impact	cm	79, 2.5 kg
V_D	m s^{-1}	*8340 at 1.942 g cm^{-3}*
P_{CJ}	GPa	*33.9*

CL-14 is an experimental very insensitive HE. CL-14 also forms upon the mechanical and/or thermal impact on *TATB* and thus also explains the extraordinary insensitivity of the latter. The alkali metal salts of CL-14 (Table C.5) are of interest as components for primary formulations.

Table C.5: Selected properties of CL-14-alkali metal salts.

	Na-CL-14	**K-CL-14**	**Rb-CL-14**	**Cs-CL-14**
CAS-Nr.	[802940-95-0]	[136869-28-8]	[802940-93-8]	[802940-94-9]
Mp (°C)	240 Expl.	265 Expl.	281 Expl.	277 Expl.
Impact (J)	15	13	10	8,4
Friction (N)	360	324	288	160

- J. Sharma, J. C. Hoffsommer, D. J. Glover, C. S. Coffey, J. W. Forbes, T. P. Liddiard, W. L. Elban, F. Santiago, Sub-Ignition Reactions at Molecular Levels in Explosives Subjected to Impact and Underwater Shocks, *Detonation Symposium*, **1985**, Albuquerque, 725–733.
- W. P. Norris, *Preparation of 5,7-diamino-4,6-dinitrobenzofuroxan an insensitive high density explosive*, US Patent 5039812, **1991**, USA.
- M. N. Sikder, S. K. Chougule, A. K. Sikder, B. R. Gandhe, Synthesis, Characterisation, and Thermal and Explosive Properties of Alkali Metal Salts of 5,7-Diamino-4,6-Dinitrobenzoforoxan (CL-14), *Energ. Mater.* **2004**, *22*, 117–126.

CL-15

2,3,6,7-Bis(furazan)-1,4,5,8-tetranitro-1,4,5,8-tetraazadecalin

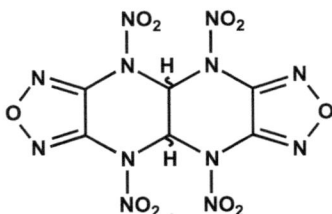

1,4,5,8-Tetranitro-1,4,5,8-tetraazadifurazano-(3,4-c)(3,4-h)decaline		
Aspect		Yellow crystals
Formula		$C_6H_2N_{12}O_{10}$
CAS		[97288-73-8]
m_r	g mol^{-1}	402.15
ρ	g cm^{-3}	1.987
$\Delta_f H$	kJ mol^{-1}	−774.04
$\Delta_{ex}H$	kJ mol^{-1}	−1246.3
	kJ g^{-1}	−3.107
	kJ cm^{-3}	−6.173
$\Delta_c H$	kJ mol^{-1}	−1872.9
	kJ g^{-1}	−4.657
	kJ cm^{-3}	−9.254
Λ	wt.-%	−11.94
N	wt.-%	41.80
Mp	°C	112
Impact	J	5

CL-15 is prone to decomposition at ambient temperature but can be stored for months without any decomposition at T < −15 °C.

- R. L. Willer, *1,4,5,8-Tetranitro-1,4,5,8-tetraazadifurazano-[3,4-c][3,4-h]decalin*, US 4503229A, **1985**, USA.

CL-20, hexanitrohexaazaisowurtzitane

Hexanitrohexaazaisowurtzitan

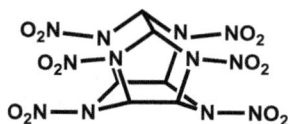

2,4,6,8,10,12-Hexanitro-2,4,6,8,10,12-hexaza
[5.5.0.0^{3,11}.0^{5,9}]tetracyclododecane
HNIW

Formula		$C_6H_6N_{12}O_{12}$
REACH		LPRS
EINECS		603-913-8
CAS		[135285-90-4]
m_r	g mol^{-1}	438.188

Polymorphs:		β	γ	ε
ρ	g cm^{-3}	1.985	1.916	2.044
$\Delta_f H$	kJ mol^{-1}	431 ± 13		377.4 ± 13
Λ	wt.-%	−10.95	=	=
N	wt.-%	38.36	=	=
$\Delta_{ex}H$	kJ mol^{-1}			−2731
	kJ g^{-1}			−6.234
	kJ cm^{-3}			−12.219
$\Delta_c H$	kJ mol^{-1}			−3595.6
	kJ g^{-1}			−8.207
	kJ cm^{-3}			−16.774
Mp	°C	163(dec)		167(dec)
Impact	J	3.5		5.25
Friction	N	64		158
V_D	m s^{-1}	9208 @ 1.942 g cm^{-3} (ε-CL-20/Estane 95.5/4.5)		
Trauzl	cm^3	*517*		

CL-20 is the most-dense commercially available CHNO explosive. It forms many poly-morphs with the ε-polymorph showing the highest density, the lowest sensitiveness, and the greatest thermodynamic stability at ambient temperature. The different poly-morphs can be identified with great certainty by FT-Raman (*Goede*). ε-CL-20 has a mechanical sensitivity comparable to that of PETN. When crystallized from ethylene glycol or diethylene glycol CL-20 is only moderately friction sensitive (*Maksimosvki*). So far, the highly expensive synthesis of CL-20 has prevented a broader use of CL-20. To meet with IM standards (e.g. AOP39), the use of Cl-20 requires considerable phleg-matization which on the other side also lowers its performance significantly. Hence, so far CL-20 has been used only in experimental formulations (*Müller, Bohn*), which require high performance but tolerate non-compliance with IM standards. CL-20 is also a candidate material for future primary formulations requiring both high performance and high sensitivity (*Sandstrom*).

- R. L. Simpson, P. A, Urtview, D. L. Ornellas, G. L. Moody, K. J. Scribner, D. M. Hoffman, CL-20 Performance exceeds that of HMX and its sensitivity is moderate, *Propellants Explos. Pyrotech.* **1997**, *22*, 249–255.
- D. Müller, New Gun propellant with CL-20, *Propellants Explos. Pyrotech.* **1999**, *24*, 176–181.
- P. Goede, N. V. Latypov, H. Ostmark, Fourier Transform Raman Spectroscopy of the Four Crystallographic Phases of *α,β,γ*, and *ε* 2,4,6,8,10,12-Hexanitro-2,4,6,8,10,12-hexaazatetracyclo[5.5.0.0^{5,9}.0^{3,11}]dodecane (HNIW, CL-20), *Propellants Explos. Pyrotech.* **2004**, *29*, 205–208.
- M. A. Bohn, M. Dörich, J. Aniol, H. Pontius, P. B. Kempa, V. Thome, Reactivity between ε-CL20 and GAP in Comparison to β-HMX and GAP, *ICT-Jata*, **2004**, Karlsruhe, V-4.
- J. Sandstrom, A. A. Quinn, E. Erickson, *Non-toxic, heavy-metal free sensitized explosive percussion primers and methods of preparing the same*, US 8206522 B2, **2012**, USA.
- P. Maksimovski, P. Tchórznicki, CL-20 Evaporative Crystallization Under Reduced Pressure, *Propellants Explos. Pyrotech.* **2015**, *41*, 351–359.

Co-crystals
Cokristalle

A co-crystal is a crystal with a primitive cell, containing two or more molecules that do not maintain a covalent bond with each other. The performance and sensitivity of crystalline materials very much depends on the type (polymorph) and quality (shape, defects, etc.) of the individual crystals. Hence, co-crystallization of common energetic materials opens up the opportunity to design new materials with tailored performance and sensitivity. Table C.6 displays the properties of unambiguously characterized co-crystals.

Table C.6: Properties of selected CL-20-based energetic co-crystals.

1	2	Stoichiometry (mol mol^{-1})	Density (g cm^{-3})	Impact$_{50}$ (%) Co	1	2	V_D (m s^{-1}) Co	1	2
CL-20	TNT	1/1	1.84	210	100	585	8402	9800	6886
CL-20	HMX	2/1	1.945	189	100	189	9484	9800	9322

Due to improved hydrogen bonding energetic materials, co-crystals often display sensitiveness that is at least lower than the most sensitive starting component but a performance approaching the most powerful component. Hence, co-crystals could bridge future capability gaps provided means are found to produce such at a required scale.

– R. A. Wiscons, A. J. Matzger, Evaluation of the Appropriate Use of Characterization Methods for Differentiation Between Cocrystals and Physical Mixtures in the Context of Energetic Materials, *Cryst. Growth Des.* **2017**, *17*, 901–906.
– McBain, V. Vuppuluri, I. E. Gunduz, L. J. Groven, S. F. Son, Laser ignition of CL-20 (hexanitrohexaazaisowurtzitane) cocrystals, *Combust. Flame*, **2018**, *188*, 104–115

Colored smoke compositions
Farbrauchsätze

Colored smoke formulations are pyrotechnics that upon combustion yield an aerosol composed from tiny water droplets with the hydrophobic organic dye particles dispersed in but mainly sitting on the surface of those droplets. The formulations consist of a dye capable to undergo sublimation (40–50 wt.-%) and a so-called heating bed composition. The latter consists of a mixture from potassium chlorate, $KClO_3$, and lactose monohydrate. Lactose monohydrate, $C_{12}H_{22}O_{11} \cdot H_2O$, is a fuel with a large negative heat of formation ($\Delta_f H$ = –2723 kJ mol^{-1}).

The stoichiometric reaction of lactose with $KClO_3$ requires 8 mol $KClO_3$ per mol lactose monohydrate:

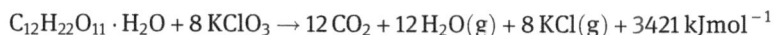

$$C_{12}H_{22}O_{11} \cdot H_2O + 8\,KClO_3 \rightarrow 12\,CO_2 + 12\,H_2O(g) + 8\,KCl(g) + 3421\,kJmol^{-1}$$

However, the combustion of this balanced mixture (ξ(Lactose) = 26 wt.-%) yields temperatures of up to T ~ 2000 °C which would incinerate any organic compound getting in close vicinity. It is hence that heating bed compositions contain a large amount of surplus fuel to lower the calculated combustion temperature down to ~900 °C. While this binary material alone still burns too hot, when it is mixed with the dye in a 1:1 ratio yields manageable combustion temperatures in the 300–500 °C ballpark depending on the particular dye and overall stoichiometry. Often those formulations are further modified up to 20 wt.-% of endothermic components (e.g. ammonium oxalate, kaolin, magnesium carbonate, and sodium hydrogen carbonate) to absorb heat and further release gases and water vapor adding to the overall performance.

The requirements for smoke dyes are as follows:
– The dye must be stable in the temperature range up to minimum 300 °C against decomposition and oxidation. Therefore, it should not bear any oxygen rich substituents like peroxy, nitro or sulfonate groups but rather substituents like amino and/or hydroxyl groups.
– The dye should have a boiling or sublimation point between T = 100–300 °C and possess a low latent heat (heat of fusion and heat of vaporization). Intermolecular hydrogen bonding or ionic bonding is hence not desirable. In addition, the molecular weight of the fuel should be below m_r = 400 g mol^{-1}.

- The dye, its technical congeners and impurities, and its potential oxidation and pyrolysis products should be safe from both occupational health point of view and in regard to the environment.

Figure C.5 shows structures, trade names, and color aspect of typical smoke dyes which are acceptable with regard to occupational health, Table C.7 displays their properties.

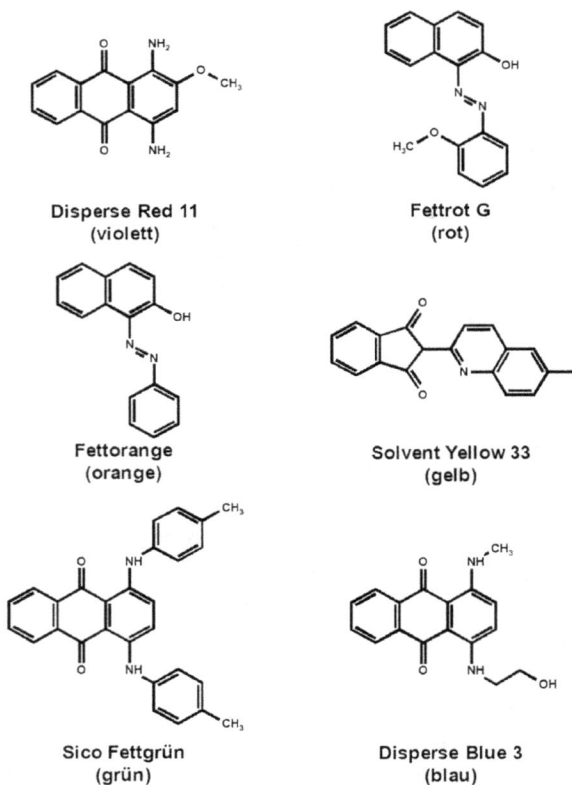

Disperse Red 11
(violett)

Fettrot G
(rot)

Fettorange
(orange)

Solvent Yellow 33
(gelb)

Sico Fettgrün
(grün)

Disperse Blue 3
(blau)

Figure C.5: Currently used colored smoke dyes.

- 1,4-Diamino-2-methoxy-anthraquinone = Disperse Red 11
- 1-(2-Methoxyphenylazo)-2-naphthol = Fettrot G, Solvent Red 1
- 1-Hydroxy-1-phenylazonaphthalene = Fettorange
- 6-Methyl-2-phthalochinolin = Solvent Yellow 33
- 1,4-Dip-*p*-toluidino-anthraquinone = Sico Fettgrün, Solvent Green 3
- 1-Methylamino-4-hydroxyethylanthraquinone = Disperse Blue 11

Table C.7: Properties of smoke dyes.

Name	Formula	CAS	$\Delta_x H$ (kJ mol^{-1})	Mp; *Sbl* (°C)
1,4-Diamino-2-methoxy-anthraquinone	$C_{15}H_{12}N_2O_3$	2872-48-2	35.29 (sm)	242;
1-(2-Methoxyphenylazo)-2-naphthole	$C_{17}H_{14}N_2O_2$	1229-55-6	142.4 (sbl)	183; *226*
1-Hydroxy-1-phenylazonaphthalene	$C_{16}H_{12}N_2O$	842-07-9	116.7 (sbl)	134; *197*
6-Methyl-2-phthaloquinoline	$C_{19}H_{13}NO_2$	6493-58-9		236; *243*
1,4-Di-*p*-toluidino-anthraquinone	$C_{28}H_{28}N_2O_2$	80094-92-4		214; *272*
1-Methylamino-4-hydroxyethylanthraquinone	$C_{17}H_{16}N_2O_3$	2475-46-9		187

Two typical smoke formulations are depicted in Table C.8.

Table C.8: Typical smoke formulations.

	Orange	Violet
Potassium chlorate (wt.-%)	25	23.5
Saccharose (wt.-%)	25	15.5
Fettorange (wt.-%)	48	
Disperse Red (wt.-%)		38.0
Terephthalic acid (wt.-%)		7.6
Sodium bicarbonate (wt.-%)		5.1
Magnesium carbonate (wt.-%)		10.2
Kaolin (wt.-%)	2	
Polyvinylalcohol (wt.-%)		2.0

Current research in the field of colored smokes addresses replacement of objectionable dyes and potassium chlorate. For quite some time, $KClO_3$ is under suspicion to act as a source of chlorine under the particular combustion conditions:

$$KClO_3 + \text{Aromatic hydrocarbon}, \Lambda \ll 0 \xrightarrow{\;300-600°C, \text{anaerobic}\;}$$

Chlorinated aromatic compounds

Hence the *de-novo* formation of the infamous polychlorinated biphenyls, dibenzo-dioxins (PCDD) and dibenzofuran (PCDF) is considered feasible. Recent research by *Springer et al.* with the violet smoke material (Table C.8) confirm the formation of 100 pg TEQ g^{-1} upon combustion of the smoke composition. (TEQ = toxicity equivalent of all PCDD and PCDF congeners). To relate this quantity to common figures the daily uptake by a grown-up human of all PCDDs and PCDFs by regular food without causing any detrimental health effects is in the 30–120 pg TEQ ballpark.

Both *Reed* (1982) and, more recently, *Klapötke* (2020) have demonstrated the potential of high nitrogen compounds to replace KClO₃/lactose in colored smoke formulations, reducing the risk for *de-novo* formation of PCBs, PCDBs, and PCDFs.

– G. Krien, Thermoanalytische Untersuchungen von Farbstoffen für pyrotechnische RauchPhrases, *Az.3.0-3/3715/75*, BICT, Swisttal-Heimerzheim, **1975**.
– C. G. Choufhry, O. Hutzinger, *Mechanistic Aspects of the Thermal Formation of Halogenated Organic Compounds including Polychlorinated Dibenzo-p-Dioxins*, Gordon & Breach, New York, **1983**, 194 pp.
– H. Hagenmaier, H. Brunner, R. Haag, M. Kraft, Die Bedeutung katalytischer Effekte bei der Bildung und Zerstörung von polychlorierten Dibenzodioxinen und polychlorierten Dibenzofuranen, *VDI-Berichte Nr. 634*, VDI-Verlag, **1987**, 557–584.
– G. Krien, Thermoanalytische Untersuchungen an Rauchfarbstoffen, *Thermochimica Acta* **1984**, *81*, 29–43.
– A. L. Brooks, F. A. Seiler, R. L. Hanson, R. F. Henderson, In Vitro Genotoxicity of Dyes Present in Colored Smoke Munitions, *Environ. Mol. Mutagen.* **1989**, *13*, 304–313.
– M. L. Springer, T. Rush, H. M. Beardsley, K. Watts, J. Bergmann, Demonstration of the Replacement of the Dyes and Sulfur on the M18 Red and Violet Smoke Grenades, *ESTCP Project WP-0122*, US ARMY Environmental Center, September **2008**.
– T. M. Klapötke et al., Guanidinium 5,5'-Azotetrazolate: A Colorful Chameleon for Halogen-Free Smoke Signals, *Angew. Chem. Int. Ed.* **2020**, *59*, 12326–12330.
– R. Reed Jr., M. L. Chan, *High Nitrogen Smoke Compositions*, US-Patent 5061329, USA, **1991**, filed **1982**.

Combustion, burn rate
Abbrand, Abbrandgeschwindigkeit

The burning/combustion of an explosive is the subsonic propagation of a luminous/incandescent exothermic reaction front supported by radioactive heat transfer, heat conduction, heat transport, and mass transport in case of porous media. The burn front of an energetic pellet (e.g. solid rocket grain, powder grain, illuminant candle, and obscurant wedge.) propagates perpendicular to the surface area in parallel layers of thickness x(mm) per unit time (*Piobert's law*). The linear burn rate

$$u = \frac{dx}{dt} \cdot \left[mm\, s^{-1} \right]$$

at ambient pressure, p [MPa], and temperature, T [K], is described by Vieille's law

$$u = a \cdot p^n$$

or differential terminology:

$$dx = ap^n dt.$$

a is the temperature sensitivity of the combustion, p the ambient pressure, and n the pressure exponent.

The influence of energy content, q_i, heat flow, Q, and heat of radiation, I, *on the burn rate in accordance with Eisenreich et al.* can be determined at different ambient temperatures T_∞

$$u = \frac{Q+I}{\rho \cdot \left(c_p \cdot [T_s - T_\infty] + L - \sum_i q_i \right)}$$

where L represents the latent heat, q_i the heat of reaction, c_p the specific heat, ρ the density, and T_s the temperature of the reaction zone.

On the surface of a non-inhibited grain, the combustion proceeds with the velocity u′ which is different to u (in which u′ > u). At low and atmospheric pressure, both velocities can be observed. A cylindrical propellant grain ignited at the top after some time shows stationary combustion with a cone shape burn front and the cone angle 2α:

$$\frac{u}{u'} = \sin\alpha$$

Solid grains of cuboid geometry burn by maintaining their rectangular geometry. However, bulk grains with square channels in it upon combustion quickly show a smoothened and finally circular-shaped burning surface (Figure C.6).

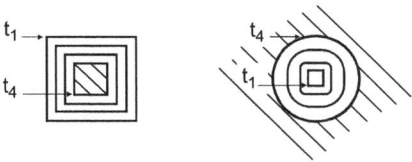

Figure C.6: Development of burning surface area of initially cubcid external (left) and internal (right) burning grains.

Upon ignition of an energetic material depending on the substrate a particular time passes until a stationary burn rate u = u_{stat} (mm s^{-1}) is obtained. The burn rate u(t) as function of time can be approached with the following expression containing the empirical constant c:

$$u(t) = u_{stat}\left(1 - e^{-ct^2}\right)$$

If m_0 is the starting mass of an energetic grain and m_t the mass at some time later t, then

$$\mu(t) = 1 - \frac{m_t}{m_0}$$

describes the burnt fraction, $\mu(t)$ at time t. The burnt fraction can also be written as

$$dm = m_0 d\mu$$

If we use S [m^2] to designate the surface area of the grain and ρ [kg m^{-3}] its density then one obtains:

$$dm = S\rho dx$$

$$m_0 d\mu = S\rho dx$$

$$d\mu = \frac{S\rho}{m_0} dx$$

With the Vieille's law $dx = ap^n dt$,
   one finally obtains

$$\frac{d\mu}{dt} = \frac{S\rho a}{m_0} p^n$$

Expanding the right side with the starting surface area S_0 yields

$$\frac{d\mu}{dt} = \frac{S}{S_0} \frac{S_0 \rho a}{m_0} p^n$$

If we use $A = \dfrac{S_0 \rho a}{m_0}$, A is called vivacity or combustion coefficient.

Altering combustion rate

The subsonic combustion rate of an energetic material can be altered by various mechanisms as is indicated below with given examples for various propellant and pyrolant systems:
-   catalysts for the decomposition of an oxidizer
    -   Fe_2O_3 for AP decomposition
-   catalysts for the decomposition of a fuel
    -   transition metal phthalocyanines on HTPB decomposition
-   catalysts for the oxidation in the condensed phase
    -   bismuth salts and carbon particles
-   catalysts for the combustion in the gas phase
    -   water gas catalysts
        -   low-temperature shift catalysts ($CuO/ZnO/Al_2O_3$)
        -   high-temperature shift catalysts ($Fe_2O_3/Cr_2O_3/MgO$)
-   increase in combustion surface temperature (*Isert, Hahma, Mueller*)
    -   High decomposition temperature of either fuels or oxidizers
        -   Mg versus Mg_2Si
        -   Polytetrafluoroethylene versus graphite fluoride
-   increase surface area and porosity to favor filtering combustion
    -   Embed open porosity combustible or inert structures
        -   Expanded graphite, microballoons

- increase thermal diffusivity
  - Embed carbon fibers, carbon nanotubes, graphene, nanodiamonds.

- H. Mache, *Die Physik der Verbrennungserscheinungen*, Walter deGruyter, **1918**, 133 pp.
- T. Vahlen, *Ballistik*, 2. Aufl., Walter deGruyter, **1942**, 267 pp.
- R. E. Kutterer, *Ballistik*, 3. Aufl. Vieweg, **1959**, 304 pp.
- N. Eisenreich, W. Eckl, T. Fischer, V. Weiser, S. Kelzenberg, G. Langer, A. Baier, Burning Phenomena of the Gun Propellant JA2, *Propellants Explos. Pyrotech.* **2000**, *25*, 143–148.
- A. Hahma, Thermomechanical Combustion Enhancer and The Effect of Combustion Catalysis on the Burn Rate and Radiation Output of Magnesium-Teflon-Viton Composition, *Int. J. Ener. Mater. and Chem. Propulsion*, **2020**, DOI: 10.1615/IntJEnergeticMaterialsChemProp.2020033664.
- G. T. Muller, A. Gany, Burning Phenomena of a Polymeric Fuel Containing Expandable Graphite, *Propellants Explos. Pyrotech.* **2020**, https://doi.org/10.1002/prep.202000138.
- S. Isert, C. D. Lane, I. E. Gunduz, S. F. Son, Tailored Burning Rates Using reactive Wires in Composite Solid Rocket Propellants, *Proc. Combust. Inst.* **2017**, 36, 2283–2290.

Commercial explosives
Gewerbliche Sprengstoffe

In the civilian field, HEs are used for the exploration of mineral deposits (seismic investigations), mining operations, for road works and tunneling, demolition activities, metal forming and welding (see *welding and cladding*), and for the synthesis of special materials (*Nanodiamonds*).

Civilian that are commercially utilized HEs are usually considered for immediate use in a single climatic zone. This is why commercial explosives may not need to meet stringent military requirements regarding service life and applicable climatic zones. In addition, (with few exceptions) those explosives often have lower density and consequently lower performance (detonation pressure and metal acceleration capability) than military used ones.

Top commercial explosives are often less expensive in production as they are based on basic industrial raw explosives such as ammonium nitrate and petroleum. It is hence that those formulations due to their low sensitiveness also possess an advantageous HD classification such as HD 1.6 or HD 5.1 and consequently ease up logistic effort too.

Generally, commercial HEs are divided into
- powdered explosives
  - free flowing
  - in cartridge form
- gelled explosives

and

– water-based explosives
  – pumpable

Free-flowing explosives are, for example, the ANFO explosives (see ANFO). The free-flowing explosives in cartridge form comprise the permitted explosives. As both methane and coal dust in a mine can form highly explosive mixtures, initiation must be avoided at all cost. Therefore, the permitted explosives yield lower detonation plume temperatures. They also contain endothermic additive such as NaCl which has a high heat of sublimation ($\Delta_{sub}H$: 230 kJ mol^{-1}) which aids in cooling the plume. Fine dispersed NaCl is also prepared in situ by a salt-pair reaction of the type shown later occurring with the detonation

$$NaNO_3 + NH_4Cl \longrightarrow NaCl + N_2 + 2H_2O + 0,5O_2$$

Important gelled explosives are is blasting gelatin.

The water-based HEs comprise the blasting slurries containing about:

55 wt.-% ammonium nitrate

15 wt.-% water

30 wt.-% nitroaromatics (e.g. TNT)

A further development of slurries are the emulsion explosives. These are highly concentrated aqueous solutions of ammonium nitrate which are mixed on site with petroleum, emulsifiers, and additives to sensitize the final mix (e.g. microballoons). The logistic advantage of the latter is that there is no item bearing HD. 1.

A typical composition reads:

76 wt.-% ammonium nitrate

17 wt.-% water

7 wt.-% oil, emulsifiers, and sensitizers

– A. Neubauer, H. Steinicke, Praxisrelevante Beispiele zum Explosivumformen in der industriellen Fertigung – ein Überblick, *Sprenginfo*, **2016**, *38*, 26–30.
– J. N. Johnson, C. L. Mader, S. Goldstein, Performance Properties of Commercial Explosives, *Propellants Explos. Pyrotech.* **1983**, *8*, 8–18.
– R. Meyer, Die Entwicklung der Wettersprengstoffe, *Nobel Hefte*, **1969**, *35*, 109–116.

Composition A

Composition A is a term used for pressable composition based on RDX and solid or pasty hydrocarbons. While older formulations (A and A2) even contained bees-wax, this is today obsolete mainly due to its low softening point. Composition A containing additional aluminum is known as *Hexal* (→).

Table C.9: Components of various composition A.

	A3	A3 Typ 11	A4	A5	A5 Typ 11	A6
RDX (wt.-%)	91.0	90.8	97.0	98.5	98.0	86
Bees-wax (wt.-%)	9.0	–	3.0	–	–	14
Stearic acid (wt.-%)	–	–	–	1.5	1.6	–
Polyethylene (wt.-%)	–	9.2	–	–	–	–
ρ (g cm^{-3})	1.65					

Table C.10: Properties and sensitiveness of composition A3.

	A3		
ρ (g cm^{-3})	1.62	1.65	1.67
V_D (m s^{-1})	8200	8300	8520
P_{CJ} (GPa)	25.7	30.0	
Trauzl (cm^3)	432		
$\Delta_{ex}H$ (kJ g^{-1})	5062		
Impact (J)	18.8		
$\varnothing_{cr}$ (mm)	<2.2		

Composition B

Composition B is a melt-castable formulation based on RDX and TNT in a mass ratio of about 60/40. Compositions bearing higher mass fractions of RDX are designated cyclotol. However, above 65 wt.-% RDX, the viscosity of the mix impedes proper casting. The experimental densities of the higher cyclotols are hence significantly lower than the TMD. Composition B modified with additional aluminum is termed *hexotonal* ($\rightarrow$) (e.g. S17, SSM887, HBX, Torpex, and Trialen) and is used as an underwater explosive.

Table C.11: Compositions of composition B.

	B	B-2, B-3	B-4		Cyclotol	
SSM	TR 8510		TR 8520		TR 8560	
RDX (wt.-%)	60	60	60	65	70	75
TNT (wt.-%)	40	40	39,5	35	30	25
Wax (wt.-%)	+1	–	–	–	–	–
Calcium silicate (wt.-%)	–	–	0.5	–	–	–
ρ (g cm^{-3})	1.68	1.68		1.71	1.71	1.71

Table C.12: Properties of composition B.

B		B-2, B-3	Cyclotol			
			65/35	70/30	75/25	77/23
ρ (g cm^{-3})	1.72	1.72	1.72	1.73	1.70	1.754
V_D (m s^{-1})	7920	7900	7975	8060	8035	8250
P_{CJ} (GPa)	29.5	29.5				31,6
Trauzl (cm^3)	390					
$\Delta_{ex}H$ (kJ g^{-1})		5.00	5.04	5.08	5.13	
$\varnothing_{cr}$ (mm)	4.3 (at 1.71 g cm^{-3})				8.1	6
19 mm E (MJ kg^{-1})	1.33					

Composition C

Composition C explosives are based on RDX and various plastic polymer binders.

Table C.13: Composition and properties of various composition C.

	C	C2	C3	C4	C5
RDX (wt.-%)	88.3	78.7	77	91	91
Inert binder (wt.-%)	11.7				
Energetic binder (wt.-%)		21.3*	23$^{#}$		
Polyisobutylene (wt.-%)				2.1	
Di(2-ethylhexyl sebacate) (wt.-%)				5.3	
Zinc stearate (wt.-%)					1
Motor oil (wt.-%)				1.6	
Polydimethylsiloxane (wt.-%)					8
ρ (g cm^{-3})			1.59	1.59	
V_D (m s^{-1})				8040	8000
Impact (J)				20	
Friction (N)					160

*DNT /TNT /MNT/NC /(60/25/13.5/1.5 wt.-%), $^{#}$DNT /MNT/TNT /CE /NC (43.6/21.7/17.4/13.0/4.3).

Original composition C4 (often referred to as C4 only) has unsatisfactory properties such as embrittlement at low temperatures and bleeding out of components at high temperatures. Hence, C4 has been replaced in many armed forces with formulations containing silicone oils (polydimethylsiloxane) (*Lingens*) (composition C5).

- P. Lingens, *Plastische von Hand leicht verformbare Sprengmasse*, DE2027709B2, **1978**, Germany.
- C. Collet, *An international Review on Green Insensitive Plastic Explosives*, O-217, NATO-Munitions Safety Information Analysis Center (MSIAC), Brussels, Belgium, April **2020**, 23 pp. Accessible at: https://www.msiac.nato.int/products-services/publications/o-217-an-international-review-on-green-insensitive-plastic-explosives.

Compositions I & II

Compositions I and II are eutectic mixtures of nitrates with dicyandiamide melting at about 100 °C. Comp II can be modified with up to 30 wt.-% tetryl.

Table C.14: Composition of compositions I and II.

	I	II
Ammonium nitrate (wt.-%)	65.5	60.0
Sodium nitrate (wt.-%)	10.0	24.0
Calcium nitrate (wt.-%)	14.5	–
Guanidinium nitrate (wt.-%)	–	8.0
Dicyandiamide (wt.-%)	10.0	8.0
ρ (g cm^{-3})	1.808	1.765

Conventional Weapons Effects Calculations

The Conventional Weapons Effects Calculations (CONWEP)-program package allows to estimate weapons effects in accordance with handbook *TM-5-855-1 Fundamentals of Protective Design for Conventional Weapons*. CONWEP is available only for users authorized by the US-DOD.

– conwep.erdc.usace.army.mil

Cook-off
Thermische Selbstzündung

The term "cook-off" generally describes the unintentional deflagration of an energetic material upon heating. Typical examples include the premature deflagration of a gun propellant cartridge in a barrel overheated through excessive firing or the deflagration of an explosive containing article by heating through direct or nearby fire. The cook-off testing in the context of insensitive munitions testing is looking primarily at accidental ignition through direct (fast cook-off) and nearby fires (slow cook-off). Accordingly, test procedures are defined in NATO AOP 39. Depending on the explosive and the particular system holding it cook-off can either yield a burn/deflagration of a detonation (DDT).

– *NATO Standard AOP-4240 Fast Heating Munition Procedures*, NATO Standardization Office, Brussels, November **2018**, 46 pp.
– *STANAG 4382 PPS (ed. 2) Slow Heating, Munitions Test Procedures*, NATO Standardization Office, Brussels, April **2003**, 9 pp.
– B. W. Asay, Cookoff, in B. W. Asay (Eds.) Shock Wave Science and Technology Reference Library, Volume 5, Non-shock Initiation of Explosives, Springer, **2010**, pp. 403–482.

Copper
Kupfer

Aspect		Copper red
Formula		Cu
GHS		02, 07
H-phrases		228-318-355
P-phrases		210-241-261-305+351-338-405-501
UN		3089
EINECS		231-159-6
CAS		[7440-50-8]
m_r	g mol^{-1}	63.546
ρ	g cm^{-3}	8.960
Mp	°C	1085
Bp	°C	2570
c_p	J g^{-1} K^{-1}	0.385
$\Delta_m H$	kJ mol^{-1}	13.14
$\Delta_v H$	kJ mol^{-1}	300.7
κ	W m^{-1} K^{-1}	400
c_L	m s^{-1}	4760
Λ	wt.-%	0
$\Delta_c H$	kJ mol^{-1}	−157.318
	kJ g^{-1}	−2.476
	kJ cm^{-3}	−22.182

Copper powder is occasionally used as a color imparter and fuel in blue burning flares as well as blue and "black" strobe compositions.

- C. Jennings-White, Blue Strobe Light Pyrotechnic Compositions, *Pyrotechnica,* **1992**, XIV, 33–45.

Copper ammonium nitrate
Tetrakis(ammin)kupfer(II)dinitrat

Tetraamminecopper dinitrate		
Aspect		Light blue crystals
Formula		$CuH_{12}N_6O_6$
CAS		[31058-64-7]
m_r	g mol^{-1}	255.678
ρ	g cm^{-3}	1.91
$\Delta_f H$	kJ mol^{-1}	−812.95
$\Delta_{ex} H$	kJ mol^{-1}	−3816
	kJ g^{-1}	−14.925
	kJ cm^{-3}	−28.507
Λ	wt.-%	0
N	wt.-%	32.87
Mp	°C	160 (dec)

Copper ammonium nitrate starts to decompose at 135 °C with the release of ammonia to give bis(ammin)copper(II) dinitrate [184850-98-4], ρ = 2.290 g cm^{-3}, Δ = +14.44 wt.-%, $\Delta_f H$ = −586 kJ mol^{-1}). Copper ammonium nitrate serves as oxidizer in igniter formulations and gas generators.

Copper(I) oxide
Kupfer(I)oxid

Aspect		Red powder
Formula		Cu_2O
CAS		[1317-39-1]
GHS		09, 07
H-phrases		400-410-302
P-phrases		264-270-273-301+312-330-501a
UN		3077
EINECS		215-270-7
m_r	g mol^{-1}	143.091
ρ	g cm^{-3}	6.0
$\Delta_f H$	kJ mol^{-1}	−170.7
Λ	wt.-%	+11.18
Mp	°C	1244
Bp	°C	1800
$\Delta_m H$	kJ mol^{-1}	−64.8
c_p	J K^{-1} mol^{-1}	69.9

Cu_2O serves as a burn rate catalyst and an oxidizer in pyrotechnics.

Copper(II) oxide
Kupfer(II)oxid

Aspect		Black fine powder
Formula		CuO
CAS		[1317-38-0]
GHS		09, 07
H-phrases		400-410-302
P-phrases		264-270-273-301+312-330-501a
UN		3077
EINECS		215-269-1
m_r	g mol^{-1}	79.545
ρ	g cm^{-3}	6.49
$\Delta_f H$	kJ mol^{-1}	−156.1

Λ	wt.-%	+20.11
Mp	°C	1326
$\Delta_{sub}H$	kJ mol^{-1}	−462
c_p	J K^{-1} mol^{-1}	42.3

CuO serves as a burn rate catalyst and an oxidizer in pyrotechnics.

Copper-5-nitrotetrazolate, DBX-1

Kupfer-5-nitrotetrazolate

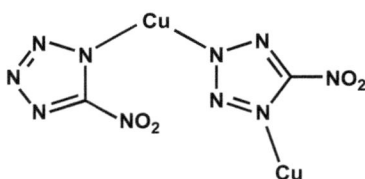

Aspect		Red bead-like crystals
Formula		$C_2Cu_2N_{10}O_4$
CAS		[957133-97-0]
m_r	g mol^{-1}	355.20
ρ	g cm^{-3}	1.955
$\Delta_f H$	kJ mol^{-1}	99.78
Λ	wt.-%	−9.00
N	wt.-%	39.43
Impact	mJ	40
Friction	g	10/0
T_{5ex}	°C	350

Copper-5-nitrotetrazolate was first prepared and patented as a primary explosive by German chemist *Dr. Ing. Edmund Ritter von Herz* (1891–1964) in 1932. Compared to lead azide, copper-5-nitrotetrazolate is better compatible with a great number of components but it is also very prone to hydrolysis.

- S. Tappan, J. Patrick Ball, Jill C. Miller, DBX-1 (copper(I)-5-nitrotetrazolate) reactions at sub-millimeter diameters, *International Pyrotechnics Seminar*, Valencia, Spain, **2013**.
- J. W. Fronabarger, M. D. Williams, W. B. Sanborn, J. G. Bragg, D. A Parrish, M. Bichay, DBX -1 – A lead Free Replacement for Lead Azide, *Propellants Explos. Pyrotech.* **2011**, *36*, 541–550.
- E. Ritter von Herz, *Verfahren zur Herstellung von Nitrotetrazol*, DE562511, **1932**, Germany.

Cordite
Kordit

Cordite is a term designating UK legacy gun propellants.

Table C.15: Various cordite types.

Name	Nitrocellulose (wt.-%)	Nitroglycerin (wt.-%)	Centralite I (wt.-%)	Dibutyl phthalate (wt.-%)
F4/3*	55	22	3.5	19.5
CSC	50	35	9	6
SC	49.5	41.5	9	0
HSC	49.5	47	3.5	0

*+1 wt.-% $K_3[AlF_6]$.

Coruscatives
Koruskativstoffe

The term coruscatives (derived froim Latin: *cruscare = flash, gleam*) was coined by Swiss born engineer and inventor *Fritz Zwicky (1898–1974)* for those combinations of materials that react with very little permanent and transient gas production. *Zwicky* primarily considered binary and ternary systems forming alloys such as those depicted later in Table C.16. He speculated that liners made of shaped charges could bring increased penetration. However, experimental work by *Held* showed that neither increased jet tip velocities nor increased slug velocities are observed with coruscatives as the only liner material or when coruscative liners are placed between HE and common liner from electrolytic copper.

Table C.16: Ignition and combustion temperature of some coruscatives.

System/alloy	Stoichiometry (wt.-%)	Gas (mL g^{-1})	Heat of reaction (kJ g^{-1})	T_i (°C)	T_{flame} (°C)
MgTe	16:84	7	1.420	544	n.d.
Mg_2Si	63:37	–	0.514	560	1040
MgS	37:63	36	4.150	227	580
Mg_2Sn	29:71	3	0.225	450	770
Mg_2Pb	19:81	2	0.125	485	790
Mg_3Sb_2	23:77	5	0.765	438	n.d.

Table C.16 (continued)

System/alloy	Stoichiometry (wt.-%)	Gas (mL g^{-1})	Heat of reaction (kJ g^{-1})	T_i (°C)	T_{flame} (°C)
Mg_3P_2	54:46	10	3.340	414	1090
ZnS	67:33	24	1.890	222	945
AlSb	18:82	3	0.552	555	960
AlP	46:54	20	3.500	438	954
Al_2Te_3	14:86	8	1.338	532	1160
TiTe	27:73	8	0.815	433	870
TiSbPb	48:23:29	–	1.045	570	1010

- F. Zwicky, *Coruscative Ballistic Device*, US3135205, USA, **1964**.
- A. Reichel, Hochenergetische Gaslose Reaktionen, *14. Arbeitstagung – Wehrtechnik*, Mannheim, Mannheim, **1968**.
- M. Held, Untersuchungen an Koruskativstoffen, *ICT -JATA*, Karlsruhe, **1985**, V-40.

Crackling compositions and microstars
Cracklingsätze, Mikrosterne

Crackling compositions belong to the group of strobe compositions. With these compositions, the light emission is always associated with a concomitant report. Crackling compositions are most typically manufactured into small so-called micro-stars of 2–4 mm diameter. Typical formulations and their sensitivities are given in Table C.17:

Table C.17: Composition and sensitivity of crackling compositions.

	1	2
Lead(II,IV) oxide, $2PbO \cdot PbO_2$ (wt.-%)	89	–
Bismuth(III) oxide, Bi_2O_3 (wt.-%)	–	75
Magnalium, MgAl (wt.-%)	11	15
Copper(II) oxide, CuO (wt.-%)	–	10
Nitrocellulose (wt.-%)	2	2
Frequency (Hz)	1.25	2.5
Sound pressure (dB)	134	132
Friction (NoGo)(N)	160	320
Impact (NoGo) (J)	3	8

- C. J. White, Lead-Free Crackling Microstars, *Pyrotechnica*, **1992**, *14*, 30–32.
- T. Shimizu, Studies on Mixtures of Lead Oxides with Metals (Magnalium, Aluminum or Magnesium), *Pyrotechnica*, **1990**, *13*, 10–18; Plate I-III.

Cranz, Julius Carl (1858–1945)

Carl Cranz was a German mathematician, physicist, and an internationally renowned ballistics expert. After studies of philosophy and theology, he studied mathematics and physics at Tübingen and Berlin with German mathematician *Paul Du Bois-Reymomd* (1831–1889) and completed his dissertation on a new theoretical model to explain the deviation of gun projectiles in 1883. Between 1890 and 1936 he developed a 4-volume monography on ballistics. His most important scientific disciples were *Hubert Schardin* and *Richard Emil Kutterer*. Together with Schardin, he developed a high-speed camera (*Cranz-Schardin camera*) to record 24 frames at a rate of 5 MHz on a static film.

- Carl Cranz Ballistik, Lehrbuch in vier Bänden (1922–1936), Springer Verlag, Berlin.
- Bd. I Außenballistik, 1923
- Bd. II Innere Ballistik, 1926
- Bd. III Experimentelle Ballistik, 1927
- Bd. IV Ergänzungen, 1936
- C. Cranz, H. Schardin, Kinematographie auf ruhendem Film und mit extrem hoher Bildfrequenz, *ZS für Physik*, **1929**, *56*, 147–183.
- Entry in the German National Library: http://d-nb.info/gnd/116717742

Crawford bomb
Crawford Bombe

The Crawford bomb is a standardized combustion chamber with an optical window designed to determine the interior ballistic properties of solid rocket propellants. In the Crawford bomb, the propellant is burnt under a very high pressure of nitrogen. Hence, the combustion related change of pressure can be neglected and the combustion conditions are considered isobaric.

- B. L. Crawford Jr., C. Hugget, F. Daniels, R. E. Wilfong, Direct Determination of Burning Rates of Propellant Powders, *Anal. Chem.* **1947**, *19*, 630–633.

Critical diameter
Kritischer Durchmesser

The critical diameter, $\varnothing_{cr}$ (mm), of a HE is the threshold diameter at which at given density, ρ (g cm^{-3}), temperature, T (K), and confinement – for example, in a metal tube or freestanding in air – an HE is still able to undergo a stationary detonation (i.e. a constant detonation velocity at different points of a charge of that diameters).

Confined explosives show smaller critical diameters. The higher the acoustic impedance, Z [kg m^{-2} s^{-1}], of confinement, the smaller the critical diameter

$$Z = \rho\ c_L$$

It is hence that sleeves made of gold (19.3 x 2.3) are used in detonation calorimetry
to effect near ideal conditions. The critical diameter further correlates with the
width of the chemical reaction zone (CRZ) that is the volume in which the actual
energy release behind the shock wave takes place. So, the wider the CRZ, the wider
is the critical diameter. In general, near-ideal HEs such as HMX or PETN have small
critical diameters whereas non-ideal HEs and those formulated with additional
components typically show larger critical diameters (Figure C.7/Table C.18).

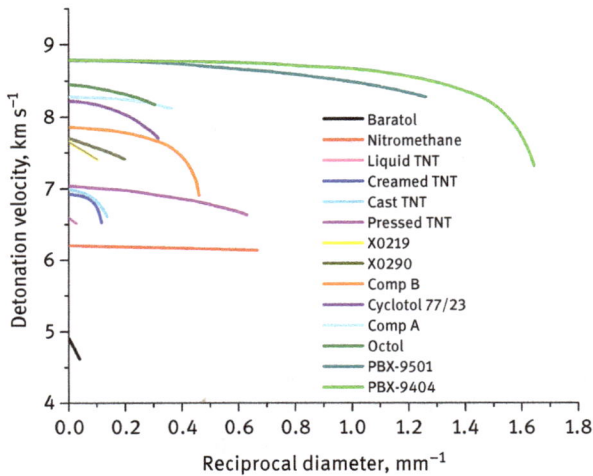

Figure C.7: Critical diameter and detonation velocity for selected high explosives.

Table C.18: Critical diameter for selected high explosives.

High explosive	Density (g cm^{-3})	Critical diameter (mm)	$V_D\varnothing_{cr}$ (m s^{-1})	$V_D\varnothing_{\infty}$ (m s^{-1})
Baratol (76)	2.619	43.2	4607	4874
Nitromethane	1.128	1.42	6129	6213
Liquid TNT	1.443	36.4	6507	6574
Creamed TNT	1.615	8.5	6508	6942
Cast TNT	1.62	7.3	6594	6999
Pressed TNT	1.62	1.6	6628	7045

Table C.18 (continued)

High explosive	Density (g cm^{-3})	Critical diameter (mm)	$V_D\emptyset_{cr}$ (m s^{-1})	$V_D\emptyset_{\infty}$ (m s^{-1})
X-0219	1.915	10	7420	7627
X-0290	1.895	5.1	7420	7706
Composition B	1.70	2.2	7859	6930
Cyclotol 77/23	1.74	3.1	7712	8210
Composition A	1.687	2.8	8108	8274
Octol	1.814	3.3	8167	8481
PBX-9501	1.832	0.8	8271	8802
PBX-9404	1.846	0.6	7308	8776

The critical diameter is typically determined with elongated conical samples (half angle $\angle = 5°$) or stepped cylinders. With conical samples there is always the tendency to overdrive the detonation and thus to obtain too small diameters. Hence, using stepped cylinder with a l/d ratio of ≥4 avoids any overdriven detonation.

- P. W. Cooper, *Explosives Engineering*, Wiley-VCH, **1996**, pp. 277, 284–293.
- H. Badners, C. O. Leiber, Method for the Determination of the Critical Diameter of High Velocity Detonation by Conical Geometry, *Propellants Explos. Pyrotech.* **1992**, *17*, 77–81.
- A. W. Campbell, The diameter effect in high-density heterogeneous explosives, *6th Symposium on Detonation*, White Oak, Auhust 24–27 **1976**, 642–652.

Cryolithe

Kryolith

Sodium hexafluoroaluminate		
Formula		Na_3AlF_6
GHS		08, 07, 09
H-phrases		332-372-411
P-phrases		260
WGK		1
UN		3260
REACH		LPRS
EINECS		237-410-6
CAS		[15096-52-3]
m_r	g mol^{-1}	209.941
ρ	g cm^{-3}	2.97
Tp1	°C	572

Tp2	°C	880
Mp	°C	1012
$\Delta_f H$	kJ mol^{-1}	−3310
Λ	wt.-%	0

Cryolithe was once an important non-hygroscopic flame color agent for yellow fire-work stars. It is also occasionally used to reduce the muzzle flash of gun propel-lants. *Hahma* could show that aluminum powder coated with cryolithe shows improved ignition and combustion over pristine aluminum which is due to eutectic (mp 960 °C) of forming from 10.5 wt.-% Al_2O_3 with cryolithe.

- A. Hahma, *Method of improving the burn rate and ignitability of aluminum fuel particles and aluminum fuel so modified* US7785430, **2002**, Sweden.
- A. Hahma, A. Gany, K. Palovuori, Combustion of activated aluminum, *Combust. Flame* **2006**, *145*, 464–480.

CTPB, carboxy-terminated polybutadiene
Carboxylterminiertes Polybutadien

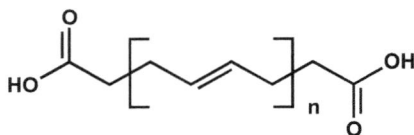

Formula		$C_{10}H_{14.958}O_{0.194}$
CAS		[68441-48-5]
m_r	g mol^{-1}	138.293
ρ	g cm^{-3}	0.916
$\Delta_f H$	kJ mol^{-1}	−10
$\Delta_c H$	kJ mol^{-1}	−6062.9
	kJ g^{-1}	−43.843
	kJ cm^{-3}	−40.160
Λ	wt.-%	−315.67
Mp	°C	280 (dec)

CTPB has molecular weight between 2300 and 3000. It was the binder of choice for rocket propellants in the 1950s. However, with the advent of HTPB by end of the 1960s, it has lost significance as the latter has far superior mechanical and ageing properties than CTPB.

- H. G. Ang, S. Pisharath, *Energetic Polymers*, Wiley-VCH, **2012**, pp. 4, 13.

Cyanuric triazide, 2,4,6-triazidotriazine
2,4,6-Triazido-1,3,5-triazin

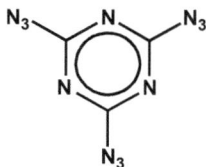

TAT		
Formula		C_3N_{12}
CAS		[5637-83-2]
m_r	g mol^{-1}	204.113
ρ	g cm^{-3}	1.54
Δ_fH	kJ mol^{-1}	916.30
$\Delta_{ex}H$	kJ mol^{-1}	−930.8
	kJ g^{-1}	−4.560
	kJ cm^{-3}	−7.023
Δ_cH	kJ mol^{-1}	−2096.6
	kJ g^{-1}	−10.272
	kJ cm^{-3}	−15.818
Λ	wt.-%	−47.03
N	wt.-%	82.35
Mp	°C	94
Friction	N	<10
Impact	J	10 at 2 kg BM machine
V_D	m s^{-1}	5500 at 1.02 g cm^{-3}
Trauzl	cm^3	415

TAT is a primary explosive. Its high volatility and low stability, however, impedes practical use. Its detonation in copper-lined containers yields fullerene.

- T. Utschig, M. Schwarz, G. Miehe, E. Kroke, Synthesis of carbon nanotubes by detonation of 2,4,6-triazido-1,3,5-triazine in the presence of transition metals, *Carbon* **2014**, *42*, 823–828.

Cyclosarin, GF

Methylfluorphosphon-säureisocyclohexylester

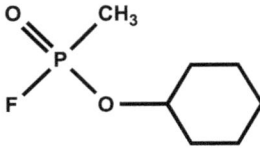

Cyclohexyl methylphosphonofluoridate		
Aspect		
Formula		$C_7H_{14}O_2PF$
CAS		[329-99-7]
m_r	g mol^{-1}	180.157
ρ	g cm^{-3}	1.133
Δ_fH	kJ mol^{-1}	−1200 (estimated)
Mp	°C	−30
Bp	°C	239
P_{vap}	Pa	5.87 @ 20 °C
$\Delta_{sub}H$	kJ mol^{-1}	90.7
Λ	wt.-%	−186.50
LCt_{50}	ppm	>10.000
ICt_{50}	ppm min^{-1}	80

Cyclotrimethylentrinitrosamine

1,3,5-Trisnitrosohexahydro-1,3,5-triazin, R-Salz

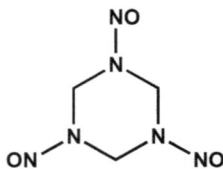

TMTA, TRDX, TNX		
Formula		$C_3H_6N_6O_3$
REACH		LPRS
EINECS		237-766-2
CAS		[13980-04-6]
m_r	g mol^{-1}	174.118
ρ	g cm^{-3}	1.588
Δ_fH	kJ mol^{-1}	285.85

$\Delta_{ex}H$	kJ mol^{-1}	−1183.5
	kJ g^{-1}	−6.797
	kJ cm^{-3}	−10.793
Δ_cH	kJ mol^{-1}	−2323.9
	kJ g^{-1}	−13.347
	kJ cm^{-3}	−21.195
Λ	wt.-%	−55.13
N	wt.-%	48.27
Mp	°C	105.6
Impact	J	125
V_D	m s^{-1}	7600 T ρ = 1.50 g cm^{-3} (with 2.5 wt.-% phenanthrene and 1 wt.-% **DPA)**
V_D	m s^{-1}	*7740 at ρ = 1.588 g cm^{-3}*
P_{CJ}	GPa	*25.83*
Trauzl	cm^3	370

Cyclotrimethylenetrinitrosamine is a very insensitive melt-castable HE. As its production does not require concentrated nitric acid, it was used as "*R-Salz*" as an *Ersatzsprengstoff* in wartime Germany. Eutectic mixtures with benzoid aromatic compounds and up to 75 wt.-% RDX melt considerably below 100 °C and thermal stability comparable to *pentolite* (PETN/TNT), however, are in contrary to pentolite insensitive toward bullet impact.

- S. A. Rothstein, P. Dube, S. R. Anderson, An Improved Process Towards Hexahydro-1,3,5-trinitroso-1,3,5-triazine (TNX), *Propellants Explos. Pyrotech.* **2017**, *42*, 126–130.
- *ARAED-TR-95014*, Picatinny Arsenal, October **1995**, 31 pp.

Cylinder test
Zylindertest

The cylinder test is a procedure to determine the detonation performance of a HE. The HE is confined in a copper cylinder (length = 12 inch, 30.48 cm) with an internal diameter 0.5 inch (12,7 mm) (r_i) and wall thickness 0.1 inch (2.54 mm). Upon detonation the copper wall is laterally accelerated. The wall velocity at an expansion ratio of the external diameter ($r_a - r_{a0}$ = 19 mm) is referred to as a measure of the working capacity, E_{19} [kJ cm^{-3}].

- M. Sućeska, *Test Methods for Explosives*, Springer, **1995**, 188–191.
- B. E. Fuchs, Picatinny Arsenal Cylinder Expansion Test and a Mathematical Examination of the Expanding Cylinder, *ARAED-TR-95014*, Picatinny Arsenal, October **1995**, 31 pp.

D

D2-wax
D2-Wachs

D2-wax is an energetic binder for high explosives. Its composition is
- 84 wt.-% Montan wax/ozokerit
- 14 wt.-% nitrocellulose
- 2 wt.-% lecithin

Dautriche method
Dautriche-Methode

The Dautriche method serves the determination of the detonation velocity of a high explosive using a calibrated detonation cord. Therefore, the explosive under investigation is prepared in a steel tube (500 length × 30 mm diameter) at the desired density. At one end of the tube, a blasting cap is inserted. In a distance of 300 mm, two other blasting caps crimped to a piece of calibrated detonation cord (V_{Dcord}) are inserted laterally into the tube as indicated (Figure D.1). After initiation of the terminal blasting cap, the passing shock wave consecutively initiates the first and second lateral blasting cap which trigger the detonation cord from both ends. The location where the wave fronts of the detonation cords collide leave a dent in the witness plate made of lead. The initiation delay of both blasting caps (which is based on the sought-after detonation velocity of the explosive) yields a deviation of that dent from the midpoint of the

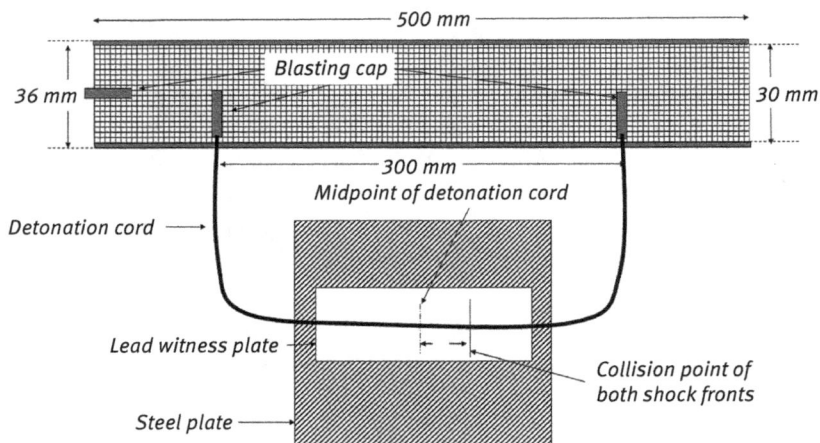

Figure D.1: Dautriche method.

https://doi.org/10.1515/9783110660562-004

detonation cord x (mm). Knowing the detonation velocity of the detonation cord, V_{Dcord}, allows to determine the detonation velocity of the explosive using the expression:

$$V_D = V_{Dcord} \cdot \frac{1}{2x}$$

The mean error is reported to be in the <4% ball park.

Dautriche, Henri-Joseph (1876–1915)

Dautriche was a French engineer who developed a method to determine the detonation velocity. He perished on February 16, 1915, in an accident involving chlorate-based explosives in Chedde, France.

DBX

DBX also designated hexamonal, were melt-castable high explosives for use in depth charges to fight submarines, hence the abbreviation **d**epth **b**omb explosive (Table D.1).

Table D.1: Composition and properties of DBX.

Components		Parameters	
Ammonium nitrate (wt.-%)	21	ρ (g cm^{-3})	1.76 (cast)
RDX (wt.-%)	21	V_D (m s^{-1})	6800
Trinitrotoluene (wt.-%)	40	c_p (J g^{-1} K^{-1})	1.323 (1.68)
Aluminum (wt.-%)	18	k (J cm^{-1} s^{-1} K^{-1})	5.53 10^{-3}
		Dp (°C)	176

DDT, deflagration-to-detonation transition

A laminar combustion process of an energetic material can experience pressure oscillations eventually resulting in a change of combustion regime to convective or pulsating combustion. Upon further pressure increase in the combustion zone, this process can lead to detonation. This process is summarized as DDT. Depending on the type of material and system, a low-velocity detonation (LVD) or high-velocity detonation (HVD) may result. Figure D.2 shows the qualitative phenomenology of DDT.

Fast DDT typically occurs with primary explosives and many oxygen-balanced high explosives such as nitroglycerine (NGl) or pentaerythritol tetranitrate (PETN).

In addition, DDT is observed in other energetic materials having intentional or unwanted (accidental) open porosity which favors convective burn. The increasing gas pressure at the burn front increases the burn rate resulting in a further pressure rise which can yield further mechanical damage to the energetic material (structural collapse) and eventually lead to LVD, which may transition into HVD depending on the boundary conditions (density, diameter, temperature, etc.).

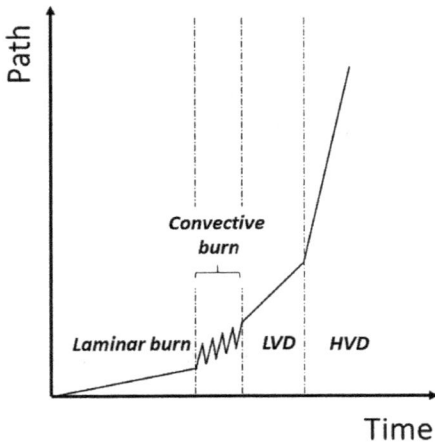

Figure D.2: Deflagration-to-detonation transition (DDT) after *Leiber*.

DDT is an important underlying phenomenon in the accidental/unintentional reaction of energetic materials and particularly ammunition.

Fast heating (fast cook-off) of confined explosive materials (e.g. in rocket motors, and warheads) yields a distinct temperature gradient affecting mechanical stress, which eventually leads to delamination of the energetic material from its container. This causes an increase in free surface area and therefore after ignition a rapid transition from laminar to convective burn may occur and DDT is immanent.

In general, the reverse phenomenon (transition from detonation-to-deflagration) may occur too particularly when the boundary conditions are in the ballpark of critical diameter/density.

- J. M. McAfee, *The Deflagration to-Detonation Transition* in W. Asay (Eds.) *Shock Wave Science and Technology Reference Library*, Volume 5, Non-Shock Initiation of Explosives, Springer Verlag, **2010**, 483–533.
- C.-O. Leiber, Stabile Bereiche beim Übergang von der Verbrennung zur Detonation, *Beiträge zum 18. Sprengstoffgespräch*, **1996**, 121–130.
- R. F. Vetter, Cook-Off in Fuel Fire, *16th JANAF Combustion Meeting, Vol.I*, **1979**, 199–204.

Decanitrodiphenyl

Decanitrobiphenyl

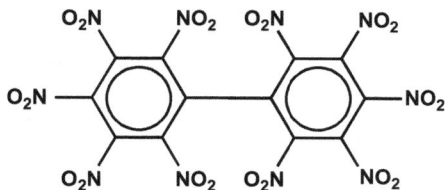

Aspect		Yellow prisms
Formula		$C_{12}N_{10}O_{20}$
CAS		[84647-88-1]
m_r	g mol^{-1}	604.18
ρ	g cm^{-3}	2.212 (calculated)
$\Delta_f H$	kJ mol^{-1}	80
$\Delta_{ex} H$	kJ mol^{-1}	−3712.2
	kJ g^{-1}	−6.144
	kJ cm^{-3}	−13.591
$\Delta_c H$	kJ mol^{-1}	−4802.2
	kJ g^{-1}	−7.948
	kJ cm^{-3}	−17.582
Λ	wt.-%	−10.59
N	wt.-%	23.17
Mp	°C	243 (dec)

Decanitrobiphenyl is practically insoluble in typical solvents. It is only stable at temperatures below 0 °C under an inert atmosphere. Even weak nucleophiles (e.g. moisture) attack decanitrobiphenyl. The molecule lacks stabilization through resonance due to the tilt between the phenyl planes impeding it.

– T. Nielsen, W. P. Norris, R. L. Atkins, W. R. Vuono, Nitrocarbons. 3. Synthesis of Decanitrobiphenyl, *J. Org. Chem.* **1983**, *48*, 1056–1059.

Deflagration

A deflagration is an exothermic reaction with a flame front (reaction zone) moving at subsonic speed supported exclusively by thermal radiation, heat conduction, heat transport, as well as diffusion. In a deflagration reaction, the reaction products flow away from the reaction zone (in contrary to a detonation reaction). In the p–v diagram (Figure D.5, p. 216), any connecting line between the state A (p_0/v_0) and any point on the branch of the product Hugoniot $\mathfrak{H}_1$ fulfilling the condition $v > v_0$ and $p < p_0$ are considered deflagrations. However, only the tangent intersecting point E indicates a stable deflagration.

- I. Glassman, R. A. Yetter, *Combustion*, 4th Ed., Academic Press, London, **2008**, pp. 147–260.

DEGDN, diethylene glycol dinitrate

Diethylenglykoldinitrat

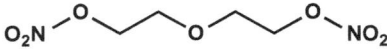

Formula		$C_4H_8N_2O_7$
REACH		LRS
EINECS		211-745-8
CAS		[693-21-0]
m_r	g mol^{-1}	196.117
ρ	g cm^{-3}	1.485
$\Delta_f H$	kJ mol^{-1}	−429
$\Delta_{ex}H$	kJ mol^{-1}	−974.8
	kJ g^{-1}	−4.970
	kJ cm^{-3}	−7.381
$\Delta_c H$	kJ mol^{-1}	−2288.4
	kJ g^{-1}	−11.669
	kJ cm^{-3}	−17.328
Λ	wt.-%	−40.79
N	wt.-%	14.28
Mp	°C	2
Bp	°C	160
$P_{vap(20 °C)}$	Pa	0.48
DSC-onset	°C	190
Friction	N	n.a.
Impact	J	0.2
Trauzl	cm^3	410

Next to NGI, DEGDN is the most important energetic plasticizer in nitrocellulose-based doublebase and triplebase gun propellants. DEGDN was used for the first time in wartime Germany in the diglycol powders (Both G- and gudol powder).

Delay (element)

Verzögerungs(element)

A delay serves to decelerate the speed of reaction of an ignition or initiation chain to a well-defined interval. Delay elements cover velocity ranges between $u < 1$ mm s^{-1} up to and beyond $u > 1000$ m s^{-1} (see *Hivelite*). Until the beginning of World War I, pressed black powder blended with inert materials like clay was the main delay formulation used in delays. Today, pressed black powder still is an important material for delays and fuses in the fireworks field and some technical pyrotechnics. However, in ammunition, a great many number of more complex formulations are now used in delay elements. There are either gasless (Table D.3) or gassy delay (Table D.2) compositions. Gasless delay formulations mostly contain two complementary oxidizers such as KClO$_4$

Table D.2: Gassy delay formulations.

Component	RL-58	SR74	SR112
Ammonium perchlorate	87.38		
Barium nitrate		60	
Potassium nitrate			60
Manganese dioxide	2.91		
Anthracene	6.80		
Stearic acid	2.91		
Tetranitrooxanilide		40	
Tetranitrocarbazole			40
u (mm s^{-1}) at 100 kPa	0.9	0.6	7.5
u (mm s^{-1}) at 1 MPa		13	

Table D.3: Gasless delay formulations.

Component	CP-22-5	SR92	AZM-961
Barium chromate	50		59
Bismuth(IIII) oxide		66.3	
Chromium(III) oxide		22.1	
Potassium perchlorate	10		14
Boron, amorphous		11.6	
Tungsten	40		
Zirconium–nickel 70-30			26
u (mm s^{-1}) at 100 kPa	0.23	26	12.7
u (mm s^{-1}) at 1 MPa		28	

and a metallate (e.g. chromate and tungstate) together with a metallic fuel selected from the group of molybdenum titanium, tungsten, zirconium, and their hydrides and alloys. Figure B.6 depicts the burn rate of $W/BaCrO_4/KClO_4$ as a function of stoichiometry. The above-mentioned formulations are pressed in steps in metal tubes. It is also still practiced to fill thick-walled tubes of lead or similar ductile materials with delay formulations and to machine the tube having an initial length of 30 cm to a length of 10 m. By this machining, the delay formulation is nicely consolidated inside the tube and firmly adheres to the metal tube which can be cut thereafter into desired segments.

Modern mechanical robust and precise delay elements can be printed on combustible or inert substrates. Therefore, thermite-type compositions are blended with energetic binders to make inks processable with common inkjet printers. The dried ink is later protected with polymer coatings.

- M. A. Wilson, R. J. Hancox, Pyrotechnic Delays and Thermal Sources, *Pyrotechnic Chemistry*, **2004**, pp. 8-1–8-22.
- T. M. Bell, D. M. Williamson, S. M. Walley, C. G. Morgan, C. L. Kelly, L. Batchelor, An Assessment of Printing Methods for Producing Two-Dimensional Lead-Free Functional Pyrotechnic Delay-Lines for Mining Applications, *Propellants Explos. Pyrotech.* **2020**, *45*, 53–76.

Density
Dichte

The density, ρ [g cm^{-3}], is a parameter of paramount importance when it comes to influencing the performance of high explosives. The density correlates linearly with the detonation velocity, V_D:

$$\rho \sim V_D$$

In addition, its square correlates with the detonation pressure, P_{CJ}:

$$\rho^2 \sim P_{CJ}$$

The *Kamlet–Jacobs equations* ($\rightarrow$ *detonation*) read correspondingly:

$$V_D = 3.9712\,(1 + 1.3 \cdot \rho)f^{0.5}$$

$$P_{CJ} = 240.86 \cdot \rho^2 \cdot \phi$$

The crystal density of a high explosive can be derived from its X-ray diffraction data using the equation:

$$\rho = \frac{M_r \cdot Z}{V_E \cdot N_A}$$

where M_r is the molecular mass, Z the number of formula units per unit cell, V_E the volume of the unit cell, and N_A the Avogadro constant.

The cell volume of organic covalent and ionic molecules, V, typically increases linearly with increasing temperature:

$$V_{298} = V_T \cdot (1 + a \cdot (298 - T))$$

With the linear thermal expansion coefficient

$$\alpha = \frac{1}{V} \cdot \left(\frac{dV}{dT}\right)_p$$

α varies considerably (see Tab T.7, p. 661) and even polymorphs of the same compound may show intersecting V/T plots. Hence, the following formula (with α set to 1.5×10^{-4}, for experimental thermal expansion coefficients found with a variety of cmpounds and formulations see Table T.7→TMA. Thermal Mechanical Analysis) to extrapolate the density at ambient temperature is just an educated guess but cannot replace a measurement at a particular temperature:

$$\rho_{298K} = \frac{\rho_T}{1 + \alpha(298 - T)}, \quad a = 1.5 \times 10^{-4} \, K^{-1}$$

The ambient temperature density of organic covalent and ionic compounds containing B, C, H, N, O, F, Cl, Br, I, S, and P can be estimated reasonably with the additivity method developed by the late Herman L. Ammon.

- C. C. Sun, Thermal Expansion of Organic Crystals and Precision of calculated Crystal Density A Survey of Cambridge Crystal Data, *J. Pharm. Sci.* **2007**, *96*, 1043–1052.
- B. M. Rice, J. J. Hare, E. F. C. Byrd, Accurate Prediction of Crystal Densities Using Quantum Mechanical Molecular Volumes, *J. Phys. Chem. A* **2007**, *111*, 10874–10879.
- H. L. Ammon, Updated Atom/Functional Group and Atom_code Volume Additivity Parameters for the Calculation of Crystal Densities of Single Molecules, Organic Salts, and Multi-Fragment Materials Containing H, C, B, N, O, F, S, P, Ck, Br and I, *Propellants Explos. Pyrotech.* **2008**, *33*, 92–102.

Detasheet®

Detasheet was a PETN-based sheet explosive marketed by *DuPont*. The product has been adopted by *Ensign–Bickford* in the meantime and is sold under the trade name *Primasheet®* as an olive-tinted material in rolls with thicknesses varying between 1 and 10 mm. Primasheet 1000 is based on PETN/NC, Primasheet 2000 is based on RDX /NC (→ *Nipolit*).

Detonating cord

Sprengschnur

A detonating cord serves both transfer of detonation as well as explosive charge itself either as a single line or as an accumulated coil of lines. A detonating cord consists of an inner core of crystalline explosive (PETN, RDX, and HMX) which is supported by an external woven thread and impermeable polymer coating to lend mechanical stiffness and protection against environmental effects. Detonating cords require a detonator as an initiator. Typical charge densities range from 15 to 100 g m^{-1}. A common application of detonating cords is for the canopy disintegration of jet planes to enable the safe ejection of aviators (Figure D.3).

Figure D.3: Canopy of a BAE hawk showing the detonating cord *WIkIPEDIA Capt. Alex Scott.* https://commons.wikimedia.org/wiki/Category:Detonating_cord#/media/File:View_from_the_cockpit_of_a_Hawk_TMK1.jpg

Detonation

The rapid and violent form of exothermic reaction called detonation differs from other forms in that all the important energy transfer is by mass flow in strong compression waves, with negligible contributions from other processes like heat conduction which are so important in flames. The leading part of a detonation front is a strong shock wave propagating into the explosive. This shock heats the material by compressing it, thus triggering chemical reaction, and balance is attained such that the chemical reaction supports the shock. In this process, material is consumed 10^3 to 10^8 times faster than in a flame, making detonation easily distinguishable from other reaction processes. The very rapid energy conversion in explosives is the property that makes them useful. If the reaction rates are essentially infinite and chemical equilibrium is attained, one has a steady-state detonation whose propagation rate is governed solely by thermodynamics and hydrodynamics.

Berthelot was the first to recognize the occurrence of a shock wave (w > c) in a reacting phase as a necessary characteristic of a detonation. Though *Riemann* in 1860 had correctly interpreted the gas dynamic development of a shock wave, it was not until *Becker*'s seminal work in 1917 that the shock wave was understood and explained correctly in terms of physics. The shock wave induces a strongly exothermic chemical reaction (CRZ = chemical reaction zone) which provides the energy required to support the shock front.

By this unique phenomenology, the detonation is different from the relatively slow (w < c) combustion and *deflagration* processes in which the CRZ is propagating only by means of heat conduction, heat transport, radiation of heat, and transport of matter.

ZND model

The ZND model is a one-dimensional model for the detonation of an ideal high explosive. It was developed and proposed in World War II independently from *Zel'dovich, von Neumann,* and *Döring* .

The model assumes a finite reaction velocity for chemical reactions. Initially, an infinitely narrow shock wave compresses the high explosive to a high-density state, termed von Neumann spike (Figure D.4, Point F) or jump plane. However, in the jump plane, no reaction has yet taken place and this marks the beginning of the CRZ which ends with the Chapman–Jouguet (CJ) state. Following the CJ state, the detonation products merely expand (rarefaction wave). Any exothermic process behind the CJ state (e.g. the reaction of the detonation products with atmospheric oxygen or aluminum) cannot contribute to the propagation of the shock front (w).

In real explosives, the shock front propagates with the detonation velocity w. To facilitate a mathematical treatment, both the shock front and CRZ are considered stationary, whereas the unreacted explosive is assumed to be non-stationary (Figure D.4). The unreacted explosive (p_1, v_1, T_1, U_1) flows from the right with the detonation velocity w into the CRZ. The fumes of the detonation products (p_2, v_2, T_2, U_2) exit the CRZ through the so-called CJ plane with the velocity u < w.

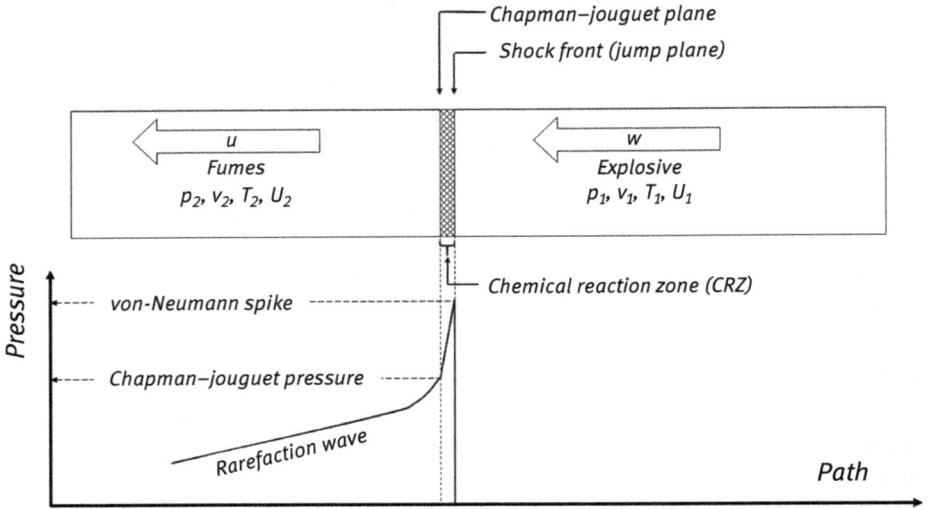

Figure D.4: Schematic depiction of a detonation and related pressure.

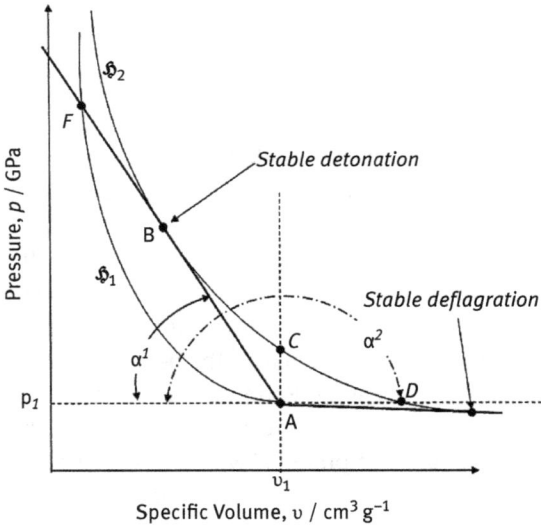

Figure D.5: Hugoniot for unreacted explosive ($\mathfrak{H}_1$) and fumes ($\mathfrak{H}_2$).

As this is a thermodynamically open system (energy and mass exchange), the conservation equations for the three quantities, mass, momentum, and energy, take the form of equilibrium equations.

Mass

$$p_2 \cdot u = p_1 \cdot u$$

Momentum

$$p_1 + p_1 \cdot w^2 = p_2 + p_2 \cdot u^2$$

Energy

$$p_1 \cdot v_1 + \frac{w^2}{2} + U_1 = p_2 \cdot v_2 + \frac{u^2}{2} + U_2$$

Based on hydrodynamics, the detonation velocity, w, and fume velocity, u,

$$w = v_1 \cdot \sqrt{\frac{p_2 - p_1}{v_1 - v_2}}$$

$$u = (v_1 - v_2)\sqrt{(p_2 - p_1) \cdot (v_1 - v_2)}$$

As the compression by the shock yields a heat release, the state functions before and behind the shock front can no longer be described with the known adiabatic equation

$$p_2 \cdot v_2{}^\gamma = p_1 \cdot v_1{}^\gamma,$$

with $\gamma = c_p/c_v$,
    but are described by the *Hugoniot* equation

$$U_2 - U_1 = c_{v2}T_2 - c_{v1}T_1 - Q = 0.5(p_1 + p_2) \cdot (v_1 - v_2)$$

All possible thermodynamic states an unreacted explosive can assume after being shocked are given by the curve $\mathfrak{H}_1$ in the p–v diagram. The possible states of the detonation fumes due to the heat released (Q) are described by curve $\mathfrak{H}_2$ which is positioned higher position relative to $\mathfrak{H}_1$ (Figure D.4).

    The initial state of the unreacted explosive is given by point **A** with (p_1,v_1). An exothermic reaction of the explosive yields a state on the curve $\mathfrak{H}_1$. The connecting line between the initial state and any state on the curve $\mathfrak{H}_2$ yields a distinct angle a with the v-axis. Now the following relationship can be established

$$\tan \alpha \sim \frac{p_2 - p_1}{v_1 - v_2}$$

Based on equation (1) it follows that

$$w \sim \sqrt{\tan \alpha}$$

Any state on $\mathfrak{H}_2$ left to point **C** has $v_1 > v_2$ and $p_2 > p_1$. Hence, tan α assumes large values. So, there is a considerable compression in the shock front leading to a pressure rise. The velocity of the fumes is positive and in the same direction as the shock front. That is why the flow of fumes of the rarefaction wave follow the detonation front. The detonation velocity, w, is given by the sum of the velocity of the fumes, u and the sonic velocity, c:

$$w = u + c$$

The tangent (*Rayleigh line*) leading from point **A** to the upper Hugoniot (point **B**) characterizes the stable detonation. **B** is called CJ point. The corresponding pressure is the CJ pressure, P_{CJ}. The point of intersection between the same line (v_1–A) with the Hugoniot of the unreacted explosive, $\mathfrak{H}_1$, (Point **F**) yields the pressure prevailing at the *von Neumann peak*.

Any state on the Hugoniot, $\mathfrak{H}_2$ between points **C** and **D** would yield a negative tan α and hence no real change of state can occur from v_1 into this branch of the Hugoniot.

For any point on $\mathfrak{H}_2$, which is below **D**, w assumes relatively smaller values as compared to the range above **C**. In addition, the fume velocity assumes negative values. This is indicative of the fumes venting away from the reaction zone. The second possible tangent from point **A** to the lower branch of $\mathfrak{H}_2$ in point **E** characterizes a stable deflagration. Figure D.4 also easily shows effect of density on the detonation velocity and pressure. As the density of an explosive is increased (decrease of specific volume), any tangent from $\mathfrak{H}_1$ to $\mathfrak{H}_2$ becomes steeper, that is the angle α becomes larger, and following the above equation, w increases. Likewise, with the tangent touch point climbing upward on $\mathfrak{H}_2$, the corresponding P_{CJ} pressure rises too.

According to *Leiber* many liquid and solid high explosives show two or even three different discrete detonation velocities or better three different range of velocities. At first, there is the "normal" detonation velocity which can be described with the above reasoning (ZND model). This phenomenon is called the HVD. Then frequently, there is lower detonation regime where the relationship is still w > c. This is called the *LVD*. And finally, for some explosives, there is an even slower velocity regime observed with w < c. This latter one is called slow velocity detonation (*SVD*). It is a characteristic that there are distinct velocity gaps between the regimes for the corresponding materials. For NGI, the following phenomena are observed:

HVD	7000–8900 m s^{-1}
No stable detonation	3000–5000 m s^{-1}
LVD	1000–2000 m s^{-1}
No stable detonation	900–1000 m s^{-1}
SVD	777–800 m s^{-1}

There are hints that the acoustic impedance $Z = \rho \cdot c_L$ of the confining material influences the type of detonation. Black powder, for example, shows some unusual "reaction behavior." When a black powder-filled safety fuze (Bickford type) is compressed firmly between metal clamps or accidentally pinched between a door and a frame, it has been observed to yield LVD. That is regular laminar combustion (0.008 m s^{-1}) at the location of clamping/damage transitions to LVD (>1000 m s^{-1}) which has led to numerous accidents (*Leiber*).

For the semiempirical calculation of detonative properties of high explosives, Kamlets equations can be used.

The detonation velocity is

$$V_{CJ} = A \left[NM^{1/2}(-\Delta_{det}H)^{1/2} \right]^{1/2} (1 + B\rho_0)$$

A = 3.9712 (empirical constant)
N = Molar amount of gaseous products per gram of explosive
M = The Mass of gaseous products per molar amount of gaseous products
$\Delta_{det}H$ = Detonation enthalpy [kJ g^{-1}]
B = 1.30 (empirical constant)
ρ_0 = Density of unreacted explosive

The detonation pressure P_{CJ} is

$$P_{CJ} = KN(M - \Delta_{det}H)^{1/2}\rho_0^2$$

K = 240.86 (empirical constant)

In the following, we will condense the expression and write $\phi = NM^{0,5} (-\Delta_{det}H)^{0,5}$.

In case experimental detonation velocities have not been determined, the pressure can be assessed in good approximation by the following expression:

$$P_{CJ} = \rho V_D^2/4$$

Papers

- R. Becker, Zur Theorie der Detonation, *Z. Elektrochem.* **1917**, *23*, 40–49.
- R. Becker, Stoßwelle und Detonation, *Z. Physik*, **1922**, *8*, 321–362.
- W. Döring, Über den Detonationsvorgang in Gasen, *Ann. Phys.* **1943**, 421–436.
- W. Döring, Die Geschwindigkeit und Struktur von intensiven Stoßwellen in Gasen, *Ann. Phys.* **1949**, 133–150.
- W. Döring, Über die Detonationsgeschwindigkeit des Methans und Dizyans im Gemisch mit Sauerstoff und Stickstoff, *Z. Elektrochem.* **1950**, *54*, 231–239.
- R. Schall, Die Stabilität langsamer Detonationen, *Z. Angew. Phys.* **1954**, *6*, 470–475.
- C. O. Leiber, Die Detonation als reaktive Mehrphasenströmung, *Rheol. Acta* **1975**, *14*, 92–100.

Monographs

- H. D. Gruschka, F. Wecken, *Gasdynamic Theory of Detonation*, Gordon and Breach, 198 S, **1971**.
- W. Fickett, W. C. Davis, *Detonation*, University Press, 386 S, **1979**.
- R. Chéret, *Detonation of Condensed Explosives*, Springer, 427 S., **1993**.
- P. W. Cooper, *Explosives Engineering*, Wiley-VCH, 460 S, **1996**.
- C. L. Mader, *Numerical Modeling of Explosives and Propellants*, CRC Press, 439 S., **1998**.
- C. O. Leiber, *Assessment of Safety and Risk with a Microscopic Model of Detonation*, Elsevier, 594 S., **2004**.
- M. H. Keshavarz, T. M. Klapötke, *Energetic Compounds – Methods for the Prediction of Their Performance*, 2nd Edition, De Gruyter, Berlin, **2020**, 144 pp

Detonative ignition
Detonative Anzündung

Detonative ignition also termed *shock ignition* was first studied by *Hardt* in the 1970s on intermetallic and thermite systems. He recognized that the shock energy required to ignite a pyrotechnic material, E_i, is equivalent to the autoignition energy for regular thermal ignition, H_i:

$$E_i \sim H_i$$

$$E_i = \frac{1}{2} \cdot \frac{P^2}{U_s^2 \cdot \rho_0^2}$$

$$H_i = c_p \cdot (T_i - T_0)$$

Hence, the ignition conditions for shock ignition can de derived from common thermal investigation. Figure D.6 depicts the schematic streak of a shock wave in a consolidated pyrotechnic pellet. The shock wave eventually fades out, but after a certain delay, a combustion reaction starts. However, this combustion proceeds much faster (blue) than a pellet regularly ignited by a thermal stimulus (green). This is because the whole bulk of the energetic material is ignited simultaneously. Hence the mass consumption rate is much higher than with a regular one-point thermal ignition. Figure D.7 shows a stack of cylindrical obscurant pellets based on red phosphorus/metal/nitrate upon shock ignition and dissemination by an internal detonation cord. While a regular thermal central ignition yields a combustion time of 25–35 s, shock ignition effects a complete consumption of the same payload amount in 2–3 s only.

- A. P. Hardt, R. H. Martison, Initiation of Pyrotechnic Mixtures by Shock, *8th Symposium on Explosives and Pyrotechnics*, The Franklin Research Center, Philadelphia, PA, **1974**, 53.
- A. P. Hardt, S. L. McHugh, S. K. Weinland, Chemistry and Shock Initiation of Intermetallic Reactions, *11th International Pyrotechnics Seminar*, 7–11 July, **1986**, Vail, CO, 255.

- A. P. Hardt, Shock Initiation of Thermite, *13th International Pyrotechnics Seminar*, 11–15 July, **1988**, Grand Junction, CO, 425.
- S. A. Sheffield, A. C. Schwarz, Shock Wave Response of Titanium Subhydride-Potassium Perchlorate, *8th International Pyrotechnics Seminar*, 12–16 July, **1982**, Steamboat Springs, CO, 972.

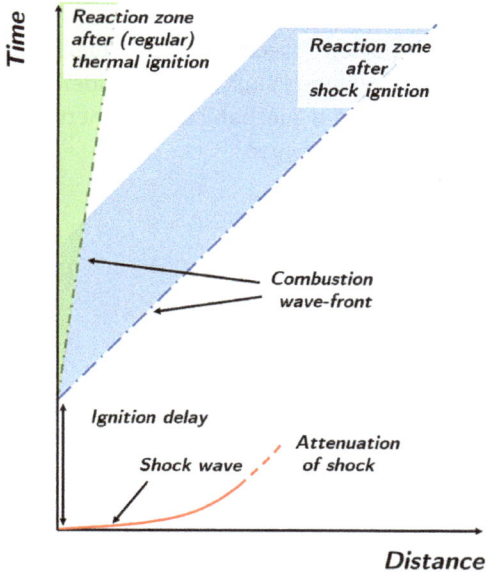

Figure D.6: Schematic showing a shock ignition (blue reaction zone) of a pyrotechnic compared to regular thermal ignition of a pyrotechnic (green reaction zone).

Figure D.7: Sequence showing the effect of detonative ignition of 700 g RP-based pyrotechnic obscurant.

Detonator

Detonator, initiator

A detonator serves the direct initiation of a sensitive secondary high explosive charge. (Less sensitive high explosives such as TNT and truly shock insensitive high explosives like TATB, NTO, or others require an additional booster charge in between.)

Detonators contain an output charge of a powerful secondary high explosive (PETN or RDX) and an adjacent primary explosive charge such as lead azide. Next to the primary explosive comes a charge containing a mechanically or thermally very sensitive pyrotechnic formulation that serves as a trigger.

Detonators can be triggered

mechanically:
- stab initiators
- percussion initiators

thermally:
- pyrotechnic initiators

electrically:
- hot-wire initiators
- electric match initiators

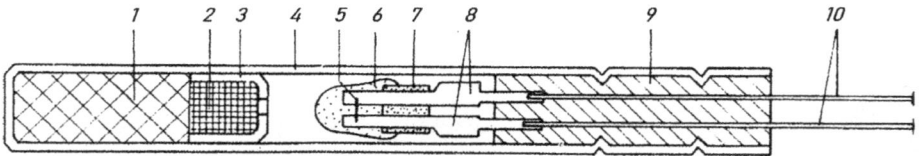

Figure D.8: Cross section of an electric detonator.

Sequence showing the effect of detonative ignition of 700 g RP-based pyrotechnic obscurant:
1 Secondary charge
2 Primary explosive
3 Internal cap
4 Outer cap
5 Bridgewire
6 Pyrotechnic composition
7 Isolation

8   Terminals
9   Filler
10 Leads

- exploding bridgewire initiators
- exploding foil initiators

optically:

- laser detonators (see *laser initiation*)

In a detonator, the initial trigger stimulus is amplified to effect the detonation of the secondary charge.

Exploding bridgewire and exploding foil initiators do not contain any primary explosive but rely entirely on generating a primary shock wave by electric signals.

Dextrin

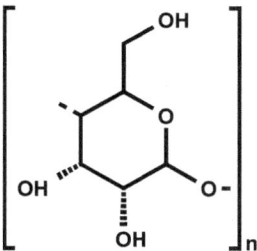

Aspect		Yellowish powder
Formula		$(C_6H_{10}O_5)_n$
CAS		[9005-84-9]
m_r	g mol^{-1}	162.142
ρ	g cm^{-3}	1.038
$\Delta_f H$	kJ mol^{-1}	−954.7
$\Delta_c H$	kJ mol^{-1}	−2835.6
	kJ g^{-1}	−17.488
	kJ cm^{-3}	−18.153
Λ	wt.-%	−118.41
c_p	J g^{-1} K^{-1}	1.22
Dp	°C	228

Dextrin is obtained by partial hydrolysis of potato or corn starch. It serves as a water-soluble binder mainly in the fireworks field and is also used as a crystallization aid for the α-polymorph of lead azide.

Diacetone diperoxide, DADP

Diacetondiperoxid

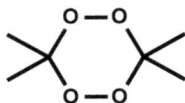

Formula		$C_6H_{12}O_4$
CAS		[1073-91-2]
m_r	$g\ mol^{-1}$	148.157
ρ	$g\ cm^{-3}$	1.33
$\Delta_f H$	$kJ\ mol^{-1}$	−598
$\Delta_{ex}H$	$kJ\ mol^{-1}$	−454.3
	$kJ\ g^{-1}$	−3.066
	$kJ\ cm^{-3}$	−4.078
$\Delta_c H$	$kJ\ mol^{-1}$	−3478.1
	$kJ\ g^{-1}$	−23.476
	$kJ\ cm^{-3}$	−31.223
Λ	wt.-%	−151.18
$\Delta_{subl}H$	$kJ\ mol^{-1}$	71.3
Dp	°C	>150
Friction	N	0.1 (BAM)
Impact	J	4−6 (BAM)
V_D	$m\ s^{-1}$	5300 at $1.18\ g\ cm^{-3}$
Trauzl	cm^3	250

DADP is less impact sensitive than its congener triacetone triperoxide (TATP) but shows similar friction sensitivity. DADP forms upon decomposition of TATP in solution as well as from solid TATP in the presence of acid traces. DADP in principle could be used as a primary explosive. However, its high vapor pressure ($p_{25\ °C}$ ~ 18 Pa) causes sublimation and recrystallization of highly sensitive needle-like crystals ruling out safe handling. DADP is insoluble in water but soluble in many organic solvents.

- A. E. Contini, A. J. Bellamy, L. N. Ahad, Taming the Beast Measurement of the Enthalpies of Combustion and Formation of Triacetone Triperoxide (TATP) and Diacetone Diperoxide (DADP) by Oxygen Bomb Calorimetry, *Propellants Explos. Pyrotech.* **2012**, *37*, 320–328.

Diamino-1,2,4,5-tetrazine-1,4-di-*N*-oxide, LAX-112, TZX

3,6-Diamino-1,2,4,5-tetrazin 1,4-dioxid

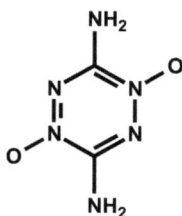

Formula		$C_2H_4N_6O_2$
CAS		[153757-93-8]
m_r	g mol^{-1}	144.093
ρ	g cm^{-3}	1.86
$\Delta_f H$	kJ mol^{-1}	164.01
$\Delta_{ex}H$	kJ mol^{-1}	−654.0
	kJ g^{-1}	−4.539
	kJ cm^{-3}	−8.442
$\Delta_c H$	kJ mol^{-1}	−1522.7
	kJ g^{-1}	−10.568
	kJ cm^{-3}	−19.656
Λ	wt.-%	−44.41
N	wt.-%	58.32
Mp	°C	266 (dec)
Friction	N	>350
Impact	cm	>177 (tool type 12)
V_D	m s^{-1}	8780 at 1.81 g cm^{-3}
P_{CJ}	GPa	33.5 at 1.81 g cm^{-3}

LAX-112 is both friction and impact insensitive and has thermal decomposition higher than RDX. It was a candidate IM material in the 1990s.

– M. A. Hiskey, D. E. Chavez, D. L. Naud, S. F. Son, H. L. Berghout, C. A. Bolme, Progress in High-Nitrogen Chemistry in Explosives, Propellants and Pyrotechnics, *IPS-Seminar*, **2000**, Grand Junction, 3–14.

Diamino-1,2,4-triazoliumnitrate, DtrzN

3,5-Diamino-1,2,4-triazoliumnitrat

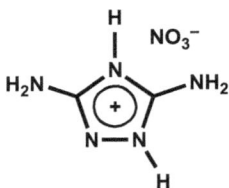

Formula		$C_2H_6N_6O_3$
CAS		[261703-47-3]
m_r	g mol^{-1}	162.107
ρ	g cm^{-3}	1,649
Δ_fH	kJ mol^{-1}	−76
$\Delta_{ex}H$	kJ mol^{-1}	−654.033
	kJ g^{-1}	−4.035
	kJ cm^{-3}	−6.653
Δ_cH	kJ mol^{-1}	−1568.5
	kJ g^{-1}	−9.676
	kJ cm^{-3}	−15.956
Λ	wt.-%	−39.48
N	wt.-%	51.84
Mp	°C	159
Dp	°C	241
Friction	N	288
Impact	J	40

DtrzN serves as a component in gas-generating composition and is used in eutectics based on ammonium nitrate and 3,5-diamino-1,2,4-triazolium nitrate as EILS (→ *AAD*) and insensitive TNT replacement.

- T. M. Klapötke, F. A. Martin, N. T. Mayr, J. Stierstorfer, Synthesis and Characterization of 3,5-Diamino-1,2,4-triazolium Dinitramide, *Z. Anorg. Allg. Chem.* **2010**, *636*, 2555–2564.
- P. W. Leonard, D. E. Chavez, P. R. Bowden, E. G. Francois, Nitrate Salt Based Melt Cast materials, *Propellants Explos. Pyrotech.* **2018**, *43*, 11–14.

1,1'-Diamino-2,2',3,3',5,5',6,6'-octranitrodiphenyl, CL-12

2,2',3,3',5,5',6,6'-Octanitro[1,1'-biphenyl]-4,4'-diamin

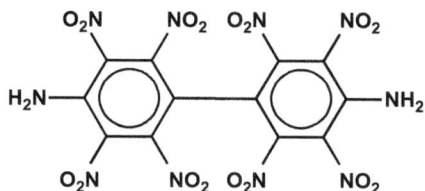

Aspect		Yellow crystals
Formula		$C_{12}H_4N_{10}O_{16}$
CAS		[84642-53-5]
m_r	g mol^{-1}	544.22
ρ	g cm^{-3}	1.82
$\Delta_f H$	kJ mol^{-1}	338.99
$\Delta_{ex}H$	kJ mol^{-1}	−3045.2
	kJ g^{-1}	−5.596
	kJ cm^{-3}	−10.184
$\Delta_c H$	kJ mol^{-1}	−5632.9
	kJ g^{-1}	−10.350
	kJ cm^{-3}	−18.838
Λ	wt.-%	−29.4
N	wt.-%	25.74
Mp	°C	310 (dec)
Friction	N	>350
Impact	cm	95 (2.5 kg)
V_D	m s^{-1}	8300 at 1.82 g cm^{-3}
P_{CJ}	GPa	31.7 at 1.82 g cm^{-3}

CL-12 is hydrolytically very sensitive. It is a precursor for the synthesis of decanitrodiphenyl.

–   T. Nielsen, W. P. Norris, R. L. Atkins, W. R. Vuono, Nitrocarbons. 3. Synthesis of Decanitrobiphenyl, J. Org. Chem. 1983, 48, 1056–1059.

1,3-Diamino-2,4,6-trinitrobenzene, DATB

1,3-Diamino-2,4,6-trinitrobenzol

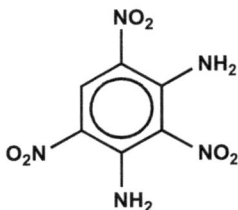

Aspect		Yellow crystals
Formula		$C_6H_5N_5O_6$
REACH		LPRS
EINECS		216-626-4
CAS		[1630-08-6]
m_r	g mol^{-1}	243.136
ρ	g cm^{-3}	1.838
$\Delta_f H$	kJ mol^{-1}	−98.74
$\Delta_{ex}H$	kJ mol^{-1}	−1137.0
	kJ g^{-1}	−4.676
	kJ cm^{-3}	−8.595
$\Delta_c H$	kJ mol^{-1}	−2976.7
	kJ g^{-1}	−12.243
	kJ cm^{-3}	−22.503
Λ	wt.-%	−55.93
N	wt.-%	28.80
c_p	J mol^{-1} K^{-1}	70.2
$\Delta_{subl}H$	kJ mol^{-1}	140
$T_{1 \to 2}$	°C	217
Mp	°C	286 (incipient dec)
Friction	N	>350
Impact	cm	>320 (tool type 12)
V_D	m s^{-1}	7659 at 1.816 g cm^{-3}
P_{CJ}	GPa	25.1 at 1.78 g cm^{-3}
$\varnothing_{cr}$	mm	5.3 at 1.816 g cm^{-3}

DATB an insensitive and temperature-resistant high explosive. Similar to its structural progenitor TATB, DATB is only slight soluble in DMF or DMSO (3–5 wt.-%).

2,6-Diamino-3,5-dinitropyrazine-1-oxide, LLM-105

2,6-Diamino-3,5-dinitropyrazin-1-oxid

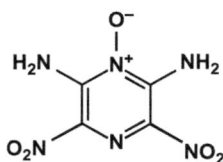

ANPZO, LLM-105, PZO, N-PEX-1		
Aspect		Citron yellow crystals
Formula		$C_4H_4N_6O_5$
CAS		[194486-77-6]
m_r	g mol^{-1}	216.113
ρ	g cm^{-3}	1.913
$\Delta_f H$	kJ mol^{-1}	−12.97
$\Delta_{ex}H$	kJ mol^{-1}	−1021.1
	kJ g^{-1}	−4.725
	kJ cm^{-3}	−9.039
$\Delta_c H$	kJ mol^{-1}	−2132.8
	kJ g^{-1}	−9.869
	kJ cm^{-3}	−18.879
Λ	wt.-%	−37.02
N	wt.-%	38.89
Dp	°C	310
Friction	N	>360
Impact	J	21
V_D	m s^{-1}	8560 at 1.913 g cm^{-3}
P_{CJ}	GPa	33.4 at 1.913 g cm^{-3}

ANPZO possesses a high density and high thermal stability surpassing TATB. However due to its poor solubility in common solvents it is difficult to convert the needle-type crystal into other morphologies. This and the complicated synthesis make ANPZO an expensive high explosive restricting research into small-scale applications.

- E.-C. Koch, *Insensitive Explosive Materials II – 2,6-Diamino-3,5-Dinitro-1,4-Pyrazine-1-Oxide ANPZO, L-159*, MSIAC, October **2009**, 24 S.; Weblink https://www.msiac.nato.int/products-serv ices/publications/l-159-insensitive-explosive-materials-2-26-diamino-35-dinitro-14 accessed on September 10 2020.

2,6-Diamino-3,5-dinitropyrazine, ANPZ

2,6-Diamino-3,5-dinitropyrazin

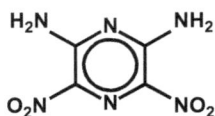

Aspect		Citron crystals
Formula		$C_4H_4N_6O_4$
CAS		[52173-59-8]
m_r	g mol^{-1}	200.114
ρ	g cm^{-3}	1.84
$\Delta_f H$	kJ mol^{-1}	−22.59
$\Delta_{ex}H$	kJ mol^{-1}	−835.0
	kJ g^{-1}	−4.173
	kJ cm^{-3}	−7.678
$\Delta_c H$	kJ mol^{-1}	−2123.2
	kJ g^{-1}	−10.610
	kJ cm^{-3}	−19.522
Λ	wt.-%	−47.97
N	wt.-%	42.00
Dp	°C	343
Friction	N	>360
Impact	cm	>177
P_{CJ}	GPa	29.6 at 1.84 g cm^{-3}

ANPZ is a synthetic precursor to the important insensitive explosive *LLM-105*.

3,3'-Diamino-4,4'-azoxyfurazan, DAAF

4-[2-(4-Amino-1,2,5-oxadiazol-3-yl)-1-oxidodiazenyl]-1,2,5-oxadiazol-3-amin

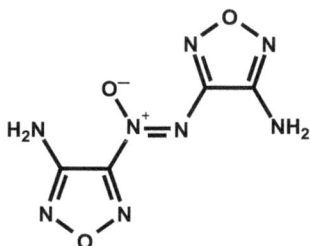

LAX -117, LAX-120, LAX-133, RX-64		
Aspect		Yellow orange crystals
Formula		$C_4H_4N_8O_3$
CAS		[78644-89-0]
m_r	g mol^{-1}	212.13
ρ	g cm^{-3}	1.745
Δ_fH	kJ mol^{-1}	+444
$\Delta_{ex}H$	kJ mol^{-1}	−1069.0
	kJ g^{-1}	−5.039
	kJ cm^{-3}	−8.793
Δ_cH	kJ mol^{-1}	−2589.7
	kJ g^{-1}	−12.208
	kJ cm^{-3}	−21.304
Λ	wt.-%	−52.8
N	wt.-%	52.83
Mp	°C	Under pressure
Dp	°C	248
Friction	N	>360
Impact	J	>50
V_D	m s^{-1}	8,057 at 1.700 g cm^{-3}
P_{CJ}	GPa	30.6 at 1.685 g cm^{-3}
$\varnothing_{cr}$	mm	1.25 at 1.60 g cm^{-3}

DAAF is a friction- and impact-insensitive high explosive that does not undergo cook-off. It has a superior performance when compared with TATB and is shock sensitive than the latter. DAAF performs particularly well at low temperatures and possesses a very low critical diameter qualifying it for use in geometrically challenging initiation chains.

– E.-C. Koch, Insensitive High Explosives II 3,3-Diamino-4,4-azoxyfurazan (DAAF), *Propellants Explos. Pyrotech.* **2016**, *41*, 526–538.
– E. G. Francois, D. E. Chavez, M. M. Sandstrom, The Development of a New Synthesis Process for 3,3'-Diamino-4,4'-azoxyfurazan (DAAF), *Propellants Explos. Pyrotech.* **2010**, 35, 529–534.

1,5-Diamino-4-*H*-1,2,3,4-tetrazolium perchlorate
1,5-Diamino-4-H-1,2,3,4-tetrazoliumperchlorat

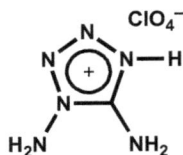

1*H*-Tetrazole-1,5-diamine perchlorate		
Formula		$CClH_5N_6O_4$
CAS		[857282-98-5]
m_r	g mol^{-1}	200.5412
ρ	g cm^{-3}	1.909 (100 K)
	g cm^{-3}	1.85 (298 K)
$\Delta_f H$	kJ mol^{-1}	117
$\Delta_c H$	kJ mol^{-1}	−1174.5
	kJ g^{-1}	−5.857
	kJ cm^{-3}	−10.835
Λ	wt.-%	0
N	wt.-%	41.91
Mp	°C	113-114
Dp	°C	192
Friction	N	60 (BAM)
Impact	J	7 (BAM)

DATP has a high ambient pressure burn rate (u = 14 mm s^{-1} and yields up to u = 149 mm s^{-1} at 10 MPa) and has been investigated as solid propellant ingredient by *Sinditski et al.*

- V. P. Sinditskii, V. Y. Egorshev, T. Y. Dutova, M. D. Dutov, T. L. Zhang, J. G. Zhang, Combustion of Derivatives of 1,5-Diaminotetrazole, *Comb. Explos. Shock.* **2011**, *47*, 36–44.
- J. C. Gálvez-Ruiz, G. Holl, K. Karaghiosoff, T. M. Klapötke, K. Löhnwitz, P. Mayer, H. Nöth, K. Polborn, C. J. Rohbogner, M. Suter, J. J. Weigand, Derivatives of 1,5-Diamino-1H-tetrazole: A New Family of Energetic Heterocyclic-Based Salts, *Inorg. Chem.* **2005** *44*, 4237–4253.

1,1-Diamino-2,2-dinitroethylene, FOX-7, DADNE
1,1-Diamino-2,2-dinitroethylen

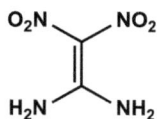

Aspect		Yellow-green crystals
Formula		$C_2H_4N_4O_4$
REACH		LPRS
EINECS		604-466-1
CAS		[145250-81-3]
m_r	g mol^{-1}	148.08
ρ	g cm^{-3}	1.907
Δ_fH	kJ mol^{-1}	−134
$\Delta_{ex}H$	kJ mol^{-1}	−724.4
	kJ g^{-1}	−4.892
	kJ cm^{-3}	−9.329
Δ_cH	kJ mol^{-1}	−1224.7
	kJ g^{-1}	−8.271
	kJ cm^{-3}	−15.772
Λ	wt.-%	−21.61
N	wt.-%	37.84
Dp	°C	225
Friction	N	>350 (BAM)
Impact	J	15–30 (BAM)
V_D	m s^{-1}	8342 at 1.78 g cm^{-3}
P_{CJ}	GPa	28.8 at 1.78 g cm^{-3}
$\varnothing_{cr}$	mm	1.25
Koenen	mm	6, type F

DADNE is an insensitive high-density explosive. Its performance just undershoots RDX. It is currently in use in booster charges.

– O. Pham-Schoenwetter, B. Donner, A. Hahma, *Method for Synthesizing 1,1-Diamino-2,2-dinitroethylene (FOX-7) or a Salt thereof*, US2018/0002270, **2018**, Germany.

Review papers after 2007

– E.-C. Koch, *Insensitive Explosive Materials VIII 1,1-Diamino-2,2-dinitroethylene – DADNE, L-178*, MSIAC, November **2012**, 35 S.; Weblink https://www.msiac.nato.int/products-services/publica tions/l-178-insensitive-explosive-materials-viii-11-diamino-22 accessed on September 10 2020.

Review papers before 2007

- H. Dorsett, Computational Study of FOX -7, A New Insensitive Explosive, *DSTO-TR-1054*, Weapons Systems Division, Aeronautical and Maritime Research Laboratory, September **2000**.
- J. Lochert, FOX -7, A New Insensitive Explosive, *DSTO-1238*, Weapons Systems Division, Aeronautical and Maritime Research Laboratory, November **2001**.
- J. Bellamy, FOX -7 (1,1-Diamino-2,2-dinitroethene), *Struct. Bond.* **2007**, *125*, 1–33.

Diaminoguanidinium-5,5'-azotetrazolate, DAGZT

Diaminoguanidinium-5,5'-azotetrazolat

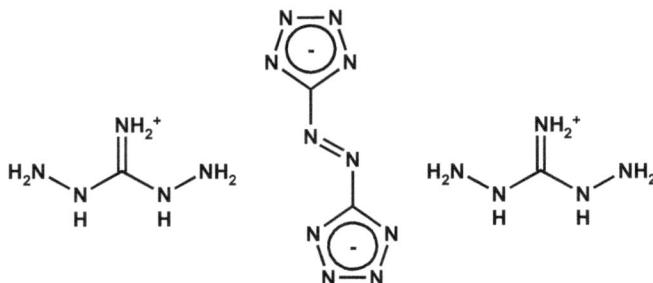

Aspect		Yellow crystals
Formula		$C_4H_{16}N_{20}$
CAS		[862107-18-4]
m_r	g mol^{-1}	344.30
ρ	g cm^{-3}	1.599
$\Delta_f H$	kJ mol^{-1}	708.8
$\Delta_{ex}H$	kJ mol^{-1}	−1040.3
	kJ g^{-1}	−3.023
	kJ cm^{-3}	−4.833
$\Delta_c H$	kJ mol^{-1}	−4569.7
	kJ g^{-1}	−13.272
	kJ cm^{-3}	−21.222
Λ	wt.-%	−74.4
N	wt.-%	81.36
Mp	°C	195 (dec)
Friction	N	>350
Impact	J	4

DAGZT is used as a high nitrogen source in gas-generating compositions and as a flame expander in spectral decoy flares.

- Hammerl, M. A. Hiskey, G. Holl, T. M. Klapötke, K. Polborn, J. Stierstorfer, J. J. Weigand, Azidoformamidinium and Guanidinium 5,5'-Azotetrazolate Salts, *Chem. Mater.* **2005**, *17*, 3784–3793.

Diazodinitrophenol, DDNP

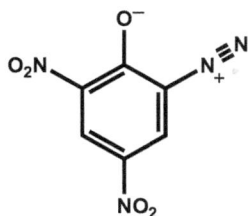

Dinol, Diazol		
Aspect		Yellow powder
Formula		$C_6H_2N_4O_5$
REACH		LPRS
EINECS		225-134-9
CAS		[4682-03-5]
m_r	g mol^{-1}	210.106
ρ	g cm^{-3}	1.719
$\Delta_f H$	kJ mol^{-1}	194
$\Delta_{ex}H$	kJ mol^{-1}	−1034.0
	kJ g^{-1}	−4.921
	kJ cm^{-3}	−8.460
$\Delta_c H$	kJ mol^{-1}	−2841.0
	kJ g^{-1}	−13.522
	kJ cm^{-3}	−23.244
Λ	wt.-%	−60.92
N	wt.-%	26.67
Dp	°C	144
Mp	°C	150
Friction	N	0.1–1 N
Impact	J	1–2
V_D	m s^{-1}	6900 at ρ = 1.6 g cm^{-3}
		6600 at ρ = 1.5 g cm^{-3}

DDNP is a powerful primary explosive exceeding by far the performance of both mercury fulminate and lead azide. Upon exposure to light it darkens and eventually becomes more thermally sensitive than the pristine material. Its maximum initiation capacity is available at densities between ρ = 1.2–1.3 g cm^{-3}. Sometimes too low mechanical sensitivity can cause malfunctions.

Dibenz[b,f]-1,4-oxazepine, CR

Dibenz[b,f]z-1,4-oxazepin

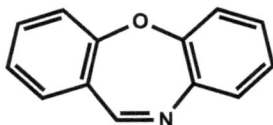

Aspect		Yellow powder
Formula		$C_{13}H_9NO$
REACH		LPRS
EINECS		607-782-8
CAS		[257-07-8]
m_r	g mol^{-1}	195.217
ρ	g cm^{-3}	1.328
Λ	wt.-%	−241.77
N	wt.-%	7.18
Mp	°C	72
$\Delta_f H$	kJ mol^{-1}	+182 (gas phase)
$\Delta_c H$	kJ mol^{-1}	−6584.0
	kJ g^{-1}	−33.727
	kJ cm^{-3}	−44.789
P_{vap}	Pa	7.9×10−3 at 20 °C
IC_{50}	ppm	0.7

CR is a more powerful irritating agent compared to both CN and CS. Thanks to its thermal stability, it is predominantly disseminated in pyrotechnics. CR is only slightly soluble in water but very soluble in diethyl ether and alcohol.

Dicyandiamide

Dicyandiamid, Cyanoguanidin

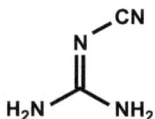

DCD		
Formula		$C_2H_4N_4$
REACH		LPRS
EINECS		207-312-8
CAS		[461-58-5]
m_r	g mol^{-1}	84.081
ρ	g cm^{-3}	1.404
$\Delta_f H$	kJ mol^{-1}	23
$\Delta_c H$	kJ mol^{-1}	−1381.7
	kJ g^{-1}	−16.433
	kJ cm^{-3}	−23.072
Λ	wt.-%	−114.17
N	wt.-%	66.64
Dp	°C	252
Mp	°C	211

DCD serves as a flame expander in pyrotechnics and is an industrial precursor for guanidinium nitrate.

- E.-C. Koch, Insensitive High Explosives III. Nitroguanidine, *Propellants Explos. Pyrotech.* **2019**, *44*, 267–292.

Diergol

Diergol is a German term to describe a propellant system consisting of two (*di*) individual components such as a fuel and an oxidizer. Both fuel and oxidizer can be either pure substances or mixtures.

- A. Dadieu R. Damm E. W. Schmidt, *Raketentreibstoffe*, Springer, Vienna, **1968**, pp. 99.

DIME

DIME is the acronym for **D**ense **I**nert **M**etal **E**xplosive. DIME formulations contain a high mass fraction (>50 wt.-%) of a very fine (d_p ~ 100 μm) heavy metal powder mostly tungsten alloys (e.g. *rWNiCo*: tungsten (91–93 wt.-%), nickel (3–5 wt.-%), and cobalt (2–4 wt.-%)). Upon detonation of DIME, the accelerated heavy metal particles cause a strong impulse in the close vicinity of the charge. However, the large surface/volume ratio of the particles results in a high aerodynamic drag and consequently, a strong deceleration of the particles resulting in a very quick reduction of the impulse. Thus, the blast performance of a DIME warhead is very localized and thus appropriate for military operation in urban terrain scenarios. Figure D.9 depicts the total and blast impulse for both a conventional blast explosive PBXN-109 and the insensitive DIME formulation B2277A (Table D.2).

Table D.4: Composition and properties of B227A.

Components		Parameter	
Tungsten (wt.-%)	79.5	ρ (g cm^{-3})	5.44
RDX (wt.-%)	15.0	V (m s^{-1})	4400 at 50 mm
Binder (wt.-%)	5.5	Dp (°C)	217
		$\varnothing_{cr}$ (mm)	> 5 < 10

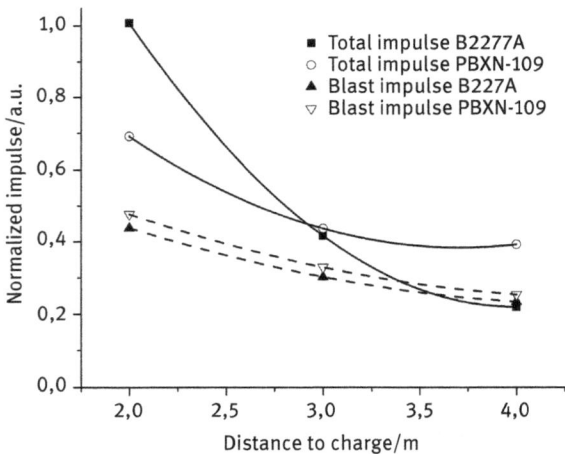

Figure D.9: Normalized impulse of two blast explosives.

DIME formulations have been developed as a so-called *multi-phase blast explosives* for special use with miniaturized bombs having carbon fiber-reinforced composite structures for reduced collateral effects (e.g. GBU 39 A/B and small diameter bomb). Non-lethal injuries sustained from detonating DIME charges are suspected to cause cancer. This is believed to be due to the high content of cobalt in the heavy metal alloys. That is why cobalt was replaced by tungsten heavy alloys to avoid these health issues in the future.

- J. Grundler, G. Guerke, J. Corley, A Method to characterize HE charges according to their potential to produce debris throw, *US DOD Explosives Safety Seminar*, New Orleans, **2000**.
- J. F. Kalinich, C. A. Emond, T. K. Dalton, S. R. Mog, G. D. Coleman, J. E. Kordell, A. C. Miller, D. E. McClain, Embedded Weapons-Grade Tungsten Alloy Shrapnel Rapidly Induces Metastatic High-Grade Rhabdomyosarcomas in F344 Rats, *Environ. Health Perspect.* **2005**, *113*, 7129–734.
- C. Collet, A. Combe, J. Groux, M. Werschine, Development and Characterization of New Families of Explosives Able to Generate Specific Effects, *IMEMTS*, **2012**, Las Vegas.

Dimethylhydrazine, UDMH

1,1-Dimethylhydrazin, Aerozin

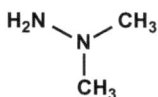

H_2N—$N(CH_3)$—CH_3
|
CH_3

Formula		$C_2H_8N_2$
GHS		02, 05, 06, 08, 09
H-phrases		225, 301, 314, 331, 350, 411
P-phrases		210, 261, 273, 280, 301+310
UN		1163
REACH		LRS
EINECS		200-316-0
CAS		[57-14-7]
m_r	g mol^{-1}	60.099
ρ	g cm^{-3}	0.784 at 25 °C
$\Delta_f H$	kJ mol^{-1}	48.8
$\Delta_c H$	kJ mol^{-1}	−1979.4
	kJ g^{-1}	−32.935
	kJ cm^{-3}	−25.821
Λ	wt.-%	−212.97
N	wt.-%	46.61
c_p	J g^{-1} K^{-1}	164
Mp	°C	−58
Bp	°C	63
Dp	°C	234

UDMH is an important liquid fuel for rocket propulsion and a strong carcinogen.

DINCH – 1,2-cyclohexanedicarboxylic acid diisononylester
Diisononylcyclohexan-1,2-dicarboxylat

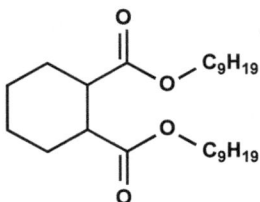

Formula		$C_{26}H_{48}O_4$
REACH		LRS
CAS		[166412-78-8]
m_r	g mol^{-1}	424.70
ρ	g cm^{-3}	0.944–0.954 at 25 °C
Λ	wt.-%	−271.24
Mp	°C	−54

DINCH is a plasticizer that is currently used as a non-toxic replacement for phthalates.

DINGU
2,6-Dinitro-2,4,6,8-tetraazabicyclo[3.3.0]octa-3,7-dion

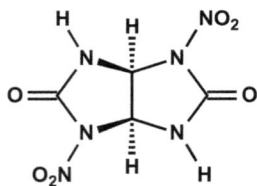

Tetrahydro-1,4-dinitroimidazo[4,5-d] imidazole-2,5(1H,3H)-dione		
Formula		$C_4H_4N_6O_6$
CAS		[55510-04-8]
m_r	g mol^{-1}	232.112
ρ	g cm^{-3}	1.94
$\Delta_f H$	kJ mol^{-1}	−176
$\Delta_{ex}H$	kJ mol^{-1}	−1059.0
	kJ g^{-1}	−4.562
	kJ cm^{-3}	−8.851
$\Delta_c H$	kJ mol^{-1}	−1969.7

	kJ g^{-1}	−8.486
	kJ cm^{-3}	−16.463
Λ	wt.-%	−27.57
N	wt.-%	36.21
Dp	°C	130
Mp	°C	249 (after Dec)
Friction	N	360
Impact	J	7.5
Trauzl	cm^3	445
V$_D$	m s^{-1}	7580 at 1.46 g cm^{-3}
P$_{CJ}$	GPa	*28.5*

DINGU is not stable in contact with alkaline components and does hydrolyze quickly in aqueous solution. Successive nitration of DINGU yields *sorguyl*.

– J. Boileau, E. Wimmer, M. Pierrot, R. Gallo, Structure du dinitroglycolurile (DINGU), *Propellants Explos. Pyrotech.* **1984**, *9*, 180.

5,7-Dinitro-1-picrylbenzotriazole, BTX

5,7-Dinitro-1-pikrylbenzotriazol

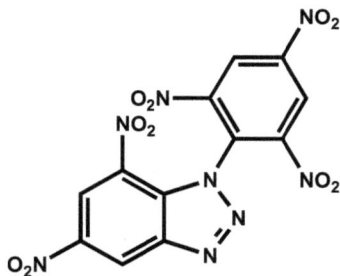

Formula		C$_{12}$H$_4$N$_8$O$_{10}$
CAS		[50892-90-5]
m$_r$	g mol^{-1}	410,211
ρ	g cm^{-3}	1.74
Δ$_f$H	kJ mol^{-1}	299.16
Δ$_{ex}$H	kJ mol^{-1}	−1967.1
	kJ g^{-1}	−4.795
	kJ cm^{-3}	−8.344
Δ$_c$H	kJ mol^{-1}	−5592.9
	kJ g^{-1}	−13.310

	kJ cm^{-3}	−23.159
Λ	wt.-%	−60.92
N	wt.-%	26.67
Mp	°C	263
Dp	°C	>263
Friction	N	>355
Impact	cm	35 with type 12 tool
V_D	m s^{-1}	7170
P_{CJ}	GPa	23.4

BTX is a temperature-resistant high explosive that has been suggested for use in Exploding bridgewire detonators (EBW).

- R. H. Dinegar, L. A. Carlson, M. D. Coburn, BTX – A Useful High Temperature EBW Detonator Explosive, *Int. Det. Symp.*, White Oak, **1976**, 460–465.

2,4-Dinitroanisole, DNAN
2,4-Dinitroanisol

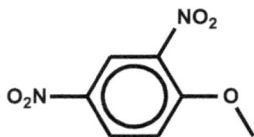

Aspect		Light yellow crystals
Formula		$C_7H_6N_2O_5$
GHS		08, 07
H-phrases		341, 302
P-phrases		264, 281, 301+312, 308+313, 405, 501a
REACH		LPRS
EINECS		204-310-9
CAS		[119-27-7]
m_r	g mol^{-1}	198.135
ρ	g cm^{-3}	1.546
$\Delta_f H$	kJ mol^{-1}	−186
$\Delta_{ex}H$	kJ mol^{-1}	−823.3
	kJ g^{-1}	−4.155
	kJ cm^{-3}	−6.424
$\Delta_c H$	kJ mol^{-1}	−3426.1

	kJ g^{-1}	−17.292
	kJ cm^{-3}	−26.734
Λ	wt.-%	−96.9
N	wt.-%	14.14
Mp	°C	94.5
Dp	°C	347
Friction	N	>360
Impact	J	>50
V$_D$	m s^{-1}	6032 at 1.546 g cm^{-3}
P$_{CJ}$	GPa	13.55

DNAN was first used in wartime Germany as a low-performance *Ersatzsprengstoff*. Nowadays, DNAN has great importance as a melt-cast base for insensitive high explosive, for example, PAX-21, *IMX-101*, or *IMX-104*. Upon temperature cycling, DNAN shows a 15% volume increase ruling out any use in high-performance (SC, EFP, etc.) applications.

- P. Samuels, L. Zunino, K. Patel, B. Travers, E. Wrobel, C. Patel, Irreversible Growth of DNAN Based Formulations, *IMEMTS*, **2012**, Las Vegas.
- P. L. Coster, C. A. Henderson, S. Hunter, W. Marshall, C. R. Pulham, Explosives at extreme conditions Polymorphism of 2,4 dinitroanisole, *NTREM*, **2014**, 164–179.
- A. Provatas, C. Wall, Ageing of Australian DNAN -Based Melt-Cast Insensitive Explosives, *Propellants Explos. Pyrotech.* **2016**, 41, 555–561.

1,3-Dinitrobenzene

1,3-Dinitrobenzol

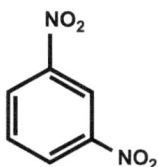

Aspect	Yellowish crystals
Formula	C$_6$H$_4$N$_2$O$_4$
GHS	06, 08, 09
H-phrases	300, 330, 373, 400, 410, 312
P-phrases	260, 284, 301+310, 320, 405, 501a
UN	3443

REACH		LPRS
EINECS		202-776-8
CAS		[99-65-0]
m_r	g mol^{-1}	168.109
ρ	g cm^{-3}	1.575
$\Delta_f H$	kJ mol^{-1}	−27.20
$\Delta_{ex} H$	kJ mol^{-1}	−738.8
	kJ g^{-1}	−4.395
	kJ cm^{-3}	−6.922
$\Delta_c H$	kJ mol^{-1}	−2905.8
	kJ g^{-1}	−17.285
	kJ cm^{-3}	−27.224
Λ	wt.-%	−95.17
N	wt.-%	16.66
Mp	°C	89.9
Bp	°C	291
Friction	N	>360
Impact	J	39
Trauzl	cm^3	242−250
V_D	m s^{-1}	6100 at 1.50 g cm^{-3}

1,3-Dinitrobenzene was used as an *Ersatzsprengstoff* in wartime Germany. Despite having a moderate performance and being practically insensitive, its high toxicity has made it obsolete.

4,6-Dinitrobenzofuroxane, DNBF

4,6-Dinitrobenzofurazoxan

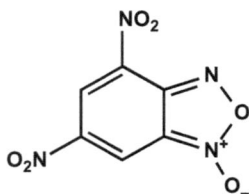

Aspect		Yellow powder (technical grade: red)
Formula		$C_6H_2N_4O_6$
REACH		LRS
EINECS		700-179-1
CAS		[5128-28-9]
m_r	g mol^{-1}	226.105

ρ	$g\ cm^{-3}$	1.79
$\Delta_f H$	$kJ\ mol^{-1}$	192
$\Delta_{ex} H$	$kJ\ mol^{-1}$	−1200.6
	$kJ\ g^{-1}$	−5.310
	$kJ\ cm^{-3}$	−9.505
$\Delta_c H$	$kJ\ mol^{-1}$	−2839.0
	$kJ\ g^{-1}$	−12.556
	$kJ\ cm^{-3}$	−22.475
Λ	wt.-%	−49.53
N	wt.-%	24.78
Mp	°C	174.5
Bp	°C	273 (dec)
Dp	°C	221
Friction	N	>355
Impact	J	4−4.75
V_D	$m\ s^{-1}$	7700 at $1.77\ g\ cm^{-3}$
P_{CJ}	GPa	26

While DNBF itself is not attractive as a secondary high explosive due to high cost and low performance, both the potassium (→*KDNBF*) and the barium salts are interesting primary explosives. Table D.5 shows the sensitiveness and explosion temperatures of the alkali salts of DNBF.

Table D.5: Selected properties of some DNBF – alkali metal salts.

	NaDNBF	KDNBF	RbDNBF	CsDNBF
CAS	[29307-66-2]	[29267-75-2]		
Color	Red brown	Gold orange	Adobe red	Yellow brown
ρ ($g\ cm^{-3}$)	1.72	2.21		
$\Delta_f H$ ($kJ\ mol^{-1}$)	−397	−654		
Mp (°C)	172 Expl.	208 Expl.	188 Expl.	165 Expl.
Impact (J)	0.6	0.7	0.7	0.6
Friction (N)	5.4	2.8	1.0	2

- V. P. Sinditskii, V. Y. Egorshev, V.V. Serushkin, A. V. Margolin, H. W. Dong, Study on Combustion of Metal-Derivatives of 4,6-Dinitrobenzofuroxan, *Theory and Practice of Energetic Materials*, C. Lang, F. Changgen (Eds.), Vol IV, China Science and Technology Press, **2001**, 69−77.
- M. L. Chan, C. D. Lind, P. Politzer, Shock Sensitivity of Energetically Susbtituted Benzofuroxans, *Int. Det. Symp.*, Portland, USA, **1989**, 566- 572.

Dinitrochlorobenzene, 1-chloro-2,4-dinitrobenzene
Chlordinitrobenzol

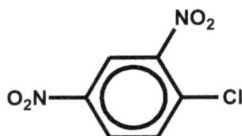

Aspect		Light-yellow crystals
Formula		$C_6ClH_3N_2O_4$
GHS		06, 08, 09
H-phrases		301, 311, 331, 373, 400, 410
P-phrases		260, 301+310, 302+352, 361, 405, 501a
UN		3441
REACH		LRS
EINECS		202-551-4
CAS		[97-00-7]
m_r	g mol^{-1}	202.554
ρ	g cm^{-3}	1.70
$\Delta_f H$	kJ mol^{-1}	−24.27
$\Delta_{ex}H$	kJ mol^{-1}	−776.3
	kJ g^{-1}	−3.832
	kJ cm^{-3}	−6.515
$\Delta_c H$	kJ mol^{-1}	−2715.0
	kJ g^{-1}	−13.404
	kJ cm^{-3}	−22.787
Λ	wt.-%	−71.09
N	wt.-%	13.83
Mp	°C	53
Bp	°C	315
Friction	N	>360
Impact	J	>50
Trauzl	cm^3	225
Koenen	mm	1, no reaction

DNCB is an important intermediate in the synthesis (*Ullmann reaction*) of polynitroaromatic compounds such as *HNS* and *PNP* .

Dinitrochlorhydrin, glycerinechlorohydrin dinitrate, DNCH

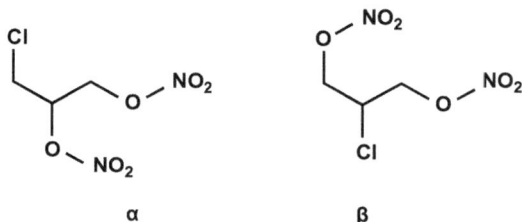

		α	β
CAS		2612-33-1	not listed
Aspect		Clear yellowish liquid	
Formula		$C_3ClH_5N_2O_6$	
m_r	g mol^{-1}	200.536	
		α	β
ρ	g cm^{-3}	1.54	
Λ	wt.-%	−15.96	
N	wt.-%	13.97	
$\Delta_f H$	kJ mol^{-1}	−255 (gas phase),	−256 (gas phase)
$\Delta_{ex} H$	kJ mol^{-1}	−928.20	
	kJ g^{-1}	−4.629	
	kJ cm^{-3}	−7.128	
$\Delta_c H$	kJ mol^{-1}	−1589.5	
	kJ g^{-1}	−7.926	
	kJ cm^{-3}	−12.207	
Mp	°C	5	16.2
Bp	°C	190 (dec)	
Impact	J	7	
Trauzl	cm^3	475	

DNCH occurs in two different isomers, α and β, with a viscosity lower than that of NGI. DNCH is non-hygroscopic and immiscible with water. It is an energetic plasticizer for NC.

Dinitrodiphenylamine, NDPA

Dinitrophenylamin

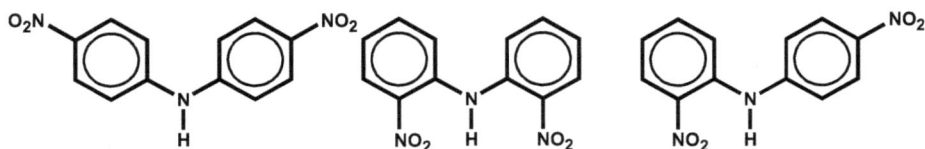

CAS		[1821-27-8]	[18264-71-6]	[612-36-2]
Aspect		Yellow crystals		
Formula		$C_{12}H_9N_3O_4$		
m_r	g mol^{-1}	259.22		
Ω	wt.-%	−151.22		
N	wt.-%	16.21		
ρ	g cm^{-3}		1.575	—
Mp	°C	217-218	172-173	219-220
$\Delta_f H$	kJ mol^{-1}	3.77	−27.20	5.44
$\Delta_c H$	kJ mol^{-1}	−6012.2	−5981.5	−6013.9
	kJ g^{-1}	−23.194	−23.075	−23.200

NDPA theoretically yields six isomers from which only the 2,2'-, 2,4'-, and 4,4'- have been prepared. NDPA is used preferably over MNA as a long-time stabilizing agent due to its lower reactivity.

Dinitroglycerole nitrolactate

Dinitroglycerolnitrolactat

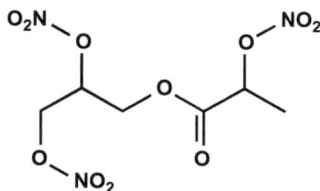

Propanoic acid, 2-(nitrooxy)-, 2,3-bis(nitrooxy)propyl ester		
Formula		$C_6H_9N_3O_{11}$
CAS		[1480889-31-3]
m_r	g mol^{-1}	299.151
ρ	g cm^{-3}	1.47
Δ_fH	kJ mol^{-1}	−618
$\Delta_{ex}H$	kJ mol^{-1}	−1451.4
	kJ g^{-1}	−4.852
	kJ cm^{-3}	−7.132
Δ_cH	kJ mol^{-1}	−3029.4
	kJ g^{-1}	−10.127
	kJ cm^{-3}	−14.886
Λ	wt.-%	−29.42
N	wt.-%	14.05
Bp	°C	190 (dec)

Low-sensitivity gelling agent with low water solubility and good thermal stability.

- M. H. Keshavarz, S. Moradi, B. E. Saatluo, H. Rahimi, A. R. Madram, A simple accurate model for prediction of deflagration temperature of energetic compounds, J. *Therm. Anal. Calorim.* **2013**, *112*, 1453–1463.

Dinitroglycol, EGDN

Dinitroglykol

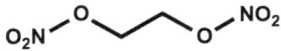

$O_2N{-}O{-}\diagup\diagdown{-}O{-}NO_2$

Ethylene glycol dinitrate, nitroglykol

Formula		$C_2H_4N_2O_6$
CAS		[628-96-6]
m_r	g mol^{-1}	152.064
ρ	g cm^{-3}	1.492
Δ_fH	kJ mol^{-1}	−241.00
$\Delta_{ex}H$	kJ mol^{-1}	−1039.6
	kJ g^{-1}	−6.836
	kJ cm^{-3}	−10.20
Δ_cH	kJ mol^{-1}	−1117.7
	kJ g^{-1}	−7.350
	kJ cm^{-3}	−10.967
Λ	wt.-%	0
N	wt.-%	18.42
Mp	°C	−22
Bp	°C	199 at ambient pressure (101.3 kPa)
	°C	105.5 at 2.53 kPa
P	Pa	10.313
$\Delta_{vap}H$	kJ mol^{-1}	65.1
Dp	°C	217
Friction	N	>353
Impact	J	0.2
Trauzl	cm^3	620
Koenen	mm	24

EGDN is more volatile than NGI and hence is not suitable as an ingredient in gun propellants. It is however used to lower the melting point of blasting gelatine.

- M. A. C. Haertel, T. M. Klapötke, J. Stierstorfer, L. Zehetner, Vapor Pressure of Linear Nitrate Esters Determined by Transpiration Method in Combination with VO-GC/MS, *Propellants Explos. Pyrotech.* **2019**, *44*, 484–492.

2,5-Dinitroimidazole, 2,4-DNI

2,5-Dinitro-1H-imidazol

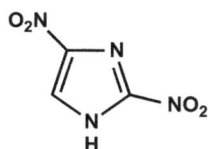

Formula		$C_3H_2N_4O_4$
CAS		[5213-49-0]
m_r	g mol^{-1}	158.073
ρ	g cm^{-3}	1.763
$\Delta_f H$	kJ mol^{-1}	7.91
$\Delta_{ex} H$	kJ mol^{-1}	−766.2
	kJ g^{-1}	−4.847
	kJ cm^{-3}	−8.546
$\Delta_c H$	kJ mol^{-1}	−1474.3
	kJ g^{-1}	−9,327
	kJ cm^{-3}	−16.443
Λ	wt.-%	−30.36
N	wt.-%	35.44
$T_{1 \rightarrow 2}$	°C	228
Mp	°C	280 (dec)
Ea	kJ mol^{-1}	205 ± 15
Friction	N	360
Impact	J	16

Dinitroimidazole is an inexpensive insensitive high explosive. However, it suffers compatibility issues due to its acidic proton.

- M. Anniyappan, S. H. Sonawane, S. J. Pawar, A. K. Sikder, Thermal decomposition and kinetics of 2,4-dinitroimidazole: An insensitive high explosive, *Thermochim. Acta* **2015**, *614*, 93−99.

2,4-Dinitrophenyl-1-ethyl nitrate

β-(2,4-Dinitrophenyl)ethylnitrat

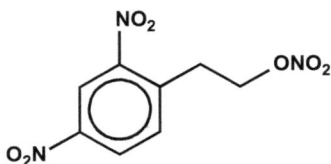

Formula		$C_8H_7N_3O_7$
CAS		[29627-24-5]
m_r	g mol^{-1}	257.16
ρ	g cm^{-3}	1.55
$Δ_fH$	kJ mol^{-1}	−200
$Δ_{ex}H$	kJ mol^{-1}	−923.5
	kJ g^{-1}	−4.458
	kJ cm^{-3}	−6.910
$Δ_cH$	kJ mol^{-1}	−3948.6
	kJ g^{-1}	−15.355
	kJ cm^{-3}	−23.800
Λ	wt.-%	−77.77
N	wt.-%	16.34
Mp	°C	33−35
Dp	°C	150

Dinitroethylphenyl nitrate is a good gelling agent for NC. Due to its low explosion temperature, it is suitable for low erosion propellants. Further advantageous traits are low volatility and insolubility in water. Propellants based on it possess good stability.

- G. Knoffler, *Rauchschwaches Pulver und/oder Treibmittel*, DE 1056989, **1959**, Germany.
- V. G. Sinyavskii, V. Kovaleva, Mono- and dinitrophenylethyl nitrates, their synthesis and properties, *Zurnal organiceskoj chimii* **1970**, *6*, 1692–1696.

3,4-Dinitropyrazole, DNP

3,4-Dinitro-1H-pyrazol

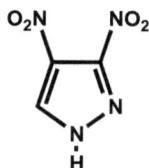

Aspect		Light-yellow crystals
Formula		$C_3H_2N_4O_4$
CAS		[38858-92-3]
m_r	g mol^{-1}	158.074
ρ	g cm^{-3}	1.788
Δ_fH	kJ mol^{-1}	185.2
$\Delta_{ex}H$	kJ mol^{-1}	−913.0
	kJ g^{-1}	−5.776
	kJ cm^{-3}	−10.859
Δ_cH	kJ mol^{-1}	−1651.4
	kJ g^{-1}	−10.447
	kJ cm^{-3}	−18.682
Λ	wt.-%	−30.36
N	wt.-%	35.44
Mp	°C	88.5
Dp	°C	216
Friction	N	>360
Impact	J	20
V_D	m s^{-1}	8115 at 1.79
P_{CJ}	GPa	29.4 at 1.79

DNP is currently investigated as an insensitive TNT replacement as melt-cast base. Though DNP/RDX (60/40) is more powerful than comp B, it is more shock sensitive in the SSGT than comp B.

- J. Ritums, C. Oscarson, M. Liljedahl, P. Goede, K. Dudek, U. Heiche, Evaluation of 3(5),4-Dinitropyrazole (DNP)as New Melt Cast Matrix, *ICT Jata*, **2014**, V2.
- J. Morris, D. Price, A. DiStasio, K. Maier, Energetic Ingredients Research for Freedm Program, *IMEMTS*, Rome, **2015**.
- C. Lizhen, S. Liang, C. Duanlin, W. Jianglong, Crystal Structure of 3,4-dinitropyrazole, $C_3H_2N_4O_4$, *Z. Kristallogr. NCS* **2016**, *231*, 1099–1100.

Dinitrotoluene, DNT

Dinitrotoluol

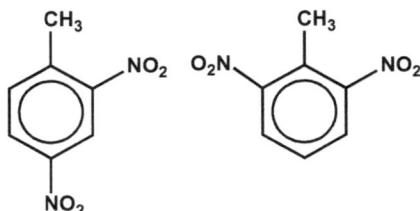

-Dinitrotoluene		2,4	2,6
Aspect		Monoclinic prisms	Yellow needles
Formula		$C_7H_6N_2O_4$	
GHS			06, 08
H-phrases		301-311-331-350-341-361f-373-412	
P-phrases		260-301+310-361-302+352-405-501a	
UN		3454	
REACH		LPRS, SVHC	LPRS
EINECS		204-450-0	210-106-0
CAS		[121-14-2]	[606-20-2]
m_r	g mol^{-1}	182.136	
ρ	g cm^{-3}	1.3208 at 71 °C (mp)	1.2833 at 111 °C
ρ	g cm^{-3}	1.519 at 25 °C	1.536 at 25 °C
$\Delta_f H$	kJ mol^{-1}	−71.55	−55.23
$\Delta_{ex} H$	kJ mol^{-1}	−766.7	−778.7
	kJ g^{-1}	−4.209	−4.275
	kJ cm^{-3}	−6.394	−6.567
$\Delta_c H$	kJ mol^{-1}	−3540.6	−3556.9
	kJ g^{-1}	−19.439	−19.529
	kJ cm^{-3}	−29.528	−29.997
Λ	wt.-%	−114,20	
N	wt.-%	15.38	
$\Delta_{sub} H$	kJ mol^{-1}	99.57	98.32 ($\Delta_v H$)
Mp	°C	71	66
Bp	°C	300	285
Dp	°C	360	360
Friction	N	>360	
Impact	J	>50	
Trauzl	cm^3	240	
Koenen	mm	1 mm (type A)	
V_D	m s^{-1}	5900 at 1.52 g cm^{-3} in 60 mm steel tube	

DNT is an important insensitive energetic plasticizer used in single-base propellants such as M1 and OD6320. It is also an important intermediate in the production of TNT. At infinite diameter, the detonation velocity for variable density can be expressed by $V_D = 1.96 + 2.913\,\rho$.

– D. Price, J. O. Erkman, A. R. Clairmont Jr., D. J. Edwards, Explosive characterization of dinitrotoluene, *Combust. Flame*, **1970**, *14*, 145–148

Dinitroxyethylnitramine, DINA

Dinitroxynitroazapentan

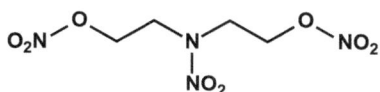

Formula		$C_4H_8N_4O_8$
CAS		[4185-47-1]
m_r	g mol^{-1}	240.13
ρ	g cm^{-3}	1.675
$\Delta_f H$	kJ mol^{-1}	−276
$\Delta_{ex}H$	kJ mol^{-1}	−1317.6
	kJ g^{-1}	−5.487
	kJ cm^{-3}	−9.163
$\Delta_c H$	kJ mol^{-1}	−2441.4
	kJ g^{-1}	−10.167
	kJ cm^{-3}	−17.030
Λ	wt.-%	−26.65
N	wt.-%	23.33
Mp	°C	52.5
Bp	°C	207 (dec.)
Friction	N	360
Impact	J	7.5
Trauzl	cm^3	445
V_D	m s^{-1}	7580 at 1.46 g cm^{-3}
P_{CJ}	GPa	28.5

DINA is a NENA compound and was prepared already in the early 1940s for use as a gelating agent with the albanite gun propellants.

– J. Halfpenny, R. W. H. Small, The Structure of, 2,2'-Dinitroxydiethylnitramine (DINA), *Acta Cryst.* **1978**, *B34*, 3452–3454.

Dioctyl adipate, DOA

Diethylhexyladipat

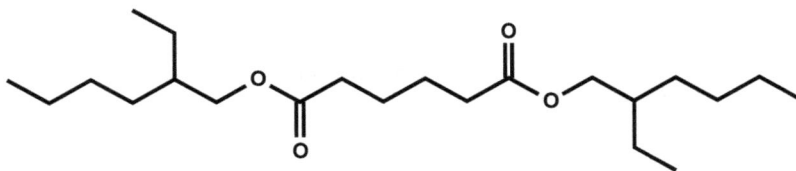

Aspect		Colorless liquid
Formula		$C_{22}H_{42}O_4$
REACH		LPRS
EINECS		203-90-1
CAS		[103-23-1]
m_r	g mol^{-1}	370.567
ρ	g cm^{-3}	0.866
$\Delta_f H$	kJ mol^{-1}	−1215
$\Delta_c H$	kJ mol^{-1}	−13444.8
	kJ g^{-1}	−36.282
	kJ cm^{-3}	−33.561
Λ	wt.-%	−263.37
Mp	°C	−67.8
Bp	°C	215 at 0.67 kPa
Spec		DOD-D-23443

DOA is used as a plasticizer in a series of plastic-bonded explosives (e.g. PBXN-109).

DIPAM, 3,3'-diamino-2,2',4,4',6,6'-hexanitrodiphenyl

Diaminohexanitrodiphenyl

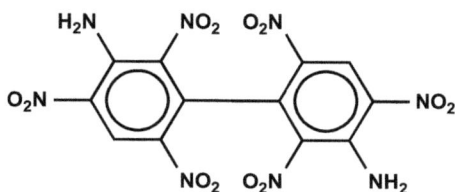

Aspect		Yellow needles
Formula		$C_{12}H_6N_8O_{12}$
REACH		LPRS
EINECS		241-258-6
CAS		[17215-44-0]
m_r	g mol^{-1}	454.226
ρ	g cm^{-3}	1.79
$\Delta_f H$	kJ mol^{-1}	−28.45
$\Delta_{ex} H$	kJ mol^{-1}	−2170.4
	kJ g^{-1}	−4.778
	kJ cm^{-3}	−8.553
$\Delta_c H$	kJ mol$^-$	−5551.3
	kJ g^{-1}	−12.222
	kJ cm^{-3}	−21.876
Λ	wt.-%	−52.84
N	wt.-%	24.67
Mp	°C	306 (dec.)
Impact	cm	128 (type 12 tool)
V_D	m s^{-1}	7490 at 1.79 g cm^{-3}
P_{CJ}	GPa	24.2

DIPAM is a temperature-resistant high explosive.

– M. J. Kamlet, C. V. Sickman, *Process for producing dimethoxyhexanitrobiphenyl and diaminohexanitrobiphenyl*, US Patent 3320320, USA, **1967**.

Dipentaerythritole hexanitrate, DIPEHN

Dipentaerythritolhexanitrat

Dipenta		
Formula		$C_{10}H_{16}N_6O_{19}$
REACH		LPRS
EINECS		236-135-9
CAS		[13184-80-0]
m_r	g mol^{-1}	524.263
ρ	g cm^{-3}	1.630
Δ_fH	kJ mol^{-1}	−979.47
$\Delta_{ex}H$	kJ mol^{-1}	−2706.9
	kJ g^{-1}	−5.163
	kJ cm^{-3}	−8.416
Δ_cH	kJ mol^{-1}	−5242.4
	kJ g^{-1}	−10.000
	kJ cm^{-3}	−16.299
Λ	wt.-%	−27.47
N	wt.-%	16.03
Mp	°C	75
Bp	°C	265 (dec)
Impact	J	4
V_D	m s^{-1}	7410 at 1.589 g cm^{-3}
Trauzl	cm^3	392

DIPEHN is a common side product in the synthesis of PETN. While it is less sensitive than PETN, it is thermally less stable than it and also has a low melting point, and hence needs to be separated from PETN.

- W. Brün, Die sprengtechnischen Eigenschaften des Dipentaerythrithexanitrats, *Z. Schieß. Spreng.* **1932**, *27*, 73–76.

Diphenyl, biphenyl

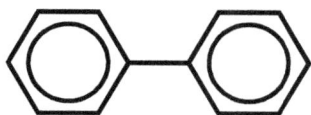

Formula		$C_{12}H_{10}$
GHS		09, 07
H-phrases		400-410-315-335
P-phrases		261-305+351+338-302+352-321-405-501a
UN		3077
EINECS		202-163-5
REACH		LRS
EINECS		202-163-5
CAS		[92-52-4]
m_r	g mol^{-1}	154.21
ρ	g cm^{-3}	0.989 at 77.1 (molten)
	g cm^{-3}	1.04 at 25 °C
$\Delta_f H$	kJ mol^{-1}	100.50
$\Delta_c H$	kJ mol^{-1}	−6251.9
	kJ g^{-1}	−40.542
	kJ cm^{-3}	−42,164
Λ	wt.-%	−300.88
Mp	°C	71
Bp	°C	255.9
$\Delta_{sub} H$	kJ mol^{-1}	82
$\Delta_{vap} H$	kJ mol^{-1}	62
$\Delta_{sm} H$	kJ mol^{-1}	19
P	Pa	7 at 20 °C
c_p	J mol^{-1} K^{-1}	198

Diphenyl is a pearlescent powder with an unpleasant smell. It is used as a carbon source in both IR obscurants and black body decoy flares and has been proposed as a melt-cast binder in firework rocket propellants.

- Z. Deluga, E. Plachta, W. Rembiszewski, B. Florczak, B. Zygmunt, E. Daniluk, M. Koch, M. Gilewicz, M. Maziejuk, *Mieszanina pirotechniczna do wytwarzania dymo maskujacych, zwlascza w podczerwieni*, PL 175254 B1, **1998**, Poland.
- B. Cook, Novel Powder Fuel for Firework Display Rocket Motors, *J. Pyrotech.* **1998**, *7*, 59–64.

Diphenylamine, DPA

Diphenylamin

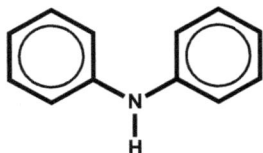

Formula		$C_{12}H_{11}N$
GHS		06, 08, 09
H-phrases		301-311-331-373-400-410
P-phrases		260-301+310-361-302+352-405-501a
UN		3077
REACH		LRS
EINECS		204-539-4
CAS		[122-39-4]
m_r	g mol^{-1}	169.226
ρ	g cm^{-3}	1.158 at 25 °C
		1.05 at 64 °C
$\Delta_f H$	kJ mol^{-1}	130
$\Delta_c H$	kJ mol^{-1}	−6424.3
	kJ g^{-1}	−37.964
	kJ cm^{-3}	−43.962
Λ	wt.-%	−278.91
N	wt.-%	8.28
Mp	°C	53.85
Bp	°C	301.9
$\Delta_{sm} H$	J mol^{-1}	17.9
$\Delta_{vap} H$	J mol^{-1}	55.2
c_p	J K^{-1} mol^{-1}	238.61

DPA was the first stabilizer for nitrocellulose proposed by *Nobel* in 1888 for use in *Ballistite* gun propellant. Though other stabilizers have evolved since then (e.g. *centralites* and *akardites*), DPA remains an important stabilizing ingredient in certain gun propellants. The first DPA reaction product found in stored gun propellants is *N*-nitrosodiphenylamine CAS: [86-30-6] which by itself is a good stabilizer for NC.

Diphenylarsine chloride, DA

Diphenylarsinchlorid

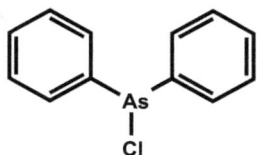

Diphenylchlorarsin, Clark I, sneezing gas		
Aspect		Pure: colourless crystals, Technical: brown liquid
Formula		$AsC_{12}ClH_{10}$
REACH		LPRS
EINECS		211-921-4
CAS		[712-48-1]
m_r	g mol^{-1}	264.586
ρ	g cm^{-3}	1.422 at 25 °C
		1.388 at 50 °C (molten)
$\Delta_f H$	kJ mol^{-1}	−96
$\Delta_c H$	kJ mol^{-1}	−6334
	kJ g^{-1}	−23.939
	kJ cm^{-3}	−34.042
Λ	wt.-%	−181.41
Mp	°C	44
Bp	°C	333 (extrapolated) (dec)
$D_{vap}H$	kJ mol^{-1}	63.2
P_{vap}	Pa	$6.7 \cdot 10^{-2}$
ICt_{50}	ppm min^{-1}	15
LCt_{50}	ppm min^{-1}	15,000

– P. O. Dunstan, Thermochemistry of complexes of pyridine and picolines with phenyldihaloarsines, *Thermochim. Acta* **1992**, *197*, 201–210.

Diphenyl-*N*-nitrosamine, DPNA

N-Nitrosodiphenylamin

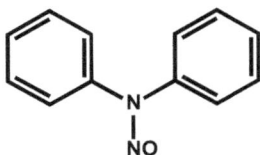

Aspect		Brown crystals
Formula		$C_{12}H_{10}N_2O$
REACH		LPRS
EINECS		201-663-0
CAS		[86-30-6]
m_r	g mol^{-1}	198.224
ρ	g cm^{-3}	1.251 at 25 °C
$\Delta_f H$	kJ mol^{-1}	227
$\Delta_c H$	kJ mol^{-1}	−6378.4
	kJ g^{-1}	−32.178
	kJ cm^{-3}	−40.255
Λ	wt.-%	−226.0
N	wt.-%	14.133
Mp	°C	67.2
Dp	°C	115
Bp	°C	215

N-Nitrosodiphenylamine is the first reaction product formed from *diphenylamine* with nitrous acid in stored gun propellants. By itself it is also a good stabilizer and also used industrially as an antioxidant to slow down the curing of rubber. In the presence of mineral acids, *N*-nitrosodiphenylamine can isomerize to 4-nitrosodiphenylamine CAS: [156-10-5].

– Banerjee, C. J. Brown, J. F. P. Lewis, N-Nitrosodiphenylamine, *Acta Cryst.* **1982**, *B38*, 2744–2745.

Diphenylurethane
N,N-Diphenylcarbaminsäureethylester

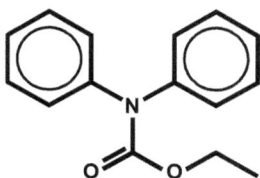

Ethyl diphenylcarbamate		
Formula		$C_{15}H_{15}NO_2$
REACH		LPRS
EINECS		210-047-0
CAS		[603-52-1]
m_r	g mol^{-1}	241.29
ρ	g cm^{-3}	1.146
$\Delta_f H$	kJ mol^{-1}	−280.75
$\Delta_c H$	kJ mol^{-1}	−7765.8
	kJ g^{-1}	−32.185
	kJ cm^{-3}	−36.884
Λ	wt.-%	−235.39
N	wt.-%	5.81
Mp	°C	72
Bp	°C	360

Diphenylurethane is an important stabilizer for nitrocellulose.

– A.W. Archer, Separation and Identification of minor components in smokeless powders by thin-layer chromatography, *J. Chromatography*, **1975**, *108*, 401–404.

Dipicryl-1,3,4-oxadiazole, DPO

2,5-Dipikryl-1,3,4-oxadiazol

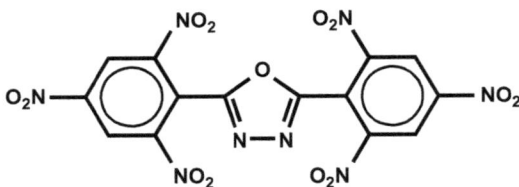

5,5'-Bis(2,4,6-trinitrophenyl)-1,3,4-oxadiazole		
Aspect		Light-yellow crystals
Formula		$C_{14}H_4N_8O_{13}$
CAS		[22358-64-1]
m_r	g mol^{-1}	492.232
ρ	g cm^{-3}	1.87
$\Delta_f H$	kJ mol^{-1}	84
$\Delta_{ex}H$	kJ mol^{-1}	−2385.4
	kJ g^{-1}	−4.846
	kJ cm^{-3}	−9.062
$\Delta_c H$	kJ mol^{-1}	−6164.9
	kJ g^{-1}	−12.525
	kJ cm^{-3}	−23.421
Λ	wt.-%	−55.26
N	wt.-%	22.77
Mp	°C	335
T_{5ex}	°C	>370
V_D	m s^{-1}	6965 at 1.605 g cm^{-3}
		7747 at 1.87 g cm^3
P_{CJ}	GPa	19.5 at 1.605 g cm^{-3}
		28.55 at 1.87 g cm^{-3}
$\varnothing_{cr}$	mm	12 at 1.605 g cm^{-3}
Impact	J	6

DPO is a high-temperature-resistant, friction-insensitive high explosive with an impact and shock sensitivity comparable to PETN.

– M. E. Sitzmann, 2,5-Dipicryl-1,3,4-oxadiazole: A shock-sensitive explosive with high thermal stability (thermally sTable substitute for PETN), *J. Energ. Mater.* **1988**, *6*, 129–144.
– D. L. Sheng, L. K. Cheng, B. Yang, Y. H. Zhu, M. H. Xu, Performances of new heat-resistant insensitive booster explosive 2,5 dipikryl-1,3,4-oxadiazole, *Hanneng Calliao*, **2011**, *19*, 184–188.

DMA, dynamic mechanical analysis

DMA, Dynamisch mechanische Analyse

The DMA serves the determination of strain and lengthening.

Therefore, a sample body is subjected to a dynamic load (typically sinusoidal) and tension and elongation, that is, the plastic–elastic deformation is measured. Two elastic moduli are obtained, the storage modulus, E', and the loss modulus, E''. For this test, the sample is pretensioned with a constant elongation/strain on which the actual dynamic load is superimposed. The measurement is done with a stepwise increase in temperature.

In accordance with STANAG the following parameters are chosen (Table D.6):

Table D.6: DMA – parameter in accordance with STANAG.

Parameter	Wert
Mode	Elongation
Frequency (Hz)	1, 10, 100
Pretension	0.3%, maximum 60 N
Dynamic load	0.05%, maximum 40 N
Heating rate (K min^{-1})	2

The temperature at the maximum of the loss modulus, E'', is of greatest importance as this is the temperature of the glass transition, T_g. That is, below this temperature, brittle fracture has to be expected. Hence this temperature should be as low as possible and should be below the lowest required service temperature. A brittle fracture in say gun propellant grains spontaneously increases the combustion surface area and hence can result in a steep pressure rise and a consecutive transition from laminar to convective combustion (DDT). With high explosives, brittle fracture can affect premature ignition (e.g. set back in bore premature)

- W. deKlerk, Mechanical Analysis – Different Methods and Applications on Energetic Materials, *IMEMTS*, Rom, **2015**.
- STANAG 4540, ed. 1, September 2001 Explosives, Procedures for Dynamic Mechanical Analysis (DMA) and Determination of Glass Transition Temperature.

DNDA

DNDA is the acronym for open chain dinitrodiaza compounds which are used as plasticizers for temperature-independent gun propellants (Ger: *TU-Pulver*) developed by *Langlotz* and *Müller*. The famous DNDA-5,7 is a blend from 2,4-dinitro-2,4-diazapentan, (DNDA-5), 2,4-dinitro-2,4-diazahexan (DNDA-6), and 3,5-dinitro-3,5-diazaheptan (DNDA-7)in the about mass proprtions 40 wt.-%:45 wt.-%:15 wt.-%. DNDA-5,7 can be obtained by nitration of a mixture of *N,N'*-dimethylurea and *N,N'*-diethylurea, hydrolysis of both dinitrodimethylurea and dinitrodiethylurea, and subsequent condensation reaction with methyl- and ethylnitramine. A typical TU-Pulver is given as follows:

N11 /M1000
- 35 wt.-% nitrocellulose
- 40 wt.-% RDX
- 23 wt.-% DNDA-5,7
- 2 wt.-% stabilizer and additives

DNDA yields a cold brittleness of the powder resulting in an *erosive burn* at low temperatures. This compensates the temperature dependence of Vieille's law and assists in retaining the ambient temperature vivacity.

- G. Pauly, R. Scheibel, Burning Behavior of Nitramine Gun Propellants under the Influence of Pressure Oscillations, *Propellants Explos. Pyrotech.* **2010**, *35*, 284–291.
- H. G. Emans, L. Lichtblau, R. Schirra, *Verfahren zur Herstellung von DNDA*, EP000001317415B1, Germany, **2005**.
- W. Langlotz, D. Müller, *Treibladungspulver für Rohrwaffen*, DE 19757469A1, **1997**, Germany.

DNDA-5
2,4-Dinitro-2,4-diazapentan

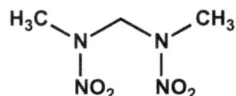

2,4-Dinitro-2,4-diazapentane		
Formula		$C_3H_8N_4O_4$
CAS		[13232-00-3]
m_r	g mol^{-1}	164.121
ρ	g cm^{-3}	1.389
$\Delta_f H$	kJ mol^{-1}	−31.76
$\Delta_{ex} H$	kJ mol^{-1}	−851.5

	kJ g^{-1}	−5.188
	kJ cm^{-3}	−7.206
$\Delta_c H$	kJ mol^{-1}	−2292.1
	kJ g^{-1}	−13.966
	kJ cm^{-3}	−19.399
Λ	wt.-%	−58.49
N	wt.-%	34.14
T_g	°C	−47
Mp	°C	56
$\Delta_m H$	kJ mol^{-1}	−20.48
Dp	°C	227

−   L. Goodman, Condensations of primary aliphatic nitramines with formaldehyde, *J. Am. Chem. Soc.* **1953**, *75*, 3019–3020.

DNDA-6

2,4-Dinitro-1,4-diazahexan

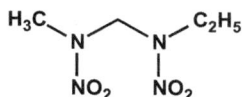

2,4-Dinitro-2,4-diazahexane

Formula		C$_4$H$_{10}$N$_4$O$_4$
CAS		[168983-72-0]
m_r	g mol^{-1}	178.148
ρ	g cm^{-3}	1.323
$\Delta_f H$	kJ mol^{-1}	−79.50
$\Delta_{ex} H$	kJ mol^{-1}	−866.2
	kJ g^{-1}	−4.862
	kJ cm^{-3}	−6.433
$\Delta_c H$	kJ mol^{-1}	−2923.7
	kJ g^{-1}	−16.412
	kJ cm^{-3}	−21.713
Λ	wt.-%	−80.83
N	wt.-%	31.45
T_g	°C	−60
Mp	°C	33
$\Delta_m H$	kJ mol^{-1}	−20.93
Dp	°C	222

−   V. A. Tartakovsky, A. S. Ermakov, N. V. Sigai, O. N. Varfolomeeva, Synthesis of *N,N'*-dialkylmethylenebis(nitramines) from *N*-alkylsulfamates, *Russ. Chem. Bull.* **2000**, *49*, 1079–1081.

DNDA-7

3,5-Dinitro-3,5-diazaheptan

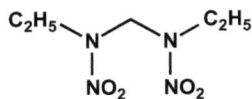

$$C_2H_5-N(NO_2)-CH_2-N(NO_2)-C_2H_5$$

3,5-Dinitro-3,5-diazaheptane		
Formula		$C_5H_{12}N_4O_4$
CAS		[134273-34-0]
m_r	g mol^{-1}	192.173
ρ	g cm^{-3}	1.271
Δ_fH	kJ mol^{-1}	−135.10
$\Delta_{ex}H$	kJ mol^{-1}	−876.2
	kJ g^{-1}	−4.559
	kJ cm^{-3}	−5.795
Δ_cH	kJ mol^{-1}	−3547.5
	kJ g^{-1}	−18.460
	kJ cm^{-3}	−23.462
Λ	wt.-%	−99.91
N	wt.-%	29.15
T_g	°C	−10
Mp	°C	76
Δ_mH	kJ mol^{-1}	−32.84
Dp	°C	229

- G. Bunte, H. Schuppler, H. Krause, Thermal Analytical Characterization of DNDA -5, DNDA-6 and DNDA-7 and certain Binary and Ternary Mixtures, *ICT -Jata*, Karlsruhe, **2004**, P-174.

DODECA

2,2',2'',2''',4,4',4'',4''',6,6',6'',6'''-Dodecanitro-1,1' 3'1'' 3'',1'''-quaterphenyl

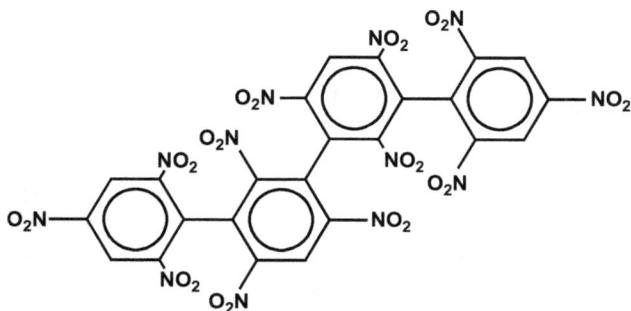

Aspect		Creme crystals
Formula		$C_{24}H_6N_{12}O_{24}$
CAS		[23242-92-4]
m_r	g mol^{-1}	846.37
ρ	g cm^{-3}	1.81
$\Delta_f H$	kJ mol^{-1}	211.71
$\Delta_{ex}H$	kJ mol^{-1}	−4263.7
	kJ g^{-1}	−5.038
	kJ cm^{-3}	−9.118
$\Delta_c H$	kJ mol^{-1}	−10513.7
	kJ g^{-1}	−12.422
	kJ cm^{-3}	−22.484
Λ	wt.-%	−51.04
N	wt.-%	19.86
Mp	°C	>425

DODECA is an extremely temperature-resistant high explosive. Due to the steric repulsion of the nitro groups, DODECA is a bent molecule in the crystalline state with dihedral angles between the phenyl and phenylene units of 90° each.

– S. Zeman, M. Roháč, Z. Friedl, A. Růžička, A. Lyčka, Crystallography and Structure–Property Relationships in 2,2',2'',2''',4,4',4'',4''',6,6',6'',6'''-Dodecanitro-1,1'3'1'' 3'',1'''- Quaterphenyl (DODECA), *Propellants Explos. Pyrotech.* **2010**, *35*, 339–346.

Donarit

Donarit is the trade name for a mining explosive based on ammonium nitrate with low NGl content.

Döring, Werner (1911–2006)

Döring was a German theoretical physicist. As a disciple of *Richard Becker,* he became introduced to detonation physics. In the 1940s, Döring developed, independently but coincidentally with *Zeldovitch* and *von Neumann*, a theory that describes gas detonations which nowadays is known as *ZND model* (Zeldovitch–von Neumann–Döring).

- W. Döring, Über den Detonationsvorgang in Gasen, *Ann. Phys.* **1943**, 421–436
- W. Döring, Die Geschwindigkeit und Struktur von intensiven Stoßwellen in Gasen, *Ann. Phys.* **1949**, 133–150.
- W. Döring, Über die Detonationsgeschwindigkeit des Methans und Dizyans im Gemisch mit Sauerstoff und Stickstoff, *Z. Elektrochem.* **1950**, *54*, 231–239.
- Entry in the catalogue of the German National Library: http://d-nb.info/gnd/137378165

Double-base propellant
Zweibasiges Treibladungspulver

Double-base propellants are formulations containing about 40–70 wt.-% *nitrocellulose* and between 20 and 45 wt.-% *blasting oil* (*NGL/DEGDN*). For a binary system from NC and NGl, Figure D.10 shows the calculated performance. While the heat of explosion, Q_{ex}, increases linearly up to 90% NGl with increasing mass content of NGl, the force only increases up to a content of 55 wt.-% NGl (equaling an oxygen balance of − 13.6 g O_2/100 g) as with higher NGl contents the number of moles decreases. This is due to formation of less but higher oxidized products.

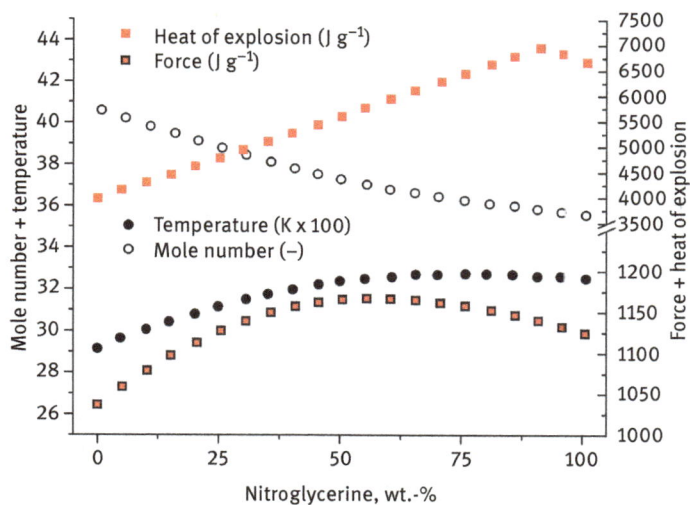

Figure D.10: Calculated thermochemical performance data for a binary NC/NGl system after *Volk*.

– F. Volk. H. Bathelt, Influence of Energetic Materials on the Energy Output of Gun-Propellants, *Propellants Explos. Pyrotech.* **1997**, *22*, 120–124.

Douda, Bernard E. (1930–2020)

Figure D.11: *Dr. Bernard E. Douda* (Photo courtesy of Sara Pliskin, PhD).

Bernard Douda was an American chemist and international pyrotechnics expert (Figure D.11). He studied at Cornell College and received a BS in chemistry in 1951. After serving for the United States Navy in the Korean War he was transferred to Crane Naval Ammunition Depot, IN, as a staff chemist. There he began research into the then developing field of pyrotechnic illuminants spectroscopy and its modelling. He received a MS in chemistry and finally completed his dissertation at the University of Indiana in 1971 on a subject of flare spectroscopy. His corresponding publications have built the foundation for today's understanding of the chemistry and physics of pyrotechnic flares. Douda was a visionary, to suggest homogeneous energetic materials as replacements for multicomponent pyrotechnics as early as 1964. Then he prepared tris-glycin-κ-N,N',N''-strontium diperchlorate a non-hygroscopic (sic!) crystalline material serving altogether as fuel, oxidizer, and color agent. The pure compound burns with an intensive red colored flame with no smoke. Douda further developed plastic-bonded pyrotechnic illuminants and in his final service years also contributed to the historic account of the US Navy research conducted in the field of infrared decoy flares. In 1983, together with Bill Cronk, Joseph H. McLain, Allen Tulis, and Bob Blunt, he founded the International Pyrotechnics Society (IPS) and became president of the IPS from 1988 to 1990. He is recipient of the Dr. Fred Saalfeld Award of the United States Navy (2010) and eponym of a young scientist award of the IPS since 2012. Douda, after 60 years of service for the United States Navy, retired in 2011 at the age of 81.

- B. E. Douda, *Glycine-Strontium Perchlorate Compounds Synthesis, Characterization and Discussions, RDTN No.26*, U.S. Naval Ammunition Depot Crane, IN, **1963**, 33 pp.
- B. E. Douda, *Unique Chemical Compound; Synthesis and Characterization, RDTN No. 52*, Naval Ammunition Depot Crane, IN, **1963**, 16 pp.
- B. E. Douda, *Theory of Colored Flame Production, RDTN No. 71*, Naval Ammunition Depot Crane, IN, **1964**, 60 pp.
- B. E. Douda, Emission Studies of selected pyrotechnic flames, *J. Opt. Soc. Am.* **1965**, *55*, 787–793.
- B. E. Douda, *Pyrotechnic Compound tris(glycine)strontium(II) perchlorate*, US Patent 3,296,045, USA, **1967**.
- B. E. Douda, R. M. Blunt, E. J. Blair, Visible Radiation from illuminating-flare flames: strong emission features, *J. Opt. Soc. Am.* **1970**, *60*, 1116–1119.
- B. E. Douda, E. J. Blair, Visible Radiation from illuminating-flare flames. II. Formation of the sodium resonance continuum, *J. Opt. Soc. Am.* **1970**, *60*, 1257–1261.
- B. E. Douda, E. J. Blair, Radiative transfer model of a pyrotechnic flame, *J. Quant. Spect. Rad. Transf.* **1974**, *14*, 1091–1105.
- B. E. Douda, R. J. Exton, Optically thick line widtths in pyrotechnic flares, *J. Quant. Spect. Rad. Transf.* **1975**, *15*, 615–617.
- B. E. Douda, Green chemical light spectrum, *Am. J. Opt. Phys. Opt.* **1975**, *52*, 123–124.
- B. E. Douda, *The Genesis of Infrared Decoy Flares – The Early Years from 1950–1970*, Naval Surface Warfare Center, Crane Division, Crane, Indiana, USA, **2009**.
- E.-C. Koch, B. E. Douda -on his 80th birthday, *Propellants Explos. Pyrotech.* **2010**, *35*, 203–204.
- E.-C. Koch, B. E. Douda receives US Navy 2010 Dr. Fred Saalfeld Award for Outstanding Lifetime Achievements in Science, *Propellants Explos. Pyrotech.* **2011**, *36*, 7.
- S. Pliskin, E.-C. Koch, Bernard E. Douda (1930–2020), *Propellants Explos. Pyrotech.* **2020**, *45*, 693–694.

DSC, differential scanning calorimetry
Dynamische Differenzkalorimetrie

DSC is a thermoanalytical method to determine temperature-dependent physical and or chemical processes that are associated with an enthalpy change in the sample. The sample size is typically in the two-figure milligram range explaining the low sensitivity of the method.

In a DSC experiment, both a sample in a metal or ceramic crucible (open or closed) and a standard reference (e.g. an empty crucible) are subjected to a predetermined temperature/time program (mostly heating but also cooling and cycling). The heat capacity of the sample as well as endothermal and exothermal processes going on in the sample result in a characteristic heat flow into and from the crucible which can be determined. Typical parameters that can be determined comprise:

- Inherent (extensive) properties
  - Heat capacity, cp
  - Crystallinity
- Reversible processes
  - Glas transition
  - Phase change
- Irreversible processes
  - Endothermal (decomposition)
  - Exothermal (oxidation, decomposition)

A similar method – however determining the temperature change only without measuring the heat flow – is the differential thermal analysis. The latter is often coupled with measuring the mass loss (thermogravimetry).

- R. C. Mackenzie, *Differential Thermal Analysis*, 2 Volumes, Academic Press, London, **1970**.
- STANAG 4515, ed. 2, Explosives, Thermal Analysis using Differential Thermal Analysis (DTA), Differential Scanning Calorimetry (DSC), Heat Flow Calorimetry (HFC) and Thermogravimetric Analysis (TGA); Brussels, **2015**, 41 pp.

DTTO and *iso*-DTTO

[1,2,3,4]Tetrazino[5,6-e]-1,2,3,4-tetrazin-1,3,6,8-tetraoxid (DTTO)
[1,2,3,4]Tetrazino[5,6-e]-1,2,3,4-tetrazin-2,4,6,8-tetraoxid (iso-DTTO)

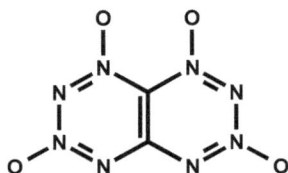
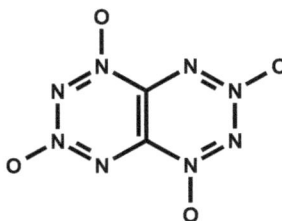

DTTO   *iso*-**DTTO**

Acronym		DTTO /TTTO	*iso*-DTTO/*iso*-TTTO
Formula		$C_2N_8O_4$	
CAS		[244613-32-9]	[244613-33-0]
m_r	g mol^{-1}	200.07	
$\rho_{calc.}$	g cm^{-3}	1.97 ± 0.10	2.00 ± 0.10
$\Delta_f H$	kJ mol^{-1}	+959.6	+955
$\Delta_{ex}H$	kJ mol^{-1}	−17585	
	kJ g^{-1}	−8.789	
	kJ cm^{-3}	−17.315	
$\Delta_c H$	kJ mol^{-1}	−1746.6	
	kJ g^{-1}	−8.730	
	kJ cm^{-3}	−17.198	
Λ	wt.-%	0	
N	wt.-%	56.00	
Mp	°C	183-186 (dec.)	n.a.
V_D*	m s^{-1}	9560	9670
P_{CJ}*	GPa	48.23	50.41

*with Cheetah 7.0 at TMD.

Predictions call for high density and high positive heat of formation for both DDTO and *iso*-DTTO. Hence, both explosives could yield detonation properties in excess of HMX add even CL-20. The recently synthesized DTTO, however, is extremely sensitive to hydrolysis imposing on any practical use (see also *sorguyl* or *hexanitrobenzene*).

- K. O. Christe, D. A. Dixon, M. Vasiliu, R. I. Wagner, R. Haiges, J. A. Boatz, H. L. Ammon, Are DTTO and iso-DTTO worthwile Targets for Synthesis? *Propellants Explos. Pyrotech.* **2015**, *40*, 463–468.
- M. S. Klenov, A. A. Guskov, O. V. Anikin, A. M. Churakov, Y. A. Strelenko, I. V. Fedyanin, K. A. Lyssenko, V. A. Tartakovsky, Synthesis of Tetrazino-tetrazine 1,3,6,8-Tetraoxide (TTTO), *Anwt. Chem. Int. Ed.* **2016**, *55*, 11472–11475.

DU, depleted uranium
Abgereichertes Uran

Depleted uranium (^{238}U > 99.7 wt.-%, ^{235}U = 0.2–0.3 wt.-%) due to its high density (ρ = 19.16 g cm^{-3}) and hardness serves both as armor in vehicles and as self-sharpening projectile (by adiabatic shear) in kinetic energy ammunition with calibers greater than 20 mm. Debris formed upon impact and penetration of DU projectiles is pyrophoric and yields complimentary thermal effects. The chemical toxicity of uranium as well as its radioactivity

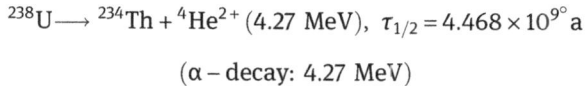

$$^{238}\text{U} \longrightarrow {}^{234}\text{Th} + {}^{4}\text{He}^{2+} \, (4.27 \text{ MeV}), \ \tau_{1/2} = 4.468 \times 10^{9°}\,a$$

$$(\alpha - \text{decay: } 4.27 \text{ MeV})$$

and the debated relation to medical conditions to exposed personel have led to repeated requests for a moratorium of its use.

– L. S. Magness and T. G. Farrand, Deformation Behavior and its Relationship to the Penetration Performance of High Density KE-Penetrator Materials, *Proc. 1990 Army Science Conference*, Durham, May **1990**, pp. 149–164.
– L. BakerJr., J. G. Schnizlein, J. D. Bingle, The ignition of uranium, *J. Nucl. Materials*, **1966**, *20*, 22–38.

Duttenhofer, Max (1843–1903)

Figure D.12: Max von Duttenhofer (1890).

Max Duttenhofer was a German entrepreneur who run the *Pulverfabrik Rottweil (Rottweil gunpowder works)* from 1863 to 1903 (Figure D.12). Before becoming involved with explosives, he had studied pharmacy and made his dissertation. As director of the *Pulverfabrik Rottweil*, his scientific education led him to rigorously promote research into the underlying chemistry of gun powders and ballistics. It was hence that he was more successful than many of his national and international competitors. In 1884, coincidently with *Vieille*, but idependently his company invented the first smokeless gun powder (R.C.P = *Rottweil Chemisches Pulver*) which was produced by nitration of grains from partially carbonized alder buckthorn wood (the same raw material Duttenhofer used for producing regular black powder). Gelling of the nitrated charcoal with ethyl acetate and extrusion yielded strands that were cut in size, dried, and finally coated with graphite. Though the R.C.P gave a blueing of KI–I$_2$ starch paper (*Abel test*) after 4 min only, a sample stored at ambient temperature produced in 1886 was still in mint condition in 1939. In 1896, Duttenhofer was awarded nobility and became "von Duttenhofer" by the King of Baden.

– O. Schmid. *Vom braunen prismatischen Pulver zum R.C.P.*, I.G. Farbenindustrie AG, Rottweil, **1939**, 80 pp.

Dynamite
Dynamit

The trade name dynamite was chosen by *Alfred Nobel* for the first plastic explosive formulation based on NGI adsorbed on diatomaceous earth (kieselguhr). Many other NGI-based compositions and even formulations not containing any NGI at all are still being referred to as dynamites.
- **Guhr dynamite** NGl adsorbed in diatomaceous earth. This first dynamite type is no longer used today.
- **Blasting gelatine** Blasting oil (NGl/DEGN) gelatinised with nitrocellulose
- **Mixed dynamite** containing wood meal, inorganic nitrates, and relatively low mass percentages of nitroglycerin (4–12 wt.-%)
- **Gelatine dynamite** Mixtures of Blasting gelatine with certain amounts of dinitroglycol and aromatic nitrocompounds to lower the freezing point of blasting oil
- **Permitted dynamites** Gelled or non-gelled formulations containing alkali salts to lower the deronation temperature (see also *commercial explosives*).

E

ECL®

Extruded composite low sensitivity (ECL®) gun propellant is a registered trademark of Nitrochemie Wimmis AG, Switzerland. ECL® is NGl free. It contains NC, a nitramine (RDX, or HMX) in mass percentages between 1 and 25 wt.-% as well as an inert plasticizer, for example, phthalic esters of long-chain linear C_9C_{11} alcohols (Table E.1). ECL® are insensitive towards bullet impact or fragment impact (see *Insensitive Munitions*). In addition, they show a reduced influence of temperature on ballistic performance when compared to common single-base GPs (Table E.2). ECL are used for propelling charges for calibers between 30 and 120 mm.

Table E.1: Typical ECL® composition.

Components	wt.-%
Nitrocellulose, 13.2 wt.-% N	68.23
RDX	24.00
Akardite II	1.73
Potassium sulfate	0.38
Manganese(II) oxide	0.10
Calcium carbonate	0.19
Phthalic ester	1.44
Graphite	0.14
Camphor	3.84

Table E.2: Sensitiveness and performance of a ECL® seven-hole grain GP-loaded cartridge.*

Parameter	Wert
Bulk density (g L^{-1})	1024
Heat of explosion (J g^{-1})	3580
Autoignition temperature (°C)	179
Bullet impact	V[#]
Fragment impact	V[#]
Impact (J)	19

*Determined on a 30 × 173 mm cartridge containing 174 g GP in accordance with STANAG 4439. [#]Reaction type in accordance with STANAG 4439.

https://doi.org/10.1515/9783110660562-005

- U. Schaedeli, H. Andres, K. Ryf, D. Antenen, B. Vogelsanger, *Antrieb zur Beschleunigung von Geschossen*, EP1857429B1, **2013**, Schweiz.
- B. Vogelsanger, U. Schädeli, D. Antenen, ECL – A New propellant family with improved safety and performance properties, *ICT -JATA*, **2007**, Karlsruhe, V-15.

Ednatol

Ednatol designates a melt-castable high explosive based on ethylene dinitramine and TNT. Typical ednatol formulations and their properties are given in Tables E.3 and E.4.

Table E.3: Ednatol formulations.

	60/40	55/45	50/50
Ethylene dinitramine (wt.-%)	60	55	50
Trinitrotoluene (wt.-%)	40	45	50
Theoretical density (g cm^{-3})	1.710	1.705	1.701

Table E.4: Sensitiveness and performance of ednatol 55/45.

	55/45
ρ (g cm^{-3})	1.62
Ω (wt.-%)	−51
V_D (m s^{-1})	7340
P_{CJ} (GPa)	21.78
Trauzl (cm^3)	360
$\Delta_{ex}H$ (kJ cm^{-3})	−7.976
Impact (J)	19

- U.S. Army Material Command (USAMC), *Engineering Design Handbook; Explosives Series, Properties of Explosives of Military Interest, AMCP 706-177 (AD-764-340)* Headquarters USAMC, Washington, DC, **1971**, pp. 130–132

EI(D)S⁻

Extremely insensitive (detonating) substances (EI(D)S) are high explosives that yield no positive result in <u>any</u> of the UN Test Series 7a–f tests. Articles/munitions containing EI(D)S which do not yield a positive outcome in any of the UN Test Series 7g–k are then categorized under HD 1.6. High explosives that have been successfully classified as EI(D)S typically contain non-ideal high explosives (e.g. ammonium perchlorate) and often constitute mixtures from ideal high explosives and oxidizers. Selected EI(D)S and their properties are given in Table E.5.

Table E.5: EI(D)S formulations.

Name	Country	Composition	ρ (g cm^{-3})	V_D (m s^{-1})	P_{CJ} (GPa)
RDC35	GB	TATB/Kel-F-800 = 95/5	1.90	7710	28.9
AFX-930	USA	RDX/NGu/Al/HTPB = 32/37/15/16	1.60	6700	
XF13333	F	NTO/TNT/Al/Wachs = 48/31/14/7	1.70	7143	22.6
FOXIT	FIN	I-RDX/AP/Al/HTPB = ?/?/?/?	1.80	5500	

- *Recommendations on the Transport of Dangerous Goods – Manual of Tests and Criteria –* 6th rev. Ed. United Nations, **2009**, pp. 157–175.
- E.-C. Koch, Extrem Insensitive Detonierende Stoffe, EIDS, Testverfahren und Materialien, *Seminar – Insensitive Munition,* Bundesakademie für Wehrverwaltung und Wehrtechnik, Mannheim, **2008**.

Eicosanitrododecahedrane, ENDH

Eicosanitrododecahedran

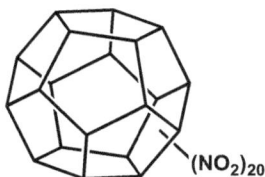

(NO$_2$)$_{20}$

Formula		C$_{20}$N$_{20}$O$_{40}$
m_r	g mol^{-1}	1160.333
ρ	g cm^{-3}	2.2–2.4
$\Delta_f H$	kJ mol^{-1}	−353.55
$\Delta_{ex} H$	kJ mol^{-1}	−7581.9
	kJ g^{-1}	−6.534
	kJ cm^{-3}	−14.375
$\Delta_c H$	kJ mol^{-1}	−7516.9
	kJ g^{-1}	−6.478
	kJ cm^{-3}	−14.252
Λ	wt.-%	0
N	wt.-%	24.14
V_D	m s^{-1}	10186 at 2.20 g cm^{-3}
P_{CJ}	GPa	48.55 at 2.20 g cm^{-3}
V/V$_0$ = 7.20	kJ cm^{-3}	−11.80 (106% CL-20, 124% HMX)

ENDH, $C_{20}(NO_2)_{20}$, is a sought-after target molecule that has not been synthesized yet. It belongs to the small group of homoleptic nitrocarbon compounds. The dodecahedrane framework is predicted to provide ENDH and its derivatives densities far beyond $\rho_{20\,°C} = 2.4$ g cm^{-3}. Even pretty conservative density predictions ($\rho = 2.2$ g cm^{-3}) for ENDH call for performance values exceeding HMX and even CL-20 by far.

- O. Sandus, Detonation Performance Calculations on Novel Explosives, *6th Ann. Working Group Meeting on Synthesis of HE Density Materials*, **1987**.

Eimite

Eimite is an extrudable gun propellant igniter material based on nitrocellulose-bound pyrotechnics having a boron-containing glaze (Table E.6).

Table E.6: Eimite – composition and properties.

	Eimite
Nitrocellulose, 13.5% N (wt.-%)	40
Potassium nitrate (wt.-%)	27.6
Magnesium$_{atomized}$ (wt.-%)	16.7
Sulfur (wt.-%)	9.8
Resorcin (wt.-%)	5.9
$\Delta_{ex}H$, exp. (J g^{-1})	530
T_{ad} (°C)	2317

- H. Hassmann, Evaluation of Eimite as a Substitute for Black Powder in Artillery Primers, *Picatinny Arsenal, Technical Report* 2525, **1957**.

Ellern, Herbert (1902–1987)

Ellern was born in Nuremberg, Germany. Following a move to Munich, he studied chemistry at the University of Munich (today LMU) from 1921 to 1925. He completed his dissertation in 1927 with Arthur Rosenheim (1865–1942) on a subject related to the oxidizing properties of hydroxylamine (NH_2OH), then at the Friedrich Wilhelms University, Berlin (today Humboldt University). Ellern immigrated to the United States at the beginning of the 1930s and started to work for the Universal Match Corporation in St. Louis Missouri in 1937. He wrote a first monograph on pyrotechnics in 1961 which was followed by an extensively revised and enlarged version in 1968. Even today, a little more than 50 years after the first appearance of its second edition, the "Ellern" is still an indispensable reference work for anyone working in the field of pyrotechnics. In the 1980s, Ellern was teaming up with explosive chemist Alexander Hardt (1930–1989) on a new edition and modernized version of his book. However, both Ellern's and Hardt's consecutive untimely passing impeded realization of this project.

– H. Ellern, D. E. Olander, Spontaneous explosion of a normally stable complex salt, *J. Chem. Educ.* **1955**, *32*, 24.
– H. Ellern, *Modern Pyrotechnics – Fundamentals of Applied Physical Pyrochemistry*, Chemical Publishing Corp. New York **1961**, 320 pp.
– H. Ellern, *Military and Civilian Pyrotechnics*, Chemical Publishing Company, New York, **1968**, 464 pp.
– H. Ellern, *Matches* in M. Grayson, D. Eckroth (Eds.), *Kirk Othmer Encyclopedia of Chemical Technology*, Volume 15, **1981**. pp. 1–8.
– Entry in the Catalogue of the German National Library:http://d-nb.info/gnd/125217706

Ellingham diagram
Ellingham Diagramm

In the 1940s, the British Physicist Harold Johann Thomas Ellingham (1897–1975) developed diagrams that visualize how Gibbs free energy

$$dG = dH - TdS$$

for the oxidation of a metal changes with temperature. Depending on the particular author, those diagrams show the reaction of 1 mol O_2 or ½ mol O_2 in the range between 0 and 2500 °C (Figure E.1).

The change of entropy, ΔS, upon oxidation of a metal is mainly determined by the removal of oxygen, and hence is negative, $DS < 0$ and about equal for all metals. The slope of the line is

$$d\Delta G/dT = -\Delta S$$

hence is positive for all metals and about parallel.

Upon oxidation of carbon to carbon monoxide,

$$C + \tfrac{1}{2}O_2 \longrightarrow CO$$

the molar amount is doubled and hence $\Delta S > 0$ and the slope is negative. Upon oxidation of carbon to carbon dioxide, the mole number does not change

$$C + \tfrac{1}{2}O_2 \longrightarrow CO_2$$

Hence, the entropy change is negligible negative, $\Delta S \sim 0$, with a slightly negative slope.

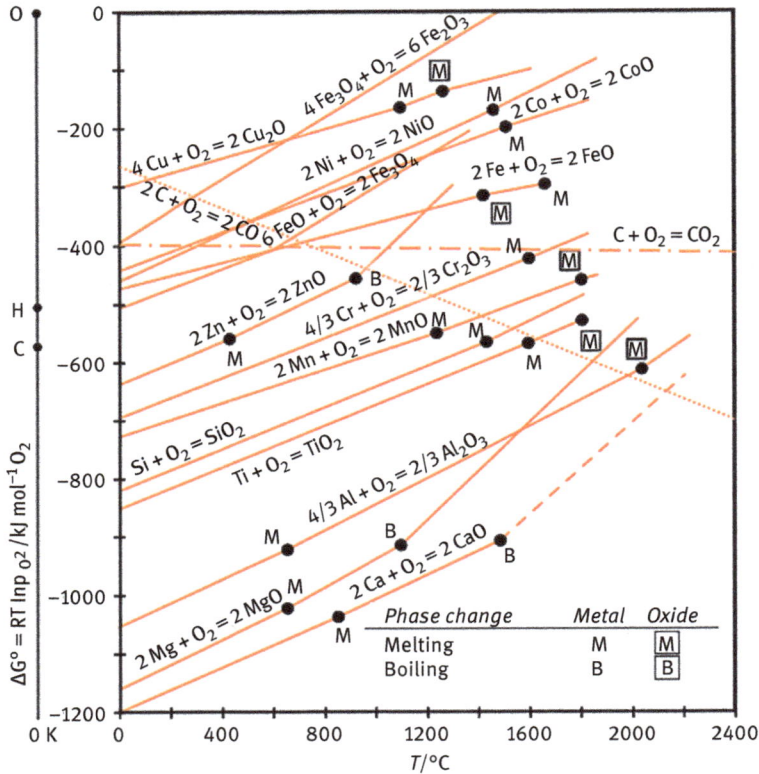

Figure E.1: Ellingham diagram modified after Wikipedia.

Above the vaporization point of metals, the slope increases due to higher decrease of entropy. At the vaporization temperature of the metal oxide, the slope flattens again due to increase in entropy. The lower the line of the corresponding metal, the more stable the resulting metal oxide. The metal of the lower line can reduce any oxide above. The Ellingham diagram is very often used to assess and visualize the energetics of thermite and delay systems.

– J. H. McLain, *Pyrotechnics*, The Franklin Instutute Press, Philadelphia, **1981**, pp. 85–87.
– D. Swanepoel, O. Del Fabro, W. W. Focke, C. Conradie, Manganese as Fuel in Slow-Burning Pyrotechnic Time Delay Compositions, *Propellants Explos. Pyrotech.* **2010**, *35*, 105–113.
– https://en.wikipedia.org/wiki/Ellingham_diagram#/media/File:Ellingham_Richardson-diagram_english.svg

Enerfoil

Enerfoil, also termed *firesheet*, was an experimental film-type igniter. It consists of polytetrafluorethylene carrier foils (d = 40–70 µm) coated on both sides (by physical vapor deposition) with magnesium (d = 2–20 µm) as fuel and bearing thin top layers of aluminum (d = 5–10 µm) as protective barriers against oxidation. Single foils showed ambient pressure linear burn rates in the u = 400–500 mm s^{-1} range, while piles or coils of foil would show burn rates up to u = 1000 mm s^{-1}. Enerfoil is very insensitive and shows a better long-term stability than common Mg/PTFE mixtures (also due to absence of problematic Viton®). It was found that interior ballistic igniting properties of enerfoil were insufficient for use with military gun propellants.

- E.-C. Koch, *Metal-Fluorocarbon Based Energetic Materials*, Wiley-VCH, **2012**, 258–263.
- F. Schötzig, *personal communication*, **2017**.

Energetic ionic liquids, EILS
Energetische Ionische Flüssigkeiten

EILS are a developing field of energetic materials with potential applications in high explosives, propellants, and pyrotechnics. Since the first public account of EILS by *Christe et al.* in 2006, the international research in the field has dramatically increased. The large interest in EILS stems from the enormous range of properties available by a huge number of possible combinations from energetic and non-energetic anions and cations (Table E.7). Per definition ionic liquids are ionic compounds (*vulgo* = salts) with melting points below T = 100 °C (Mp <–40 °C = *low-temperature EILs*; Mp > 25 °C = *melt-castable EILS*).

Ionic liquids contain cations and/or anions with relatively low symmetry and good charge distribution. Hence, facilitated by the absence of hydrogen bridges low-interactive forces result between the ions thereby favoring low melting points.

Typical organic cations of EILs contain nitrogen atoms, which favor the stabilization of the positive charge (Figure E.2). Typical heterocyclic anions of EILS are shown in Figure E.3.

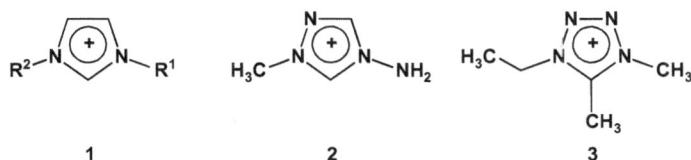

Figure E.2: Typical organic cations in EILS 1,3-dialkyl-imidazolium (**1**), 4-amino-1-methyl-1,2,4-triazolium (**2**) and 1-ethyl-4,5-dimethyltetrazolium (**3**).

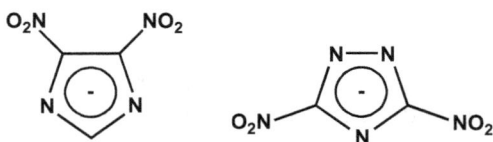

Figure E.3: Typical organic anions in EILS 4,5-dinitro-imidazolate (**4**), 3,5-dinitro-1,2,4-triazolate (**5**).

Apart from heterocyclic anions, EILS very often contain "classical" inorganic anions such as nitrate, dinitramide, chlorate, perchlorate, and bis-trifluorosilylamide.

Table E.7: Particular properties of EILS and potential applications.

Property	Potential application
Melting point < 100 °C	TNT replacements, for example, *1-amino-3-methyl-1,2,3-triazolium nitrate*
Glass transition temperature < −40 °C	Plasticizer in GP and/or PBX, for example, *4-amino-1-methyl-1,2,4-triazolium nitrate*
Low vapor pressure	No VOC issues in production
	No safety issues with regards to flammable or explosive vapors
	Low tendency to migrate upon temperature cycling
Countless possible combinations between energetic and non-energetic anions and cations	Ca. 10,000 different EILS currently possible

For substances to qualify as potential EILS, they must satisfy certain requirements:
- Sufficient oxygen content $\Lambda_{CO} \geq 0$ wt.-% to form CO at minimum
- Stable against hydrolysis
- Low hygroscopicity (DRH ≥80% RH)
- Compatibility with metallic fuels and other common energetic materials

On top of that, EILS candidate materials must meet certain physical and sensitiveness criteria (Tables E.8 and E.9).

Table E.8: Physical requirements EILS (1).

Physical properties	Value	Thermodynamic properties	Value
Melting point (°C)	a) 80–100	$\Delta_f H$ (kJ mol^{-1})	>0
	b) < −40	$\Delta_c H$ (kJ g^{-1})	>25
Density (g cm^{-3})	>1.4		
Surface tension (N m)	0.1		
Viscosity (Pa s)	As low as possible		

Table E.9: Sensitiveness requirements EILS.

Sensitiveness	Value	Stability	Value
Impact sensitiveness (J)	>50	TGA – 75 °C isotherm, 24 h	<1 % loss
Friction sensitiveness (N)	>120	TGA – 10 K min^{-1}	Exo-onset > 120 °C

Apart from EILS satisfying the above criteria, there is also interest in ionic liquids that may react with typical liquid oxidizers (e.g. IRFNA, N_2O_4, or H_2O_2) in a hypergolic fashion. Though hypergolicity is a property of a system containing two different partners, those ionic liquids are nonetheless termed hypergolic-ionic liquids. The hypergolic reactivity is favored by presence and number unsaturated side chains on an organic cation and by electron-deficient anions such as the dicyandiamide $(N(CN)_2)^-$ species.

– A. Brand, T. Hawkins, G. Drakem I. M. K. Ismail, G. Warmouth, L. Hudgens, *Energetic Ionic Liquids as TNT Replacements*, AFRL. June **2005**.
– C. Bigler Jones, R. Haiges, T. Schroer, K. O. Christe, Oxygen-Balanced Energetic Ionic Liquid, *Anwt. Chem.* **2006**, *118*, 5103–5106.
– U. Schaller, Flares Based on Ionic Liquids, *WPC*, **2011**, Reims.
– U. Schaller, V. Weiser, M. Bohn, T. Keicher, H. Krause, Energetische Ionische Liquide, *ICT*, November **2011**.
– Y. Zhang, H. Gao, Y. Joo, J. M. Shreeve, Ionic Liquids as Hypergolic Fuels, *Anwt. Chem. Int. Ed.* **2011**, *50*, 9554–9562.
– K.-T. Han, S. Braun, F. Cisek, B. Wanders, Triazole Salts as a new Component for propellants, *ISL*, March **2012**.
– E. Sebastiao, C. Cook, A. Hu, M. Murugesu, Recent developments in the field of energetic ionic liquids, *J. Mater. Chem. A.* **2014**, *2*, 8153–8173.
– J. Glorian, S. Hagenbach, K.-T. Han, S. Braun, F. Ciszek, B. Baschung, Targeting Energetic Azole-Based Salts Via Compuational Chemical Calculations, *50th ICT-Jata*, **2019**, Karlsruhe, P 51.

Enthalpy of formation, heat of formation
Bildungsenthalpie, Bildungswärme

The heat released upon formation of 1 mol of a chemical substance from the elements at constant pressure is called enthalpy of formation. Per definition heat released upon formation (exothermal reaction) is indicated with a minus. In monographs, the standard enthalpy of formation ($\Delta_f H°$) at 25 °C (298.15 K) and ambient pressure 0.1 MPa is listed.

Table E.10 displays the enthalpy of formation and other pertinent properties of important compounds occurring as fuels, oxidizers, or reaction products.

Table E.10: Enthalpy of formation and other pertinent properties of compounds.

Compound	m_r (g mol^{-1})	ρ (g cm^{-3})	Mp (°C)	Bp (°C)	$\Delta_f H$ (kJ mol^{-1})	Cp at 298 K (J mol^{-1} K^{-1})
$AlBr_3$	266.71	3.205	97.5	257	−511.3	100.5
Al_4C_3	143.96	2.95	2230	Dec	−208.8	116.1
$Al_2O_3 \cdot CaO$	158.04	3.64	1600	Dec	−2326.2	120.8
$Al_2O_3 \cdot 3\,CaO$	270.20	3.02	1535	Dec	−3587.8	209.7
$AlCl_3$	133.34	2.44	192	Sub	−705.6	91.1
AlF_3	83.98	3.197	454 (pt*)	1275 sub	−1510.4	75.1
AlI_3	407.69	3.98	191	360	−302.9	98.9
AlN	40.99	3.09	>2400	–	−318.0	30.1
Al_2O_3	101.96	4.05	2050	–	−1675	79
$Al(OH)_3$	78.00	3.98	300 (-H_2O)	–	−1276.1	93.15
AlP	57.96	2.424	2550	–	−164.4	42
$AlPO_4$	121.95	2.56	580, 705, 1047 (pt) 2000	–	−1733.4	93
Al_2S_3	150.16	2.02	1100	–	−724	112.9
$Al_2(SO_4)_3$	342.15	2.83	450 (dec)		−3440.8	259.4
$AlSb$	148.73	4.279	1080		−50.4	46.4
$Al_2O_3 \cdot SiO_2$	162.05	3.14	1816		−2590.3	122.8
$3\,Al_2O_3 \cdot 2\,SiO_2$	426.05	3.00	1840		−6820.5	325.4
$AsCl_3$	181.28	2.16	−19.8	131.4	−305.1	133.5
AsF_3	131.92	3.01	−5.95	58	−821.3	126.5
AsF_5	169.91	7.71	−79.8	−52.8	−1,237	
As_2O_3	197.84	3.87	312	459	−654.8	97
As_2S_3	246.04	3.49	170 (pt), 312	707	−167.4	116.5
BBr_3	250.54	2.643	−46	91.3	−238.5	128.03
BCl_3	117.19	1.434	−107,2	12.4	−403	62.4
BF_3	67.81	2.99	−131 (pt), −128.7	−99.9	−1136.6	50
$LiBF_4$	93.75	1.967			−1876	
HBO_2	43.82	2.486	176		−789	
BI_3	391.52	3.35	49.9	210	+71.1	70.7

Table E.10 (continued)

Compound	m_r (g mol^{-1})	ρ (g cm^{-3})	Mp (°C)	Bp (°C)	$\Delta_f H$ (kJ mol^{-1})	Cp at 298 K (J mol^{-1} K^{-1})
B_2O_3	69.62	1.805	460	2,066	−1271.9	62.59
BP	41.785	2.90	1227		−79	30.25
B_2S_3	117.81	1.93	563		−252.2	117.1
$BaBr_2$	297.16	4.781	857	2028	−757.7	77.0
$BaCl_2$	208.24	3.888	920 pt, 963	2026	−858.6	75.1
BaF_2	175.34	4.893	967 pt, 1207 pt, 1368	2270	−1208	72.2
BaI_2	391.15	5.15	711		−605.4	77.5
BaO	153.34	5.685	2015		−553.5	47.27
$Ba(OH)_2$	171.35	4.50	284 pt, 508		−946.3	101.6
$Ba_3(PO_4)_2$	601.93	4.1	1727		−4174	77.8
BaS	169.40	4.25	2227		−460.2	49.4
$BaSiO_3$	213.42	4.399	1605		−1623.6	90
$BaTiO_3$	233.21	5.85	120 pt, 1460 pt, 1616		−1659.8	102.5
$BaWO_4$	385.19	5,04	1490		−1703	133.8
$3BeO·B_2O_3$			1495		−3134	
$BeBr_2$	168.83	3.465	473 Sbl		−355.6	66.06
$BeCl_2$	79.92	1.899	403 pt, 415	487	−496.2	62.4
BeF_2	47.01	1.986	227 pt, 552	1167	−1026.8	51.8
BeI_2	262.82	4.325	480	486	−188.7	68.9
BeO	25.01	3.020	2507		−608.4	25.57
$Be(OH)_2$	43.03	1.924	>1000		−902.9	65.7
BeS	41.08	2.36			−234.3	34.1
$BeSO_4$	105.07	2.443	590 pt, 635 pt		−1200.8	86.1
$BeWO_4$					−1513.4	97.28
$BiBr_3$	448.69	5.72	158 pt, 218	453	−276.1	100.8
$BiCl_3$	315.34	4.75	233	447	−379.1	100.4
BiF_3	265.98	8.25	649	900	−909.2	85.8
BiI_3	589.69	5.778	408	540	−150.6	105.6
BiO	224.98	7.2	180 (dec)		−209	

Table E.10 (continued)

Compound	m_r (g mol^{-1})	ρ (g cm^{-3})	Mp (°C)	Bp (°C)	$\Delta_f H$ (kJ mol^{-1})	Cp at 298 K (J mol^{-1} K^{-1})
Bi(OH)$_3$	260	4.36	100-H$_2$O		−711	
Bi$_2$S$_3$	514.15	6.78	>763		−143.1	122.1
HBr	80.92	3.6443*	−86.9	−66.77	−36.2	29.12
CBr$_4$	331.63	3.42	90.1	102	+50.2	91.2
CCl$_4$	153.82	1.5867	−23	76.8	−132.8	133.9
CF$_4$	88.0	1.960*	−184	−128	−933.2	61.1
CI$_4$	519.63	4.34	171 (dec)	307	392.21	
CO	28.01	1.250*	−205	−191	−110.5	29.14
CO$_2$	44.01	1.101*	−56.6		−393.5	37.13
COF	47.0085				−171.5	
COF$_2$	66.007	2.045*	−114	−83	−623.8	
Ca$_3$(BO$_3$)$_2$	237.86	3.10	1487		−3429.1	187.9
CaBr$_2$	199.90	3.354	742	1783	−683.3	75.1
CaCl$_2$	110.99	2.152	782	2206	−795.8	72.86
CaF$_2$	78.08	3.18	1151 pt, 1418	2505	−1225.2	68.6
CaI$_2$	293.89	3.956	754	1110	−536.8	77.2
CaO	56.08	3.4	2927	3570	−635.1	42.1
Ca(OH)$_2$	74.09	2.23	450 (dec)		−986.1	87.5
Ca$_3$P$_2$	182.19	2.51	−1600 (dec)		−506.3	116.3
Ca$_3$(PO$_4$)$_2$	310.18	3.14	1150 pt, 1470 pt, 1810		−4120	227.8
CaS	72.14	2.58	2525		−473.2	47.4
CaSiO$_3$	116.16	3.07	1125 pt, 1544		−1634.9	85.27
CaTiO$_3$	135.98	4.10	1257 pt, 1960		−1660.6	97.65
CaWO$_4$	287.93	6.06	1555		−1645.2	124.44
HCl	36.46	1.639*	−114.8	−84.9	−92.31	29.12
Cr	51.9961	7.14	1907	2482	0	23.34
CrBr$_3$	291.72	4.63	958 Sbl	−	−432.6	96.4
CrCl$_3$	158.36	2.76	1150	945 Sbl	−556.4	91.8
CrF$_3$	108.99	3.8	1100−1200 Sbl		−1174	78.74

Table E.10 (continued)

Compound	m_r (g mol^{-1})	ρ (g cm^{-3})	Mp (°C)	Bp (°C)	$\Delta_f H$ (kJ mol^{-1})	Cp at 298 K (J mol^{-1} K^{-1})
CrI$_3$	305.80	4.92	868, 1100 (dec)		−156.9	73.7
CrO$_3$	99.99	2.7	193	−	−589.5	69.3
CsBF4	219.71	3.2	220 pt, 550		−1887.82	
CsBr	212.81	4.44	636	1300	−405.4	52.2
Cs$_2$CO$_3$	325.82	4.24	610		−1147.3	123.8
CsCl	168.36	3.988	470 pt, 645	1290	−442.8	52.4
CsF	151.90	4.115	703	1250	−554.7	52
CsI	259.81	4.51	445 pt, 627	1280	−346.6	52.6
Cs$_2$O	281.81	4.25	400 (dec)		−346	76
CsOH	149.9	3.675	220 pt, 227	990	−416.1	67.9
Cs$_2$SO$_4$	361.87	4.243	667 pt, 1005		−1443	135.1
Cs$_3$PO$_4$	493.71				−1909	
CsPF$_6$	277.82				−2379	
CuBr$_2$	223.35	4.77	498	900	−138.5	75.7
CuCl$_2$	134.45	3.386	370 pt, 488 pt, 628	993 (dec)	−218	71.9
CuF$_2$	101.54	4.23	836	1670	−538.9	65.6
Cu(OH)$_2$	97.56	3.93			−450	95
CuP$_2$	125.49	4.2			−121	71.1
Cu$_3$P	221.61	7.147	1022		−151.5	87.8
Cu$_2$S	159.15	5.6	110 pt, 440 pt, 1127		−81.2	76.9
CuS	95.61	4.671	507 (dec)		−53.1	47.82
CuSO$_4$	159.60	3.606	200.560 (dec)		−771.4	98.8
HF	20.01	0.901*	−83.36	19.46	−268.5	29.14
FeBr$_3$	295.57		120		−268.2	96.9
FeCO$_3$	115.86	3.85			−740.5	82.06
FeCl$_3$	162.21	2.904	303 (dec)	319	−399.4	96.6
FeF$_3$	112.84	3.52	926 Sbl		−1041.8	91
Fe$_2$P	142.67	6.56	1370		−160.2	74.99

Table E.10 (continued)

Compound	m_r (g mol^{-1})	ρ (g cm^{-3})	Mp (°C)	Bp (°C)	$\Delta_f H$ (kJ mol^{-1})	Cp at 298 K (J mol^{-1} K^{-1})
FeS_2	119.98	5	743 (dec)		−171.5	62.1
$Fe_2(SO_4)_3$	399.88	3.097	1178 (dec)		−2583	265
Fe_2SiO_4	203.78	4.34	1205		−1438	132.8
H_2O	18.01	1	0	100	−285.83	75.28
HDO	19.0214	1.054	2.04	100.7	−286	
D_2O	20.0276	1.106	3.81	101.4	−294.6	84.4
$HfBr_4$	498.11	4.90	420		−767.3	17.6
HfC	190.5	12.6	3890		−251	34.4
$HfCl_4$	320.3		315		−990.4	120.5
HfF_4	254.48	7.13	962 Sbl		−1930.5	100.4
HfI_4	686.11	5.5	449		−493.7	144.3
HfN	192.5	13.7	3305		−373.6	39.4
HfO_2	210.49	9.68	1700 pt, 2900		−1144.7	60.2
HI	127.91	5.789*	−50.79	−35.54	−26.4	29.2
KBF_4	125.908	2.505	278 pt, 570		−1887	115
KBO_2	81.91		950	1401	−994.96	67.04
KBr	119.01	2.75	734	1383	−393.8	52.31
KCN	65.12	1.52	622	1625	−113.47	66.35
K_2CO_3	138.21	2.428	250, 428, 622 pt, 901		−1150.18	114.24
$KHCO_3$	100.12	2.17	100–200 (dec)		−963	48.1
KCl	74.56	1.984	772	1437	−436.68	51.71
KF	58.097	2.49	857	1502	−568.61	48.97
KI	166.01	3.126	685	1345	−327.9	52.78
KNO_2	85.11	1.915	88 pt, 440 (dec)		−369.8	74.94
K_2O	94.20	2.32	740		−361.5	74.42
KOCN	81.12	2.056	700–900		−418.65	
KOH	56.11	2.044	249 pt, 410	1327	−424.68	64.90
KPF_6	184.07	2.591	575 (dec)		−2350.61	

Table E.10 (continued)

Compound	m_r (g mol^{-1})	ρ (g cm^{-3})	Mp (°C)	Bp (°C)	$\Delta_f H$ (kJ mol^{-1})	Cp at 298 K (J mol^{-1} K^{-1})
K_3PO_4	212.266	2.61	1640		−1988.24	164.85
K_2HPO_4	174.18	1.5	315 pt, 400 (dec)		−1775.77	141.29
KH_2PO_4	136.09	2.338	171 pt, 253		−1568.33	116.57
K_2S	110.27	1.740	777 pt, 948		−376.56	74.68
KSCN	97.18	1.886	141.4 pt, 175.1	500 (dec)	−203.4	
K_2SO_3	158.27				−1126.75	123.43
K_2SO_4	174.27	2.662	583 pt, 1069		−1437.79	131.19
$KHSO_4$	136.17	2.322	164.2 pt, 180.5, 218		−1158	
$K_2S_2O_3$	190.315	2.59			−1173.61	
K_2SiF_6	220.28	2.66			−2956	
K_2TiF_6	240.09	3.012	780			
K_2TiO_3	174.10	3.1	1515		−1610	
K_2WO_4	326.05	4.208	388 pt, 933		−1510.3	194.56
KPF_6	184.07	2.591	575 (dec)		−2350.6	
$LiAlO_2$	65.9	3.41	1609.8		−1188.7	67.83
$LiBF_4$	93.74	0.852	195 (dec)		−1876	
$LiBO_2$	49.75	1.397	785 pt, 844		−1019.22	60.37
LiBr	86.85	3.464	550	1265	−350.9	48.94
Li_2CO_3	73.89	2.111	350 pt, 410 pt, 720		1216.04	96.23
LiCl	42.39	2.068	610	1383	−408.27	48.03
LiF	25.94	2.64	848	1717	−616.93	41.92
LiI	133.84	4.06	469	1176	−270.08	50.28
Li_3N	48.96	1.84	115 (dec)		+10.8	
Li_2O	29.88	2.013	1560	2563	−598.73	54.09
LiOH	23.95	1.46	471	1624	−484.93	49.58
Li_2SO_4	109.94	2.221	575 pt, 859		−1436.49	120.96
Li_2SiF_6	155.96	2.8			−2880	
Li_2SiO_3	89.96	2.52	1201		−1649.5	100.48

Table E.10 (continued)

Compound	m_r (g mol^{-1})	ρ (g cm^{-3})	Mp (°C)	Bp (°C)	$\Delta_f H$ (kJ mol^{-1})	Cp at 298 K (J mol^{-1} K^{-1})
Li_2SiO_4	119.84	2.392	1255		−2330	146.63
Li_2TiO_3	109.78	3.42	1212 pt, 1547		−1670.7	109.9
Li_2ZrO_3	153.10	3.51			−1760.2	109.54
$MgBr_2$	184.12	3.72	711	1156	−524.26	73.16
$MgCO_3$	84.32	3.037	900 dec (-CO_2)		−1095.8	75.52
$MgCl_2$	95.22	2.316			−1279	159.1
MgF_2	62.31	3.13	1263	2262	−1124.24	61.54
MgI_2	278.12	4.43	633	981	−366.94	74.85
MgO	40.31	3.576	2831	3600	−601.24	37.11
$Mg(OH)_2$	58.33	2.4	269 dec (-H2O)		−924.66	77.22
$Mg_3(PO_4)_2$	262.88	2.74	1348		−3780.66	213.11
MgS	56.38	2.84	>2000		−345.72	45.58
$MgSO_4$	120.37	2.96	1127		−1284.9	96.20
$MgSiO_3$	100.40	3.11	630 pt, 985 pt, 1577		−1548.92	81.95
Mg_2SiO_4	140.71	3.275	1898		−2076.94	118.72
$MgTiO_3$	120.21	4.05	1630		−1576.2	91.9
$MnBr_2$	214.75	4.385	698		−385.8	75.3
$MnCl_2$	125.84	2.977	650	1231	−481.3	73.0
MnF_2	92.93	3.891	750 pt, 900		−849.9	67.8
MnF_3	111.93	3.54	285 (dec), Sbl		−1071.1	91.2
MnI_2	308.75	5.01	638	1017	−266.1	75.3
MnO	70.94	5.18	1842		−385.2	44.1
MnP	85.91	5.39	1190		−113	46.86
MnS	87.00	3.99	1430		−214.2	50
MnS_2	119.07	3.46			−223.8	70.1
$MnSO_4$	151.00	3.181	700 (dec)		−1065.3	100.2
$MnSiO_3$	131.02	3.72	1286 (dec)		−1320.9	86.4
Mn_2SiO_4	201.96	4.043	1346 (dec)		−1730.5	129.9

Table E.10 (continued)

Compound	m_r (g mol^{-1})	ρ (g cm^{-3})	Mp (°C)	Bp (°C)	$\Delta_f H$ (kJ mol^{-1})	Cp at 298 K (J mol^{-1} K^{-1})
MnTiO$_3$	150.82	4.54	1404		−1358.6	100.1
MnWO$_4$	302.79	7.2			−305	124.3
MoBr$_2$	255.76	4.88			−121.4	
MoBr$_3$	335.67				−171.5	
MoBr$_4$	415.58				−188.3	
MoCl$_2$	166.85	3.714	727	1427	−285.8	74.5
MoCl$_3$	202.3	3.74	300 (dec)		−403.3	94.8
MoCl$_4$	237.75	3.192	272 (dec)		−479.5	118.3
MoCl$_5$	273.21	2.928	194	268	−423.6	155.6
MoF$_6$	209.93	2.543	17.5	35	−1586	170
MoO$_2$	127.94	6.47			−588.9	55.98
MoO$_2$Cl$_2$	198.84	3.31	170	250	−717.1	104.4
MoOF$_4$	187.93	3	98	186	−1380	127
Mo$_2$S$_3$	288.06	5.91	1807		−407.1	109.3
MoS$_2$	160.04	4.8	450 Sbl	1750	−276.1	63.6
MoS$_3$	192.13		350 (dec)		−309.6	82.6
NH$_3$	17.03	0.77147*	−77.7	−33.4	−45.9	35.7
N$_2$O	44.01	1.9775*	−90.91	−88.56	+82.05	38.8
NO	30.006	1.3402*	−163.6	−151.7	+90.3	29.8
NO$_2$	46.01	1.4494*	−11.25	21.1	+33.1	36.6
NH$_4$Br	97.95	2.431	137	452 Sbl	−270.3	96
NH$_4$F	37.04	1.0092			−464	65.3
NH$_4$I	144.94	2.515	551		−202.1	81.7
NaBF$_4$	109.79	2.47	311 pt, 408		−1844	
NaBO$_2$	65.80	2.464	966	1447	−975.7	66.63
Na$_2$B$_2$O$_4 \cdot$10H$_2$O	381.37	1.73	320 dec - H$_2$O		−6262	615
NaBr	102.9	3.202	747	1390	−361.4	51.89
Na$_2$CO$_3$	105.99	2.532	450 pt, 851		−1130.8	111

Table E.10 (continued)

Compound	m_r (g mol^{-1})	ρ (g cm^{-3})	Mp (°C)	Bp (°C)	$\Delta_f H$ (kJ mol^{-1})	Cp at 298 K (J mol^{-1} K^{-1})
$NaHCO_3$	84.01	2.238	270 dec - CO_2		−950.8	87.6
$NaCl$	58.44	2.163	800	1461	−411.1	50.5
NaF	41.99	2.79	993	1695	−575.4	46.9
NaI	149.89	3.667	661	1304	−287.9	52.2
$NaNO_2$	68.99	2.168	284	320 (dec)	−359	69
Na_2O	61.98	2.27	750.970 pt, 1132	1950 (dec)	−418	68.9
$NaOH$	40.00	2.13	299 pt, 323	1554	−425.9	59.6
NaH_2PO_4	119.977				−1537	117
Na_2HPO_4	141.959				−1748	135
Na_3PO_4	169.94	2.5	1583		−1917.4	150
$NaHSO_4$	120.06	2.435	315 (dec)		−1126	
Na_2SO_4	142.04	2.663	185.241 pt, 884		−1387.8	128.2
Na_2SiF_6	188.06	2.679			−2833	
Na_2SiO_3	122.06	2.4	1089		−1561.5	111.9
$NiBr_2$	218.51	5.098	963	904 Sbl	−211.9	75.4
$NiCO_3$	118.72	4.388	400-CO_2		−694.5	86.2
$NiCl_2$	139.62	3.55	1031/1001	993/965 Sbl	−304.9	71.68
NiF_2	96.71	4.63			−657.7	64.03
NiI_2	312.52	5.834	797		−78.2	77.4
NiO	74.71	7.45	250.292 pt, 1955		−239.7	44.31
Ni_2O_3	165.378	4.83	600 (dec)		−489.53	
$Ni(OH)_2$	92.72	4.1	230 dec - H_2O		−528.1	59.51
Ni_2P	148.39	7.2	1112		−184.1	64.81
Ni_3P	207.08	5.99			−220	87.79
Ni_5P_2	355.50	7.28	1185		−435	152
NiS	90.77	5.5	379 pt, 976		−87.9	47.12
NiS_2	122.8	4.45	1007		−131.4	70.6
Ni_3S_2	240.26	5.82	556 pt, 789		−216.3	117.74

Table E.10 (continued)

Compound	m_r (g mol^{-1})	ρ (g cm^{-3})	Mp (°C)	Bp (°C)	$\Delta_f H$ (kJ mol^{-1})	Cp at 298 K (J mol^{-1} K^{-1})
Ni_3S_4	304.39	4.81	353 (dec)		−301.2	164.8
$NiSO_4$	154.77	3.68	848		−872.9	137.96
$NiTiO_3$	154.59	5.097	1775		−1202.4	99.3
PBr_3	270.70	2.852	−40.5	172.9	−184.5	134.7
PBr_5	430.494	3.6	84 (dec)		−269.91	
PCl_3	137.333	1.5778	−93.6	75.5	−319.7	131.38
PCl_5	208.24	2.12	164 Sbl		−445.5	142.52
PF_3	87.97	3.907*	−151	−101	−958.4	58.7
PF_5	125.97	5.805	−83	−75	−1594.4	84.9
PH_3	34.00	1.5307*	−133.8	−87.77	+5.4	37.1
P_2H_4	65.979		−99	63	−5.02	
PI_3	411.69	4.18	61.5	227	−18	78.00
PO	46.973	g			−23.55	
PO_2	62.973	g			−276.6	41.04
P_4O_6	109.945	2.135	23.8	175.3	−2263.8	238.49
P_4O_{10}	283.89	2.3	422		−3009.8	211.5
$POCl_3$	153.33	1.675	1.2	105.3	−597.1	138.79
H_3PO_2	66.00	1.49	26.5	130 Sbl	−608.8	
H_3PO_3	82.00	1.65	73.6	250 (dec)	−971.5	
H_3PO_4	98.00	1.88	42.35		−1279	106.23
PbB_4O_7	362.44	5.85	160 dec - H_2O		−2857	151.66
$PbBr_2$	367.01	6.667	373	916	−277.4	79.6
$PbCO_3$	267.2	6.6	315 (dec)		−699.1	87.4
$PbCl_2$	278.10	5.85	501	950	−359.4	77.1
$PbCl_4$	349.00	3.18	−15	>50 Expl	−553.4	100.5
PbF_2	245.19	8.37	260 pt, 855	1290	−677	72.3
PbI_2	461.00	6.16	410	847	−175.4	77.6
PbO	223.19	8.0	890	1472	−218.1	45.8

Table E.10 (continued)

Compound	m_r (g mol^{-1})	ρ (g cm^{-3})	Mp (°C)	Bp (°C)	$\Delta_f H$ (kJ mol^{-1})	Cp at 298 K (J mol^{-1} K^{-1})
PbS	239.25	7.5	1114		−98.6	49.43
PbSO$_4$	303.25	6.29	866	1170	−923.1	86.4
PbSiO$_3$	283.27	6.49	766		−1144.9	90.06
PbTiO$_3$	303.09	7.82	490	1286	−1198.7	104.4
PbWO$_4$	455.04	8.52	1123		−1121.7	119.9
RbBF$_4$	172.27	2.82	590		−1880	
RbBr	165.37	3.35	693	1340	−394.6	52.8
RbCO$_3$	230.94	3.545	303	873	−1136.0	117.6
RbHCO$_3$	146.48		175 (dec)		−963	
RbCl	120.92	2.76	718	1390	−435.4	52.3
RbF	104.47	3.557	795	1410	−557.7	50.5
RbI	212.37	3.55	656	1300	−333.9	52.5
Rb$_2$O	186.94	3.72	270, 340 pt, 505		−339.0	74.1
RbOH	102.48	3.203	94, 245 pt, 301		−418	
RbHSO$_4$	182.54	2.892	208		−1145	
Rb$_2$SO$_4$	266.99	3.613	653 pt, 1060	1700	−1435.6	113.1
SF$_6$	146.05	6.516*	−51	−64	−1222.15	96.6
H$_2$S	34.08	1.539*	−85.6	−60.4	−20.7	34.22
S$_4$N$_4$	184.27	2.2	179	185 Expl	+460.2	
SO	48.059				+5	
SO$_2$	64.06	2.9262*	−72.7	−10.8	−296.8	39.9
SO$_3$	80.06	1.9229	16.8	44.8	−395.8	50.7
H$_2$SO$_4$	98.07	1.834	10.36	330	−814	138.9
H$_2$SO$_4 \cdot$H$_2$O	116.09	1.788	8.62	290	−1127	215.1
SbBr$_3$	361.46	4.148	96.6	288	−259.4	112.7
SbCl$_3$	228.11	3.14	73.4	219	−381.1	110.5
SbCl$_5$	299.02	2.336	4	140 (dec)	−440.2	158.9
SbF$_3$	178.75	4.379	292	376	−915.4	90.2

Table E.10 (continued)

Compound	m_r (g mol^{-1})	ρ (g cm^{-3})	Mp (°C)	Bp (°C)	$\Delta_f H$ (kJ mol^{-1})	Cp at 298 K (J mol^{-1} K^{-1})
SbI$_3$	502.46	4.917	170	401	−100.4	98.1
Sb$_2$O$_3$	291.50	5.76	656	1425	−708.6	101.4
Sb$_2$O$_5$	323.50	6.7	380 (dec)		−971.9	117.6
SiBr$_4$	347.72	2.814	5.4	152.8	−457.3	146.4
SiC	40.1	3.22	2100 pt	2700 Sbl	−73.2	26.9
SiCl$_4$	169.9	1.483	−70.4	57.57	−577.4	145.3
SiF$_4$	104.08	4.69	−90.3	−86	−1614.9	73.6
SiI$_4$	535.70	4.108	120.5	287.5	−189.5	108
Si$_3$N$_4$	140.28	3.17	1900 (dec)		−744.8	99.5
SiO	44.09	2.13	1702		−99.58	29.9
SiO$_2$	60.08	2.648	575, 806 pt, 1550		−910.9	44.6
SiS$_2$	92.21	2.02	1090	1250 Sbl	−213.4	77.5
SmCl$_2$	221.26	4.56	680 pt, 859	1960	−815.5	82.4
SmCl$_3$	256.71	4.465	682		−1025.9	99.5
SmF$_2$	188.35	6.16	1417	2400	−1180	
SmF$_3$	207.35	6.928	490 pt, 1306	2323	−1778	96.2
SmI$_2$	404.16	5.47	793	1580	−590	
SmI$_3$	531.06	850	−641.8			
Sm$_2$O$_3$	348.7	8.347	2325		−1827.4	115.8
SnBr$_2$	278.51	4.923	231	853	−243.5	79
SnBr$_4$	438.33	3.34	33	215	−405.8	136.5
SnCl$_2$	189.6	3.951	247	614	−328	78
SnCl$_4$	260.5	2.226	−33.3	114.1	−511.3	165.3
SnF$_2$	156.69	4.85	213	850	−648.5	72.4
SnF$_4$	194.68	4.78		701	−1188	
SnI$_2$	372.5	5.285	320	717	−143.9	78.5
SnI$_4$	626.22	473	144.5	354.5	−215.3	132
SnO	134.69	6.446	270 pt, 1080		−286.8	47.8

Table E.10 (continued)

Compound	m_r (g mol^{-1})	ρ (g cm^{-3})	Mp (°C)	Bp (°C)	$\Delta_f H$ (kJ mol^{-1})	Cp at 298 K (J mol^{-1} K^{-1})
SnO_2	150.69	7.02	410 pt, 1630	1800 Sbl	−580.8	52.6
SnS	150.75	5.2	602 pt, 882	1230	−107.9	49.3
SnS_2	182.82	3.9	765		−153.6	70.1
$SnSO_4$	214.75	4.15	>360-SO_2		−887	150.6
$SrBr_2$	247.44	4.216	645 pt, 657	2143	−718	76.9
$SrCO_3$	147.63	3.736	924 pt, 1289 dec-CO_2		−1219.8	84.3
$SrCl_2$	158.53	3.094	872	2056	−828.9	75.6
SrF_2	125.62	4.24	1148, 1211 pt, 1477	2486	−1217.1	70
SrI_2	341.43	4.549	538		−561.5	78
$Sr(NO_2)_2$	179.63	2.997	200 pt		−762.7	
SrO	103.62	4.7	2665	3090	−592	45.4
$Sr(OH)_2$	121.63	3.625	375, 710 dec(-H_2O)		−968.9	74.9
$SrHPO_4$	183.6	3.544	340 (dec)		−1822	
$Sr_3(PO_4)_2$	452.8	4.53	1767		−4129	
SrS	119.68	3.7	>2000		−468.6	48.7
$SrSO_4$	183.68	3.96	1156 pt, 1605		−1459	101.7
$SrSiO_3$	163.7	3.65	1580		−1634	88.5
Sr_2SiO_4	267.32	4.506	>1750		−2305	143.3
$SrTiO_3$	183.51		2080		−1672.4	98.4
$SrWO_4$	335.47	6.187	1535		−1639.7	125.5
$SrZrO_3$	226.84	5.45	2750		−1767.3	103.4
TiB_2	69.52	4.52	2920	3977 (dec)	−323.8	44.1
$TiBr_2$	207.68	4.41	>400		−405.4	78.7
$TiBr_3$	287.63	4.4	400 (dec)		−550.2	101.7
$TiBr_4$	367.54	3.25	38.2	230	−618	131.5
TiC	59.91	4.93	3020	4820	−184.5	33.8
$TiCl_2$	118.81	3.13	1310 Sbl		−515.5	69.8
$TiCl_3$	154.26	2.64	475 (dec)		−721.7	97.1

Table E.10 (continued)

Compound	m_r (g mol^{-1})	ρ (g cm^{-3})	Mp (°C)	Bp (°C)	$\Delta_f H$ (kJ mol^{-1})	Cp at 298 K (J mol^{-1} K^{-1})
TiCl$_4$	189.71	1.726	−24.3	136.5	−804.2	145.2
TiF$_3$	104.9	3.4	950 (dec)		−1435.5	92
TiF$_4$	123.89	2.798	283 Sbl		−1649.3	114.3
TiI$_2$	301.71	4.99	400		−266.1	86.2
TiI$_4$	555.52	4.3	106 pt, 150	377.2	−375.7	125.7
TiN	61.91	5.22	2930		−337.9	37.1
TiO	63.90	4.93	991 pt, 1750	3227	−542.7	40
TiO$_2$	79.90	3.84	642 pt, 1560		−938.7	
Ti$_2$O$_3$	143.80	4.6	200 pt, 2130	3000	−1520.9	95.8
TiS	79.96	4.12	1927		−272	48.1
TiS$_2$	112.03	3.22	147		−407.1	67.9
TiSi$_2$	104.06	3.90	1480 (dec)		−134.3	65.5
Ti$_5$Si$_3$	323.76	4.3	2130		−579	187
WBr$_5$	583.40		276	333	−311.7	155.5
WC	195.86	15.7	2976		−40.2	35.4
W$_2$C	379.71	16.06	2857	6000	−26.4	76.6
WCl$_2$	254.76	5.436	500 (dec)		−257.3	77.8
WCl$_4$	325.66	4.624	300 (dec)		−443.1	129.5
WCl$_5$	361.12	3.875	253	286	−513	155.6
WCl$_6$	396.57	3.52	177, 240 pt, 292	348	−593.7	175.4
WF$_6$	297.84	12.9	2.3	17.06	−1721.7	119
WI$_4$	691.47	5.2			0	
WO$_2$	215.85	12.11	1500	1730	−589.7	55.7
WOCl$_4$	341.66	3.95	211	223	−671.1	146.3
WOF$_4$	275.84	5.07	101	188	−1394.4	133.6
WS$_2$	247.98	7.5	1250		−259.4	63.5
XeF$_4$	207.287		117		−215	90
XeF$_2$	169.287	4.32	120 Sbl		−163	

Table E.10 (continued)

Compound	m_r (g mol^{-1})	ρ (g cm^{-3})	Mp (°C)	Bp (°C)	$\Delta_f H$ (kJ mol^{-1})	Cp at 298 K (J mol^{-1} K^{-1})
Xe	131.29	5.8971*	−111	−107	0	20.79
YbBr$_2$	332.86	5.91	677	1800	−552	
YbBr$_3$	412.77	5.117	956	1470	−775	
YbCl$_2$	243.95	5.08	720	1900	−800	82.9
YbCl$_3$	279.40	3.98	875		−959.9	95.3
YbF$_2$	211.04	7.985	1407	2380	−1184	
YbF$_3$	230.04	8,2	1157	2200	−1569.8	94.56
Yb$_2$O$_3$	394.08	9.215	1092 pt, 2450	4070	−1814.5	115.4
Yb$_2$(SO$_4$)$_3$	778.39	3.286	900 (dec)		−3890	275
ZnBr$_2$	225.19	4.219	402	655	−329.7	65.1
ZnCO$_3$	125.38	4.44	300 (dec)		−812.9	80.1
ZnCl$_2$	136.28	2.93	290	732	−415	71.3
ZnF$_2$	103.37	4.95	820 pt, 872	1500	−764.4	65.6
ZnI$_2$	319.18	4.736	446	750	−208.2	65.7
Zn$_3$N$_2$	224.12	6.22	600 (dec)		−22.6	109.3
ZnO	81.37	5.66	1975		−50.5	41.1
Zn(OH)$_2$	99.38	3.082	125 dec(-H$_2$O)		−642.0	72.4
Zn$_3$(PO$_4$)$_2$	368.05	3.83	942		−2899.5	234.1
ZnS	97.44	4.079	1020 pt		−205.2	45.4
ZnSO$_4$	161.43	3.546	740		−982.8	99.1
ZnSb	187.12	6.383	537 (dec)		−151	52.05
Zn$_3$Sb$_2$	439.61	6.327	405, 455 pt, 566		−30.5	
ZnSiO$_3$	141.45	3.52	1429		−1262	84.8
Zn$_2$SiO$_4$	222.82	3.9	1512		−1644	121.3
ZrB$_2$	112.84	5.64	3060	4193 (dec)	−322.6	48.4
ZrBr$_2$	251.04	5	400 (dec)		−404.6	86.7
ZrBr$_3$	330.95	4.52	310		−636.0	99.5
ZrBr$_4$	410.86	3.98	450	357 Sbl	−760.7	124.8

Table E.10 (continued)

Compound	m_r (g mol^{-1})	ρ (g cm^{-3})	Mp (°C)	Bp (°C)	$\Delta_f H$ (kJ mol^{-1})	Cp at 298 K (J mol^{-1} K^{-1})
ZrC	103.23	6.51	3530	5100	−196.6	37.9
ZrCl$_2$	162.13	3.6	650 (dec)		−431	72.6
ZrCl$_3$	197.58	2.28	300 (dec)		−714.2	96.2
ZrCl$_4$	233.03	2.80	437	331 Sbl	−979.8	119.8
ZrF$_2$	129.221		902	2264	−962.32	65.67
ZrF$_3$	148.219				−1441.6	83.64
ZrF$_4$	167.21	4.43	450 pt	903 Sbl	−1911.3	103.4
ZrI$_2$	345.03		600 (dec)		−259.4	94.1
ZrI$_3$	471.93	5.18	275 (dec)		−397.5	103.9
ZrI$_4$	598.84	4.793	500	431 Sbl	−488.7	127.8
ZrN	105.23	6.97	2982		−365.2	40.4
ZrO$_2$	123.22	5.82	1205, 1377 pt, 1687	4270	−1097.5	56.1
Zr(OH)$_4$	159.25	3.25			−1720	
ZrS$_2$	155.35	3.87	1450		−577.4	68.8
Zr(SO$_4$)$_2$	283.34	3.71			−2499	
ZrSi$_2$	147.39	4.88	1925		−159.4	64.4
ZrSiO$_4$	183.30	4.6	1540		−2023.8	98.8

* for gases g/l

- *D'Ans Lax Taschenbuch für Chemiker und Physiker*, R. Blachnik (Eds.) *Band 3, Elemente, anorganische Verbindungen und Materialien, Minerale*, Springer, **1998**, 1463 pp.
- *D'Ans Lax Taschenbuch für Chemiker und Physiker*, C. Syrowietz (Eds.) *Band 2, Organische Verbindungen*, Minerale, Springer, **1983**, 1128 pp.

EOR and EOD

EOR and EOD stand for explosive ordnance reconnaissance and explosive ordnance disposal. The procedures to determine, identify, and dispose-off of unexploded ammunitions are described in NATO STANAG 2143.

- Explosive Ordnance Reconnaissance und Explosive Ordnance Disposal EOR /EOD, *STANAG 2143 Ed. 5*, NATO, **2005**.

Eprouvette

The eprouvette (French: *épreuve* = test) is an apparatus to gauge the ballistic performance of black powder. While various designs have evolved over the centuries, nowadays, the eprouvette is a vertically oriented mortar with a conical projectile fitting closely into the barrel and being directly connected to a frame riding in two slide rails coaligned to the bore axis of the mortar. Upon firing of the mortar, the projectile assumes a maximum height that is secured by ratchets. The projectile height at given charge mass is a comparative measure of ballistic performance.

- W. Hintze, Untersuchungsberichte 1. Zwei-Komponenten-Schwarzpulver, 2. Einfluß des Kohlenstoffgehaltes der Holzkohle auf die Schwarzpulvereigenschaften, *Explosivstoffe*, **1968**, *16*, 25–48.

Equation of state, EOS
Zustandsgleichung

Chapman–Jouguet (C-J) detonation theory after *Mader* implies that the performance of an explosive is determined by thermodynamic states: the C-J state and the C-J adiabat. Thermochemical codes use thermodynamics to calculate these states and hence obtain a prediction of explosive performance. The allowed thermodynamic states behind a shock are intersections of the Rayleigh line (expressing conservation of mass and momentum), and the shock Hugoniot (expressing conservation of energy). C-J theory states that a stable detonation occurs when the Rayleigh line is tangent to the shock Hugoniot.

This point of tangency can be found, given that the equation of state (EOS) $P = P(V,T)$ of the products is known. The chemical composition of the products changes with the thermodynamic state, so thermochemical codes must simultaneously solve for state variables and chemical concentrations. This problem is relatively straightforward, given that the EOS of the gaseous and solid products is known.

One of the most difficult parts of this problem is accurately describing the EOS of the gases. The Becker–Kistiakowski–Wilson (BKW) gaseous EOS has a long and venerable history in the explosives field. The BKW-EOS is the most used and best calibrated of those used to calculate detonation properties assuming steady state and chemical equilibrium. Experimental data are required in evaluating any model of the detonation process. Data that relate to the detonation process are detonation velocity as a function of density, C-J pressures and temperatures, and state points on the shock Hugoniot and expansion isentrope of the detonation products:

$$\frac{p \cdot v}{RT} = 1 + X \cdot e^{\beta \cdot X}$$

With

$$X = \frac{\kappa \sum x_i k_i}{v \cdot (T + \Theta)^\alpha}$$

P represents the pressure, v the molar volume of the products, x_i the mole fraction, κ_i the covolume.

Empirical constants:

Table E.11: Empirical BKW constants.

	BKWR	BKWS	BKWC
α	0.51	0.50	0.50
β	0.402	0.398	0.403
κ	12.31	10.50	10.86
Θ	3856757	6620	5441

Caution must be employed when comparing predicted C-J properties from an equilibrium code and experimental data. According to *Zeldovich–von Neumann–Doering* theory, the C-J state is attained behind a planar detonation wave traveling through an infinite medium after an infinite period of time. Of course, real experiments are always carried out on finite systems. Finite systems size reduces the effective time available for equilibration: the developing C-J state is disrupted by rarefaction waves. Thus, the C-J state is never exactly attained in any real experiment.

– Ch.L. Mader, *Numerical Modeling of Detonation*. University of California Press, Berkeley and Los Angeles, California, **1979**, 44–88.

Ernst-Mach Institut, EMI

The Ernst Mach Institut (EMI) operates under the legal name Fraunhofer-Institut für Kurzzeitdynamik, with facilities at Freiburg im Breisgau, Efringen-Kirchen, and Kandern. The institute was founded in 1960 and since then deals with defence related research such as terminal ballistics, effect of blast waves on structures and shock initiation of high explosives. The budget provided by the German Ministry of Defence (BMVg) in the year 2015 amounted to 27 Mio. EUR.

The current director of the EMI is shock physics expert *Prof. Dr. Ing. habil. Stefan Josef Hiermaier* (1966*).

– Webseite https//www.emi.fraunhofer.de/en.html

Erosion

Erosion, $E = (\Delta_m/m_L)$, is the loss of metal from the gun barrel caused by both corrosion and wear by the firing process, where Δ_m is the mass loss of the gun barrel (kg) and m_L is the mass of the gun propellant (kg).

Erosion is a system property and hence it is dependent on
- barrel type and material
- chrome plating
- ammunition
- gun propellant
- rate/rhythm of fire
- ambient temperature, relative humidity

Already, *Grune and Kegler* in 1968 found that the erosion increases with the charge mass, however, after reaching a maximum value $(\Delta_m/m_L)_{max}$, the erosion decreases. *Stein* found that the erosion of a large caliber gun barrel is related to charge mass, explosion temperature, initial velocity of projectile, and gas pressure by

$$(\Delta_m/m_L) \sim m_L{}^{1.5} \cdot T_{ad}{}^7 \cdot v_0{}^{1.4} \cdot p_{max}{}^5$$

In a erosion bomb, the mass loss of an erosion nozzle correlates with the third power of the explosion temperature

$$(\Delta_m/m_L)_{max} = c \, T_{ad}{}^3$$

with c being an empirical constant.

Apart from the temperature and gas pressure, the composition of the combustion gases and consequently the initial propellant composition play an important role in determining the erosion. The combustion products attack the barrel and yield oxidation, carbide formation, and hydrogen brittleness with following simplified equations:

$$Fe + CO_2 \longrightarrow FeO + CO$$

$$4\,Fe + CO \longrightarrow FeO + Fe_3C$$

$$Fe + H_2 \rightarrow FeH_2$$

In view of the above, *Kimura* gauged the decrease of relative erosivity in the following order with high nitrogen amounts actually leading to renitridation of steel and hence mass increase:

$$CO_2 > CO > H_2O > H_2 > 0 > N_2$$

Lawton has formulated an equation to relate erosion to chemical composition of the combustion gases:

$$E = 114 \exp\{0.0207\,[(CO) - 3.3\,(CO_2) + 2.4\,(H_2) - 3.6\,(H_2O) - 0.5\,(N_2)]\}$$

Additives to the GP may also help to reduce erosion. At first, additives (e.g. TiO_2 or WO_3) may absorb thermal energy by virtue of their high specific heat. At second, additives may protect the barrel against thermal radiation by being optically opaque.

- J. Lavoie, C.-F. Petre, C. Dubois, Erosivity and Performance of Nitrogen-Rich Propellants, *Propellants Explos. Pyrotech.* **2018**, *43*, 879–892.
- W. Langlotz, F. Schötzig, A. B. Wellm, P. Bott, Method for the Examination of the Erosivity of Gun Propellants, *48th ICT -Jata*, Karlsruhe, **2017**, V-13.
- K. Ryf, B. Vogelsanger, D. Antenen, A. Skriver, A. Huber, Moderne Pulverentwicklungen, *CCG Seminar WB.235*, **2002**.
- B. Lawton, Thermo-chemical erosion in gun barrels, *Wear* **2001**, *251*, 827–838.
- J. Kimura, Hydrogen Gas Erosion of High-Energy LOVA Propellants, *16th International Symposium on Ballistics*, San Francisco, CA, USA, **1996**.

Erosive burning
Erosiver Abbrand

Erosive burn is a phenomenon to account for the deviation of the burn rate, u of a solid energetic grain from Vieille's law:

$$u = u_{p,t} + u_e$$

with the Vieille's law $u_{p,t} = a \cdot P^n$ and the rate of erosion, u_e after *Lenoir*:

$$u_e = \frac{b \cdot c_p \cdot \mu^c \cdot Pr^{0.667} \cdot k}{L^{0.2} \cdot e^f} \cdot G^d$$

with the constants b, c, d, f, the heat capacity, c_p, the viscosity of the combustion products, μ, the Prandtl number, Pr, the mass flow, G, and the distance of the burning area of the energetic grain from the head of the grain, L.

Erosive burn predominantly occurs with internally perforated grains (hollow cylinder, star, dogbone, etc.) and is seen at the aftend of the grain. High L/D ratio also favor occurrence of erosive burning. The grain at the lower end is exposed to the mass flow of combustion products of the upper area of the burning grain. It is observed that velocity of this mass flow determines the erosive burn. At very low mass flow rates, the heat feedback to the condensed phase can be interfered leading to negative erosion, that is a lower burn rate than regular. At higher mass flow rate, the convective heat transport effects a higher heat feeback and consequently increases the burn rate which as positive erosive burn. At lower temperature, erosive burn is more pronounced than at higher temperatures.

– J. M. Lenoir, G. Robillard, A Mathematical Method to predict the effect of erosive Burning in Solid-Propellant Rockets, *Combustion Symposium*, **1957**, 663–667.
– M. K. King, Erosive Burning of Solid Propellants, *J. Propul. Power* **1993**, *9*, 785–805.
– N. Kubota, *Propellants and Explosives*, Wiley-VCH, 3rd Ed., Weinheim, **2015**, pp. 402–408.

Erythritol tetranitrate, ETN

Tetranitroerythrit

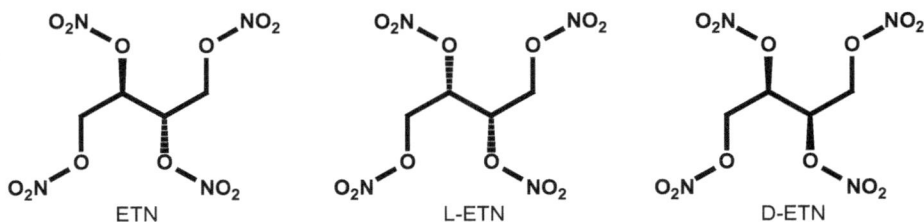

Formula		$C_4H_6N_4O_{12}$		
REACH		LPRS		
EINECS		230-734-9		
CAS		[7297-25-8]		
m_r	g mol^{-1}	302.11		
ρ	g cm^{-3}	1.773		
$\Delta_f H$	kJ mol^{-1}	−499.99		
$\Delta_{ex} H$	kJ mol^{-1}	−1818.6		
	kJ g^{-1}	−6.02		
	kJ cm^{-3}	−10.673		
$\Delta_c H$	kJ mol^{-1}	−1931.6		
	kJ g^{-1}	−6.394		
	kJ cm^{-3}	−11.336		
Λ	wt.-%	+5.30		
N	wt.-%	18.54		
P_{vap}	Pa	3.2 10^{-3}		
$\Delta_{sub} H$	kJ mol^{-1}	129.1		
		ETN	L-ETN	D-ETN
Mp	°C	62.5	liquids at ambient temperature	
	–		do not crystallize down to −80 °C	
DSC onset	°C	163.4	154	154
ERL-Impact	J	3.6	0.74	0.74
		PETN(5.0)		
BAM-Impact	J	2 (BAM)		
BAM-Friction	N	38.1 ± 12	>360 (l)	
		PETN (63.1 ± 10.8)		

ETN is more sensitive than PETN and used primarily as a vasodilator. The diastereomer is less sensitive than either D-ETN or L-ETN, the latter of which are liquids.

- V. W. Manner, B. C. Tappan, B. L. Scott, D. N. Preston, G. W. Brown, Crystal Structure, Packing Analysis, and Structural Sensitivity Correlations of Erythritol Tetranitrate, *Cryst. Growth Des.* **2014**, *14*, 6154–6160.
- M. A. C. Härtel, T. M. Klapötke, J. Stierstorfer, L. Zehetne·, Vapor Pressure of Linear Nitrate Esters Determined by Transpiration Method in Combination with VO-GC/MS, *Propellants Explos. Pyrotech.* **2019**, *44*, 484–492.
- N. Lease, L. M. Kay, G. W. Brown, D. E. Chavez, P. W. Leonard, D. Robbins, V. W. Manner, Modifying Nitrate Ester Sensitivity Properties Using Explosive Isomers, *Cryst. Growth Des.* **2019**, *11*, 6708–6714.

Ethriol trinitrate
Ethrioltrinitrat

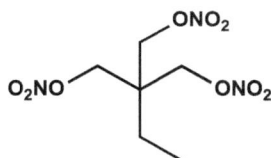

Formula		$C_6H_{11}N_3O_9$
REACH		LPRS
EINECS		220-866-5
CAS		[2921-92-8]
m_r	g mol^{-1}	269.168
ρ	g cm^{-3}	1.50
$\Delta_f H$	kJ mol^{-1}	−479.86
$\Delta_{ex}H$	kJ mol^{-1}	−1340.1
	kJ g^{-1}	−4.979
	kJ cm^{-3}	−7.468
$\Delta_c H$	kJ mol^{-1}	−3453.3
	kJ g^{-1}	−12.830
	kJ cm^{-3}	−19.245
Λ	wt.-%	−50.52
N	wt.-%	15.62
Mp	°C	51
V_D	m s^{-1}	6440 at 1.48 g cm^{-3}
Trauzl	cm^3	415

ETTN is used as an energetic plasticizer in GPs. However, it does not possess the ability to gel and hence needs to be formulated with gelling agents. ETTN serves also as a vasodilatating medicament.

- J. Mohammad, M. Kamalvand, M. H. Keshavarz, S. Farrashi, Assessment of the Strength of Energetic Compounds Through the Trauzl Lead Block Expansions Using their Molecular Structures, *Z. Anorg. Allg. Chem.* **2015**, *641*, 2446–2451.

Ethyl nitrate

Ethylnitrat

Formula		$C_2H_5NO_3$
REACH		LPRS
EINECS		210-903-3
CAS		[625-58-1]
m_r	g mol^{-1}	91.067
ρ	g cm^{-3}	1.108
Δ_fH	kJ mol^{-1}	−190.37
$\Delta_{ex}H$	kJ mol^{-1}	−424.3
	kJ g^{-1}	−4.659
	kJ cm^{-3}	−5.162
Δ_cH	kJ mol^{-1}	−1311.2
	kJ g^{-1}	−14.399
	kJ cm^{-3}	−15.954
Λ	wt.-%	−61.49
N	wt.-%	15.38
Impact	cm	2, (Nitroglycerine: 1 cm)
Mp	°C	−94.6
Bp	°C	87.7
V_D	m s^{-1}	5800 at 1.10 g cm^{-3}
Trauzl	cm^3	420

Ethyl nitrate was investigated as monergol but due to its extreme sensitiveness is no longer considered.

Ethylenediammonium dinitrate, EDDN, EDAD

Ethylendiammoniumdinitrat, PH-Salz

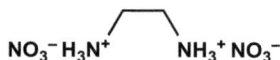

$$NO_3^- H_3N^+ \diagup\diagdown NH_3^+ NO_3^-$$

Formula		$C_2H_{10}N_4O_6$
CAS		[20829-66-7]
m_r	g mol^{-1}	186.125
ρ	g cm^{-3}	1.597
Δ_fH	kJ mol^{-1}	−653.54
$\Delta_{ex}H$	kJ mol^{-1}	−752.0
	kJ g^{-1}	−4.041
	kJ cm^{-3}	−6.477
Δ_cH	kJ mol^{-1}	−1562.7
	kJ g^{-1}	−8.396
	kJ cm^{-3}	−13.408
Λ	wt.-%	−25.79
N	wt.-%	30.10
Mp	°C	186
Friction	N	>355 (BAM)
Impact	J	10 (BAM)
V_D	m s^{-1}	6915 at 1.50 g cm^{-3}
		4650 at 1.00 g cm^{-3}
Trauzl	cm^3	345–375
Koenen	mm	2; type A

Ethylenediammonium dinitrate (often falsely named ethylenediamine dinitrate) forms a eutectic with ammonium nitrate melting at T = 105 °C. Additional calcium nitrate hydrate, $Ca(NO_3)_2 \cdot H_2O$, inhibits volume contraction upon solidification. Mixtures with calcium nitrate and aluminized formulations were used in wartime Germany as *Ersatzsprengstoff (replacement explosives)* to overcome shortening of organic nitrocompounds. Nowadays, EDD is used as an experimental component in very insensitive melt-cast formulations both as eutectic with AN (EA) and in ternary mixtures with additional potassium nitrate, KNO_3, (EAK) which was the base for the first-generation IMX explosive formulations (*see IMX*).

– C. George, R. Gilardi, J. L. Flippen-Andersen, Structures of 1,2-Ethanediammonium Dinitrate (1) and 1,3-Propanediammonium Dinitrate (2), *Acta Cryst.* **1991**, *C47*, 2713–2715.

Ethylenedinitramine, EDNA, haleite

N,N'-Dinitroethylendiamin

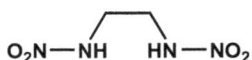

$$O_2N-NH \qquad HN-NO_2$$

Formula		$C_2H_6N_4O_4$
REACH		LPRS
EINECS		208-018-2
CAS		[505-71-5]
m_r	g mol^{-1}	150.094
ρ	g cm^{-3}	1.709
$\Delta_f H$	kJ mol^{-1}	−103.81
$\Delta_{ex} H$	kJ mol^{-1}	−788.9
	kJ g^{-1}	−5.256
	kJ cm^{-3}	−9.198
$\Delta_c H$	kJ mol^{-1}	−1541.5
	kJ g^{-1}	−10.270
	kJ cm^{-3}	−17.973
Λ	wt.-%	−31.98
N	wt.-%	37.33
Mp	°C	177
Bp	°C	265 (dec)
Impact	J	
V_D	m s^{-1}	7570 at 1.49 g cm^{-3}
		5650 at 1.00 g cm^{-3}
Trauzl	cm^3	366–410

EDNA was tested by Picatinny Arsenal in the 1930s for the first time as a high explosive. It was used in World War II together with TNT as *ednatol* but is no longer used today.

Ethyl-*N*-methylcarbanilate

Methylphenylurethan

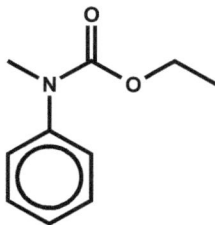

Formula		$C_{10}H_{13}NO_2$
GHS		07
H-phrases		302-312-315-319-332-335
P-phrases		261-280-305+351+338
CAS		[2621-79-6]
WGK		3
m_r	g mol^{-1}	179.219
ρ	g cm^{-3}	1.09
$\Delta_f H$	kJ mol^{-1}	−384.09
$\Delta_c H$	kJ mol^{-1}	−5409.1
	kJ g^{-1}	−30.182
	kJ cm^{-3}	−32.898
Λ	wt.-%	−218.72
N	wt.-%	7.82
Mp	°C	242.44
Bp	°C	116 at 1.1 kPa

Ethyl-*N*-methylcarbanilate is a non-energetic stabilizer and a gelling agent for nitrocellulose.

Ethyl ethylphenylcarbamate

Ethylphenylurethan

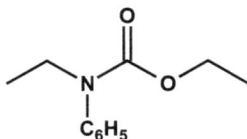

Formula		$C_{11}H_{15}NO_2$
REACH		LRS
EINECS		213-796-1
CAS		[1013-75-8]
m_r	g mol^{-1}	193.246
ρ	g cm^{-3}	1.043
$\Delta_f H$	kJ mol^{-1}	−420.49
$\Delta_c H$	kJ mol^{-1}	−6052.0
	kJ g^{-1}	−31.318
	kJ cm^{-3}	−32.665
Λ	wt.-%	−227.68
N	wt.-%	7.25
Mp	°C	35
Bp	°C	255

Ethyl is a gelling aid for nitrocellulose.

Ethyltetryl, 2,4,6-trinitrophenylethylnitramine

Ethyltetryl

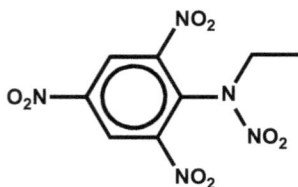

Formula		$C_8H_7N_5O_8$
REACH		LPRS
EINECS		227-961-0
CAS		[6052-13-7]
m_r	g mol^{-1}	301.172
ρ	g cm^{-3}	1.63

$\Delta_f H$	kJ mol^{-1}	-17.99
$\Delta_{ex} H$	kJ mol^{-1}	-1480.9
	kJ g^{-1}	-4.917
	kJ cm^{-3}	-8.015
$\Delta_c H$	kJ mol^{-1}	-4130.6
	kJ g^{-1}	-13.715
	kJ cm^{-3}	-22.356
Λ	wt.-%	-61.09
N	wt.-%	23.25
Mp	°C	95.8
Impact	J	5 (BAM)
Friction	N	>355
V_D	m s^{-1}	7300 at 1.60 g cm^{-3}
Trauzl	cm^3	327

Ethyltetryl and *Tetryl* in mass proportions 70/30 forms a eutectic melting between 85 and 88 °C.

– E.-C. Koch, Insensitive high explosives: IV. Nitroguanidine – Initiation & detonation, *Defence Technology*, **2019**, *15*, 467–487.

Ex-98

EX-98 is an oxygen-rich extrudable propellant igniter based on an insensitive double-base propellant and a perchlorate-based pyrotechnic (Table E.12).

– S. T. Peters, EX-98 An Igniter Material Designed for LOVA Gun Propellants, *ICT-JATA*, Karlsruhe, **2002**, Germany, V27.

Table E.12: Composition and properties of Ex-98.

	EX-98
Nitrocellulose, 13.5% N (wt.-%)	17.4
Triethylene glycol dinitrate (wt.-%)	13.7
Potassium perchlorate, Class 3 (wt.-%)	61.9
Magnesium, Typ II Gran 16 (wt.-%)	6.4
Centralite I (wt.-%)	0.6
Water (wt.-%)	1.0 max
ρ (g cm^{-3})	1.82
$\Delta_{ex} H$ (J g^{-1})	-5439
Λ (wt.-%)	$+10.16$
f, exp. (J g^{-1})	618
T_{ex} (°C)	3435

Exit velocity, w

Ausströmgeschwindigkeit

w is the velocity of which the deflagration products from a rocket motor stream through the throat into the environment. w is primarily determined by the ratio of combustion chamber temperature (T_0) to mean molecular mass ($\bar{M}$) of the combustion products and at second by the ratio between pressure in the combustion chamber (p_0) and the pressure in the environment (p_e).

If the gas velocity before the throat is neglected, the exit velocity can calculated as follows:

$$w = \sqrt{2c_p \cdot (T_0 - T_e)}$$

with the indices 0 for the combustion chamber and e for the exit plane.

With

$$c_p = \frac{\gamma}{\gamma - 1} \cdot \frac{R_0}{\bar{M}} \text{ and } \frac{T_e}{T_0} = \left(\frac{p_e}{p_0}\right)^{\frac{\gamma}{\gamma-1}}$$

it follows that

$$w = \sqrt{\frac{2\gamma}{\gamma - 1} \cdot \frac{R_0}{\bar{M}} \cdot T_0 \left[1 - \left(\frac{p_e}{p_0}\right)^{\frac{\gamma}{\gamma-1}}\right]}$$

where R_0 is the universal gas constant, γ the istropic exponent, $\bar{M}$ the mean molecular mass of combustion gases.

- H. G. Mebus, *Berechnung von Raketentriebwerken*, C.F. Wintersche Verlagsbuchhandlung, **1957**, 120 pp.

EXPLO5 thermochemical code

EXPLO5 is thermochemical computer program that predicts the performance of high explosives, propellants, and pyrotechnic mixtures based on chemical formula, heat of formation, and density. As such, EXPLO5 is a useful tool in synthesis, formulation, and numerical modelling of energetic materials. The first version of the code was developed in 1991 by Muhamed Sućeska, and since then the code has been continuously improved. Today, EXPLO5 is used in more than 80 research laboratories in 35 countries.

EXPLO5 consists of four modules that solves individual tasks, and common database of reactants and products:

Ideal detonation module

The ideal detonation calculation is based on the C-J theory and uses either BKW or Exp-6 EOS for gas phase products, and Murnaghan EOS for condensed. It calculates equilibrium composition and thermodynamic parameters of state of detonation products along the shock adiabat of detonation products, the expansion isentrope of detonation products and evaluates the coefficients in the JWL EOS, detonation energy, and Gurney energy.

Kinetic detonation module

The module is based on the Wood–Kirkwood slightly divergent detonation theory. It is intended to calculation of the self-propagating detonation velocity of non-ideal as a function of unconfined charge diameter. Several different reaction rates models and radial expansion models are incorporated in the module.

Isobaric combustion module

This module calculates equilibrium composition of combustion products and thermodynamic parameters of state under constant (specified) pressure conditions. It enables calculation of theoretical rocket performance (e.g. pressure and flow velocity at the nozzle throat, exhaust and sound velocity at the nozzle exit, thrust coefficient, nozzle expansion ratio, and specific impulse).

Isochoric combustion module

The module calculates equilibrium composition of combustion products and thermodynamic parameters of state under constant volume conditions. It uses the virial or ideal gas EOS, and predicts that the force (energy) of energetic materials, along with the other thermodynamic parameters of the products (heat and temperature of isochoric combustion, maximum pressure, etc.) can handle a wide variety of different molecules and their mixtures (containing up to 48 different chemical elements: C, H, N, O, Al, Cl, Si, F, B, Ba, Ca, Na, P, Li, K, S, Mg, Mn, Zr, Mo, Cu, Fe, Ni, Pb, Sb, Hg, Be, Ti, I, Xe, U, W, Sr, Cr, Br, Co, Ag, Zn, Sn, Bi, Cs, Hf, Ge, Nb, Ta, Yb, Y, and V). Users can manipulate the database (add and remove compounds, modified data, and create user's database file). EXPLO5 comes with default database which includes around 700 of the most frequently used reactants and more than 1000 products.

- M. Sućeska, Calculation of detonation properties of C-H-N-O explosives, *Propellants, Explos. Pyrotech.* **1991**, *16* 197–202.
- M. Sućeska, EXPLO5 computer program for calculation of detonation parameters, Proc. of 32nd Int. Annual Conference of ICT, Karlsruhe **2001**, pp. 110/1–110/13.
- M. Sućeska, Calculation of detonation heat by EXPLO5 computer code results, *Proc. of 30th Int. Annual Conference of ICT*, Karlsruhe **1999**, pp. 50/1–50/14).

- M. Sućeska, Evaluation of detonation energy from EXPLO5 computer code results, *Propellants, Explos. Pyrotech.* **1999**, *28*, 280–285.
- M. Sućeska, calculation of thermodynamic parameters of combustion products of propellants under constant volume conditions using virial equation of state. Influence of values of virial coefficients, *J. Energ. Mater.* **1999**, *17*, 253–277.
- M. Sućeska, Calculation of detonation parameters by EXPLO5 computer program, *Materials Science Forum* **2004**, *465–466*, 325–330.
- M. Sućeska, C.H.Y. Serene, A. How-Ghee, Can the accuracy of BKW EOS be improved, *15th International Detonation Symposium*, San Francisco, USA, July 13–18, **2014**, pp. 1247–1256.
- M. Sućeska, M. Braithwaite, T. Klapoetke, B. Stimac, Equation of state of detonation products based on exponential-6 potential model and analytical representation of the excess Helmholtz free energy, *Propellants Explos. Pyrotech.* **2019** *44*, 1–9.

Exploding foil initiator, EFI, slapper–detonator

The EFI is a refined version of the *exploding bridgewire detonator* (EBW), originally invented and patented by *John Stroud* at LLNL in 1965. The EFI is free of any primary explosives and hence is inherently safe (Figure E.4). To function smoothly the EFI requires short (t < 10 μs) steep high voltage pulses (U ~ 2.5 kV) with high current (I ~ 100 A) which are triggered by Krytrons. The short pulse vaporizes the metal foil (copper) and the expanding Cu plasma shears off a circular piece of the Kapton™ layer which is accelerated through the barrel and impacts on the acceptor pellet which is typically a temperature-resistive high explosive such as TATB or HNS-IV. The impact on the high explosive pellets yields a shock to detonation transition initiating the material (Figure E.5). Modifications of the EFI, for example, use strong lasers to rapidly vaporize the metal.

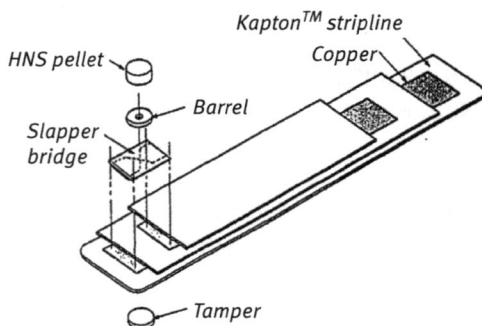

Figure E.4: Schematics of an EFI.

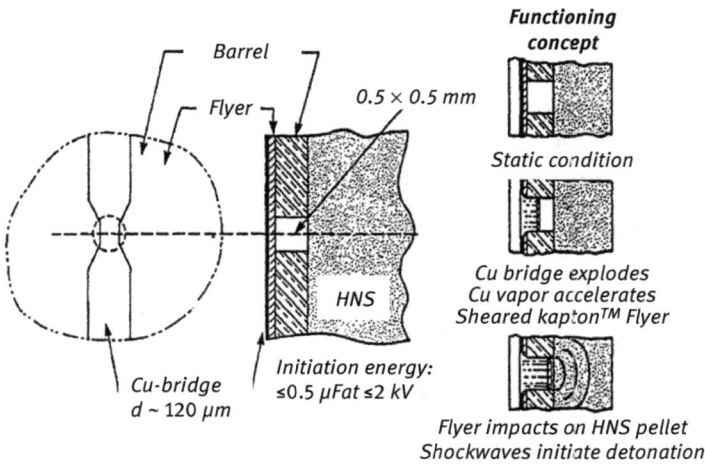

Figure E.5: Details and functioning concept of an EFI.

– J. A. Lienau, Exploding Foil Initiator Qualifications, *Technical Report RD-ST-91-16*, US Army Missile Command, August 1993.

Explosion, heat of
Explosionswärme

The heat of explosion, $\Delta_{ex}H$ (kJ g^{-1}), is the amount of heat released upon explosive decomposition (detonation or deflagration) of an energetic material in an inert atmosphere requiring that there is not any reaction of the decomposition products of the energetic material with the environment (water, oxygen, nitrogen, etc.) taking place. The heat of explosion of non-detonating materials such as gun propellants or pyrotechnics can be determined in a typical Parr-type calorimetric bomb under Ar or He. However, detonating substances have to be tested in a specially designed detonation calorimeter which can withstand the full high velocity detonation of an explosive material. Now detonating substances are peculiar in that they often show critical diameter and density under which a full C-J-type high-velocity detonation and consequently full energy release does not occur. Hence, in the case of those high explosives considerable amounts of explosive would be necessary to achieve a high velocity detonation. To overcome this issue detonating samples are therefore often confined in highly dense and inert sleeves made of gold to reduce the energy losses. After the experiment either the released heat or composition of the detonation products are analyzed to determine the heat of detonation. However, it must be noted that the detonation energy as used to describe an ideal detonation (C-J detonation) is always a little smaller than the energy determined in a calorimeter. This

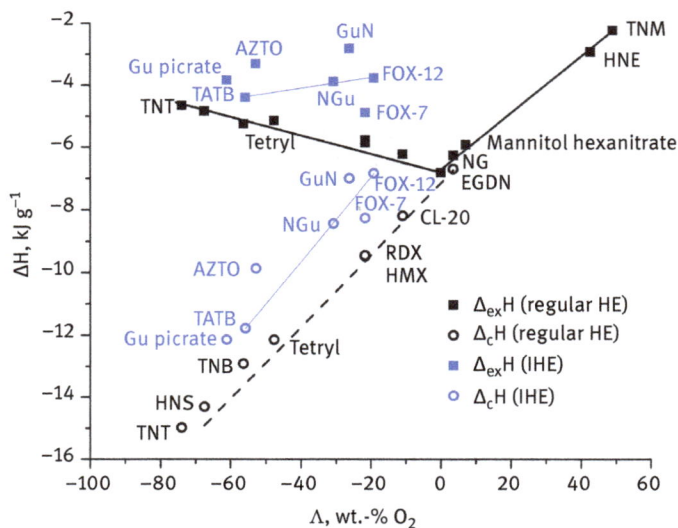

Figure E.6: Calculated Heat of explosion and calculated heat of combustion of high explosives versus oxygen balance.

is because after the C-J point exothermic processes still occur but do not contribute into the incident detonation pressure and velocity.

For common high explosives the heat of explosion correlates with the oxygen balance of the material (black solid squares in Figure E.6), however, for insensitive high explosives (blue solid squares) there is no such correlation. Figure E.6 also depicts the correlation between combustion enthalpy and oxygen balance which again outlines the differences between common and insensitive high explosives.

- V. I. Pepekin, S. A. Gubin, Heat of explosion of commercial and brisant high explosives, *Combust. Explos. Shock Wave*, **2007**, *43*, 212–218.
- D. L. Ornellas, *Calorimetric Determinations of the Heat and Products of Detonation for Explosives: October 1961 to April 1982, UCRL-52821*, Lawrence Livermore National Laboratory, April 5 **1982**, 87 pp.

Explosive
Explosivstoff

An explosive material is a substance (pure or formulation) that can undergo a spontaneous ($\Delta G < 0$) and exothermic ($\Delta H < 0$) reaction in the absence of atmospheric oxygen or any other ambient oxidizer. A necessary and innate property of explosive materials is the strong visible and ultraviolet chemiluminescence associated with the reaction. This clearly distinguishes explosives from other materials that – though may be able to undergo a spontaneous and exothermal reaction yet – are no explosives. The minimum energy required to achieve chemiluminescence by electronic excitation in the ultraviolet and visible range probably is probably around 1 ± 0.2 kJ g^{-1}, but also depends strongly on the individual thermal stability and the heat capacity of the products formed upon reaction.

Explosives can be triggered to react by means of mechanical impetus (shock wave, impact, friction) thermal excitation (flame, convective/conductive/radiative heating) and/or specific optical excitation (certain transition metal complexes can be initiated by irradiation with monochromatic light).

Besides intense flames and sparks the eponymous effects of explosives are the spontaneous release and expansion of at least temporarily gaseous products (pressure-volume work) which either constitutes the main effect (large flames) or assist to propel a container or accelerate (bullets, fragments, liners, etc.).

Explosives are identified by following procedures outlined in UN Test Series 1 (UN gap test, steel sleeve test, time pressure test, or internal ignition test). A minimum one positive outcome indicates an explosive material in any of these tests. As non-explosive materials, all materials qualify that pass UN Test Series 2 with not a single positive outcome.

Both groups in addition can be separated further into two subgroups each. That is explosives which are prepared for use in explosives and those who are made for non-explosive uses (Figure E.7). Figure E.8 finally shows a functional categorization into high explosives, propellants, and pyrotechnics and the characteristic properties.

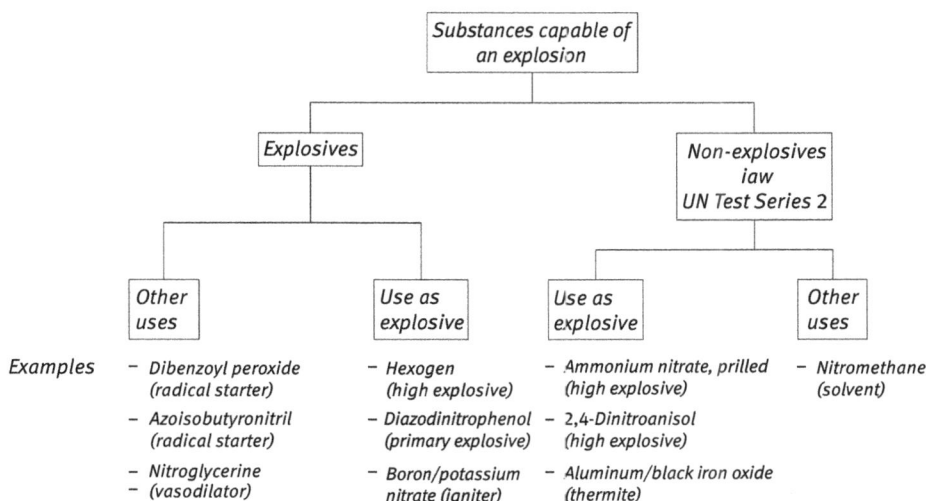

Figure E.7: Categories of explosive substances.

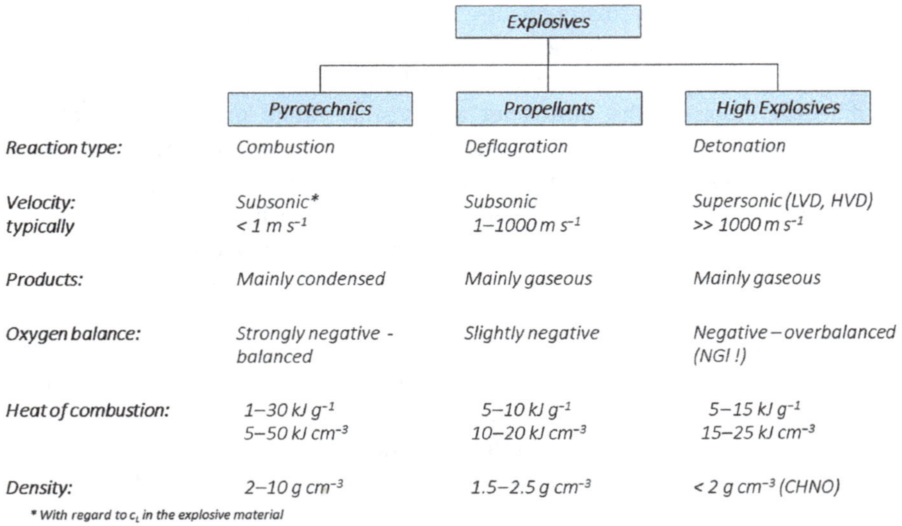

Figure E.8: Functional categories of explosives.

– *Recommendations on the Transport of Dangerous Goods, Manual of Tests and Criteria*, 5. Ed. UN, Genf, **2009**, 450 pp.

Explosophors
Explosophore

Explosphors are chemical functional groups that support the exothermal decomposition of a compound. This can be either by facilitating oxidation with oxygen:

Oxygen-containing groups

–	Peroxide	$-O-O-$
–	N-Oxide	$-N(O)-$
–	Azoxy	$-N=N(O)-$
–	Furazan	$=N-O-N=$
–	Furoxan	$=N(O)-O-N=$
–	Nitroso	$-NO$
–	Nitro	$-NO_2$
–	Nitrato	$-ONO_2$
–	Nitroxy	$=NO_2$
–	Fulminate	$=C=N-O-$

Or by other means to increase the tendency to decompose (*strain*, <u>fragment coupling</u>, <u>stability of N_2</u>).

oxygen-free groups:

– <u>Azo</u>	$-N=N-$
– <u>Azido</u>	$-N_3$
– <u>Cyclopropyl</u>	$-C_3H_5$
– *Cyclopropenyl*	$-C_3H_3$
– *Tetrahedryl*	$-C_4H_3$
– *Cubyl*	$-C_8H_7$
– <u>Acetylene</u>	$-C{\equiv}C-$

– J. P. Agrawal, R. D. Hodgson, *Organic Chemistry of Explosives*, Wiley. New York, **2007**, 384 pp.

Exsudation, Exudation

Ausschwitzen, Exsudation

Exsudation occurs upon temperature cycling of solid materials. It is related to the separation of a liquid phase at the high-temperature profile. This liquid can be an original component (e.g. a liquid plasticizer) or a eutectic that has unintentionally formed from impurities in the material. In either case, the separation of components by exsudation yields a lowered mechanical integrity of an explosive material and therefore can affect both sensitiveness and performance.

– H. W. Voigt, *Exudation Test for TNT Explosives under Confinement: Exudation Control and Proposed Standards, Technical Report ARLCD-TR-83004*, Large Caliber Weapon Laboratory, Dover, NJ, USA, February **1983**, 37 pp.

F

FactSage

FactSage is an Integrated Thermodynamic Databank System (ITDS) that provides tools both for the administration of thermochemical data of inorganic substances and for thermodynamic calculations. It offers these capabilities through a graphical user interface, which is in most parts self-explanatory.

The user of an ITDS normally has three tasks in mind when using such a package:
- Administration of thermodynamic databases
- Execution of thermodynamic application calculations (with tabular and graphical results)
- Generation of thermodynamic data from experimental information

Administration of data comprises:
- Searching the contents of (public and private) databases
- Entering, modification, and deletion of private data

Execution of thermodynamic application calculations means:
- Thermodynamic properties of pure substances in tabulated and graphical forms
- Thermodynamic properties of solution phases and their constituents in tabulated and graphical forms
- Thermodynamic properties of stoichiometric reactions both as isothermal equilibrium properties including equilibrium constants and as non-isothermal balances of the extensive properties of a reaction in tabular and graphical forms
- Complex equilibrium calculations for multiphase multicomponent systems using global conditions of elementary amounts, temperature, and total pressure, but also under constraints such as fixed extensive property balances, predefined phase occurrences, or conditions of an open system
- Phase diagrams of the classical three types with the choice of two potential axes (T, P, chemical potential or partial pressure), one potential axis and one axis with extensive property ratios (mole fractions, weight fractions), or two axes with extensive property ratios

Generation of thermodynamic data from experimental information:
This enables the user to invert complex equilibrium calculations such that the results of the calculation are used to optimize the Gibbs energy data of the various phases of a chemical system, be they stoichiometric pure substances or ideal or non-ideal solutions.

https://doi.org/10.1515/9783110660562-006

The two major modules for application calculations are the Equilib and the Phase Diagram.

Nowadays, there is a wealth of thermodynamic data that are ready to use with the software. These are compiled in user-friendly topical databases such as Pure Substances, Oxides, Salts, Light Metal Alloys, Hard Metals, Steels, Fertilisers, Non-ferrous Alloys, Multi-purpose Alloys, Nuclear Materials, Noble Metals, Solders, and also Aqueous Solutions. Several groups of scientists cooperate in the enormous task of maintaining and developing these databases. It should be noted, however, that by way of using the data administration and optimization modules of the software, the user can also establish his/her own private databases. Please visit www.FactSage.com for details.

FAE-Fuel–air explosive

FAEs were first investigated in wartime Germany for both engineering charges (clearing trenches, foxholes, bunker, and mine fields) and enhanced blast anti-aircraft ammunition.

FAE are also called volumetric explosives. This is because for the "in situ preparation," it requires to mix large volumes of flammable gases or vapors with air and/or requires dispersion of flammable droplets and particles in air. Depending on stoichiometry, pressure, temperature, size of clouds, and the type of environment/confinement, these mixtures can be ignited and yield deflagration with characteristic rates. In the past, FAE usually needed a two-step process with two igniters: a first charge was initiated for the distribution of the FAE and finally the distributed FAE-cloud was ignited. Provided sufficient volume, pressure, and temperature of the FAE, the laminar burn can progress to detonation (see DDT). Meanwhile, current FAE charges can also be directly initiated with a shock wave (e.g. with a small HE charge). However, till then, the boundary conditions for stable detonation must be satisfied, otherwise the detonation can transit back to a deflagration.

The blast effects associated with FAE are characterized by a much slower pressure decay when compared with common high explosives. This and the inherent volumetric expansion of FAE – which can extend from several 10 m^3 to some 100 m^3 – explain the increase in destructive effects compared to common HE which only act as point sources.

- M. Zippermayr, *Testimony* given towards CIOS Mission 214, "Salzburg Area", **1945**.
- H. Walter, B. Walter, B. H. Wilcox, *Myrol Vapor Detonation*, in *German Developments in High Explosives*, FIAT-Final Report 1035, 9 April **1947**, 15 pp.
- https://www.globalsecurity.org/military/systems/munitions/fae.htm
- E.-C. Koch, *Volumetric Explosives Part I, Fuel Air Explosives, L-165*, NATO-MSIAC, Brussels, July **2010**, 25 pp. https://www.msiac.nato.int/products-services/publications/l-165-volumetric-explosives-part-1-fuelair-explosives

Feistel, Fritz (1897–1957)

Feistel was a German chemist and entrepreneur. He was born in Kirchheimbolanden, (Rhineland Palatinate) served as officer in World War I, and started studies of chemistry at the Friedrich-Wilhelms University at Bonn in 1920, where he obtained his doctorate in 1926. The same year he became chemist at the chemical main laboratory of the Deutsche Pyrotechnische Fabriken AG (DEPYFAG) in Berlin. In 1933, Dr. Feistel founded the Deutsche Leucht and Signalmittelwerk (DELEU) in Berlin Reinickendorf. In 1936, the production was moved to remote Schönhagen (Teltow county). Dr. Feistel was the first to introduce zirconium as a corrosion-resistant high-energy fuel in first fires and signal tracers. His company was also the first company worldwide to develop quantitative color and light measurement of pyrotechnic illuminants and colored smokes. His son, *Fritz Feistel Jr.* (1930–1979), in 1960 founded Pyrotechnische Fabriken Fritz Feistel KG at Göllheim/Rhineland Palatinate which was later known as Piepenbrock Pyrotechnik (1981–1999), and finally was part of Diehl Group (1999–2001).

- F. Trimborn, *Explosivstoffabriken in Deutschland*, 2. Aufl. Locher, **2002**.
- H. J. Eppig, Photometric Procedures Used in Research and Production of German Pyrotechnic Ammunition, *CIOS Target Nos 3a/162 & 17/32*, June-August **1945**, London H.M. Stationery Office.
- Entry in the German National Library: http://d-nb.info/57064111X

Ferrocene
Ferrocen

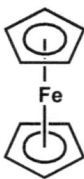

Bis(η^5-cyclopenta-2,4-dien-1-yl)iron(0), FeCp$_2$	
Aspect	Red-orange crystals
Formula	$C_{10}FeH_{10}$
GHS	02, 07, 09
H-phrases	228-302-411
P-phrases	210-240-241-280-301+312-501a
UN	1325
REACH	LRS
EINECS	203-039-3
CAS	[102-54-5]

m_r	g mol^{-1}	186.04
ρ	g cm^{-3}	1.49
$\Delta_f H$	kJ mol^{-1}	154.89
$\Delta_c H$	kJ mol^{-1}	−5,931.4
	kJ g^{-1}	−31.883
	kJ cm^{-3}	−47.506
Λ	wt.-%	−223.6
Mp	°C	173
Bp	°C	249
$\Delta_{sbl} H$	kJ mol^{-1}	72

Ferrocene is among the best investigated organometallic compounds with ~20,000 publications dealing with it ever since its first report by T. J. Kealy (1927–2012) and P. L. Pauson (1925–2013) and the independent structural characterization by G. Wilkinson (1921–1996) and E. O. Fischer (1918–2007), which earned the latter a shared Nobel Prize in 1973. Ferrocen has been tested as burn rate modifier for AP-based composite propellants. However, Ferrocen possesses a considerable vapor pressure (also responsible for the characteristic smell resembling camphor), which upon temperature cycling leads to unwanted migration. Hence, ferrocene is often modified with high molecular side chains (e.g. *Butacen* or *Catocen*) to reduce the tendency to migrate. Despite its volatility, it has been proposed as ingredient in various pyrotechnic formulations.

The isoelectronic (H–C $\cong$ N) all-nitrogen ferrocene derivative $Fe(N_5)_2$ has long been among the sought-after high-nitrogen target molecules and in view of the recent progress with N_5^-, currently merits consideration (*Christe*). In a similar way, the all-phosphorus derivative $Fe(P_5)_2$ is a surprisingly stable and non-reactive (air/water) phosphorus compound (*Baudler*) which – provided other ways of synthesis would be found – could be an ideal future fuel for pyrotechnics.

- U. A. Lehikoinen, Aluminum-Iron Oxide Incendiary Composition Containing A (Cyclopentadienyl) iron Compound, US-Patent 3,498,857, USA, **1970**.
- J. A. Boatz, K. Christe, A. Vij, Structural Isomers of bis(pentazolyl)iron(II): A Theoretical Study, *AFOSR Molecular Dynamics Contractors Conference*, Monterey, CA, 22–24 May **2005**.
- M. Baudler, S. Akpapoglou, D. Ouzounis, F. Wagestian, B. Meinigke, H. Budzikiewicz, H. Münster, On the Pentaphosphacyclopentadienide Ion, P_5^-, *Angew. Chem. Int. Ed.* **1988**, 27, 280–281.

Firecracker, banger

Knallkörper

A firecracker creates a loud sharp acoustic report upon ignition. To allow for a rapid flame spread and consequently rapid pressure built-up, the pyrotechnic formulation in a firecracker must be present as an unconsolidated material with high specific surface area. Consolidation just yields a laminar burn not sufficient to effect a deflagration and rupture of a container. Compositions used in firecrackers are black powder, flash compositions, and also dark compositions which create very little light such as the formulation given in Table F.1.

Table F.1: Dark report after Fronabarger.

Potassium ferricyanide, $K_3Fe(CN)_6$ (wt.-%)	31
Potassium perchlorate, $KClO_4$ (wt.-%)	69

- J. W. Fronabarger, *Igniter Composition Comprising a perchlorate and potassium hexacyano cobaltate*, US Patent 3,793,100, **1974**, USA.

Firedamp explosion

Schlagendes Wetter

Flammable gases (mainly methane, CH_4) can originate from coal beds and also from Upper Permian in potash underground mining. Combustible mixtures with air (4.4–16.5 Vol% methane) in the galleries are called firedamp. Upon ignition of a firedamp the combustion can quickly transition to detonation and yield a firedamp explosion. To avoid ignition and initiation of a firedamp in explosive mining, so called *permissible explosives* are used which are characterized by general low explosion temperature and considerable content of alkali halide salts (>10 wt.-%) which due to their catalytic properties as well as their high heat of vaporization inhibit flame spreading. The so-called *inverse salt-pair explosives* contain ammonium chloride and alkali nitrate salts to form alkali halides in situ:

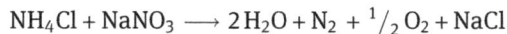

$$NH_4Cl + NaNO_3 \longrightarrow 2 H_2O + N_2 + {}^1/_2 O_2 + NaCl$$

Fire extinguisher, pyrotechnic

Feuerlöschgenerator

A pyrotechnic fire extinguisher produces aerosols that stop the combustion in C–H–O systems by a combination of three main mechanisms (Table F.2):
- chain termination reactions
- interfering with the oxygen access to the combustion zone
- interfering with the heat transport

Table F.2: Experimental fire-extinguishing compositions.

Formulation	1	2	3
Potassium bromate, $KBrO_3$	50.7		
Strontium nitrate, $Sr(NO_3)_2$		35.50	
Potassium nitrate, KNO_3 (SSA ≥ 1,500 cm^2 g^{-1})			67–82
Potassium iodide, KI		25.21	
Tripotassium cyanurate, $K_3C_3N_3O_3$	47.8		
Dicyandiamide, $C(NH_2)_2CN$ (≤15 μm)			6–25
5-Aminotetrazole, $CH(NH_2)N_4$		20.35	
Magnesium carbonate, $MgCO_3$		18.94	
Phenol formaldehyde resin (≤100 μm)			8–12
Polyvinyl alcohol, $(C_2H_4O)_n$	1.5		
u (mm s^{-1})		12.95 at 6.89 MPa	

To stop chain reactions a similar approach as in muzzle flash suppression can be used and this is why easily ionizable potassium compounds find widespread use. In addition, heavy atoms slowing down overall combustion kinetics assist in suppression (e.g. iodine or bromine compounds).

Oxygen access is terminated by use of underbalanced formulations which absorb oxygen and release heavy CO_2 which can act as a protective blanket especially in confined spaces.

Finally, heat transport can be interfered with by dissemination of conductive particles such as graphite which do not burn but show high heat capacity.

- V. N. Kozyrev, V. N. Yemelyanov, Y. I. Sidorov, V. A. Andreev, *Method and apparatus for extinguishing fires in enclosed spaces*, US5,865,257, **1999**, Russia.
- G. F. Holland, M. A. Wilson, *Chemically active fire suppression composition*, US 6,024,889, **2000**, USA.
- P. L. Posson, M. L. Clark, *Flame suppressant aerosol generant*, US 8,182,711, **2012**, USA.

Firework compositions
Feuerwerkssätze

Firework compositions serve the production of acoustic and optical effects. They can be distinguished into the following compositions:
- Bengal/gerb compositions
- Crackling compositions
- Glitter compositions
- Illuminant compositions
- Flash compositions
- Smoke compositions
- Sparkler compositions
- Strobe compositions
- Cone compositions
- Waterfall compositions
- Whistling compositions

- A.P. Hardt, *Pyrotechnics*, Pyrotechnic Publications, Post Falls ID, **2001**, 430 pp.

First fire
Anfeuerung

First fires – also known as primers – are igniter mixtures which are directly applied to the surface of the energetic material to be ignited. The first fire can be painted (Figure F.1), printed, dip coated, or consolidated directly with the main grain. Basic requirements for a first fire are: ease of ignition (low ignition temperature), high heat of combustion to transfer safely, and reliably to the main grain and high mechanical strength to resist delamination as a result of environmental stimuli such as temperature cycling, logistical and tactical vibration, drop, or expulsion at near sonic windspeeds. It has been suggested recently to use commercial alloying foils (Ni/Al) as cover and first fire for decoy flares. Table F.3 shows bright burning metal containing first fires (13) and dim first fires using antimony sulfide.

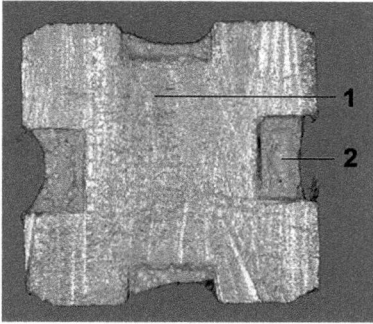

Figure F.1: Cross section of a generic 1 × 1 × 8 in. MTV grain (**1**) with machined groves and applied first fire (**2**) from *Koch 2012*.

Table F.3: First fire formulations.

Formulation	1	2	3	4
Antimony sulfide, Sb_2S_3				3.16
Charcoal, $(C_{10}H_6O_2)_n$			15	8.42
Magnesium, Mg	62	17		
Silicon (fine)			10	
Zirconium (60–200 mesh))			10	
Viton, $(C_{10}H_7F_{13})_n$	10			
Polytetrafluoroethylene, $(C_2F_4)_n$	28			
Potassium perchlorate, $KClO_4$			60	7.37
Potassium nitrate, KNO_3				78.94
Barium peroxide, BaO_2		81		
Gum acaroides, $C_6H_{5.42}O_{1.15}$			5	2.11
Calcium resinate, $CaC_{40}H_{58}O_4$		2		

– D. B. Nielson, R. L. Tanner, C. Dilg, *Flares including reactive foil for igniting a combustible grain thereof and methods of favricating and igniting such flares*, US 7,690,308, USA, **2008**.
– V. Weiser, E. Roth, S. Knapp, A. Lity, Ignition of Flare Propellants Using Al/Ni-Ananofoils at sub-atmospheric pressure, *ICT-Jata*, **2019**, P108.

Flame size

Flammengröße

The flame size of a burning material (fuel, pyrotechnic, propellant, high explosive, etc.) is determined by its oxygen balance, the ambient pressure, the specific heats of the material, components, latent heats, and by its heat of combustion. Hence, small flames are observed at

- high pressure;
- neutral or positive oxygen balance;
- high specific heats and latent heats of the combustion products; and
- low heat of combustion.

Large flames are correspondingly observed with the contra conditions.

- H. C. Hottel, W. R. Hawthorne, Diffusion in Laminar Flame Jets, *Symposium on Combustion*, **1949**, 254–266.
- W. R. Hawthorne, D. S. Weddel, H.C. Hottel, Mixing and Combustion in Turbulent Gas Jets, *Symposium on Combustion*, **1949**, 266–288.

Flame temperature, adiabatic

Flammentemperatur, adiabatische

The adiabatic flame temperature, T_{ad} [K], is the temperature that is obtained from a reaction at constant pressure in a confined system ($\delta Q = 0$). From the first law of thermodynamics for p = const. it follows that $dH = \delta Q$. The adiabatic flame temperature is reached when the sum of the integrated heat capacities $c_{p,i}(T)$ at temperature T_{ad} and the latent heats ($\Delta_{Tpj}H$) equal the amount of the reaction enthalpy.

While this correlation is valid for low temperature, C–H–N–O systems' significant deviations are observed with highly exothermic systems including metals and halogens.

$$\int_{298}^{T_{ad}} \sum_i \Delta c_{p_i}(T)dt + \sum_j \Delta_{pt_j}H - \Delta_R H^\circ = 0$$

- J. Warnatz, U. Maas, R. W. Dibble, *Verbrennung*, 3. Aufl. Springer, **2001**, pp. 50 ff.
- D. P. Dolata, T. I. Peregrin, Prediction of Flame Temperatures, Part 1: Low Temperature Reactions, *J. Pyrotech.* **1995**, *1*, 37–46.

Flare
Leuchtkörper

Pyrotechnic flares generally serve the generation of radiation in a specific part of the electromagnetic spectrum. Flare ammunition comprises means for tactical signaling and illumination in the visible and near-infrared range as well as augmentation and decoying effects in the mid-infrared range. Projectiles for signaling and illumination can be fired from proprietary projectors starting with calibers around 13 mm, flare guns, Very pistols (25.4 mm), mortars, and howitzers up to 155 mm. The burn time of free-falling stars most often does not exceed 8 s while parachute-retarded illuminating payloads can have burn times up to 180 s. Other ammunition for signaling and illumination include trip wire flares ($t_b > 40$ s) and signal mines ($t_b > 8$ s) sometimes combined with an audible report.

For an aerial illuminating payload, Figure F.2 shows the irradiance (lx) on the ground as a function of height and Figure F.3 shows the irradiance as function of lateral displacement for two heights. The irradiance B_λ (lx) at any given place with the lateral displacement, r(m) from the ground mark at a height, h(m) with an illuminating payload burning with the radiant intensity, I_λ (cd), is given by

$$B_\lambda = \frac{I_\lambda \cdot h}{\sqrt[3]{h^2 + r^2}}$$

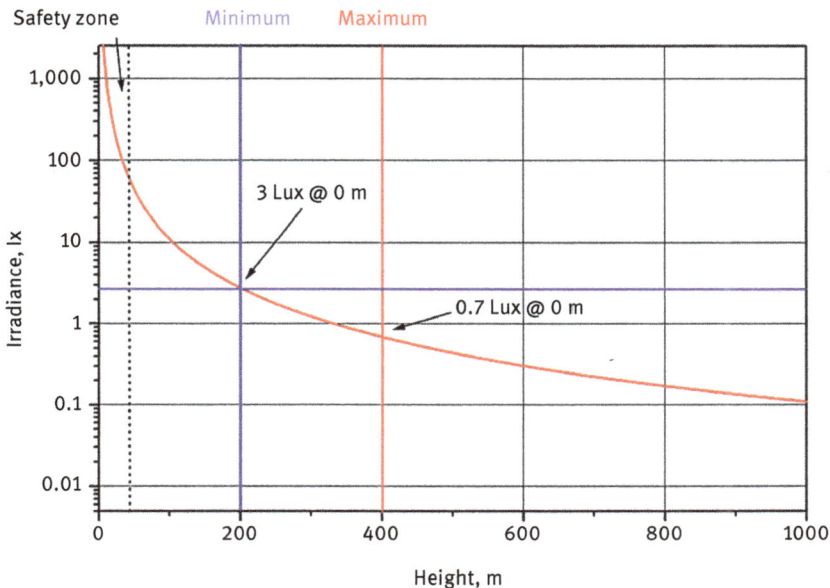

Figure F.2: Irradiance as a function of ceiling for a flare payload with $I_\lambda = 110$ kcd.

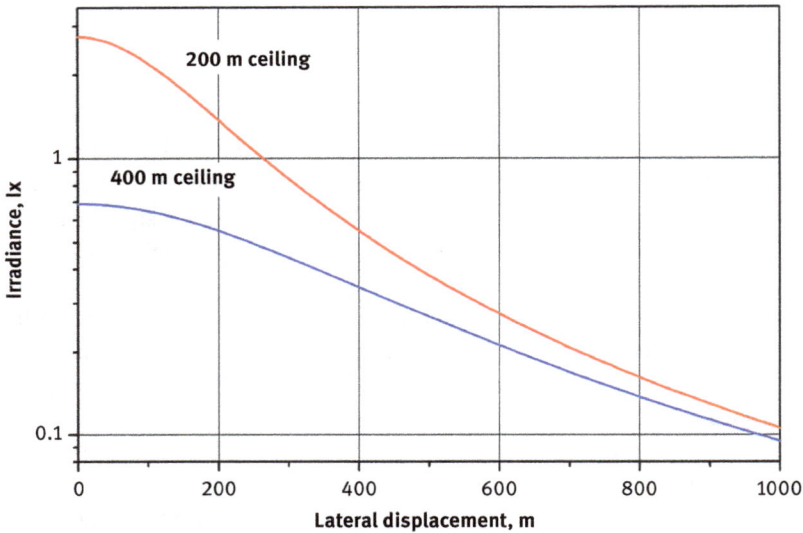

Figure F.3: Irradiance as function of lateral displacement for two heights for a flare payload with $I_\lambda = 110$ kcd.

- NN. *Engineering Design Handbook – Military Pyrotechnics Series Part One Theory and Application, AMC Pamphlet AMCP 706-185*, Headquarters, U.S. Army Materiel Command, April **1967**, 6-1–6-73.

Flare back
Nachflammer

Flare back (flareback) is the term for the afterburn of the fuel-rich propellant combustion products (H_2 and CO) in the gun barrel and emerging from the breech into a vehicle. The latter is particularly dangerous for the gun crew who might sustain burns from the flare back and due to the possibility of accidental ignition of gun propellant charges ready for use in the area. Additives in the gun propellant suppressing the muzzle flash also suppress the flare back. In naval artillery, flare backs are prevented by deluge systems which flush pressurized air or nitrogen through the barrel right after the shot has fired and before the breech is opened. In tanks and other self-propelled artillery fixed fume extractors suck the fumes off.

- R. Davis, Army's Ml09 Howitzer – Required Testing Should Be Completed Before Full-Rate Production, *Report GAO/NSIAD-92-44*, 23 January **1992**, 30 pp.
- C. R. Woodley, Modelling of Fume Extractors, *19th International Symposium of Ballistics*, 7–11 May **2001**, Interlaken, Switzerland, 273–280.

Flare composition, illuminating
Leuchtsatz

A pyrotechnic illuminating flame composition upon combustion yields a flame radiating intensively in the visible band. Typical illuminating flares contain magnesium as metallic fuel, sodium nitrate as oxidizer and an organic polymer as binder.

The radiant intensity, I_λ (cd) delivered upon combustion of a flare payload is related to its mass m_f (g), the burn time, t_b (s), and the spectral energy E_λ (cd s g^{-1}) of the flare composition:

$$I_\lambda = \frac{E_\lambda \cdot m_f}{t_b}$$

The spectral energy itself correlates with the combustion enthalpy of the flare composition, $\Delta_c H$ (J g^{-1}) in accordance with

$$E_\lambda = \frac{1}{4\pi} \cdot \Delta_c H \cdot F_\lambda$$

It has to be taken into account that

$$1\,cd = 638^{-1}\,Wsr^{-1}$$

Visible flares

With Mg/NaNO$_3$ compositions, the fraction of energy emitted in the visible band ranges between: F_λ 0.11~0.08. Figure F.4 depicts the VIS spectrum of a Mg/NaNO$_3$ flare.

Figure F.5 further shows the relationship between stoichiometry of Mg/NaNO$_3$ and flare performance (burn rate, heat of explosion, temperature, and spectral energy).

To achieve as high heat of combustion as possible illuminating flare compositions are always oxygen deficient ($\Lambda < 0$) and consequently use the atmospheric oxygen as a supplementary oxidizer.

That is the first reaction, which occurs in the condensed phase of the pyrotechnic grain, releases heat and nitrogen to inflate the plume and to volatilize the surplus magnesium:

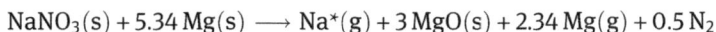

$$NaNO_3(s) + 5.34\,Mg(s) \longrightarrow Na^*(g) + 3\,MgO(s) + 2.34\,Mg(g) + 0.5\,N_2$$

The magnesium vapor then reacts with the atmospheric oxygen to give MgO:

$$2.34\,Mg(g) + atmospheric\ oxygen \longrightarrow 2.34\,MgO(sg^{-1})$$

Consequently, the plume size increases with magnesium content. The thermal excitation of sodium yielding a broad Na- doublet extending from $\lambda = 540-660$ nm is responsible for the outstanding luminosity of the flame. Condensed MgO particles in

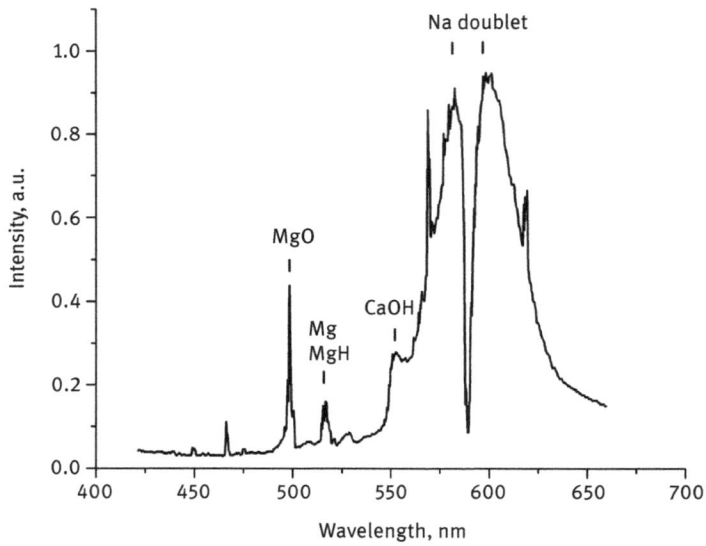

Figure F.4: VIS emission spectrum of Mg/NaNO$_3$ flare with assignments, CaOH signals due to calcium stearate pressing aid in the flare composition.

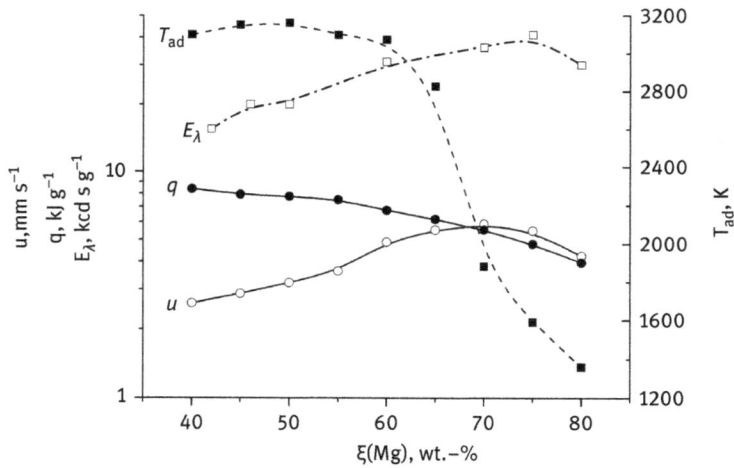

Figure F.5: Performance parameters of the binary system Mg/NaNO$_3$.

addition to having a high emissivity in the visible band further contribute to the luminous output. A typical yellow illuminating flare composition is depicted as follows:

Yellow flare
- 36 wt.-% sodium nitrate, $NaNO_3$
- 55 wt.-% magnesium, Mg
- 3 wt.-% polyolefinic resin
- 3 wt.-% chloroprene binder
- Spectral energy, $E_\lambda = 40.5$ [kcd s g^{-1}] $\cong 59.3$ [J g^{-1} sr^{-1}]

Visible flares are generally characterized by a dominant wavelength, λ_d [nm] and color saturation Σ_λ (%). For yellow flares, typical values are $\lambda_d = 587 \pm 3$ nm and $\Sigma_\lambda > 75$ %.

The undisputed position of magnesium as a flare fuel for visible flares has several reasons. Among other metallic fuels, Mg allows for the highest spectral energy and illuminant intensity (I_λ). Table F.4 shows the performance of a generic formulation:

Generic flare formulation
- 50 wt.-% metal fuel   Mass of composition 12 g
- 45 wt.-% $Ba(NO_3)_2$ Diameter   12 mm
- 5 wt.-% PVC

Table F.4: Performance of generic metal/$Ba(NO_3)_2$/PVC flare composition as a function of metallic fuel after Schmied (**1968**).

Fuel	Oxygen balance (wt.-%)	Heat of combustion (kJ g^{-1})	Burn rate (mm s^{-1})	Light intensity (kcd)	Spectral energy (kcd s g^{-1})
Magnesium (10–100 µm)	−24	−24.77	8.2	108	18.2
Al–Mg alloy (50:50)	−30	−27.91	8.9	77	11.3
Al + Mg (50:50)	−30	−27.91	7.1	67	11.6
Zirconium	−9	−12.03	14.4	57	3.6
Ti–Al alloy (50:50)	−30	−25.39	5.0	15	3.5
Aluminum	−36	−31.04	2.9	5.8	2.2
Boron	−102	−58.82	5.9	3.5	0.9
Zinc	−3	−5.36	1.6	0.8	0.4
Silicon	−48	−32.44	2.5	0.5	0.25

While other metals in comparison to Mg have higher gravimetric heats of combustion, those metals have lower or no vapor pressure at all, and hence cannot burn in a diffusion flame. This however is of paramount importance to achieve an extended luminous plume. The influence of stoichiometry and type of nitrate on the flare performance is depicted in Table F.5. It shows that $NaNO_3$ is by far the best oxidizer for illuminants.

Table F.5: Performance of generic Mg/nitrate/PVC flare composition as a function of nitrate (Schmied, **1968**).

Formulations	50 wt.-% magnesium (150–250 μm) 45 wt.-% nitrate 5 wt.-% PVC			30 wt.-% magnesium (150–250 μm) 55 wt.-% nitrate 15 wt.-% PVC		
Nitrate	Burn rate (mm s^{-1})	Light intensity (kcd)	Spectral energy (kcd s g^{-1})	Burn rate (mm s^{-1})	Light intensity (kcd)	Spectral energy (kcd s g^{-1})
$NaNO_3$	7.6	180	36	3.7	44	17.2
KNO_3	5.8	42	11	3.2	9	4.2
$Sr(NO_3)_2$	5.2	79	21.4	4.1	30	9.6
$Ba(NO_3)_2$	5.5	73	17.9	3.9	19	6.3

Generic flare formulation
- Mass of composition 12 g
- Diameter 12 mm

Figure F.6 shows the influence of grain diameter on spectral efficiency for very large cylindrical illuminant grains with Mg (150–300 μm).

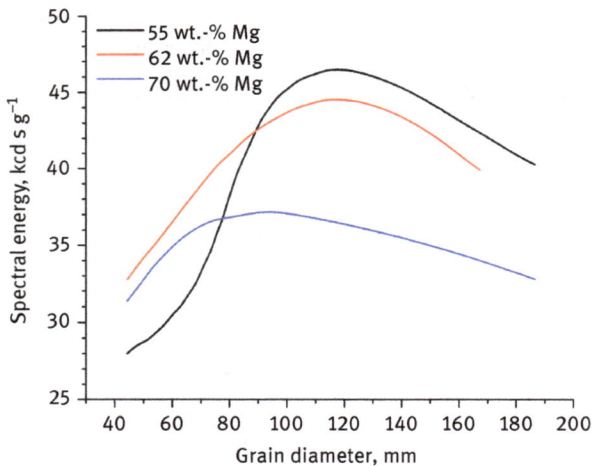

Figure F.6: Influence of grain diameter on spectral energy of an visible illuminant at three different stoichiometries after *Douda*.

Near-infrared flares

The use of night-vision goggles (NVG) – operating beyond the visible range, $\lambda > 700$ nm – necessitates the use of clandestine pyrotechnic illumination and signaling. Figure F.7 depicts the sensitivity of the human eye and the responsivity of a state-of-the-art III. generation NVG detector material.

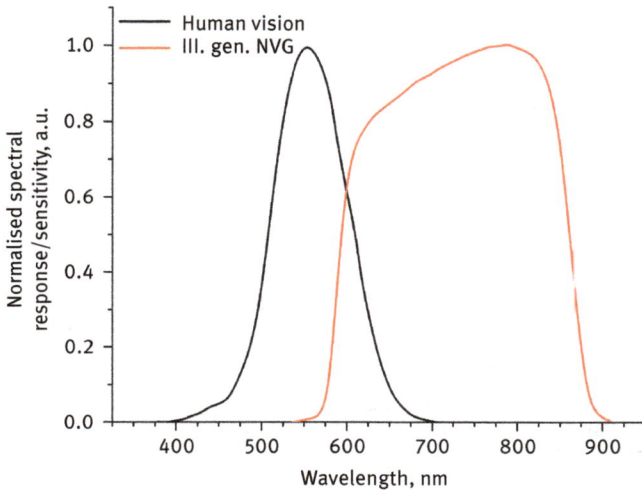

Figure F.7: Spectral sensitivity of the human eye versus III Gen NVG responsivity after *Bradley et al.* (1994).

A prime requirement for NIR-flare compositions is a high concealment index, $\chi_{NIR/VIS}$ (which is defined as ratio of radiation intensity [W sr^{-1}] emitted in the NIR (λ = 600–900 nm) versus VIS (λ = 400–700 nm) which means as low as possible visible emission. To match those requirements, NIR-flare compositions utilize fuels that give off very little visible signature upon combustion such as hexamine, elemental boron and silicon. The luminous flux of Mg/nitrate flare compositions decreases in the following about order:

$$Na > Sr > Ba > K, Rb > Cs$$

It is hence that potassium, rubidium, and cesium nitrate are preferred oxidizers in NIR flares, also because the main series transitions of K:(λ: 769, 766 nm), Rb:(λ: 795, 780 nm), and Cs:(λ: 894, 852 nm) occur in the NIR. Figure F.8 shows the emission spectrum of an NIR-flare formulation (Table F.6) invented by *Dillehay*.

Though Rb is slightly superior to K (with regards to emitting at longer wavelength and hence improved concealment index), it is however extremely pricy ruling

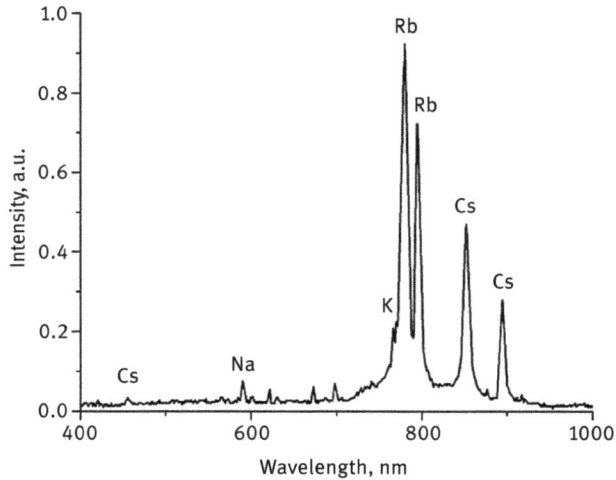

Figure F.8: Emission spectrum of a Rb and Cs containing NIR flare after *Scheutzow*.

Table F.6: Composition and performance of NIR flare (*Scheutzow*).

Components	NIR-flare	Parameter	Value
Silicon, Si,	10	I_{VIS} (cd s g^{-1})	107
Hexamine, $C_6H_{12}N_4$	16	I_{VIS} (J g^{-1} sr^{-1})	0.168
Cesium nitrate, $CsNO_3$	40	I_{NIR} (J g^{-1} sr^{-1})	11.57
Rubidium nitrate, $RbNO_3$	30	χ (–)	69
Epicon/epikure	4	u (mm s^{-1})	1

out any practical application. The current (2020) price ratio for 1 kg of the metal nitrate at 99.7% metal content and higher is approximately K : Cs : Rb / 1 : 70 : 4000.

– I. Schmied, Licht- und Farbintensitäten pyrotechnischer Phrases durch Variation der Metallkomponente, *14. Arbeitstagung – Wehrtechnik*, 5.-7. Nov. **1968**, Mannheim.
– B. E. Douda, Determination of the Amount of Energy Radiated in the visible Region by an illuminating Flare, *RDTN No. 135*, Naval Ammunition Depot Crane, IN, 24 July **1967**,
– B.E. Douda, *25 Million Candle Cast Flare Diameter and Binder Study, Volume 1 and 2*, RDTR 105, Naval Ammunition depot Crane IN, Jan **1968**.
– B. E. Douda, R. M. Blunt, E. J. Bair, Visible Radiation from Illuminating-Flare Flames: Strong Emission Features, *J. Opt. Soc. Am.* **1970**, *60*, 1116–1119.
– E.-C. Koch, *Survey on State-of-the-Art Near-Infrared Emitting Compositions for Flares and Tracers, L-155*, NATO-MSIAC, Brussels, Belgium, March **2009**, 30 pp. accessible for MSIAC member states at https://www.msiac.nato.int/products-services/publications/l-155-survey-on-state-of-the-art-near-infrared-emitting-compositions
– S. Scheutzow, *Investigations of Near and Mid Infrared Pyrotechnics*, PhD thesis, LMU, Munich, **2012**.

Flare composition, signaling
Lichtsignalsatz

In contrary to illuminating flares where the highest possible radiant intensity (cd) is the prime characteristic signaling flares both require high radiant intensity, I_λ [kcd] to be recognized but also an unambiguous color of characteristic dominant wavelength, λ_d [nm], and high saturation, Σ_λ [%]. For signaling purposes, the colors red, white, yellow, and green are used. Though formulations exist to produce blue light, its very low radiant intensity, the poor saturation, and finally the poor visibility range of short wavelength light in a blurred battlefield atmosphere make blue an unsuitable signal color. Table F.7 lists typical signal flare formulations and performance data. Though zirconium yields unsatisfactory results with $Ba(NO_3)_2$ (see Flare illuminating), *Hahma* found that it supersedes Mg as fuel in strontium containing flares when combined with a suitable auxiliary fuel to generate extra gas to expand the combustion plume (Table F.6).

Table F.7: Composition and performance of signal flare compositions.

Parameter	Red	Red	White	Yellow	Green
Magnesium, Mg (wt.-%)	35		42	32	21
Zirconium, Zr (wt.-%)		44			
Strontium nitrate, $Sr(NO_3)_2$ (wt.-%)	54	45	5		
Barium nitrate, $Ba(NO_3)_2$ (wt.-%)				49	57
Sodium nitrate, $NaNO_3$ (wt.-%)				56	
Chlorinated rubber, (wt.-%)	1				
Polyvinyl chloride, $(C_2H_3Cl)_n$ (wt.-%)	10			8	17
Polyvinylidene chloride $(C_2H_2Cl_2)_n$ (wt.-%)		8			
Binder		3	4		5
λ_d (nm)	615	618	577	587	560
Σ_λ (%)	78	86	40	80	67
E_λ (kcd s g^{-1})	44	56	35	30	12

The corresponding gaseous, atomic, and molecular emitters responsible for the emission and their properties and color coordinates are given in Table F.8. Subtractive color mixing of BaOH and SrOH yields the white flame in the abovementioned formulation. Any color hue accessible by mixing both BaOH and SrOH is given by the dotted line in Figure F.9.

Table F.8: Color values, saturation, and dominant wavelength of important optical emitters in flames.

Emitter	X	y	λ_d (nm)	Σ_λ (%)
Na	0.576	0.423	589	100
BaOH	0.066	0.606	504	90
BaCl	0.094	0.811	523	96
SrOH	0.679	0.321	615	100
SrCl	0.720	0.280	640	100
CuOH	0.290	0.666	546	88
CuCl	0.148	0.116	467	90
CuBr	0.155	0.038	459	95

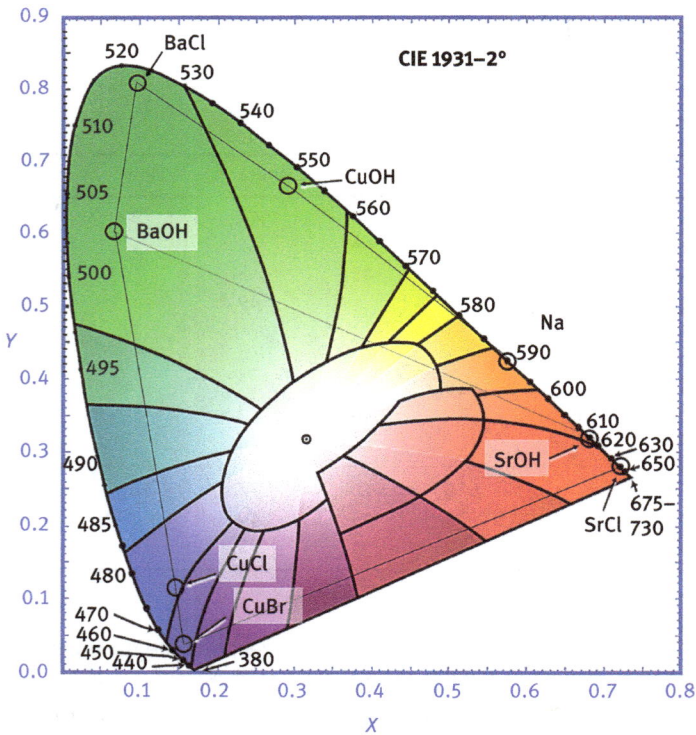

Figure F.9: 1931-2-CIE chromaticity diagram with the color positions of the selected emitters. The outer boundary limits the accessible flame colors.

- W. Meyerriecks, K. Kosanke, Color Values and Spectra of the Principal Emitters in Colored Flames, *J. Pyrotech.* **2003**, *18*, 1–22.
- E.-C. Koch, Spectral Investgation and Color Properties of Copper(I) Halides Cu*X* (X = F, Cl, Br, I) in Pyrotechnic Flames, *Propellants Explos. Pyrotech.* **2015**, *40*, 799–802.
- A. Hahma, M. Welte, O. Pham, *Verwendung von Zirconium oder eines Zirconium enthaltenden Gemischs*, DE 10 2011 116 594 A1, **2013**, Germany.

Flare, decoy
Scheinziel, Täuschkörper

Decoy flares also known as *infrared countermeasure flares* are expendable counter-measures. Based on the reaction of a pyrotechnic or pyrophoric energetic material, decoy flares produce a characteristic electromagnetic signal intended to interfere with the sensors of an infrared guided missile system.

Typically, decoy flares are used to protect aerial targets, but they are also common in the protection of naval vessels. Occasionally, flares are used as counter-measures for land-based vehicles.

There are two main uses for decoy flares
- Preemptive use → Impede lock-on
- Reactive use → Break the lock

The main performance parameters of decoy flares are
- Burning time, t_b [s]
- Spectral intensity distribution, $dI/d\lambda$
- Time/intensity Ratio, dI/dt
- Quality and dimension of spatial extension of the radiating area, dI/dx, as well as
- Kinematic behavior of the flare, dx/dt

A typical aerial decoy flare consists of an anodized aluminum casing (Figure F.10), an electrical impulse cartridge (effecting both ignition and expulsion of the actual flare payload), a sequencer or bore rider safety device with a pyrotechnic transfer charge that prevents premature ignition of the flare payload, the primed flare payload encased in a consumable ballistic cover, and finally a spacer and the terminal cap with sealant to protect the flare from moisture.

There is an enormous range of different pyrotechnic compositions and pyrophoric or flammable materials utilized in decoy flares. Despite all those different chemistries the radiant intensity I_λ (W sr^{-1}) of a flare material is directly related to the rate of combustion, u (mm s^{-1}), enthalpy of combustion $\Delta_c H$ (kJ g^{-1}), and the spectral emissivity of the combustion products, $\varepsilon_{\lambda,T}$ (–). In addition, tactical factors

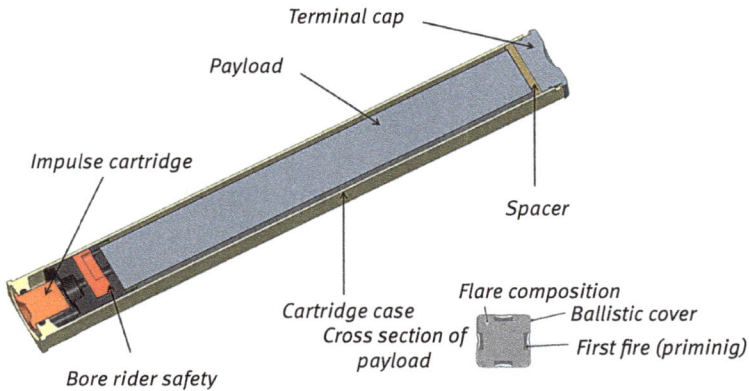

Figure F.10: Cross-sectional view of 1 × 1 × 18 in. flare cartridge and payload.

like wind speed, ambient pressure (altitude), and angle of observation (anisotropy of combustion) influence the radiant energy emitted into the steradian:

$$I_\lambda = \frac{dm}{dt} \cdot E_\lambda$$

$$E_\lambda = \frac{1}{4\pi} \cdot \Delta_c H \cdot F_\lambda \cdot \delta_w \cdot \delta_p \cdot \delta_a$$

where F_λ represents the fraction of energy emitted in the band of interest, δ_w the wind speed, δ_p the ambient pressure, and δ_a the anisotropy.

Aerial targets such as jet planes are distinct sources of radiation as depicted in Figure F.11. The hot tail pipes yield a grey body-type signature (Figure F.12a), whereas the hot combustion gases forming the plume can be best described as a selective radiator (Figure F.12b).

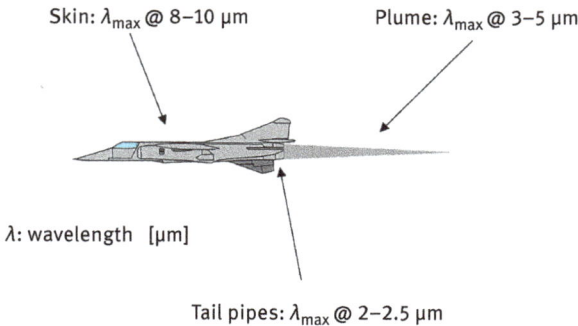

Figure F.11: Source of infrared radiation on a jet plane.

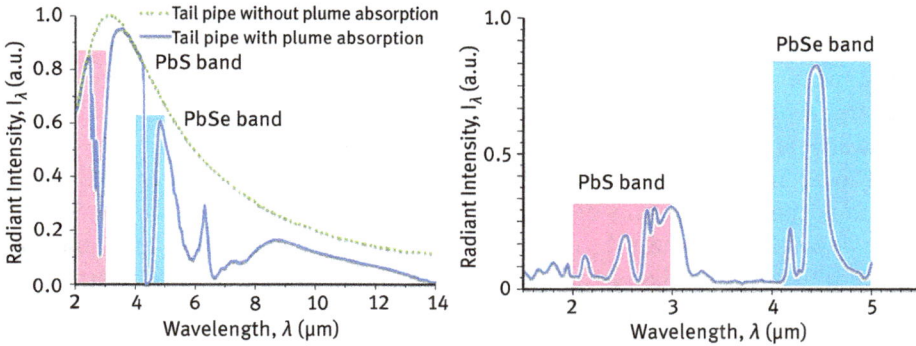

Figure F.12: Spectrum (a) tail pipe (left), (b) plume (right).

Depending on object to target angle either the tail pipes (red sector) or the hot plume (blue sector) dominate the signature of the air plane (Figure F.13).

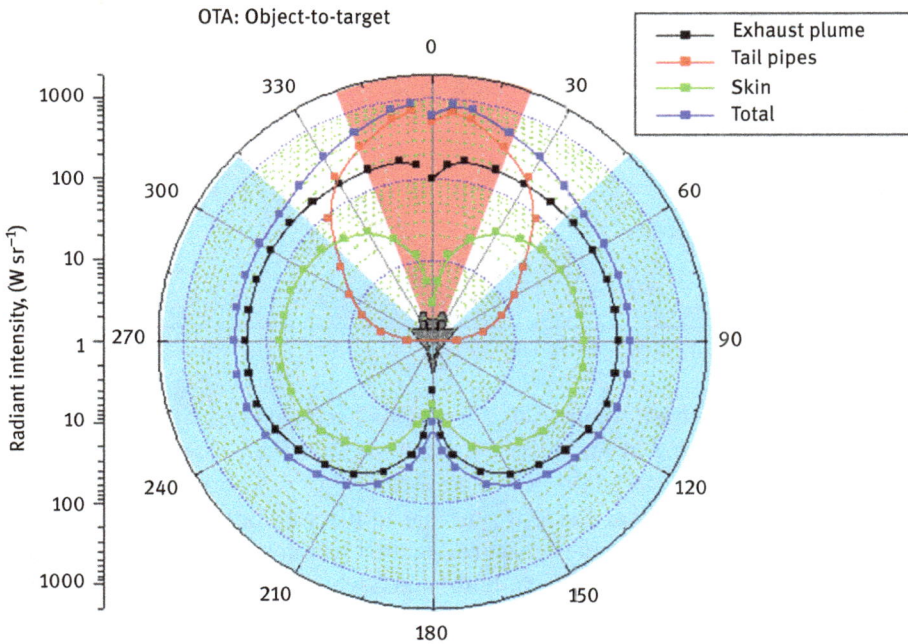

Figure F.13: Polar diagram of aerial signature.

In the first generation of infrared guided missiles, the sole detector material available was lead sulfide, PbS, having its highest responsivity in the short-wave band

(2–2.5 µm) and hence forcing the missiles to head for the tail pipes. This in turn triggered development of flare compositions both for target drones and countermeasures that would mimic the grey body-type radiation of the tail pipes. This is the largest and oldest group of infrared decoy flare compositions. Typically, but physically not correct, they are often referred to as blackbody compositions.

Those formulations (Nos. **1–4** in Table F.9) burn very hot and mainly yield a temperature-dependent intensity distribution with high amounts of radiation emitted in the low-wavelength band 1–2.5 µm. Therefore, they very often contain magnesium as energetic fuel and a carbon compound that provides hot radiating carbon particles in the combustion plume which serve as very efficient emitters.

Table F.9: Selected blackbody compositions.

	1	2	3	4	5
Magnesium, Mg (wt.-%)	60	48	38	33	
Zirconium, Zr (wt.-%)					46.8
Polytetrafluoroethylene, $(C_2F_4)_n$ (wt.-%)	25				
Polycarbon monofluoride, $(CF)_n$ (wt.-%)		19			
Viton®A (wt.-%)	10	12			
HTPB (wt.-%)				16	
Potassium nitrate, KNO_3 (wt.-%)			17		
Ammonium perchlorate, NH_4ClO_4 (wt.-%)				34	
Molybdenum oxide, MoO_3					30.1
Chromium oxide, Cr_2O_3					19.2
Graphite, C (wt.-%)	5	18			
Diphenyl, $C_{12}H_{10}$ (wt.-%)			45		
Anthracene, $C_{14}H_{10}$ (wt.-%)				15	
Spectral energy, $E_{PbS\text{-band}}$ ($J\ g^{-1}\ sr^{-1}$)	180	300	80	90	22

With the advent of detector materials such as lead selenide (PbSe) or indium antimonide (InSb) working at longer wavelength (4–5 µm), all aspect attack of aircraft became possible. In addition, the opportunity to use both PbS and PbSe in parallel in a missile seeker made it possible to discriminate blackbody sources such as first-generation countermeasures against true targets simply by evaluating the integrated intensity ratios in both bands which is different for plume and grey-bodies. Hence, spectrally matched payloads constitute the second large group of decoy flare compositions (Table F.10: Nos. **6–10**). These compositions are typically oxygen balanced and their signature mimics the hot combustion gases leading to an integrated intensity in the PbSe band at minimum five times the intensity in PbSe band.

Table F.10: Typical spectrally matched flare compositions.

	6	7	8	9	10	11
Benzotriazole, $C_6H_5N_3$ (wt.-%)	20					
Potassium benzoate, $KC_7H_5O_2$ (wt.-%)		26				
HTPB (wt.-%)			15			
Diethylenglycol dinitrate, $C_4H_8N_2O_7$ (wt.-%)					20	
Nitrocellulose (wt.-%)				50	55	
Nitroglycerine, $C_3H_5N_3O_9$ (wt.-%)					25	
Dioctyladipate, $C_{22}H_{42}O_4$ (wt.-%)				9		
Trinitrotoluene, $C_7H_5N_3O_6$ (wt.-%)						35
Ammonium perchlorate, NH_4ClO_4 (wt.-%)			85	41		
Potassium perchlorate, $KClO_4$ (wt.-%)	76	63				65
Polyacrylat, $(C_7H_{12}O_2)_n$ (wt.-%)	4					
Viton $^®$A (wt.-%)		6				
Spectral energy, PbSe band ($J\,g^{-1}\,sr^{-1}$)	27	24	20	76	80	34

Apart from classical flare grains payloads may comprise stacks of thin polymer or paper carrier slides coated with plasticized red phosphorus. Furthermore, both combustible and non-combustible carrier materials such as carbon cloth impregnated with combined pyrotechnic and pyrophoric materials have been suggested. Finally, pyrophoric metal sheets are used as IR decoys.

The mechanical and thermals sensitiveness of selected flare compositions is given in Table F.11.

Table F.11: Safety data of selected flare compositions.

Parameter	1	7	8	10	11
Friction (N)	>360	240	40	120	320
Impact (J)	10	4.5	2–3	2	15
Decomp (°C)	263				259
T_{5ex} (°C)	>360	300	230	166	–
ESD (J)	0.05	0.07	0.125	6	5
BICT gap test 50% (GPa)	n.a.	n.a.		<2.6	>2.6

- E.-C. Koch, *Pyrotechnischer Satz zur Erzeugung von IR-Strahlung*, EP 12 1713.2, **2001**, Germany.
- E.-C. Koch, Review on Pyrotechnic Aerial Infrared Decoys, *Propellants Explos. Pyrotech.* **2001**, *26*, 3–11.
- E.-C. Koch, Advanced Aerial Infrared Countermeasures, *Propellants Explos. Pyrotech.* **2006**, *31*, 3–19.
- E.-C. Koch, Experimental Advanced Infrared Flare Compositions, IPS-Seminar, Fort Collins, **2006**, pp. 71–79.

- E.-C. Koch, 2006–2008 Annual Review on Aerial Infrared Decoy Flares, *Propellants Explos. Pyrotech.* **2009**, *34*, 6–12.
- E.-C. Koch, V. Weiser, E. Roth, 2,4,6 Trinitrotoluene: A Surprisingly Insensitive Energetic Fuel and Binder in Melt-Cast Decoy Flare Compositions, Angew. *Chem. Int. Ed.* **2012**, *51*, 10038–10040.

Flash composition
Knallsatz, Blitzknallsatz

Flash compositions were formerly used in photoflash bombs to provide very short illumination for nighttime aerial photography. While flash compositions have become obsolete for this very purpose (as advanced electrooptical systems rely on image intensification and near-infrared optics), they are still indispensable for training, simulation, and stun grenades.

The main requirements for flash compositions are
- high specific radiant intensity, I (cd),
- high speed of reaction,
- high amount of condensed reaction products, and
- high combustion temperature.

Typical flash powders contain about equivalent mass percentages of potassium perchlorate and aluminum. Similar like black powder or other report compositions flash compositions work properly (undergo fast reaction) only if ignited in a non-consolidated state with high porosity. Consolidated flash compositions however upon ignition yield a laminar burn just like ordinary illuminating compositions. Table F.12 shows the various aluminum oxidizer blends.

Table F.12: Composition and performance of flash compositions with Al as fuel and various oxidizers.

Oxidizer	Al (wt.-%)	Mass (g)	Peak intensity 10^6 (cd)	Rise time (ms)	Duration (ms)	Spectral energy 10^3 (cd s g^{-1})
$LiClO_4$	46	47.5	55	1.8	22	8.4
$NaClO_4$	43	32.5	35	1.3	23	7.0
$KClO_4$	39	42.0	41	1.7	16	5.9
$Ca(ClO_4)_2$	38	33.7	34	0.6	24	9.3
$Ca(NO_3)_2$	35	18.0	20	1.6	13	3.2
$Sr(ClO_4)_2$	34	42.0	64	1.6	19	8.2
$Sr(NO_3)_2$	30	43.0	30	1.8	32	4.7
$Ba(ClO_4)_2$	30	43.0	57	1.2	19	6.2
$Ba(NO_3)_2$	26	49.0	n.a.	n.a.	n.a.	n.a.

Cegiel et al. have developed a green alternative for flash charges:

Perchlorate-free flash charge
- 34.17 wt.-% manganese(IV) oxide
- 34.17 wt.-% strontium nitrate
- 28.74 wt.-% aluminum
- 2.92 wt.-% graphite

The high content of fine aluminum powder and the unconsolidated state of flash compositions allows accidental charge accumulation and hence makes those formulations particularly sensitive towards electrostatic discharge. In addition, those formulations require very little confinement to undergo a rapid deflagration. Figure F.14 shows the static blast pressure of $Al/KClO_4$ as a function of scaled distance.

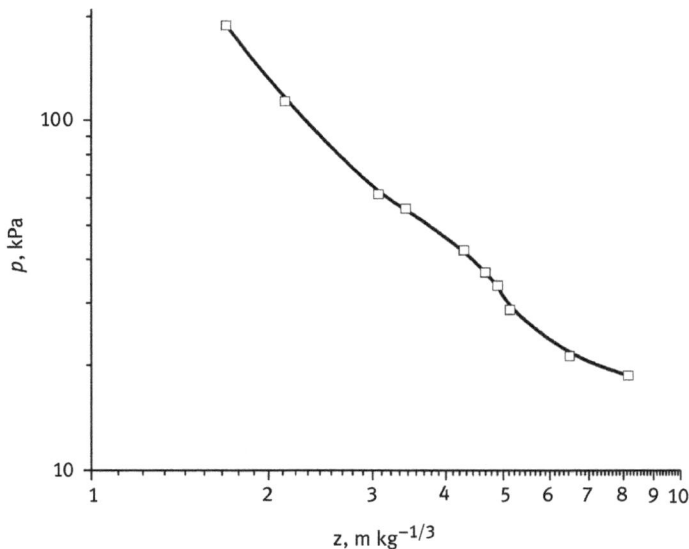

Figure F.14: Static blast pressure versus scaled distance for $Al/KClO_4$ (30/70) after *Wild* (**1978**).

- S. Lopatin, Sea-Level and High-Altitude Performance of Experimental Photoflash Compositions, *Technical Report FRL-TR-29*, Picatinny Arsenal, October **1961**, USA.
- D. Cegiel, J. Strenger, C. Zimmermann, *Perchloratfreie pyrotechnische Mischung*, DE102010052628A1, **2010**, Germany.
- M. Bishop, N. Davies, The Luminous and Blast Performance of Flash Powders, *IPS*, Adelaide, **2001**, 73–86.
- U. Krone, H. Treumann, Pyrotechnic Flash Compositions, *Propellants Explos. Pyrotech.* **1990**, *12*, 115–120.
- R. Wild, Blast waves produced by a pyrotechnic flash-mixture compared to those produced by high explosives, *Explosives Safety Seminar*, San Antonio, USA, **1978**, 727–738.

Flash point

Flammpunkt

The flash point is the lowest temperature at which a combustible air–vapor mixture can form above a volatile substance (liquid, gel, and solid). There are plenty of different test setups and methods cast/put in national and international standards such as ASTM D93 or ASTM D3828 and D3278 and EN ISO 3679 and 3680.

Floret test

The Floret test is a small-scale modification of a gap test and mainly serves to determine the detonative spreading of a sample and its sensitivity towards shock. Figure F.15a depicts the setup and Figure F.15b depicts the *postmortem* configuration indicating the particular spread of detonation which is related to sensitivity.

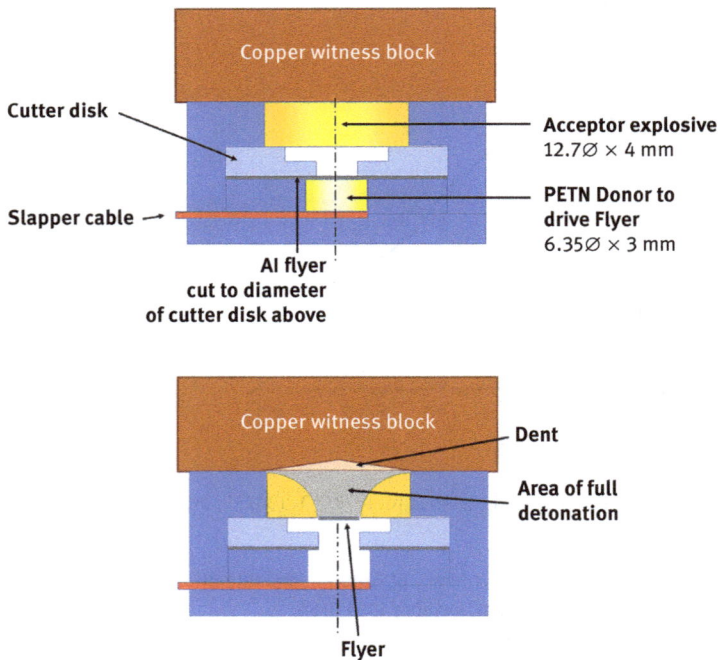

Figure F.15: (a) Floret test setup (top) and (b) *post mortem* configuration (bottom).

Thanks to its scale explosives available in small amounts only can be tested (~1 g) and compared to other materials. While the original Floret test uses a slapper detonator to detonate, the PETN pellet modifications have seen use of plastic explosive driven metal flyers. Recent modifications see introduction of instrumentation (fully instrumented floret investigation: FIFI).

- F. J. Gagliardi, R. D. Chambers, T. D. Tran, Small-scale performance testing for studying new explosives, – *VACETS International Technical Conference*, **2005** – Milpitas, CA, USA.
- M. W. Wright, Development of the floret test for screening the initiability of explosive materials, *AIP Conference Proceedings* **2012**, *1426*, 693–696.
- A. E. D. M. van der Heijden, R. H. B. Bouma, Characterization of Granular and Polymer-Embedded RDX Grades: Floret Tests, *Propellants Explos. Pyrotech.* **2016**, *41*, 360–366.
- E. G. Francois, P. R. Bowden, A. H. Mueller, M. D. W. Bowden, C. A. Scovel, *Small Scale Tests to Determine the Divergence and Spreading of Explosive Booster Materials*, *LA-UR-19-27653*, Los Alamos National Laboratory, 1. August **2019**.

FOF

FOF is the acronym for all explosive formulations developed at *FOI* Sweden.

Form function
Formfunktion

Apart from their chemical composition the geometrical shape of gun propellants, rocket propellants and pyrotechnic grains very much determines their performance (power, W).

The time dependent alteration of the surface area $S(t)/S_0$ of a geometrical shape can be described with a form function $\varphi(\mu)$. Curve progression for typical grain geometries are depicted in Figure F.16.

$\mu(t)$ is the fraction burnt at time (t) (see *Combustion, burn rate* page 186).

Depending on the shape design of a grain, the burning surface area, $S(t)$, of a grain can
- decrease (degressive) $\varphi(\mu) < 1$,
- remain constant (neutral) $\varphi(\mu) \sim 1$, or
- increase (progressive) $\varphi(\mu) > 1$.

Typical degressive geometries are solid spheres, cubes, cylinders, strips, and slotted tubes, neutral geometries are tubes and progressive geometries are multihole grains such as 7-, 19-, and 37-hole in hexagonal prisms or 4 holes in cubes.

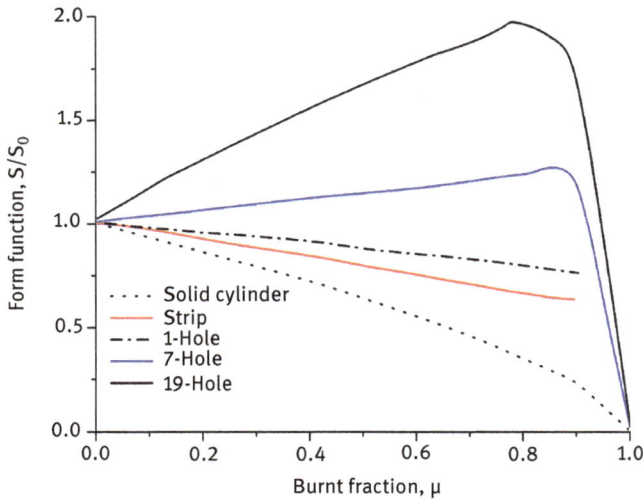

Figure F.16: Form function curves for different grain geometries.

The form function $\varphi(\mu)$ for:

Solid-cube grain

$$\Phi(\mu) = \frac{S}{S_0} = \frac{6(a-2x)^2}{6a^2} = \frac{(a-2x)^2}{a^2}$$

with $\mu(t) = 1 - \frac{m(t)}{m_0}$ it follows

$$\varphi(\mu) = (1-\mu)^{2/3}$$

The same equation can be applied to describe spherical grains.

Solid cylinders

$$\varphi(\mu) = (1-\mu)^{0.5}$$

Strips

$$\varphi(\mu) = (1-4[b/a]\mu)^{0.5} \text{ mit a = width, b = thickness (a > b)}$$

Tubes

$$\varphi(\mu) \sim 1$$

- D. Vittal, S. Singh, Form Function for Propellants in Closed Vessel Work, *Propellants Explos.* **1980**, *5*, 9–14.

Formamidinium nitroformate
Formamidiniumnitroformat

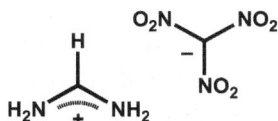

FANF		
Aspect		Light yellow crystals
Formula		$C_2H_5N_5O_6$
CAS		[2392156-77-1]
m_r	g mol^{-1}	195.091
ρ	g cm^{-3}	1.742
$\Delta_f H$	kJ mol^{-1}	−118.8
$\Delta_{ex}H$	kJ mol^{-1}	−1149.9
	kJ g^{-1}	−5.894
	kJ cm^{-3}	−10.268
$\Delta_c H$	kJ mol^{-1}	−1382.8
	kJ g^{-1}	−7.088
	kJ cm^{-3}	−12.346
Λ	wt.-%	−4.10
N	wt.-%	35.89
Dp	°C	165
Mp	°C	215
Impact	J	>100 (BAM)
Friction	N	>355
V_D	m s^{-1}	8900
P_{CJ}	GPa	32

FANF was prepared in 2018 for the first time. While comparable with RDX with regard to performance, it has a superior oxygen balance and is impressively insensitive towards impact and friction. Compared to known *hydrazinium nitroformate*, it has an onset of 40 °C higher than the latter.

- A. F. Baxter, I. Martin, K. O. Christe, R. Haiges, Formamidinium Nitroformate: An Insensitive RDX Alternative, *J. Am. Chem. Soc.* **2018**, https://pubs.acs.org/doi/10.1021/jacs.8b10200

FOX

FOX is the acronym for all explosive compounds developed at FOI/Sweden, for example, FOX-7 1,1-diamino-2,2-dinitroethylene or FOX-12, *N*-guanylurea dinitramide.

Fraunhofer Institut für Chemische Technologie, FhG-ICT

The ICT was established in 1956 by chemist Dr. *Karl Meyer* (1898–1977) as a scientific research institute for propellants and explosives at Karlsruhe University. For safety reasons it was relocated to nearby Pfinztal-Berghausen where experiments could be conducted safely in an abandoned quarry and newly erected test stands and bunkers. While the ICT dealt with propellants and explosives only for over 30 years, the fall of the iron curtain and international disarmament in the early 1990, brought some restructuring to ICT and introduced new civilian research activities such as electrochemistry, polymer technology, and environmental engineering.

Internationally renowned scientists working at ICT were the former director and chemist *Prof. Dr. Hiltmar Schubert* (1927*), the former deputy director and chemist *Dr. Fred Volk* (1930–2005), and the physicist and long-term editor of *Propellants Explosives Pyrotechnics Dr. Norbert Eisenreich* (1948–2019).

For its defence related applied research, the ICT is partially funded by the German Ministry of Defence. In 2015, the ICT received about 23 Mio. Euros.

The current director of the FHG-IC is polymer engineering expert *Prof. Dr. Ing. Peter Elsner* (*1956).

High explosives developed at ICT bear the following acronyms followed by three figures:
- HX = high explosives
- HXA = high explosives containing aluminum, figures starting with 1 contain RDX and those starting with 2 contain HMX
- GHX = castable high explosives with GAP binder
- PHX = pressable high explosives

- Webseite www.ict.fraunhofer.de
- *Entwurf zum Bundeshaushaltsplan 2015 Einzelplan 15*, Bundesministerium der Verteidigung, **2015**.

Fulminates
Fulminate

The fulminates are the ionic derivatives of the fulminic acid, $H-C\equiv N^+-O^-$. Mercury fulminate, HgCNO [628-86-4], due to its inherent instability and high toxicity has no more importance today as a primary explosive. Silver fulminate, AgCNO, [5610-59-3] though a stable and powerful primary explosive is extremely sensitive when present as crystalline needles and hence is of no technical use. However, it is used in very low amounts in toy snappers and exploding gadgets (trick matches, exploding cigarette tips) in joke articles.

– R. Matyáš, J. Pachman, *Primary Explosives*, Springer, **2013**, pp. 59–62.

Functional lifetime, service life
Funktionslebensdauer

The service life of an explosive is the time period in which unrestricted safety and functionality (performance) of the explosive in any item is guaranteed.

The service life is limited by chemical and physical ageing processes of the explosive. Beyond the service life, the explosive/the item containing the explosive should still be considered safe to use. However, then the explosive/item may no longer yield the designed performance.

The physical lifetime is the time period in which the explosive/item should still be safe to handle though a proper and safe use would no longer be possible. In addition, the physical lifetime excludes the hazards by adiabatic self-heating of the explosive.

The chemical lifetime finally seals of any safe handling of the explosive/item and hazardous adiabatic self-heating is immanent.

– K. Schneider, Funktionslebensdauer von Explosivstoffen, *Forum Explosivstoffe 2000*, WIWEB, Swisttal-Heimerzheim, March **2000**.

Furazans
Furazane

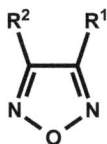

The common name furazan for 1-oxa-2,5-diazole derives from the structurally related cyclic ether tetrahydro*furan*. Furazans are attractive explosophors for energetic materials because
- They are aromatic (6 π-electrons) and hence can stabilize an attached molecule.
- The planarity of the ring makes corresponding molecules stackable and hence increases density while it also facilitates gliding along the planes avoiding bond scission and therefore reduces friction sensitiveness of a compound.

- The oxygen attached to nitrogen increases the heat of formation.
- The 3 and 4 positions are easily functionalized.

Important furazans are *DAAF* and *LLM-175*. *TATB*, upon mechanical and thermal shock in an endothermal fashion, yields furazans and furoxans contributing to the extraordinary low sensitiveness and high thermal stability of TATB.

- J. P. Agrawal, R. D. Hodgson, *Organic Chemistry of Explosives*, Wiley, **2007**, pp. 297–302.
- J. Sharma, J. C. Hoffsommer, D. J. Glover, C. S. Coffey, J. W. Forbes, T. P. Liddiard, W. L. Elban, F. Santiago, Sub-Ignition Reactions at Molecular Levels in Explosives subjected to Impact and underwater shock, *8th Detonation Symposium*, 15–19 July **1985**, Albuquerque, NM, USA, pp. 725–733.

Furoxans

Furoxane

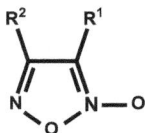

Furoxans have a 1-oxa-2,5-diazol-2-oxide framework. In contrary to furazans, they possess a higher oxygen balance. Important furoxans are *potassium dinitrobenzofuroxan, CL-14*, and *benzotrifuroxan*. CL-14 is formed as a decomposition product in shocked TATB.

- J. P. Agrawal, R. D. Hodgson, *Organic Chemistry of Explosives*, Wiley, **2007**, pp. 302–307.

G

GAP, glycidyl azide polymer

Glycidylazidpolymer

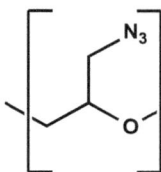

Aspect		Light yellow viscous liquid
Formula		$(C_3H_5N_3O)_n$
CAS		[143178-24-9]
m_r	g mol^{-1}	99.1
ρ	g cm^{-3}	1.29
Δ_fH	kJ mol^{-1}	113.97
$\Delta_{ex}H$	kJ mol^{-1}	−409.4
	kJ g^{-1}	−4.132
	kJ cm^{-3}	−5.343
Δ_cH	kJ mol^{-1}	−2008.1
	kJ g^{-1}	−20.266
	kJ cm^{-3}	−26.203
Λ	wt.-%	−121.09
N	wt.-%	42.40
T_g	°C	−45
Mp	°C	120–150
Dp	°C	216
Impact	J	8 (BAM)
Friction	N	>355

GAP is an important energetic binder for solid propellants (e.g. ADN) and high explosives. GAP with terminal hydroxyl groups can be cured with diisocyanates to give polyurethanes.

- P. F. Aiello, A. P. Manzara, GAP Benefits, *JANNAF Propulsion Meeting*, Indianapolis, February **1992**.
- B. Johannessen, I. M. Denenholz, A. P. Manzara, Characterization of Gap Polyol and Plasticizer, *MACH I*, **1997**.

https://doi.org/10.1515/9783110660562-007

Gas-generating compositions

Gasgeneratorsätze

Gas-generating compositions are used both in the civilian and military field. Apart from composition producing unspecified gases, that are mixtures from CO_2, H_2O, and N_2, there are formulations that produce high nitrogen levels (see *airbag*), high hydrogen concentration, or high oxygen (see *SCOG*). Hydrogen is used as fuel in aerospace applications and can be used to fuel auxiliary engines. Table G.1 shows different compositions and the hydrogen content in the product gases.

Table G.1: Hydrogen-generating compositions.

Components	1	2	3	4	5
Triaminoguanidinium nitrate (wt.-%)	45				
Triaminoguanidinium-5-tetrazolate (wt.-%)	25				
Polyester (wt.-%)	13.2				
Polymethylenpolyphenylen isocyanate (wt.-%)	1.8				
Trimethylolethane trinitrate (wt.-%)	15				
Strontium nitrate (wt.-%)		40	30	20	10
Borazane (H_2N-BH_2)$_x$ (wt.-%)		60	70	80	85
Iron(III) oxide (wt.-%)					5
Gaseous products					
Hydrogen (wt.-%)	41	11.7	13.7	15.6	16.6
Carbon monoxide (wt.-%)	1.3				
Water (wt.-%)	26				
Nitrogen (wt.-%)	31				

- J. E. Flanagan, *Solid propellant hydrogen generator*, US Patent 4234363, **1980**, USA.
- J. P. Goudon, H. Blanchard, J. Renouard, C. Vella, P. Yvart, Gaseous Hydrogen Generation from a Solid Mixture using a self-sustaining Combustion reaction, *42nd International Annual Conference of the Fraunhofer ICT*, June 28 – July 01, **2011** Karlsruhe, Germany, V-27.

Gelled propellants

Geltreibstoffe

Gelled propellants are disperse systems in which one or more miscible liquids are gelled with a thickening agent. Gels are shear-thinning fluids, that is, their viscosity decreases with increasing mechanical energy input. Gelled propellants are advantageous over liquid propellants in that they do not leak (which increases safety) or do not show slosh in flight. However, they can still be pumped as regular liquids. In contrary to solids, they show an improved tactical signature. Typical thickening agents are highly dispersed silica or urea derivatives.

- H. K. Ciezki, C. Kirchberger, A. Stiefel, Kröger, P. Caldas Pinto, J. Ramsel, K. W. Naumann, J. Hürttlen, U. Schaller, A. Imiolek, V. Weiser, Overview on the German Gel Propulsion Technology Activities: Status 2017 and Outlook, *European Conference for Aeronautics and Space Sciences (EUCASS)*, **2017**.
- H. K. Ciezki, K. W. Naumann, Some Aspects on Safety and Environmental Impact of the German Green Gel Propulsion Technology, *Propellants Explos. Pyrotech.* **2016**, *41*, 539–546.
- H. K. Ciezki, J. Hürttlen, K. W. Naumann, M. Negri, J. Ramsel, V. Weiser. Overview of the German Gel Propulsion Technology Program, *50th AIAA/ASME/SAE/ASEE Joint Propulsion Conference, AIAA Propulsion and Energy Forum*, (AIAA 2014-3794), 2014.

Glassman's condition for vapor-phase metal combustion
Glassmans Bedingung für die Gasphasenverbrennung von Metallen

Irvin Glassman (1923–2019) developed a hypothesis for the combustion of metallic and metalloid fuels in oxygen. If combustion in other atmospheres (by other oxidizers such as nitrogen, chlorine, or fluorine) is considered too then the criteria read as follows:

[1] If the boiling point (dissociation temperature) of the combustion product (MeX) is greater than that of the fuel (M), then steady-state vapor-phase combustion takes place. The contracondition calls for surface combustion of the fuel.

[2] If condition [1] is satisfied, then the boiling point (dissociation temperature) of the combustion product limits the flame temperature of the fuel.

Condition [2] derives from the fact that the heat of vaporization of a metal combustion product, $\Delta_{vap/diss}H(MeX)$ is generally greater than the heat available from the combustion process for those metals $\Delta_f H(MeX)$ to heat

$$\Delta_{vap/diss}H(MeX) \geq \Delta_f H(MeX) - \left(H^{\circ}_{Bp}(MeX) - H^{\circ}_{298}(MeX)\right) = \Delta_{avail}H$$

Typical metals burning in air in the vapor phase are Al, Be, Li, Mg, Yb, and Zn.

A clear indication of vapor-phase combustion is the presence of metal and metal oxide vapor-phase radiation giving rise to distinct electronic atomic and or molecular spectra, whereas the fuels not satisfying condition [1] primarily yield gray body radiation dependent on the surface temperature and emissivity.

- Glassman, Combustion of Metals Physical Considerations in M. Summerfield (Ed.) *Solid Propellant Rocket Research*, Academic Press, New York, **1960**, 253–269.

Glauber's exploding powder
Glaubersches Knallpulver

Iatrochemist Johann Rudolph Glauber (1604–1670) in 1648 discovered an exploding powder composed from:

    55 wt.-% potassium nitrate, KNO_3

    27 wt.-% sulfur, S_8

    18 wt.-% potassium carbonate, K_2CO_3

However, unlike other explosive materials, this powder cannot be ignited directly with a torch or safety fuse or any other similar means. But instead, the powder needs to be heated up rapidly with a Bunsen burner or torch. After fusion of the components, a fierce explosion occurs.

German Chemist *Fritz Seel* (1915–1987) assumed that sulfur reacts with potassium carbonate to give a mixture of potassium sulfides and potassium thiosulfate ($K_2S_2O_3$). The consecutive reduction of nitrate to nitrite is thought to be the basis for the explosive decomposition of nitrite and thiosulfate to give sulfite and thionitrate ($KSNO_2$), the latter of which also decomposes explosively:

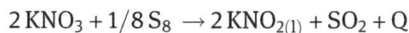

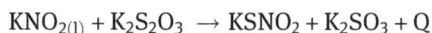

$$6/8\,S_8 + 3\,K_2CO_3 \rightarrow 2\,K_2S_n + K_2S_2O_3 + 3\,CO_2 + Q$$

$$2\,KNO_3 + 1/8\,S_8 \rightarrow 2\,KNO_{2(l)} + SO_2 + Q$$

$$KNO_{2(l)} + K_2S_2O_3 \rightarrow KSNO_2 + K_2SO_3 + Q$$

$$2\,KSNO_2 \rightarrow K_2S_2O_3 + N_2O + Q$$

–   F. Seel, Geschichte und Chemie des Schwarzpulvers – Le charbon fait le poudre – *Chemie in unserer Zeit*, **1988**, *22*, 9–16.

Glitter compositions
Glittersätze

Glitters originate as hot, however, flameless particles from a pyrotechnic combustion process and eventually burst into flames after a certain delay. There is great variety of different glitter formulations (Table G.2). In general, it is assumed that the bursting into flames is due to reactions of the hot particle with the ambient oxygen.

Table G.2: Glitter compositions.

	1	2	3	4
Color effect	Gold	Pearl	Silver	Silver
Black powder* (wt.-%)	65	65	75	
Barium nitrate (wt.-%)		11		
Potassium perchlorate (wt.-%)				48
Sulfur (wt.-%)	10	10		19
Charcoal (wt.-%)	20			9
Dextrine (wt.-%)	5	5	5	4
Aluminum (wt.-%)			10	7
Magnalium (wt.-%)		9		
Antimony sulfide (wt.-%)			10	
Sodium carbonate (wt.-%)				13

*Not pan-milled (75/15/10) KNO_3, charcoal, sulfur.

– C. Jennings-White, Glitter Chemistry, *J. Pyrotech.* **1998**, 8, 53–70.

Glycidyl nitrate polymer, polyglyn, PGN
Poly(glycidylnitrat)

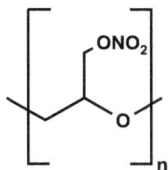

Aspect		Light yellow viscous liquid
Formula		$(C_3H_5NO_4)n$
CAS		[27814-48-8]
m_r	g mol^{-1}	119.077
ρ	g cm^{-3}	1.46
$\Delta_f H$	kJ mol^{-1}	−284.5
$\Delta_{ex} H$	kJ mol^{-1}	−524.7
	kJ g^{-1}	−4.406
	kJ cm^{-3}	−6.490
$\Delta_c H$	kJ mol^{-1}	−1610.6
	kJ g^{-1}	−13.526
	kJ cm^{-3}	−19.748
Λ	wt.-%	−60.46
N	wt.-%	11.76
T_g	°C	−31.9
Mp	°C	215

PGN cured with aliphatic isocyanates upon temperature cycling can decompose irreversibly into primary amine, CO_2, and vinyl nitrate-terminated PGN. *Sanderson* has addressed this and developed a process to produce stable non-ageing polymers from PGN.

- A. J. Sanderson, L. J. Martins, M. A. Dewey, *Process for Making stable cured PGN and energetic compositions comprising same*, US6861501, USA, **2002**.
- A. Provatas, *Energetic Polymers and Plasticisers for Explosive Formulations – A Review of Recent Advances*, *DSTO-TR-0966*, DSTO, Melbourne, April **2000**, 44 pp.

Graphite

Graphit, Grafit

Graphite is a naturally available but also synthetically accessible carbon polymorph that due to its stratified structure of graphene sheets is characterized by a low friction coefficient and considerable electrical and thermal conductivity along the sheets (ρ = 2.26 g cm^{-3}). It is a common phlegmatizing additive in many explosive materials as it reduces friction sensitivity and makes materials electrically conductive, thereby reducing their sensitiveness to ESD. Grained black powder is typically polished with graphite to make it water repellent. Graphite is used as a burn rate modifier in pyrotechnics and solid rocket propellants. Expanded graphite, due to its superior surface area, can yield a burn rate increase of pyrotechnics and solid propellant materials of nearly an order of magnitude. Graphene sheets (the single carbon layer) can be obtained by exfoliation of graphite and graphite compounds and have been suggested as lubricants for melt-cast explosives.

- E.-C. Koch, Defluorination of Graphite Fluoride Applying Magnesium, Z. Naturforsch. 2001, 56b, 512–516. http://zfn.mpdl.mpg.de/data/Reihe_B/56/ZNB-2001-56b-0512.pdf.
- E.-C. Koch, Behavior of Burn Rate and Radiometric Performance of two Magnesium /Teflon / Viton (MTV)Formulations upon Addition of Graphite, *J. Pyrotech.* **2008**, *27*, 38–41.
- A. Hahma, *Pyrotechnische Wirkmasse mit einem Abbrandbeschleunigungsmittel*, EP 2 695870A2, Germany, **2014**.

Greek fire

Griechisches Feuer

Greek fire – more correctly it should be referred to as *liquid fire* from the Greek term *hygron pyr* – is an alleged mixture of burnt lime stone (CaO), mineral oil fractions, sulfur, resin, and salpetre (KNO_3). The resins supported the tackiness of the material and probably also served as emulsifiers to prevent separation of the components. Upon contact of the material with water, the burnt lime reacts fiercely producing temperatures in excess of 400 °C sufficient to ignite sulfur and volatilize the mineral oil fractions which then burn supported by salpetre.

- V. Muthesius, *Zur Geschichte der Sprengstoffe und des Pulvers*, Hoppenstedt, Berlin, **1941**, 21–26.

GSX, gelled slurry explosive

GSX are a group of phase-stabilized non-ideal high explosives, containing ammonium nitrate, metal powders, and water. Very often, GSX is used synonymously with the following formulation DBA-22M (DBA – dense blasting agent; Table G.3).

Table G.3: Composition and performance of DBA-22M.

Component	wt.-%	Parameter	
Ammonium nitrate	50	ρ (g cm^{-3})	1.47
Aluminum	34	V_D (m s^{-1})	5000
Boric acid, water, ethylene glycol	16	$\Delta_{ex}H$ (kJ g^{-1})	5.941

DBA-22M was used in the BLU-82, also known as *daisy cutter* (NEW: 5715 kg) in the Vietnam war to swiftly produce helicopter landing zones in tropical rain forest and to clear mine fields.

- M. F. Porter, *Commando Vault (U)*, 12 October **1970**, HQ PACAF, Directorate, Tactical Evaluation, CHECO Division.

Guanidinium dinitramide
Guanidiniumdinitramid

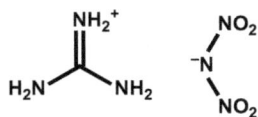

Formula		CH$_6$N$_6$O$_4$
GHS		03, 07
CAS		[170515-96-5]
m_r	g mol^{-1}	166.096
ρ	g cm^{-3}	1.67

$\Delta_f H$	kJ mol^{-1}	−157.95
$\Delta_{ex} H$	kJ mol^{-1}	−752.9
	kJ g^{-1}	−4.533
	kJ cm^{-3}	−7.570
$\Delta_c H$	kJ mol^{-1}	−1093.1
	kJ g^{-1}	−6.581
	kJ cm^{-3}	−10.990
Λ	wt.-%	−9.63
N	wt.-%	50.60
Mp	°C	139
Dp	°C	164
E_a	kJ mol^{-1}	167.4

- T. S. Kon'kova, Y. N. Matyushin, E. A. Miroshnichenko, A. B. Vorob'ev, Thermochemical properties of dinitramide salts, *Russ. Chem. Bull.* **2009**, *58*, 2020–2027.
- H. R. Blomquist, *Gas generating composition comprising guanylurea dinitramide*, US-Patent 6,117,255, USA, **2000**.

Guanidinium nitrate

Guanidiniumnitrat

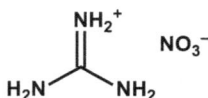

Formula		CH$_6$N$_4$O$_3$
GHS		03, 07
H-phrases		272-302-315-319-412
P-phrases		210-221-302+352-305+351+338-321-501a
UN		1467
REACH		LRS
EINECS		208-060-1
CAS		[506-93-4]
m_r	g mol^{-1}	122.084
ρ	g cm^{-3}	1.436
$\Delta_f H$	kJ mol^{-1}	−387.02
$\Delta_{ex} H$	kJ mol^{-1}	−345.7
	kJ g^{-1}	−2.831
	kJ cm^{-3}	−4.066
$\Delta_c H$	kJ mol^{-1}	−864.0
	kJ g^{-1}	−7.007
	kJ cm^{-3}	−10.163

Λ	wt.-%	−26.21
N	wt.-%	45.89
Mp	°C	216
Dp	°C	240
Friction	N	>355
Impact	J	>50
V_D	m s^{-1}	3700 at 1.436 g cm^{-3} at 60 mm Ø
		7870 at 1.660 g cm^{-3} at 52 mm Ø
P_{CJ}	GPa	26.11 at 1.666 g cm^{-3} at 60 mm Ø
Trauzl	cm^3	240
Koenen	mm	1.5, no reaction

Guanidinium nitrate (GN) is a very feeble explosive. It is used mainly as nitrogen sources in gas-generating formulations and other pyrotechnics. It is also the main precursor for nitroguanidine from which is prepared by reaction with concentrated sulfuric acid.

- E.-C. Koch, Insensitive High Explosives III. Nitroguanidine, *Propellants Explos. Pyrotech.* **2019**, 44, 267–292.

Guanidinium nitroformate
Guanidiniumnitroformat

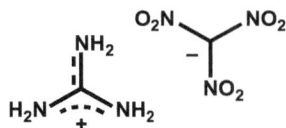

Guanidinium trinitromethanide, GNF		
Aspect		Yellow crystals
Formula		$C_2H_6N_6O_6$
CAS		[223685-92-5]
m_r	g mol^{-1}	210.106
ρ	g cm^{-3}	1.695
$\Delta_f H$	kJ mol^{-1}	−213.14
$\Delta_{ex} H$	kJ mol^{-1}	−1047.06
	kJ g^{-1}	−4.984
	kJ cm^{-3}	−8.447
$\Delta_c H$	kJ mol^{-1}	−1431.5
	kJ g^{-1}	−6.813
	kJ cm^{-3}	−11.549
Λ	wt.-%	−7.61
N	wt.-%	40.00
Dp	°C	113
Mp	°C	69 (−H_2O, hydrate)

Impact	J	30 (BAM)
Friction	N	>360
V_D	m s^{-1}	8500
P_{CJ}	GPa	28

The formation of hygroscopic guanidinium nitroformate (GNF) was first observed by *Boyer et al.* in 1989. Actually it was intended to prepare 1,1,1-triamino-2,2,2-trinitroethane, which is the higher homologue of FOX-7 (1,1-diamino-2,2-dinitroethylene). However, quantum chemical calculations then already indicated a significantly weakened C–C bond which led to the correct structural interpretation as an ionic compound. *Klapötke et al.* finally could show the ionic structure by means of crystallography.

- M. Krishnan, P. Sjøberg, P. Politzer, J. H. Boyer, Guanidinium trinitromethanide, *J. Chem. Soc. Perkin Trans. 2* **1989**, 1237–1242.
- T. S. Konk'ova, Y. N. Matyushin, Combined study of thermochemical properties of nitroform and its salts, *Russ. Chem.Bull.* **1998**, *47*, 2371–2374.
- M. Gobel, T. M. Klapötke, Potassium-, Ammonium-, Hydrazinium-, Guanidinium-, Aminoguanidinium-, Diaminoguanidinium-, Triaminoguanidinium- and Melaminiumnitroformate – Synthesis, Characterization and Energetic Properties, *Z. Anorg. Allg. Chem.* **2006**, *633*, 1006–1017.

Guanidinium perchlorate

Guanidiniumperchlorat

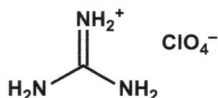

Formula		CClH$_6$N$_3$O$_4$
CAS		[10308-84-6]
m_r	g mol^{-1}	159.529
ρ	g cm^{-3}	1.74
$\Delta_f H$	kJ mol^{-1}	−311.08
$\Delta_{ex}H$	kJ mol^{-1}	−660.1
	kJ g^{-1}	−4.138
	kJ cm^{-3}	−7.200
$\Delta_c H$	kJ mol^{-1}	−889.3
	kJ g^{-1}	−5.575
	kJ cm^{-3}	−9.700
Λ	wt.-%	−5.01
N	wt.-%	26.34

Mp	°C	248
Dp	°C	240
Impact	J	10
V_D	m s^{-1}	7150 at 1.67 g cm^{-3}
Trauzl	cm^3	400

Eutectic mixtures of guanidinium perchlorate with lithium perchlorate and polymers have been suggested as impact-insensitive solid rocket propellants.

– J. J. Byrne, *Propellant containing guanidine perchlorate-lithium perchlorate eutectic in homogeneous phase with polymeric binder*, US Patent 3531338, **1970**, USA.

Guanidinium picrate
Guanidiniumpikrat, GuPi

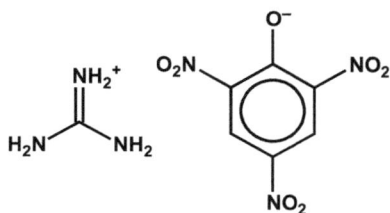

Aspect		Yellow needles
Formula		$C_7H_8N_6O_7$
CAS		[4336-48-5]
m_r	g mol^{-1}	288.177
ρ	g cm^{-3}	>1.50
$\Delta_f H$	kJ mol^{-1}	−396.60
$\Delta_{ex} H$	kJ mol^{-1}	−1109.7
	kJ g^{-1}	−3.851
	kJ cm^{-3}	−6.392
$\Delta_c H$	kJ mol^{-1}	−3501.4
	kJ g^{-1}	−12.150
	kJ cm^{-3}	−20.169
Ω	wt.-%	−61.07
N	wt.-%	29.16
Mp	°C	319 (dec)
Friction	N	>353
Impact	cm	≫20
V_D	m s^{-1}	6500 at unknown density
		7300 at 1.50 g cm^{-3}
P_{CJ}	GPa	13.99 at 1.50 g cm^{-3}

Guanidinium picrate (GuPi) is a very insensitive high explosive with a performance in the TNT ballpark. It was proposed as early as 1901 as a filler in safe explosives. Nowadays, GuPi is an energetic fuel for gas generators.

- D. Srinivas, V. D. Ghule, K. Muralidharan, Energetic salts prepared from phenolate derivatives, *New. J. Chem.* **2014**, *38*, 3699–3707.
- U. Bley, R. Hagel, J. Havlik, A. Hoschenko, P. S. Lechner, *Pyrotechnisches Mittel*, EP 1 890 986 B1, **2011**, Germany.

Guanidinium-5,5′-azotetrazolate, GAT, GZT, GUZT

Diguanidinium-5,5′-azotetrazolat

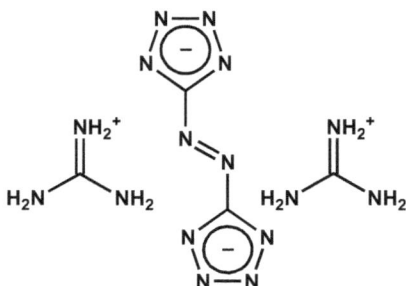

Aspect		Citron crystals
Formula		$C_4H_{12}N_{16}$
CAS		[142353-07-9]
m_r	g mol^{-1}	284.246
ρ	g cm^{-3}	1.538
Δ_fH	kJ mol^{-1}	410.03
$\Delta_{ex}H$	kJ mol^{-1}	−644.3
	kJ g^{-1}	−2.267
	kJ cm^{-3}	−3.486
Δ_cH	kJ mol^{-1}	−3699.1
	kJ g^{-1}	−13.014
	kJ cm^{-3}	−20.015
Λ	wt.-%	−78.8
N	wt.-%	78.84
Mp	°C	242 (dec)
Friction	N	>355
Impact	J	7.5

GZT is an important high nitrogen compound with considerable low toxicity. Originally developed for use in automobile airbags it has been introduced in numerous applications such as fire extinguishers, solid propellants, and also in-gredient in intumescent fire protection coatings. Consolidated GZT can be ignited with a torch whereupon the material starts to burn with a visible flame which however instantly extinguishes but the decompositions and consumption of the material proceeds at very little heat release with the formation of white fumes. Mixtures of GZT and AGTZ show lower explosion temperatures than the individual components.

- K. M. Bucerius, Stabile, stickstoffreiche Verbindung in Form des Diguanidinium-5,5′-azotetrazolat, DE-Patent 4034645C2, Germany, **1993**.

Guanylurea dinitramide, FOX-12, GuDN

Guanylharnstoffdinitramid

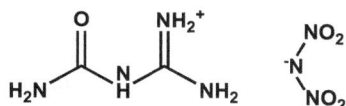

Formula		$C_2H_7N_7O_5$
CAS		[217464-38-5]
m_r	g mol^{-1}	209.121
ρ	g cm^{-3}	1.76
$\Delta_f H$	kJ mol^{-1}	−356
$\Delta_{ex}H$	kJ mol^{-1}	−792.2
	kJ g^{-1}	−3.789
	kJ cm^{-3}	−6.667
$\Delta_c H$	kJ mol^{-1}	−1431.4
	kJ g^{-1}	−6.845
	kJ cm^{-3}	−12.047
Λ	wt.-%	−19.13
N	wt.-%	46.89
Mp	°C	215
Friction	N	>355
Impact	J	50 (250–400 µm)
V_D	m s^{-1}	7966 at 1.666 g cm^{-3} at 60 mm Ø
P_{CJ}	GPa	26.11 at 1.666 g cm^{-3} at 60 mm Ø
Trauzl	cm^3	345–375
$\emptyset_{cr}$	mm	54–25 at 1.60 g cm^{-3}
E_a	kJ mol^{-1}	217–376

GuDN is an insensitive high explosive that is weaker and more sensitive than *nitro-guanidin* (NGu). GuDN is also more expensive than NGu. GuDN has been suggested as an ingredient in gas-generating compositions. Blends with TNT (*Guntol*)have been investigated as very insensitive melt-cast explosives. GuDN has been suggested as alternative to NGu in low erosion gun propellants (GPs) (e.g. navy insensitive low erosion (NILE) – GP; Table G.4). GuDN is an intermediate in the production of ammonium dinitramide.

Table G.4: Composition and performance of NILE – propellant.

Component	(wt.-%)	Performance	
GuDN	32.0	Density (g cm^{-3})	1.588
RDX	40.0	Heat of explosion (J g^{-1})	–
CAB	14.4	Force (J g^{-1})	895
ATEC	7.2	Flame temperature (K)	2175
HPC	5.3	Covolume (cm^3 kg^{-1})	0.119
Vestenamer	0.7	Temperature coefficient (mm s^{-1} MPa^{-1})	0.759
Centralite I	0,4	Pressure exponent (–)	0.809

- E.-C. Koch, *Insensitive Explosive Materials VII – Guanylurea dinitramide GuDN, L-175*, MSIAC, November **2011**, 32 pp.; Weblink http//bit.ly/2r02bth
- E.-C. Koch, Insensitive High Explosives IV: Initiation and Detonation of Nitroguanidine, *Def. Technol.* **2019**, *44*, 467–487.
- E.-C. Koch, Insensitive High Explosives, V: Ballistic Performance and Vulnerability of Nitroguanidine Based Gun Propellants, *Propellants Explos. Pyrotech.* accepted for publication, 2020.

Gum arabic
Gummiarabicum

Gum arabic is a natural occurring colorless/yellowish resin CAS-No. [9000-01-5]. It is obtained from the sap of various acacias in Asia and Africa. Gum arabic contains mainly D-galactose, L-arabinose, L-rhamnose, as well as D-glucuronic acid. The water-soluble gum was one of the main binders used in pyrotechnics in the past. However, its high content of D-glucuronic acid, CAS-No. [6556-12-3] (pK$_s$: 3.18) makes gum arabic incompatible with substances sensitive to acids such as magnesium or chlorates. The about composition of gum arabic is $C_6H_{9.778}O_{5.037}$. Its heat of formation is $\Delta_fH = -973.2$ kJ mol^{-1}. Between T = 44 and 140 °C, it loses moisture (~15% mass loss) and in air it starts to decompose exothermically starting at T = 238 °C. Its density is between 1.3 and 1.4 g cm^{-3}.

- W. Meyerriecks, Organic Fuels Composition and Formation Enthalpy Part II – Resins, Charcoal, Pitch, Gilsonite and Waxes, *J. Pyrotech.* **1999**, *9*, 1–19

Gun propellants, GP
Treibladungspulver

GPs serve the propulsion of bullets and projectiles in small, middle, and large caliber weapons. Upon combustion, these slightly oxygen-deficient propellants produce large amounts of hot gases with low molecular mass which expand and thereby propel a projectile.

The main ingredient in conventional GPs is nitrocellulose which also serves as energetic binder and base. Depending on the content of additional energetic fillers it can be distinguished between:

– Single-base GP
Contains between 80 and 98 wt.-% NC and are used predominantly for small arms caliber but also low artillery charges. Energy content is between 3 and 4 kJ g^{-1}.

– Double-base GP
Contains between 40–70 wt.-% NC and between 20 and45 wt.-%NGL or DEGN. They find application in tank, mortar, and medium caliber ammunition. Energy content with NGl between 4.5 and 5.5 kJ g^{-1}, whereas with DEGN is around 4 kJ g^{-1}.

– Triple-base GP
Similar as double base they contain NC and either NGL or DEGN but also a third component which is mostly nitroguanidine. They find application in the highest charges for artillery ammunition. Energy content is 3–4.1 kJ g^{-1}.

– Quadruple-base or seminitramine GP
These GP may contain a fourth energetic ingredient such as nitramines (e.g. RDX or HMX). Again target applications are for artillery charges. Energy content is up to 4 kJ g^{-1}.

– Insensitive (low vulnerability ammunition) GP
Insensitive GP typically do not contain any NC or at least only low levels. Alternatives to NC are polynitropolyphenylen (PNP). Classical blasting oils (NGL, DEGN) are replaced by insensitive components such as TEGDN. Further, they may contain high amounts of insensitive RDX, HMX, nitroguanidine or other low sensitivity materials (FOX-7, NTO, TEX). Energy content is up to 4 kJ g^{-1}.

– Low temperature coefficient (LTC) GPs
LTC GPs contain special plasticizers such as **DNDA-57** or **MEN-42 (proprietary blend from methyl and ethyl-NENA)**, which effect low temperature brittleness and induce erosive burn at low temperatures. For DNDA-type powders, the energy content ranges up to 4.7 kJ g^{-1}.

The performance of a GP is related to its chemical composition (Table G.5), geometry, surface treatment, and loading density.

Table G.5: Ingredients in gun propellants.

Purpose	Example	Purpose	Example
Energetic filler	Nitrocellulose (NC) Nitroguanidine (NGu) Hexogen (RDX)	*Stabilizer for nitrocellulose*	Akardite, centralite
Binder energetic	NC, CAN, GAP, BAMO, PNP, etc.	*Antierosion additive: Muzzle flash suppressor*	TiO_2, $CaCO_3$ KNO_3, K_3AlF_6
Non-energetic	CAB	*Ignition modifier*	MnO_2
Plasticizer		*Rolling addiutive*	Stearic acid (POL)
Energetic	NGl, DEGN, TEGDN, NENA, MEN, DNDA	*Burn rate modifier*	Cu, Pb, Bi salts of organic acids
Non-energetic	Camphor, DBA, akardite	*Laminar coolants*	Wax/TiO_2

Typical grain geometries are depicted in Figure G.1. A recent find is that quadruple-perforated cubes have been investigated in Germany as early as 1943. They yield a progressive burn and due to their cubic shape allow for high loading densities.

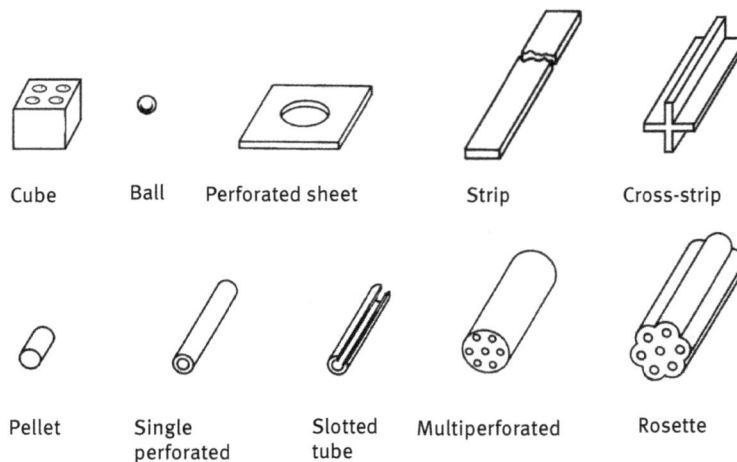

Cube   Ball   Perforated sheet   Strip   Cross-strip

Pellet   Single perforated   Slotted tube   Multiperforated   Rosette

Figure G.1: Typical grain geometries (not in proportional scale).

Depending on the type and caliber, only between 15% and 30% of the chemical energy can be transformed into kinetic energy of the projectile. The remainder is lost in gas and heat losses.

The main figures of merit of a GP are
- the heat of explosion, Q_{ex} (kJ g^{-1}),
- specific gas volume, (dm^3 g^{-1}),
- flame temperature, T_{ex} (K), and
- force, f (J g^{-1}).

The force is given by the product from gas constant, R, and the flame temperature of the propellant T_{ex}:

$$f = n \cdot R \cdot T_{ex}$$

It is related to the chemical energy of the GP, E_{ch}, by

$$E_{ch} = \frac{R \cdot T_{ex}}{M(\gamma - 1)} = \frac{f}{\gamma - 1}$$

And can be determined experimentally from combustion experiments with the GP at various loading densities in a ballistic bomb utilizing the *Abel equation*. Current research in the field of GPs aims at reducing sensitiveness and erosivity of propellants, extension of the service life, and the finally the most pressing issue: the replacement of obsolete chemicals.

Typical GPs and their composition and performance parameters are given in Table G.6.

Table G.6: Composition (wt.-%), performance, and sensitiveness of gun propellants.

	Single base	Double base		Triple base	Quadruple base		LTC	LOVA
Designation	OD6320	NK1074	L5460	PUTE 8577*	R5730	NILE	M1000	M43
NC	84.1	51.8	60	69.8	35		35	4
(N-content)	(13.15)	(13.1)	(13.2)	(13.1)	(12.5)			(12.6)
ATEC						7.2		7.6
NGl		40.5	15	22.3				
CAB						14.4		12
DEGDN			25		22			
DNDA-57							23	
DNT	9.66	1.5						
NGu				6.2	33			
GuDN						32		
HPC						6		
RDX					8	40	40	76
K$_2$SO$_4$	2							

Table G.6 (continued)

	Single base	Double base	Triple base	Quadruple base		LTC	LOVA
DBP	2.90		1.0				
Cryolithe							
DPA	0.97	1.0					
Centralite I		1.5				0.4	0.4
Akardite II		0.7	0.7	2		X	
Akardite III		3.7					
MgO		0.05					
Graphite		0.05					
T_{ex} (K)	2571	3390	3451	2700	2175	3128	2654
Q_{ex} ($J\,g^{-1}$)		4610	4603	3600			3400
f ($J\,g^{-1}$)	938	1139	1127	1018	895	1213	1010
M ($g\,mol^{-1}$)	22.791		39.27	22.05			
T_i (°C)		168		172			192
Impact (J)				3			9.6
Friction (N)				240			

*MRCA: 27 × 145 mm.

- K. Ryf, B. Vogelsanger, D. Antenen, A. Skriver, A. Huber, *Moderne Pulverentwicklungen*, Nitrochemie Wimmis, 17. September **2002**.
- K. Lapp, *Eine vergleichende Betrachtung einiger Grundformen von Progressivpulvern unter besonderer Berücksichtigung ihrer fabrikmässigen Fertigungsmöglichkeit*, Düneberg, 22 April **1943**, 12 pp.

Guntol

Guntol is the made-up name for melt-castable explosives based on GuDN and TNT. Formulations also containing aluminum are termed *guntonal*.

- P. Sjöberg, H. Östmark, A.-M. Amnéus, GUNTONAL – An Insensitive Melt Cast for Underwater Warheads, *IMEMTS*, **2010**, Munich.

Gurney model
Gurney-Modell

Solid-state physicist Ronald W. Gurney (1898–1953) during World War II developed a series of simple physical models to describe the acceleration of metal fragments by an expanding detonation plume. His model is based on a series of assumptions:

a) The detonation of a high explosive releases a specific amount of energy per unit mass unit of explosive. This energy is transformed into kinetic energy of the plume and kinetic energy of the confinement/fragments.

b) The detonation plume has a homogeneous density and a linear one-dimensional velocity profile in the space coordinates of the system.

For the fragment velocities of symmetric and asymmetric charges, the following equations can be applied:

- Sandwich

$$\frac{v}{\sqrt{2E}} = \left[\frac{M}{C} + \frac{1}{3}\right]^{-\frac{1}{2}}$$

- Cylinder

$$\frac{v}{\sqrt{2E}} = \left[\frac{M}{C} + \frac{1}{2}\right]^{-\frac{1}{2}}$$

- Sphere

$$\frac{v}{\sqrt{2E}} = \left[\frac{M}{C} + \frac{3}{5}\right]^{-\frac{1}{2}}$$

- Asymmetric sandwich (static countermass/tamper)

$$\frac{v_M}{\sqrt{2E}} = \left[\frac{1+A^3}{3(1+A)} + \frac{N}{C}A^2 + \frac{M}{C}\right]^{-\frac{1}{2}}$$

with

$$\frac{N}{C} = \frac{\text{Countermass}}{\text{Mass of explosive}}$$

it follows:

$$A = \frac{1+2\frac{M}{C}}{1+2\frac{N}{C}}$$

- Open sandwich (one metal side only)

$$\frac{v}{\sqrt{2E}} = \left[\frac{\left(1+2\frac{M}{C}\right)^3 + 1}{6\left(1+\frac{M}{C}\right)}\frac{M}{C}\right]^{-\frac{1}{2}}$$

The Gurney energy of a high explosive similarly as the V_d and P_{CJ} can be assessed with the *Kamlet* parameters, Φ, (see *Detonation*) with $\sqrt{2E} = 0.887\sqrt{\Phi} \cdot \rho_0^{0.4}$

Or may be derived from experimental data such as detonation velocity and density $\sqrt{2E} = \frac{D}{2.97}$ (Table G.7):

Table G.7: Gurney constant and detonation velocity for various high explosives.

High explosives	ρ_0 (g cm^{-3})	D (m s^{-1})	$\sqrt{2E}$ (m s^{-1})
Comp A3	1.59	8140	2630
Comp B	1.71	7890	2700
Comp C3	1.60	7630	2680
Comp C4	1.601	8190	2820
Cyclotol 75/25	1.754	8250	2790
H-6	1.76	7900	2580
HMX	1.835	8830	2800
LX-14	1.89	9110	2970
Octol 75/25	1.81	8480	2800
PBX 9404	1.84	8800	2900
PBX 9502	1.885	7670	2377
Pentolite	1.700	7530	2770
PETN	1.770	8260	2930
RDX	1.77	8700	2830
Tacot	1.61	6530	2120
TATB	1.854	7760	2440
Tetryl	1.62	7570	2500
TNT	1.63	6860	2440
Tritonal 80/20	1.72	6700	2320

- J. E. Kennedy, *The Gurney Model of Explosive Output for Driving Metal*, in *Explosive Effects and Applications*, J.A. Zukas, W. P. Walters (Eds.), Springer, **1998**, 221–257.

H

Hafnium

Aspect		Silver grey – black
Formula		Hf
GHS		02
H-phrases		250-251
P-phrases		201-222-235+410-280-420-422a
UN		2545
EINECS		231-166-4
CAS		[7440-58-6]
m_r	g mol^{-1}	178.49
ρ	g cm^{-3}	13.09
Mp	°C	2222
Bp	°C	5400
$\Delta_m H$	kJ mol^{-1}	24.06
$\Delta_b H$	kJ mol^{-1}	575.5
c_p	J mol^{-1} K^{-1}	25.74
κ	W m^{-1} K^{-1}	23
c_L	m s^{-1}	3671
$\Delta_c H$	kJ mol^{-1}	−1144
	kJ g^{-1}	−6.409
	kJ cm^{-3}	−83.898
Λ	wt.-%	−17.928

Due to its superior volumetric heat of combustion Hf has been suggested as an ingredient in igniters, composite solid propellants, and incendiary materials. The combustion properties of the refractory compounds hafnium boride, HfB$_2$, carbide, HfC, and HfSi$_2$ have been investigated recently.

- R. E. Betts, *Hafnium -potassium perchlorate pyrotechnic composition*, US3109762, **1963**, USA.
- A.C. Scurlock, K.E. Rumbel, M.L. Rice, *High density metal-containing propellants capable of maximum boost velocity*, USP 3.326.732, **1967**, USA.
- A. P. Shaw, R. K. Sadangi, J. C. Poret, C. M. Csernica, Metal_Element Compounds of Titanium, Zirconium and Hafnium as pyrotechnic Fuels, **2015**, Toulouse, 1–11.

https://doi.org/10.1515/9783110660562-008

Hafnium bomb

Hafniumbombe

The hafnium bomb is the name for speculative weapon that was discussed controversial in the 1990s. It was argued then that the hafnium bomb with its performance level could close the gap between conventional high explosives ($\Delta_{ex}H = 10^3$ J g^{-1}) and fission bombs ($\Delta_{ex}H = 10^{10}$ J g^{-1}).

The technological basis for this alleged weapon is the known metastable nuclear spin isomer ^{178m2}Hf which has a half life of 31 years and through an internal transition falls back to the ground state (^{178}Hf) by releasing the energy difference (E = 2.446 MeV) as hard gamma radiation. Based on the mass of ^{178m2}Hf, this equals an energy content of E = 1.326×10^9 J g^{-1}.

Assertions that the internal transition could be triggered by external stimuli such as soft X-rays built the basis for the idea of a bomb that would emit not only hard gamma rays but also would produce no radioactive debris. This assertion was soon rebutted and till now no peer-reviewed scientific proof of a stimulated decay of ^{178m2}Hf or any other comparable metastable nuclear-spin isomer has been reported in the literature. In addition, the interaction cross sections for any potential starting material to produce ^{178m2}Hf in the first place are so incredibly small that weighable amounts could not be produced with reasonable effort. All in all, the hafnium bomb bears the typical characteristics for pathological science and should be taken as a warning lesson to stick to scientific rigor.

- E. V. Tkalya, Induced decay of the nuclear isomer ^{178m2}Hf and the isomeric bomb, *Phys. Usph.* **2005**, *48*, 525–531.
- S. Weinberger, *Imaginary Weapons*, Nation Books, New York, **2006**, 276.

Hansen test

The Hansen test is a stability test for gun propellants that is obsolete now. Gun propellants were subjected to heating at 110 °C for 1–8 h. Every hour a separate sample would be eluted with distilled water and the hydrogen concentration (H) was plotted against time. The stability, K, is evaluated by determining K = 15−(H) · 10^4 and must be > 0.

- L. Metz, Prüfung der Beständigkeit von Nitrozellulose und rauchschwachem Pulver mit Hilfe von Wasserstoffionenmessungen nach Nic. L. Hansen, *Spreng. Schiessw.* **1926**, *21*, 186–188.

High-blast explosive

High-blast explosive (HBX) are melt-castable underwater high explosives. HBX-1 is the phlegmatized version of Torpex (→ *Hexotonal*).

Heat

Heat is the nickname for oxygen-deficient compositions ($\xi(Fe) > 80$ wt.-%) from dendritic iron powder and potassium perchlorate that are used as thermal-battery-heating elements. Due to elemental iron being prone to rust quickly in moist air, heat must be processed at controlled low humidity (RH < 10% at 20 °C). Heat is consolidated mostly to thin disks without an additional binder and hence relies completely on the interlocking of dendritic iron particles (→ *iron*). Upon ignition, heat disks react by a strong glow and concomitant release of volatile potassium chloride. However, the combustion does not affect the overall morphology of the disks and the disks also retain their electrical conductivity (due to surplus Fe content). The high heat capacity of those disks helps to maintain the temperature in a thermal battery stack. The heat of explosion q of $Fe/KClO_4$ in the 80–90 wt.-% Fe range is a linear function of Fe content and does not correlate with the experimental burn rate which has its maximum around $\xi(Fe) = 84$ wt.-% (Figure H.1).

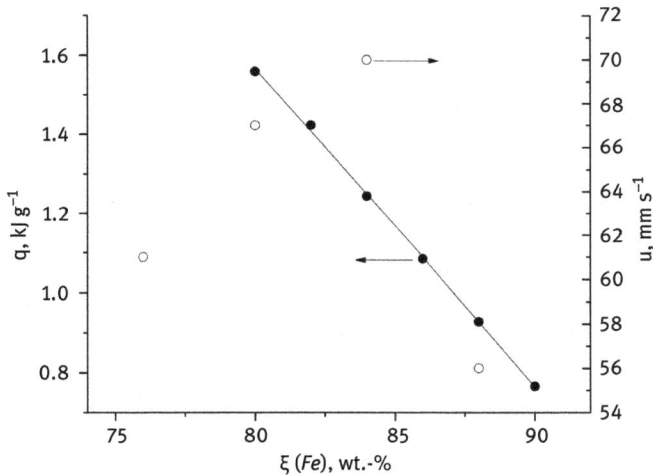

Figure H.1: Burn rate and heat of explosion of $Fe/KClO_4$.

– J. Callaway, N. Davies, M. Stringer, Pyrotechnic Heater Compositions for use in Thermal Batteries, *IPS-Seminar*, Adelaide, **2001**, pp. 153–168.
– R. A. Guidotti, J. Odinek, F. W. Reinhardt, Characterization of Fe/KClO$_4$-Heat Powders for Thermal Batteries, *IPS-Seminar*, **2002**, Westminster, 847–857.

Held's criterion

Held-Kriterium

In accordance with Held, upon the initiation of high explosives by shaped charge jets, projectiles, or fragments, the velocity correlates with their respective diameter, d as v^2d = constant $\left[mm^3\ \mu s^{-1}\right]$ (Table H.1).

Table H.1: v^2d values for various high explosives.

Type	ρ (g cm^{-3})	v^2d (mm^3 μs^{-1})
Hexanitroazobenzene	1.60	3
PBX9404	1.84	4
RDX/wax (88/12)	?	5
RDX/TNT (65/35)	?	6
LX13	1.53	12
PETN	1.77	13
Comp B	1.73	16
H6		16.5
Detasheet C3		36–53
PBX9407	1.60	40
Tetryl	1.71	41
Comp C4		64
TATB	1.80	108
PBX9502	1.89	127

However, more recent investigations by *Arnold et al.* indicate quite some scatter for v^2d for individual explosives and a linear relation of the type

$$v = A - B \cdot d$$

has been found to better match a broader span of jets and projectiles.

– W. Arnold, T. Hartmann, E. Rottenkolber, Filling the gap between hypervelocity and low velocity impacts, *J. Impact Eng.* **2020**, *139*, 10351.
– F. Bouvenot, *The Legacy of Manfred Held with Critique*, Thesis, Naval Postgraduate School, Monterey, CA, USA, **2011**, 218 pp.
– M. Held, Initiation Phenomena with Shaped Charge Jets, *Int. Detonation Symposium*, **1989**, pp. 1416–1426.

Held, Manfred (1933–2011)

Manfred Held was a German physicist and engineer. After his studies and a dissertation at the TU München, he entered then Messerschmidt Bölkow Blohm (MBB) Schrobenhausen in 1960. Under the supervision of shaped charge specialist Franz Rudolf Thomanek (1913–1990), Held matured and became an internationally respected expert for warhead detonics. His applied and fundamental research into the mode of action of explosively formed projectiles and shaped charges led him to develop the first reactive armor. He came up with a number of physically simple but robust models to assist in the design of warheads and for the investigation of the initiation and detonation of high explosives.

– N. Eisenreich, Manfred Held – a life devoted to explosive science, *Propellants Explos. Pyrotech.* **2016**, *41*, 7.
– H. Muthig, C. O. Leiber, P. Wanninger, Manfred Held 1933–2011, *Propellants Explos. Pyrotech.* **2011**, *36*, 103–104.
– M. Held, Fragmentation Warheads, S. 387–464; M. Held, Flash Radiography, pp. 555–608; M. Held, High-Speed Photography, S. 609–674, all in J. Carleone (Eds.) *Tactical Missile Warheads, Vol. 155, Progress in Astronautics and Aeronautics*, AIAA, **1993**.
– Entry in the catalogue of the German National Library: http://d-nb.info/480995087

Henkin test

The Henkin test is a standardized method to determine the reaction delay upon heating of a sample of explosive in a confined metal tube in the Wood's metal bath. As a reference value, the temperature at which explosion occurs after 5 s is taken, T_{5ex} (°C).

– Henkin 5-Second Ignition Temperature, US/202.01.016, in AOP -7 Manual of data requirements and tests for the Qualification of Explosive Materials for Military Use, 2. Ed. June 2002, NATO Standardization Agency. Brussels.

Hexachlorethane, HC

Hexachloroethan

HC, perchlorethane	
Formula	C_2Cl_6
GHS	08, 09
H-phrases	330-411
P-phrases	273-281-308+313-391-405-501a

UN		3077
REACH		LPRS
EINECS		200-666-4
CAS		[67-72-1]
m_r	g mol^{-1}	236.74
ρ	g cm^{-3}	2.091
Mp	°C	186.6
Bp	°C	185.6
$T_{p(rh\rightarrow tri)}$	°C	46
$T_{p(tri\rightarrow reg)}$	°C	71
$\Delta_{sub}H$	kJ mol^{-1}	59
$\Delta_f H$	kJ mol^{-1}	−206.27
$\Delta_c H$	kJ mol^{-1}	−727.18
	kJ g^{-1}	−3.072
	kJ cm^{-3}	−6.423
Λ	wt.-%	−6.76
c_p	J K^{-1} g^{-1}	0.722

The colorless HC has a camphor-type smell. It has been the most common organo-chlorine compound in use in obscurants together with halophilic fuels like Mg, CaSi$_2$, Ti, Fe, Zn, Al, and Si. In addition, HC has been suggested as a chlorine source in signal flare compositions.

- E.-C. Koch, Metal-Halocarbon-Pyrolants, *Handbook of Combustion*, F. Winter, M. Lackner, A. K. Agarwal, Volume 7, Wiley-VCH, **2010**, pp. 355–402.

Hexal

Hexal are pressable formulations based on **hex**ogen, **al**uminum, and montan wax (Table H.2). Hexal is used for medium caliber munition to combat aerial targets.

Table H.2: Typical hexal formulation and properties.

Composition	1	2
Hexogen (wt.-%)	66.50	56.6
Aluminum (wt.-%)	30.00	37.6
Montan wax (wt.-%)	3.50	5.8
V_D (m s^{-1})	7733 at 1.81 g cm^{-3}	
Trauzl (cm^3)	472 at 1.39 g cm^{-3}	
Koenen (mm)	6 (F)	
Friction (N)	160	

Table H.2 (continued)

Composition	1	2
Impact (J)	10	
c_p (J g^{-1} K^{-1})		1.166
Dp (°C)		168

As RDX dissolves in wax, its decomposition mostly starts soon after softening of the wax which is below T ≪ 200 °C.

- P. Langen, P. Barth, Investigation of the Explosive Properties of HMX /AL 70/30, *Propellants Explos.* **1979**, *4*, 129–131.

Hexamethylenetetramine

Hexamin, Urotropin

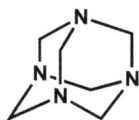

1,3,5,7-Tetraazaadamantan, urotropine, formin		
Formula		$C_6H_{12}N_4$
GHS		02, 07
H-phrases		228-317
P-phrases		210-241-261-302+352-321-501a
UN		1328
REACH		LRS
EINECS		202-905-8
CAS		[100-97-0]
m_r	g mol^{-1}	140.188
ρ	g cm^{-3}	1.331
Mp	°C	285 (subl)
$\Delta_f H$	kJ mol^{-1}	122.59
$\Delta_c H$	kJ mol^{-1}	−4198.7
	kJ g^{-1}	−29.951
	kJ cm^{-3}	−39.865
Λ	wt.-%	−205.43
N	wt.-%	39.97

Hexamethylenetetramine serves as fuel in camping and military ration packs (*Esbit*®) and is used frequently in pyrotechnics to lower the combustion temperature and to deoxidize the flame. Furthermore, it serves as flame expander for NIR (*Krone*), VIS (*Koch*), and IR decoy flares (*Hahma*). Upon combustion, the isolated carbons avoid formation of soot. Hence, it is also a popular ingredient in indoor fireworks. *Lohmann* suggested strobe compositions based on potassium, alkaline earth sulfates, and magnesium {52/25/13/10}. In addition, hexamine is the main starting material to produce hexogen and octogen.

-   E. Lohmann, *Pyrotechnischer Satz zur Erzeugung von Lichtblitzen*, DE 34 02 546 A1, **1985**, Germany.
-   U. Krone, K. Basse, Pyrotechnic Illumination, *International Pyrotechnics Seminar*, Grand Junction, CO, USA, July **2000**, pp. 141–142.
-   J. J. Sabatini, E.-C. Koch, J. C. Poret, J. D. Moretti, S. M. Harbol, Rote pyrotechnische Leucht – ohne Chlor! *Anwt. Chem.* **2015**, *127*, 11118–11120.
-   A. Hahma, *Active Material for an infra-red decoy with area effect which emits mainly spectral radiation upon combustion*, EP 2602239B1, Germany, **2020**.

Hexamethylenetriperoxidediamine, HMTD

Hexamethylentriperoxiddiamin

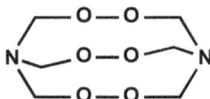

Formula		$CH_6N_2O_6$
CA		[283-66-9]
m_r	g mol^{-1}	208.171
ρ	g cm^{-3}	1.597
$\Delta_f H$	kJ mol^{-1}	−359.91
$\Delta_{ex}H$	kJ mol^{-1}	−1035.5
	kJ g^{-1}	−4.974
	kJ cm^{-3}	−7.944
$\Delta_c H$	kJ mol^{-1}	−3716.2
	kJ g^{-1}	−17.852
	kJ cm^{-3}	−28.509
Λ	wt.-%	−92.23
N	wt.-%	13.46
Mp	°C	153
T_{ex}	°C	186

Friction	N	0.1–1
Impact	J	0.5
V_D	m s^{-1}	5100 at 1.15 g cm^{-3}
P_{CJ}	GPa	26.11 at 1.666 g cm^{-3} at 60 mm diameter
Trauzl	cm^3	242
Koenen	mm	1, type F

HMTD is sensitive to hydrolysis and thermally less stable than TATP. It is hence that it is not used in the civilian or military field. Like urea nitrate, it has been suggested as an improvised explosive.

– Improvised Munitions Handbook, Department of the Army Technical Manual, *TM 31-210*, **1969**, I–17.

Hexamethylentetramine dinitrate, HDN
Hexamindinitrat

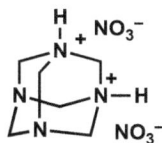

Formula		$C_6H_{14}N_6O_6$
CAS		[18423-21-7]
m_r	g mol^{-1}	266,214
ρ	g cm^{-3}	1.63
Δ_fH	kJ mol^{-1}	–377.4
$\Delta_{ex}H$	kJ mol^{-1}	–1122.5
	kJ g^{-1}	–4.217
	kJ cm^{-3}	–6.873
Δ_cH	kJ mol^{-1}	–3984.5
	kJ g^{-1}	–14.968
	kJ cm^{-3}	–24.397
Λ	wt.-%	–78.13
N	wt.-%	31.57
Mp	°C	165
Dp	°C	186
Friction	N	240
Impact	J	15
Trauzl	cm^3	220

HDN is an important intermediate in the synthesis of RDX and HMX.

Hexamite, hexanite

Hexamite, also codenamed S 18 (→ *Schiesswolle*), was an insensitive bulletproof melt-castable high explosives used in torpedo warheads, naval demolition, and depth charges in Germany.

Table H.3: Typical composition and properties.

Composition	S 18
Hexanitrodiphenylamine (wt.-%)	24
Trinitrotoluene (wt.-%)	60
Aluminum (wt.-%)	16
V_D (m s^{-1})	6900 at 1.72 g cm^{-3}
Trauzl (cm^3)	390
T_{5ex} (°C)	200–260

Hexanitroadamantane, HNA
Hexanitroadamantan

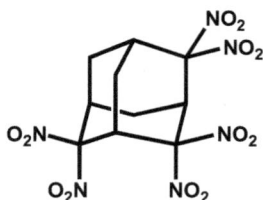

2,24,4,6,6-Hexanitrotricyclo[3.3.1.1^{3,7}]decan, HNA		
Formula		$C_{10}H_{10}N_6O_{12}$
CAS		[143850-71-9]
m_r	g mol^{-1}	406.22
ρ	g cm^{-3}	1.777
Δ_fH	kJ mol^{-1}	−160
$\Delta_{ex}H$	kJ mol^{-1}	−2149.7
	kJ g^{-1}	−5.292
	kJ cm^{-3}	−9.404

Δ_cH	kJ mol^{-1}	−5204.4
	kJ g^{-1}	−12.812
	kJ cm^{-3}	−22.766
Ω	wt.-%	−51.20
N	wt.-%	20.69
Mp	°C	198−200

Hexanitroadamantane is a colorless crystalline high explosive.

- J. Zhang, T. Hou, Y. Ling, L. Zhang, J. Luo, An efficient process for the synthesis of *gem*-dinitro compounds under high steric hindrance by nitrosation and oxidation of secondary nitroalkanes, *Tetrahedron Lett.* **2018**, *59*, 2880−2883.
- P. R. Dave, A. Bracuti, T. Axenrod, B. Liang, The Synthesis and complete proton and carbon 13 NMR spectral assignment of 2,2,4,4,6,6-hexanitroadamantane and its precursor nitro ketones by 2D NMR spectroscopy, *Tetrahedron*, **1992**, *48*, 5839−5846.

Hexanitroazobenzene, HNAB

2,2',4,4',6,6'-Hexanitroazobenzol

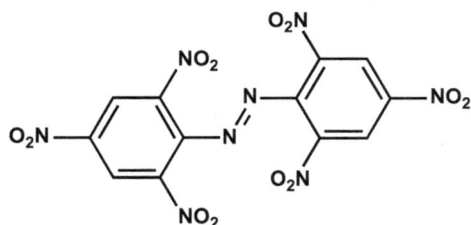

Aspect		Blood red crystals
Formula		$C_{12}H_4N_8O_{12}$
REACH		LPRS
EINECS		242-850-7
CAS		[19159-68-3]
m_r	g mol^{-1}	452.21
ρ	g cm^{-3}	1.799 (I), 1.750 (II), 1.703 (III) (polymorph)
Δ_fH	kJ mol^{-1}	284.09
$\Delta_{ex}H$	kJ mol^{-1}	−2334.2
	kJ g^{-1}	−5.162
	kJ cm^{-3}	−9.286
Δ_cH	kJ mol^{-1}	−5577.9
	kJ g^{-1}	−12.335
	kJ cm^{-3}	−22.190

Λ	wt.-%	−49.53
N	wt.-%	24.78
Mp	°C	221-222
Dp	°C	>230
Friction	N	>355
Impact	J	5 (BAM)
V_D	m s^{-1}	7600
Trauzl	cm^3	370

Hexanitroazobenzene (HNAB) forms at minimum three different crystalline poly-morphs. Due to its superior thermal stability, HNAB can be sublimed at $T \sim 230$ °C and deposited on various substrates. Crystalline layers having thickness as small as $d \sim 65$ μm are still capable to undergo full detonation ($V_D \sim 7600$ ms^{-1}), which is a unique behavior for an insensitive high explosive.

- A. S. Tappan, R. R. Wixom, R. Knepper, Critical detonation thickness in vapor-deposited hexanitroazobenzene (HNAB)films with different preparation conditions, 12th*WPC*, Colorado Springs, **2014**.
- R. Knepper, K. Browning, R. R. Wixom, A. S. Tappan, M. A. Rodriguez, M. K. Alam, Microstructure Evolution during Crystallization of Vapor-Deposited Hexanitroazobenzene Films, *Propellants Explos. Pyrotech.* **2012**, *37*, 459–467.
- J. C. Hoffsommer, J. S. Feiffer, Thermal Stabilities of Hexanitroazobenzene (HNAB)and Hexanitrobiphenyl (HNB), *NOLTR 67–74*, United States Ordnance Laboratory, White Oak, Maryland, **1967**.

Hexanitrobenzene, HNB

Hexanitrobenzol

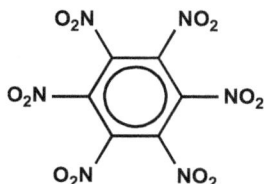

Aspect		Yellow crystals
Formula		$C_6N_6O_{12}$
CAS		[13232-74-1]
m_r	g mol^{-1}	348.099
ρ	g cm^{-3}	1.988
$\Delta_f H$	kJ mol^{-1}	146.44
$\Delta_{ex}H$	kJ mol^{-1}	−2524.1
	kJ g^{-1}	−7.251
	kJ cm^{-3}	−14.415
$\Delta_c H$	kJ mol^{-1}	−2507.1
	kJ g^{-1}	−7.202
	kJ cm^{-3}	−14.318
Λ	wt.-%	0
N	wt.-%	24.14
Mp	°C	246
Dp	°C	257
Impact	cm	10 (PA)
V_D	m s^{-1}	9500 at 2.01 g cm^{-3}
P_{CJ}	GPa	40.6 at 2.01 g cm^{-3}

HNB is one of the few homoleptic nitrocarbon compounds. Though having high density, it is very reactive towards nucleophiles, in that weak ones (e.g. moisture) lead to quick hydrolysis to the extremely sensitive 1,3,5-trihydroxy-2,4,6-trinitrobenzene. It hence has no practical importance as a high explosive.

– A. T. Nielsen, *Nitrocarbons*, Wiley-VCH, **1995**, 93–98.
– J. C. Hoffsommer, J. S. Feiffer, Thermal Stabilities of Hexanitroazobenzene (HNAB)and Hexanitrobiphenyl (HNB), *NOLTR 67-74*, United States Ordnance Laboratory, White Oak, Maryland, **1967**.

Hexanitrodiphenyl, HNDP

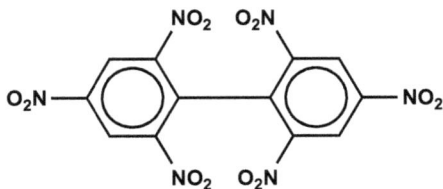

Aspect		Light yellow crystals
Formula		$C_{12}H_4N_6O_{12}$
CAS		[4433-16-3]
m_r	g mol^{-1}	424.197
ρ	g cm^{-3}	1.74
		1.839 at 150 K
$\Delta_f H$	kJ mol^{-1}	68.20
$\Delta_{ex} H$	kJ mol^{-1}	−2100.0
	kJ g^{-1}	−4.953
	kJ cm^{-3}	−8.618
$\Delta_c H$	kJ mol^{-1}	−5362.1
	kJ g^{-1}	−12.641
	kJ cm^{-3}	−21.995
Λ	wt.-%	−52.8
N	wt.-%	19.81
Mp	°C	242
Dp	°C	320
Impact	cm	85 (PA)
Trauzl	cm^3	344

HNDP is obtained via Ullmann reaction from *chlorotrinitrobenzene*. It is suitable for high-temperature initiators.

- M. Rohac, S. Zeman, R. Ales, Crystallography of 2,2',4,4',6,6'-hexanitro-1,1'-biphenyl and ist relation to Initiation Reactivity, *Chem. Mater.* **2008**, *20*, 3105–3109.

Hexanitrodiphenylamine, hexyl, HNDP

Hexanitrodiphenylamin

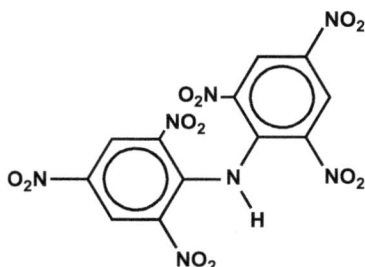

Aspect		Yellow green crystals
Formula		$C_{12}H_5N_7O_{12}$
REACH		LPRS
EINECS		205-037-8
CAS		[131-73-7]
m_r	g mol^{-1}	439.211
ρ	g cm^{-3}	1.77
$\Delta_f H$	kJ mol^{-1}	41.42
$\Delta_{ex}H$	kJ mol^{-1}	−2153.9
	kJ g^{-1}	−4.904
	kJ cm^{-3}	−8.680
$\Delta_c H$	kJ mol^{-1}	−5477.8
	kJ g^{-1}	−12.472
	kJ cm^{-3}	−22.076
Λ	wt.-%	−52.82
N	wt.-%	22.32
Mp	°C	249
Dp	°C	259
Impact	J	7.5 (BAM)
	cm	22.86 (PA 2 kg)
Friction	N	>355 (BAM)
V_D	m s^{-1}	6898 at 1.58 g cm^{-3}
V_D	m s^{-1}	7150 at 1.67 g cm^{-3}
Trauzl	cm^3	333
Koenen	mm	5, type F-G

HNDP was used as early as World War I as an insensitive high-explosive filler for bulletproof torpedo-charges (*Schießwolle* 18) based on TNT. It is nowadays used for high-temperature-resistant initiators.

Hexanitroethane, HNE

Hexanitroethan

Formula		$C_2N_6O_{12}$
REACH		LPRS
EINECS		213-042-1
CAS		[918-37-6]
m_r	g mol^{-1}	300.055
ρ	g cm^{-3}	1.998
$\Delta_f H$	kJ mol^{-1}	119.66
$\Delta_{ex}H$	kJ mol^{-1}	−886.5
	kJ g^{-1}	−2.954
	kJ cm^{-3}	−5.903
$\Delta_c H$	kJ mol^{-1}	−866.5
	kJ g^{-1}	−2.888
	kJ cm^{-3}	−5.770
Λ	wt.-%	+42.66
N	wt.-%	28.01
Mp	°C	150 (dec)
Impact	J	15.4
Friction	N	>240
Trauzl	cm^3	351

HNE decomposes already at ambient temperature causing freshly prepared crystals to become waxy and amorphous. Except for straight chain alkanes, HNE reacts with most other organic compounds and hence is not a suitable ingredient in formulations. It reacts hypergolically with strongly basic amines. It is of theoretical interest as a model compound. Careful pyrolysis under inert gas yields *tetranitroethylene*.

- P. Noble Jr., W. L. Reed, C. J. Hoffman, J. A. Gallaghan, F. G. Borgardt, Physical and Chemical Properties of Hexanitroethane, *AIAA Journal* **1963**, *1* 395–397.
- H. P. Marshall, F. G. Borgardt, P. Noble Jr., Thermal Decomposition of Hexanitroethane, *J. Phys. Chem.* **1965**, *69*, 25–29.

Hexanitrostilbene, HNS

Hexanitrostilben

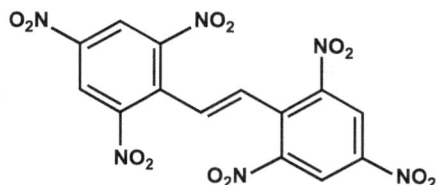

Aspect		Creme crystals
Formula		$C_{14}H_6N_6O_{12}$
REACH		LPRS
EINECS		243-494-5
CAS		[20062-22-0]
m_r	g mol^{-1}	450.235
ρ	g cm^{-3}	1.745
$\Delta_f H$	kJ mol^{-1}	78.24
$\Delta_{ex}H$	kJ mol^{-1}	−2177.6
	kJ g^{-1}	−4.837
	kJ cm^{-3}	−8.440
$\Delta_c H$	kJ mol^{-1}	−6444.8
	kJ g^{-1}	−14.314
	kJ cm^{-3}	−24.979
Λ	wt.-%	−67.52
N	wt.-%	18.67
Mp	°C	313 (dec)
Impact	J	>5 (BAM)
Friction	N	>235 (BAM)
V_D	m s^{-1}	7000 at 1.70 g cm^{-3}
V_D	m s^{-1}	6800 at 1.60 g cm^{-3}
P_{CJ}	GPa	20 at 1.60 g cm^{-3}
Trauzl	cm^3	333
Koenen	mm	5, type ?
STANAG		4230

HNS is frequently used in high-temperature-resistant initiators, detonating bolts, slapper detonators, and EFIs. As it is poorly soluble even in DMSO, it is difficult to recrystallize in different shapes and sizes. Hence, though its synthesis is simple, its cumbersome grain modification makes it comparatively expensive. HNS is used in low concentrations to affect the crystallite size in melt-cast TNT. The HNS types are distinguished as follows:

HNS-I Non-recrystallized HNS from the hypochlorite process (Shipp synthesis)
HNS-II Recrystallized HNS-I
HNS-III HNS for doping TNT
HNS-IV Recrystallized HNS-II with very high specific surface area (for EFI applications)
HNS-V With dioxane washed HNS-IV

- D. Clement, K. P. Rudolf, The Shock Initiation Threshold of HNS as a Function of its Density, *Propellants Explos. Pyrotech.* **2007**, *32*, 322–325

Hexogen, RDX, cyclonite

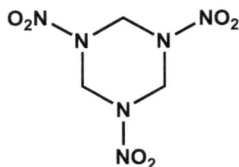

Formula		$C_3H_6N_6O_6$
REACH		LRS
EINECS		204-500-1
CAS		[121-82-4]
m_r	g mol^{-1}	222.117
ρ	g cm^{-3}	1.818
$\Delta_f H$	kJ mol^{-1}	66.94
$\Delta_{ex} H$	kJ mol^{-1}	−1284.5
	kJ g^{-1}	−5.783
	kJ cm^{-3}	−10.444
$\Delta_c H$	kJ mol^{-1}	−2107
	kJ g^{-1}	−9.486
	kJ cm^{-3}	−17.132
Λ	wt.-%	−21.61
N	wt.-%	37.84
Mp	°C	204 (dec)
Impact	J	7.5 (BAM)
Friction	N	120 (BAM)
V_D	m s^{-1}	8428 at 1.70 g cm^{-3}
	m s^{-1}	8080 at 1.60 g cm^{-3}

P_{CJ}	GPa	31.75 at 1.70 g cm^{-3}
	GPa	27.92 at 1.60 g cm^{-3}
Trauzl	cm^3	480
Koenen	mm	8, type H
Specification		STANAG 4022

Hexogen is currently the most important military high explosive in boosters and main charges. Apart from a broad variety of common crystal size types (Table H.4), insensitive crystal qualities (RS-RDX, I-RDX) have been developed too, which help to attain an altered-IM signature.

Table H.4: Hexogen – grain types.

Mesh	Mw (µm)	Passing through sieve (wt.-%)							
Grain type		1	2	3	4	5	6	7	8
8	2360				100				
12	1700			99 min					
20	850	98 ± 2							
35	500		99 ± 1		20 ± 20				100
50	300	90 ± 10	95 ± 5	40 ± 10				98 ± 2	98 min
60	250						99 + 1/−3		
80	180						97 + 3/−6		
100	150	60 ± 30	65 ± 15	20 ± 10				90 ± 8	90 min
120	125						83 + 10/−16		
170	90						65 + 15/−22		
200	75	25 ± 20	33 ± 13	10 ± 10				46 ± 15	70 + 1015
230	63						36 ± 14		
325	45					97 min	22 ± 14		50 ± 10

RDX was first made by *Lenze* in 1897 by nitration of HND. However, it was only in 1916 that *Henning* recognized its energetic properties and suggested its use in gun propellants. Finally, *Ritter von Herz*, after World War I, discovered that it is a superior and highly brisant high explosive. In World War II, RDX was produced by all bellingerent parties involved. Therefore, and also in order to circumvent shortages of individual raw materials, different production processes were invented, such as:

- **Bridgewater** (UK)
  Continuous nitration of hexamine with 99% nitric acid

- **H = Henning** (Germany)
  Nitration of HDN with nitric acid

- **SH = Schnurr and Henning** (Germany)
  Batch reaction of hexamine with 99% nitric acid

- **K = Knöffler** (Germany)
  Batch reaction of hexamine with ammonium nitrate and nitric acid

- **KA = Knöffler and Apel** (Germany) or **Bachmann** (USA)
  tReaction of HDN with ammonium nitrate and nitrosulfuric acid in the presence of acetic acid anhydride

- **E = Eble** (Germany) or **Schiessler and Ross** (Canada)
  Reaction of paraformaldehyde with ammonium nitrate in the presence of acetic acid anhydride

- **W = Wolfram** (Germany)
  Reaction of potassium amidosulfonate with formaldehyde and nitrosulfuric acid

- T. R. Gibbs, A. Popolato, *LASL Explosive Property Data*, University of California Press, Berkeley, **1980**, pp. 141–151.

Hexotonal

Hexotonals are aluminized versions of composition B (Table H.5). Those formulations can be used as enhanced blast explosibles and underwater explosives.

Table H.5: Composition and performance of hexotonal formulations.

Compositions	Alex 20	Alex 32	H-6	HBX-1	HBX-3	H-1580	S17	Torpex	Borotorpex	Trialen
Aluminum (wt.-%)	19.8	30.8	20	17	35	15.0 30 µm	40	18		30
Boron (wt.-%)									10	
RDX (wt.-%)	44	37.4	45	40	31	42.1	10	42	46	20
TNT (wt.-%)	32.2	27.8	30	38	29	42.1	50	40	44	50
CaCl$_2$ (wt.-%)			0.5	0.5	0.5		–	–		–
D2 wax (wt.-%)		4	4	5	5	5	0.8	–	–		–
Density (g cm^{-3})	1.801*	1.88	1.74	1.712	1.84	1.757	1.88	1.81	1.742	1.73

Parameter	Alex 20	Alex 32	H-6	HBX-1	HBX-3	H-1580 *	S17	Torpex	Borotorpex	Trialen
V_D (m s^{-1})	7530	7300	7194	7307	6917	7490		7495	7600	
P_{CJ} (GPa)	23	21.5		22.04		25.7			25.2	
c_p (J g^{-1} K^{-1})				1.126	0.96			1.00		
Dp (°C)						170	195			
Δ_mH (J g^{-1})			38.3	46.0	37.4		38.6	48.2		48.2

*Hexotonal.

High explosive

Sprengstoff

A high explosive is an explosive material capable of and intended to undergo detonation. High explosives that detonate directly upon exposure to thermal stimuli (flame, heat, and spark) or minor mechanical stimuli (friction, impact, and stab) are called primary explosives (*Initialsprengstoffe*). All other high explosives that require a shock wave to detonate are called secondary high explosives (*Sekundärsprengstoffe*). Among the secondary high explosives, those that are not cap sensitive or not undergoing full detonation but when exposed to a detonator but requiring the shock of a booster charge are considered main charge explosive filler, whereas those secondary explosives that can be set off with the blast from a detonator are called booster explosives.

Hivelite

Hivelite is the name for coprecipitated pyrotechnics from alkalidecaboranates ($M_2B_{10}H_{10}$) and alkali metal nitrates (MNO_3, M = K, Cs). Typically, Hivelite features Cs salts which are denser, less hygroscopic, and thermally more stable than the corresponding potassium salts. A typical composition is given thereafter:

RDM 510532

- 72.5 wt.-% cesium nitrate, $CsNO_3$
- 27.5 wt.-% cesium decaboranate, $Cs_2B_{10}H_{10}$

At the highest comparable pressing density ($\rho \sim 4$ g cm^{-3}), the coprecipitates burn at double the rate (u ~ 350 m s^{-1}) than regularly mixed compositions from micronsized salts. The sensitiveness of Hivelite approaches that of primary explosives. It is hence that Hivelite is often desensitized against friction and ESD with graphite.

Hivelite is used for very fast and reliable transfer charges, for example, in emergency rescue systems.

- L. Avrami, R. Velicky, D. Anderson, D. Downs, A Comparative Study of Very High Burning Rate Materials – Hivelite Compositions 300511 and 300435, *Technical Report ARLCD-TR-82015*, US Army. Dover NJ, **1982**.
- J. D. Glass, Evolution of an F-111 Pyrotechnic Time Delay, *11th International Pyrotechnics Seminar*, Vail, CO, 7–11 July **1986**, pp. 243–253.

Holland test

This test determines the mass loss upon storage of gun propellants at 105 °C (double- and triple-base GPs) for 72 h or 110 °C (single-base GPs). Changes occurring in the first 8 h of the test are not taken into account. This test covers all gaseous decomposition products.

- Bundeswehrprüfvorschrift TL-1376-0600 M.2.21.1

Holtex

NC/DEGDN/PETN-based explosive, similar to *Nipolit* (→).

Homburg, Axel (1936–2018)

Axel Homburg was a German mechanical engineer and entrepreneur. After his studies in the field of aerospace engines that earned him the Dr. Ing. at the TH Darmstadt in 1967, he entered the Dynamit Nobel AG on January 1, 1969. There he was responsible for the development of propellants before he was promoted to become head of research and development of explosives. After posts as director and vice president of defence technology, he became CEO of the Dynamit Nobel AG in 1988 and later became member of the board of directors in 1997. In addition to his professional duties, he was chairmain of the Fraunhofer ICT board of trustees and member of the board of trustees of the Bundesanstalt für Materialforschung und -prüfung (BAM) in Berlin. He also took over responsibility as editor for the tenth edition of the Explosives Handbook founded by Rudolf Meyer (1908–2000).

- A. Homburg, Remarks on the Evolution of Explosives, *Propellants Explos. Pyrotech.* **2017**, *42*, 851–853.
- A. Homburg, J. Köhler, R. Meyer, *Explosives*, 7th Ed. Wiley-VCH, Weinheim, **2016**, 466 pp.
- Entry in the German National Library: http://d-nb.info/gnd/106921584

HTA

During World War II, the German acronym HTA designated melt-castable high explosives based on **h**exogen, **T**NT, and **a**luminum. An important formulation was HTA-15 (aka HTA-41), which was used in the unguided air-to-air missile RM-4.

HTA -15

40 wt.-% hexogen (RDX)

45 wt.-% TNT

15 wt.-% aluminum

With the introduction of octogen (HMX) in the 1950s, nowadays HTA represents melt-castable high explosives based on **HMX, TNT, a**luminum (Table H.6).

Table H.6: Composition and performance of HTA-3, octonal.

Composition	HTA -3
Aluminum (wt.-%)	22
Octogen (HMX)(wt.-%)	49
TNT (wt.-%)	29
ρ (g cm^{-3})	1.90

Parameter	HTA-3
V_D (m s^{-1})	7866
P_{CJ} (GPa)	29
Impact (cm)	43 with 2 kg
c_p (J g^{-1} K^{-1})	1.026
Dp (°C)	212
$\Delta_m H$ (J g^{-1})	28.0

HTPB, hydroxylterminated polybutadiene

Hydroxylterminiertes Polybutadien

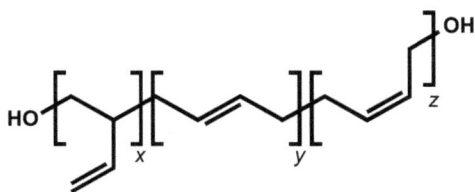

Polyvest-HT (*Evonik*), poly-Bd 45HT (*Atochem*)		
Formula		$C_{10}H_{15.4}O_{0.07}$
CAS		[69102-90-5]
m_r	g mol^{-1}	136.752
ρ	g cm^{-3}	0.90-0.92
$\Delta_f H$	kJ mol^{-1}	−51.88
$\Delta_c H$	kJ mol^{-1}	−6084.2

	kJ g^{-1}	−44.492
	kJ cm^{-3}	−40.932
Λ	wt.-%	−323.27
Tg	°C	−80
Tpp	°C	−18
Fp	°C	215

HTPB contains about 20% 1,2-vinyl (x), ~ 60% 1,4-*trans*-(y), and ~20% 1,4-*cis*-(z)-butadiene units. The degree of cross-linking is (x + y + z) = n ~ 55. HTPB is cured with both aromatic and aliphatic isocyanates (→ *isophorone diisocyanate*) to give polyurethanes with advantageous mechanical properties and high resistance towards acids and bases. HTPB has superior mechanical properties at low temperatures and is less prone to aging compared to CTPB. It is hence an important base for curable high explosives (e.g. PBX see Appendix 2) and solid rocket propellants.

- H. G. Ang, S. Pisharath, *Energetic Polymers*, Wiley-VCH, Weinheim, **2012**, pp. 5, 13, 19.
- J. P. Agrawal, *High Energy Materials*, Wiley-VCH, Weinheim, **2010**, pp. 244–249.
- M. A. Daniel, Polyurethane Binder Systems for Polymer Bonded Explosives, *DSTO-GD-0492*, Weapons Systems Division Defence Science and Technology Organisation, Edinburgh South Australia, **2006**, 34 pp.

Hugoniot, Pierre-Henri (1851–1887)

Pierre Hugoniot was a French mathematician and physicist working on ballistics, fluid dynamics, and shock physics. He developed an equation of state to describe the influence of shock on matter (→ *Hugoniot equation*).

- R. Chéret, The life and work of Pierre-Henri Hugoniot, *Shock Waves* **1992**, *2*, 1–4.

Hugoniot equation
Hugoniot-Gleichung

The Hugoniot equation describes thermodynamic changes of state induced by shock

$$h_1 - h_0 = \frac{1}{2} \cdot (v_0 + v_1) \cdot (p_1 - p_0)$$

The Hugoniot $\mathfrak{H}_1$ (e.g. in a *p–v* diagram) indicates all possible states of a substance after being subjected to a shock with the initial state (p_0/v_0). The shock itself is not described by the Hugoniot, $\mathfrak{H}_1$, but by the Rayleigh line, R, from the initial state (p_0/v_0) to the shocked state (p_1/v_1):

$$p_1 - p_0 = \frac{w^2}{v_0} - \frac{w^2}{v_0^2} \cdot v_1$$

Then a non-reactive shock relaxation of a substance proceeds through the stats indicated on the Hugoniot, $\mathfrak{H}_1$. However, if a shock causes a temperature rise over the ignition temperature of an explosive substance, the final states are no longer found on $\mathfrak{H}_1$, but are found on a $\mathfrak{H}_2$, which does not intersect (p_0/v_0). Due to the release of heat of this "reactive shock," the product Hugoniot, $\mathfrak{H}_2$, is always above the starting state Hugoniot, $\mathfrak{H}_1$ (see Figure H.2).

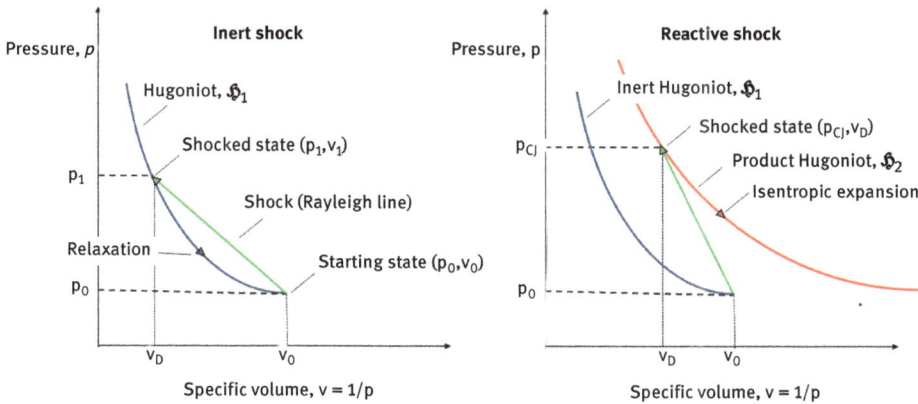

Figure H.2: Hugoniot plots in the p–v diagram for an inert (left) and reactive shock (right).

In the case of detonable substance, the pair of values (p_{cj}, v_D) indicates the tangential intersection between the Rayleigh line and $\mathfrak{H}_2$ and describes the stable detonation (see *detonation*).

The states on a Hugoniot are characterized by triples of pressure, p, specific volume, v, and particle velocity, u. Those values can be found for many reactive and non-reactive materials in the publication by *Marsh*.

– N.N. Shock Waves in Solids, J. A. Zukas (Eds.), *Studies in Applied Mechanics, Volume 49*, Springer Verlag, Heidelberg, **2004**, pp. 75–102.
– S. P. Marsh (Eds.) *LASL Shock Hugoniot Data*, University of California Press, Berkeley, USA, **1980**, 658 pp.

HWC

HWC (Hexogen Wax Carbon) is an RDX (hexogen) based explosive formulation that is in use as booster and donor in various gap tests (Table H.7).

Table H.7: Composition and performance of HWC.

Composition	HWC
Hexogen (RDX)(wt.-%)	94.5
Wax (wt.-%)	4.5
Graphite (wt.-%)	1.0
ρ (g cm^{-3})	1.707 (100% TMD)

Parameter	HWC
V_D (m s^{-1})	7890 at 95% TMD
P_{CJ} (GPa)	24.88 at 95% TMD
Friction (N)	216
Impact (J)	7.5
Dp (°C)	190

Hydrazinium diperchlorate, HP-2

Hydraziniumdiperchlorat

Constitutional formula		$[ClO_4]_2[N_2H_6]$
Sum formula		$C_2lH_6N_2O_8$
CAS		[13812-39-0]
m_r	g mol^{-1}	232.98
ρ	g cm^{-3}	2.2
$\Delta_f H$	kJ mol^{-1}	−293.29
$\Delta_{ex} H$	kJ mol^{-1}	−395.9
	kJ g^{-1}	−1.699
	kJ cm^{-3}	−3.738
Λ	wt.-%	+41.21
N	wt.-%	12.02
Mp	°C	118 (dec)
Dp	°C	170

Cryogels based on HP-2 with resorcine-formaldehyde-resin are nano-structured energetic materials and show both reduced thermal and mechanical sensitiveness than the corresponding physical mixtures of the components.

- S. Cudziło, W. Kicinski, W. Trzcinski, Thermochemical analysis of composites containing organic gels and inorganic oxidizers, *Biul. Wojskow. Akad. Tech.* **2008**, *57*, 173–183.

Hydrazinium monoperchlorate, HP

Hydraziniumperchlorat

Constitutional formula		$[ClO_4][N_2H_5]$
Sum formula		$ClH_5N_2O_4$
CAS		[13762-80-6]
m_r	g mol^{-1}	132.51
ρ	g cm^{-3}	1.939
$\Delta_f H$	kJ mol^{-1}	−177
$\Delta_{ex}H$	kJ mol^{-1}	−410.7
	kJ g^{-1}	−3.100
	kJ cm^{-3}	−6.010
Λ	wt.-%	+24.15
N	wt.-%	21.14
DSC onset	°C	220
Mp	°C	137 (dec)
Impact	J	3–6 (BAM)
Friction	N	10
V_D	m s^{-1}	4600 at 1.25 g cm^{-3} at $\varnothing$ = 12 mm
Trauzl	cm^3	362
Koenen	mm	24

HP is an extremely sensitive oxidizer that has been investigated back in the 1960s and 1970s as an alternative to AP.

- K. Klager, R. K. Manfred, L. J. Rosen, Hydrazine Perchlorate as Oxidizer For Solid Propellants, *ICT JATA 1978*, 359–382, **1978**.

Hydrazinium nitrate

Hydraziniumnitrat

Constitutional Formula		[NO$_3$][N$_2$H$_5$]
Sum Formula		H$_5$N$_3$O$_3$
CAS		[37836-27-4]
m$_r$	g mol^{-1}	95.058
ρ	g cm^{-3}	1.685
Δ$_f$H	kJ mol^{-1}	−238
Δ$_{ex}$H	kJ mol^{-1}	−376.62
	kJ g^{-1}	−3.962
	kJ cm^{-3}	−6.676
Λ	wt.-%	+8.42
N	wt.-%	44.21
DSC onset	°C	229
Mp	°C	62.1 (dec)
Impact	J	7.4 (BAM)
V$_D$	m s^{-1}	8690 at 1.60 g cm^{-3} (under confinement)
Trauzl	cm^3	408
Koenen	mm	6

HN is an oxidizer and has been tested as an eutectic ingredient with both hydrazine and water in the binary: H/HN: 55/45: mp: −45 °C, and the ternary: H/HN/H$_2$O: 53/33/13; mp: −54 °C for use with hypergolic WFNA in rocket propulsion while other formulations from H/HN/H$_2$O have been tested as liquid gun propellants.

- F. Bachmaier, J. Jimenez-Barbera, *Hydrazine nitrate – hydrazine as a monergolic propellant*, DLR **1966**, 28 pp.

Hydrazinium nitroformate, HNF

Hydraziniumnitroformat

Hydrazinium trinitromethanide		
Aspect		Yellow crystals
Constitutional formula		[C(NO$_2$)$_3$][N$_2$H$_5$]
CAS		[14913-74-7]
m$_r$	g mol^{-1}	183.081
ρ	g cm^{-3}	1.91
Δ$_f$H	kJ mol^{-1}	−76.86
Δ$_{ex}$H	kJ mol^{-1}	−936.1
	kJ g^{-1}	−5.113
	kJ cm^{-3}	−9.766
Λ	wt.-%	+13.1
N	wt.-%	38.25

Mp	°C	118 (dec)
Impact	J	2–4 (BAM)
Friction	N	18–36 (BAM)
V_D	m s^{-1}	8428 at 1.70 g cm^{-3}
P_{CJ}	GPa	31.75 at 1.70 g cm^{-3}
Trauzl	cm^3	480
Koenen	mm	8 (type ?)

HNF forms needles (l/d: 4–5). It is obtained from precipitation of hydrazine with nitroforme law:

$$N_2H_{4(l)} + HC(NO_2)_{3(l)} \longrightarrow N_2H_5{}^+ C(NO_3)_2{}^- + 84\, kJ\, mol^{-1}$$

Until 2010, HNF was produced on a 50 kg scale in the Netherlands by Aerospace Propulsion Products. Compared on theoretical results, HNF is a superior oxidizer when compared with AP and even ADN. However, HNF suffers from severely restricted thermal stability. Adiabatic self heating of pure HNF starts at 100 °C. Hence, suitable stabilizers are sought after. In addition, HNF is incompatible with typical polymeric binders of the HTPB type.

- J. Louwers, Hydrazinium Nitroformate A High Performance Next Generation Oxidizer, *J. Pyrotech.* **1997**, *6*, 36–42.
- M. A. Bohn, Thermal Stability of Hydrazinium (HNF) Assessed by Heat Generation Rate and Heat Generation and Mass Loss, *J. Pyrotech.* **2007**, *26*, 65–94.

Hydrogen peroxide

Wasserstoffperoxid, T-Stoff, Perhydrol

Aspect		Colorless yellowish liquid fuming in air
Formula		H_2O_2
GHS		03, 05, 07
H-phrases		271-302-314-332-335-412
P-phrases		280-305+351+338-310
UN		2014
EINECS		231-765-0
CAS		[7722-84-1]
m_r	g mol^{-1}	34.015
ρ	g cm^{-3}	1.447 (100% at 20 °C)
	g cm^{-3}	1.4709 (100% at 0 °C)
$\Delta_f H$	kJ mol^{-1}	−187.78
Λ	wt.-%	+47.04
c_p	J K^{-1} mol^{-1}	89.319

T_p	°C	0.42 at 34,7 Pa
Mp	°C	−0.43
Bp	°C	150
$\Delta_m H$	kJ mol^{-1}	12.50
$\Delta_v H$	kJ mol^{-1}	44.86 at 160 °C

Highly concentrated H_2O_2 was produced and used for the first time in Germany in the 1930s as rocket propellants. The Walter engine in the A4 missile to drive the fuel pumps was powered by the exothermal decomposition of it by reaction with calcium permanganate:

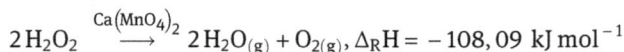

$$2\,H_2O_2 \xrightarrow{\mathrm{Ca(MnO_4)_2}} 2\,H_2O_{(g)} + O_{2(g)}, \Delta_R H = -108,09 \text{ kJ mol}^{-1}$$

Highly concentrated H_2O_2 (>85 wt.-%) is hypergolic with furfuryl alcohol. It is often stabilized against decomposition with sodium stannate, Na_2SnO_3, or organic phosphates.

– Dadieu, R. Damm, E. W. Schmidt, *Raketentreibstoffe*, Springer Verlag, Vienna, **1968**, pp. 385–404; 668–674.

Hydroxylammonium perchlorate, HAP

Hydroxylammoniumperchlorat

Constitutional		[ClO$_4$]
formula		[NH$_3$OH]
Sum formula		ClH$_4$N$_1$O$_5$
EINECS		239-650-7
REACH		LPRS
CAS		[15588-62-2]
m_r	g mol^{-1}	133.488
ρ	g cm^{-3}	2.126
$\Delta_f H$	kJ mol^{-1}	−278
$\Delta_{ex} H$	kJ mol^{-1}	−241
kJ g^{-1}	−1.809	
kJ cm^{-3}	−3.846	
Λ	wt.-%	+41.95
N	wt.-%	10.49
DRH	wt.-%	<18
DSC onset	°C	120
Mp	°C	90
Impact	cm	15 (BoM)
		(AP: 100 cm)

HAP is an extremely hygroscopic oxidizer that dissolves rapidly when exposed to air with RH > 10 % RH. It is hence not a feasible ingredient for solid propellants but has been tested as aqueous solution or solution in liquid ammonia (analogue to Diver's liquid) as monopropellant ingredient and hypergolic oxidizer for OTTO fuel. Its first decomposition step at 180 °C is to form ammonium perchlorate and oxygen.

- J. H. Robson III, *Method of making hydroxylamine perchlorate*, US Patent 2,768,874, USA, **1952**.
- J. M. Bellerby, C. S. Blackman, The interaction between Otto Fuel II and Aqueous Hydroxylammonium Perchlorate (HAP), Part I: Initial Observations and Time-to-Event Measurements, *Propellants Explos. Pyrotech.* **2004**, *29*, 262–266.

Hygroscopicity
Hygroskopizität

Hygroscopicity is the ability of substances to absorb water from the atmosphere. Figure H.3 shows the typical water content of the air as function of the temperature. While synthetic polymers can absorb and adsorb only small amounts of water, natural substances such as paper, cardboard, cork, cotton, felt, and charcoal can absorb up to 30 % of their weight. Ionic compounds having deliquescence relative humidity (DRH) below the actual relative humidity are able to absorb a multiple of their own

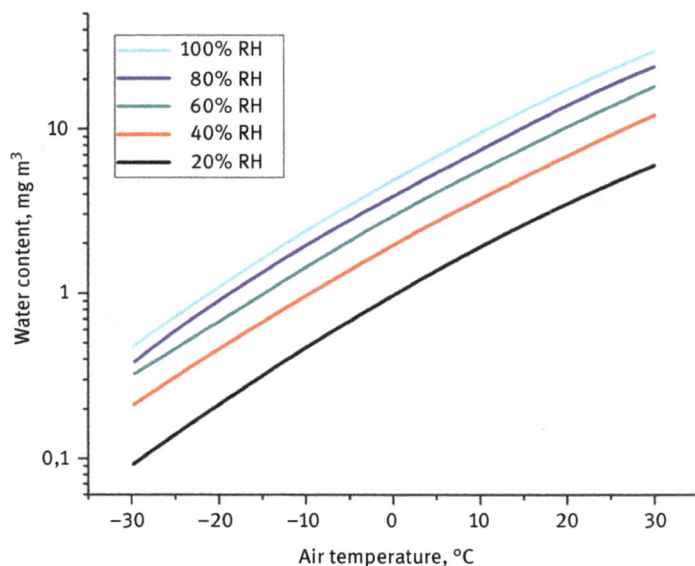

Figure H.3: Water content of the air at various relative humidities and as a function of temperature.

weight. The DRH of ionic compounds is a function of temperature and particle surface area. That is, DRH decreases with increasing temperature and decreasing particle diameter. The DRH further qualitatively correlates with the water solubility of ionic compounds. DRH values above 90 % indicate non-hygroscopic compounds, 90 > DRH > 80 indicate slightly hygroscopic compounds and DRH < 80 % RH are characteristic for hygroscopic compounds. Table H.8 shows the DRH and/or solubility for various inorganic compounds which may either be ingredients or combustion products of energetic materials.

Table H.8: Deliquescence relative humidity and water solubility of various ionic compounds in water.

	DRH at 20 °C	S (g L^{-1}) @ 20 °C		DRH at 20 °C	S (g L^{-1}) @ 20 °C
NH_4Cl	80	374	$Ca(NO_3)_2$	54	567
$LiCl$	13	832	$Sr(NO_3)_2$	86	410
KCl	84	344	$Ba(NO_3)_2$	98	82
$CsCl$	68	1862	$Mn(NO_3)_2$	3	1180 (at 10 °C)
$NaCl$	75	265	$Cu(NO_3)_2$	50	3810 (at 40 °C)
$MgCl_2$	32	352	$Pb(NO_3)_2$	98	597
$CaCl_2$	33	425	$NaClO_3$	74	1000
$MnCl_2$	57	739	$KClO_3$	97	73
$FeCl_3$	45	912	$Ba(ClO_3)_2$	94	256
$CuCl_2$	67–72	757	NH_4ClO_4	~95	178
$ZnCl_2$	<10	4300	$LiClO_4$	~70	3553
$MgSO_4$	91	660	$NaClO_4$	43	662
Na_2SO_4	86	162	$KClO_4$	99	18
K_2SO_4	98	110	$Mg(ClO_4)_2$	28	500
K_2CO_3	43	525	$Ca(ClO_4)_2$	<10	1886
Na_2CO_3	71	210	$CuSO_4$	98	11130
NH_4NO_3	65	655	$NH_4H_2PO_4$	93	368
$LiNO_3$	13	427	KH_2PO_4	96	222
$NaNO_3$	75	466	$CaHPO_4$	97	0.1
KNO_3	94	242	H_3PO_4	9	Unlimited
$CsNO_3$	>90	186	P_4O_{10}	0	Unlimited
$Mg(NO_3)_2$	56	408			

- J. Wolf, E. Lissel, Hygroskopisches Verhalten von Werkstoffen für pyrotechnische Munition, *WIWEB*, **2001**, 37 pp.

Hypergol

If a fuel/oxidizer combination ignites instantaneously upon contact without significant delay, then this substance combination is called hypergol. The intrinsic property is called hypergolicity and the behavior is called hypergolic.

The effect was probably first observed by Helmuth Walter in Germany around 1936. However, it was not until 1940 that the noun hypergol was coined by Dr. Wolfgang C. Noeggerath when working on self-igniting jet propellants combination of "C-Stoff" (methanol + hydrazine hydrate) and "T-Stoff" (80% hydrogen peroxide (H_2O_2) stabilized with 8-hydroxyquinoline) for the Me-163B (aka *Komet* or *Kraftei*).

Nowadays, hypergols should have an ignition delay under t_d = 50 ms, but ideally below t_d = 30 ms in the complete temperature band. A special case is the spontaneous ignition of substances when coming into contact with atmospheric oxygen and moisture. Those substances are called pyrophoric and the phenomenon is called *pyrophoricity*. The ignition delay in pyrophoric and hypergolic systems in general can be predicted with quantum chemical methods.

–   D. A. Newsome, G. L. Vaghjiani, D. Sengupta, An ab initio Based Structure Property Relationship for Prediction of Ignition Delay of Hypergolic Ionic Liquids, *Propellants Explos. Pyrotech.* **2015** *40*, 759–764.
–   A. Dadieu, R. Damm, E. W. Schmidt, *Raketentreibstoffe*, Springer, **1968**, 116.

Hypervelocity missile
Hyperschallflugkörper, HFK, HaFK

The hypervelocity missile, HFK (Fig H.4), was a German experimental missile that has been conjointly developed by ICT, EADS, Bayern Chemie, and Diehl in the early 2000s. It served the understanding of hypersonic missiles and upon its test flight at Meppen (WTD 91) proving ground in October 2003 the HFK E1 reached a Mach number slightly above seven marking the world record for hypersonic flight at about sea-level altitude. Tables H.9 and H.10 depict the experimental propellant and its properties.

Figure H.4: Artist depiction of HFK E1 with its common fin design and center of gravity.

Table H.9: HFK-solid propellant composition after *Menke et al.*

Ingredients	
Ammonium perchlorate (wt.-%)	20
GAP-A (wt.-%)	9
GAP-N100 (wt.-%)	12
TMETN (wt.-%)	9
ε-CL-20 (wt.-%)	47
Al (wt.-%)	3

Table H.10: HFK-solid propellant properties after *Menke et al.*

Parameter	
Density (g cm^{-3})	1.77
Oxygen balance, Λ (wt.-%)	−28.6
Adiabatic Temperature (K)	3027
I_{sp} at 70:1 expansion ratio (s)	253
u at 10 MPa (mm s^{-1})	38 ± 0.7
Pressure exponent, n (−)	≤0.45
Friction (N)	40
Impact (J)	4
Deflagration temp (°C)	189

- S. Eisele, K. Menke, About the burning behavior and other properties of smoke reduced composite propellants Based on AP/CL20 /GAP, *32nd International ICT Annual Conference*, 3–6 July **2001**, Karlsruhe, P-149.
- M. A. Bohn, M. Dörich, H. Pontius, Adiabtische Selbstaufheizung von Festtreibstoffen auf Basis e-CL20 (HNIW), Ammoniumperchlorat, GAP-Binder und energetischer Weichmacher, *34th International ICT Annual Conference*, 24–27 June **2003**, Karlsruhe, P-175
- K. W. Naumann, F. Bouteille, K. Bauer, M. Thiberge, N. Bretthauer, R. Strecker, Design and testing of the Solid Rocket Motor HFK 2000 Built for the German HFK (Hyperschall-Flugkörper) Hypervelocity Missile Program, AIAA 2005-4170, *42st AIAA/ASME/SAE/ASEE Joint Propulsion Conference & Exhibit*, 10–13 July, **2005**, Tucson, AZ, USA.

Hytemp® – polyacrylate

Copolymer von Methacylsäureethylester und Butylester

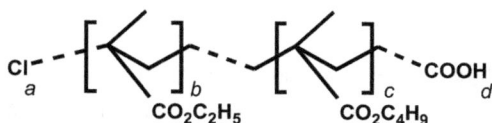

Cl⁻ ··· structure with CO₂C₂H₅ and CO₂C₄H₉ groups ··· COOH

Aspect		Opaque polymer
Formula		$\sim C_7H_{12}O_2$
m_r	g mol^{-1}	128.17
ρ	g cm^{-3}	1.1
$\Delta_f H$	kJ mol^{-1}	−500 (estimated)
$\Delta_c H$	kJ mol^{-1}	−3969.6
	kJ g^{-1}	−30.972
	kJ cm^{-3}	−34.069
Λ	wt.-%	−224.69
T_g	°C	−41 to −36
Dp	°C	>175

Hytemp is a popular binder in pressable plastic-bonded explosives (e.g. PBXN-9). Hytpem is good soluble in ketones and esters. It contains small percentages (<5 wt.-%) of terminal chlorine and carboxy groups.

– H.-D. Park, Y.G. Cheun, J.-S. Lee, J.-K. Kim, Development of a High Energy Sheet Explosive with Low Sensitivity, *36th International Annual Conference of ICT*, 28 June-1 July **2005**, Karlsruhe, P-117.

I

ICT code

The origin of the ICT-Thermodynamic Code lies in the year 1969. At that time, the Fraunhofer Institute for Chemical Technology developed a FORTRAN program for the calculation of chemical equilibria, which was expanded and improved in the following years. The MS-DOS version of this code found a wide distribution for the performance calculation of propellants. MS-DOS programs are no longer up-to-date and are to be replaced by WINDOWS applications. Therefore, it became necessary to create a WINDOWS user interface for the input and output of the ICT-Thermodynamic Code. Besides this, it is now possible to transfer the results into MS-WORD and present the results graphically.

The ICT-Thermodynamic Code is based on the method developed by the National Aeronautics and Space Administration (NASA). This method uses mass action and mass balance expressions to calculate chemical equilibria. Thermodynamic equilibria can be calculated for constant pressure conditions as well as for constant volume conditions. Up to 75 reaction products can be present; 40 of these can be in liquid or solid state. In addition to the ideal equation of state (EOS), the virial EOS can be used, especially for the high-pressure conditions of closed vessels and gun weapons. By applying the virial EOS (including the second and third virial coefficients), pressures can be calculated that are close to experimental values. The calculation of the heat of explosion is of special interest, because the experimental measurement using a calorimetric bomb is sometimes difficult due to high temperatures or erosive reaction products. Finally, the code can be used to determine the parameters of gas detonations, for example, pressure, temperature, and detonation velocity. POC: Dr. Stefan Kelzenberg, Fraunhofer ICT, stefan.kelzenberg@ict.fraunhofer.de

https://doi.org/10.1515/9783110660562-009

IDP, isodecylpelargonate

Isodecylpelargonat, Emolein

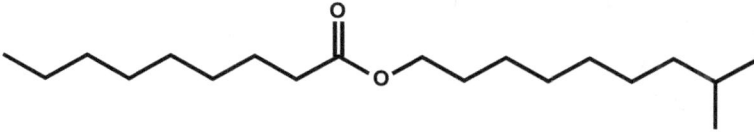

Aspect		Colorless liquid
Formula		$C_{19}H_{38}O_2$
REACH		LPRS
EINECS		203-665-7
CAS		[109-32-0]
m_r	g mol^{-1}	289.51
ρ	g cm^{-3}	0.866
$\Delta_f H$	kJ mol^{-1}	−889.10
$\Delta_c H$	kJ mol^{-1}	−12018.6
	kJ g^{-1}	−40.263
	kJ cm^{-3}	−34.867
Λ	wt.-%	−294.79
Mp	°C	−80
Bp	°C	150 at P = 333 Pa
Specification		AS 2328

IDP serves as plasticizer in HTPB-based formulations. Contrary to other non-energetic plasticizer such as dioctyl adipate, it lowers the viscosity significantly, thereby allowing higher filler levels (e.g. Rh 26 and PBXN-110).

- R. Gagnaux, Die mechanischen Eigenschaften vom Kompositsprengstoff PBXN-110. Einfluss von HTPB/IDP -Verhältnis, *35th International Annual Conference of ICT*, 29 June-2 July **2004**, Karlsruhe, V-6.

IED

Unkonventionelle Spreng und Brandvorrichtungen, USBV

IED stands for *improvised explosive device* consists of a high explosive and an initiator prepared from household products. Depending on the theater, IEDs may certainly contain components of military origin (ammunition parts (see Fig. I.1), engineering explosives, etc.). The NATO definition of IED reads: "An IED is a device which is placed or fabricated in an improvised manner, incorporating destructive,

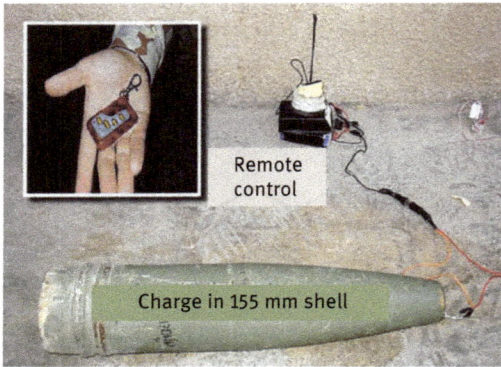

Figure I.1: Typical IED using military components.

lethal, noxious, pyrotechnic or incendiary chemicals, and designed to destroy, inca-pacitate, harass or distract. It may incorporate military stores, but is normally derived from non-military components."

Depending on the type of initiation, IEDs can be classified into

Radio-controlled IED	RCIED	
Victim-operated IED	VOIED	
Command wire IED	CWIED	Fixed
Timer-controlled IED	TCIED	
Vehicle-borne IED	VBIED	
Suicide IED	SIED	Mobile

- *Allied Joint Doctrine for Countering Improvised Explosive Devices (C-IED)*, NATO Standardization Agency, Draft, **2008**, Brussels.
- Improvised Munitions Handbook, Department of the Army Technical Manual, *TM 31-210*, **1969**, 256 pp.

Ignit

Ignit is a binary mixture from powdered iron and potassium permanganate. It was first used in British HC smoke igniters in World War I and later became adopted in Germany for the same purpose under the name Ignit. Figure I.2 shows the burn rate and heat of explosion for consolidated $Fe/K[MnO_4]$ as a function of iron content.

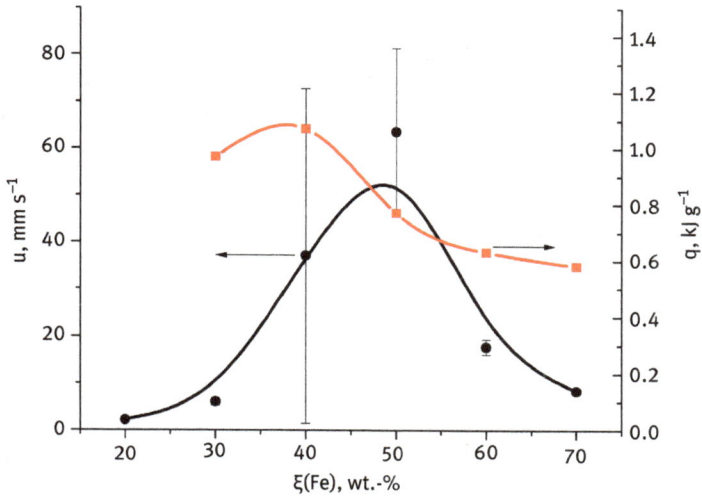

Figure I.2: Burn rate and heat of explosion of Fe (<43 μm) and K[MnO$_4$](<53 μm) consolidated at 55 MPa.

– M. J. Tribelhorn, M. G. Blenkinsop, M. E. Brown, Combustion of some iron-fueled binary pyrotechnic systems, *Thermochim. Acta* **1995**, *256*, 291–307.

Igniter composition
Anzündsatz

Igniter compositions are pyrotechnic compositions that produce hot gases and/or hot particles. Igniter compositions having a positive oxygen balance $\Lambda > 0$ typically show reduced ignition delay (t_2-t_3) (Figure I.3). Typical compositions are given hereafter in Table I.1.

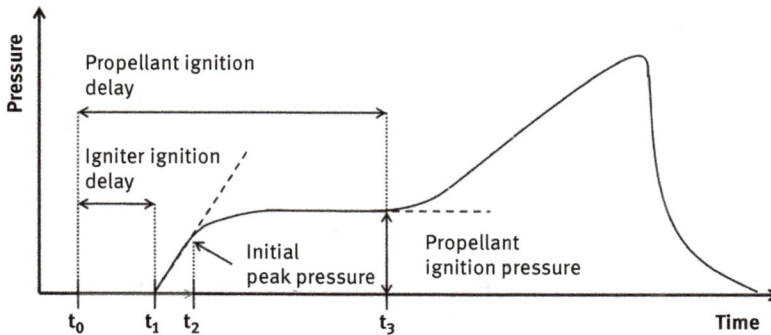

Figure I.3: Pressure–time curve for ignition of a progressive solid propellant grain after *Koch*.

Table I.1: Igniter compositions.

Components (wt.-%)	MTV	B/KNO₃	Benite	Eimite	Ex-98	Y-593	AZM-421	ZPP
Potassium perchlorate					61.9		12.9	42
Potassium nitrate		75	51	27.6		75		
Barium chromate							59.0	
Nitrocellulose		5	30	40			2.8	
Boron, amorphous		20						
Magnesium	60			16.7	6.4			
Titanium							7.6	
Zirconium							17.7	52
Charcoal			10.2			10		
PTFE	35							
Nitroglycerine					17.4			
TEGDN					13.7			
Viton® A	15							5
Sulfur			6.8	9.8		5		
Centralite			1.0		0.6			
Graphite								1
Resorcine				5.9				

Igniters
Anzünder

Igniters serve the start of a combustion/deflagration process. Therefore, igniters produce flames and/or hot particles. There are primary and secondary igniters. Primary igniters are stab, impact, friction, and electrical igniting caps. Secondary igniters are ignition boosters, ignition transfer charges, delay elements, and safety fuses. For rocket motors with a known cavity volume, V_m (cm³) the required mass, m_i (g) of an igniter can be assessed with the following formula:

$$m_i = 0.5 \cdot V_m^{0.316098}$$

- Solid Rocket Motor Igniters, *NASA SP8051*, **1971**, 112 pp.

Ignition
Anzündung

Ignition is a term describing both a threshold and the induction of a thermal, subsonic reaction (combustion and deflagration) in an energetic material. Ignition starts with an exothermic reaction of the energetic material and is completed upon a stable, that is, stationary, combustion process. An energetic material can be ignited by various means such as

- **Heat transport**
- **Heat conduction**
- **Heat radiation**
- **Illumination**
  - **Indirect heating by absorption of radiation**
  - **Selective excitation of a weak bond initiating the decomposition**
- **Friction**
- **Impact**
- **Shockwaves**
- **Hypergolic reaction**

Igniters produce flames, hot particles, and gases. The energetic materials contained in igniters are pyrotechnic compositions. The latter may contain thermally sensitive primary explosives such as *lead styphnate*. To effect ignition, the igniter must be able to heat up the energetic material beyond its ignition temperature. The ignition sensitivity of pyrotechnic compositions can be tested in the ignition gap test.

The ignition of energetic materials with shock waves has been investigated in great detail by *Hardt*. He found that the required shock energy E_i to ignite a material equals the common thermal ignition enthalpy

$$E_i \sim H_i$$

with

$$E_i = 0.5 \cdot \frac{p^2}{u_s^2 \cdot \rho_0^2}, \; H_i = c_p(T_i - T_0)$$

and T_i and T_0 (K) are ignition and ambient temperatures: c_p (J K^{-1} g^{-1}) is the specific heat; ρ_0 (g cm^{-3}) the density; p (Pa) the shock pressure; and u (m s^{-1}) the particle velocity.

The ignition of a rocket propellant with progressive burning geometry is depicted below in Figure I.10. The phase t_0–t_1 is the ignition delay of the igniter composition and correlates with the ignition sensitivity. The phase t_1–t_2 is the pressure increase up to which stationary combustion has been achieved with the igniter composition. At t_3, finally, the solid propellant has been ignited.

- E.-C. Koch, *Metal-Fluorocarbon Based Energetic Materials*, Wiley-VCH, **2012**, pp. 210–215.
- G. Klingenberg, Experimental Study on the Performance of Pyrotechnic Igniters, *Propellants Explos. Pyrotech.* **1984**, *9*, 91–107.
- A. P. Hardt, R. H. Martinson, Initiation of pyrotechnic Mixtures by Shock, *8th Symposium on Explosives and Pyrotechnics*, Philadelphia, **1974**, p. 53.

Ignition gap test
Anzündungs-Gaptest

The ignition gap test serves the determination of the safe maximum ignition distance between a standardized donor (B/KNO_3: 30/70) and a consolidated 100 mg, 5 mm ⌀ pellet of a pyrotechnic composition under investigation. The shortest possible distance is 2.5 mm. Though the test requires quite a large number of samples to be fired, it is a standard procedure in the UK as it employs a realistic configuration (donor–acceptor).

Table I.2 shows the mean ignition distance for some selected ignition compositions and Figure I.4 depicts the general setup.

Table I.2: Mean gap values for various compositions.

Composition (wt.-%)	Designation	d_{50} (mm)
Si/KNO_3 (50/50)		19
Mg/KNO_3 (50/50)		21
$Ti/KNO3$ (50/50)		31
B/KNO_3 (10/90)		36
$B/Si/KNO_3$ (20/10/70)		44
B/KNO_3 (30/70)		52
$Si/KNO_3/2$ K-SP (40/40/20)		63
$B/K_2Cr_2O_7$ (15/85)		82
3 K-SP (75/15/10)	G40	83

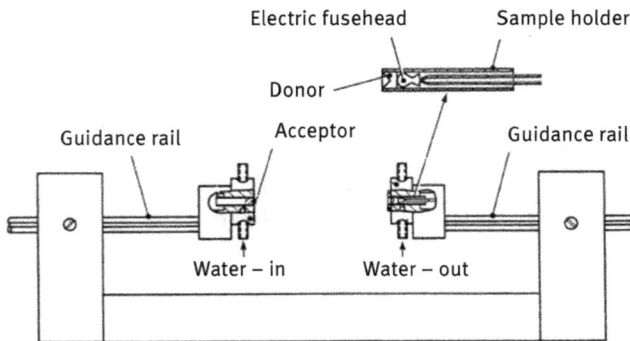

Figure I.4: Ignition gap test after *Lindsley*.

– E. A. Robinson, G. I. Lindsley, E. L. Charsley, S. B. Warrington, Assessing the Ignition Characteristics of Pyrotechnics, *10th International Pyrotechnic Seminar*, **1985**, 25.

– G. I. Lindsley, E. A. Robinson, E. L. Charsley, S. B. Warrington, A Comparison of the Ignition Characteristics of Selected Metal/Oxidant Systems, *11th International Pyrotechnic Seminar*, **1986**, 425–440.

IM signature

IM-Signatur

The IM signature describes the reaction type of a munition in response to tests conducted in accordance with AOP – 39/STANAG 4439 (see Table I.4 *Insensitive munition*). Table I.1 depicts the differences in signature for a general-purpose bomb filled with either vulnerable H-6 or insensitive PBXN-109. Color coding helps to quickly identify compliance/non-compliance with specifications (green = compliance, yellow = borderline, and red = non-compliance).

Table I.3: Response of Mk82 mod Type 2 or BLU-111/B general-purpose bomb filled with two different high explosives after *Swanson*.

High explosive	FCO	SCO	BI	FI	SR	SCJI
H-6	I	I	I	I	I	I
PBXN-109	IV	IV/V	V	V	II	II

– R. L. Swanson, Approach to IM Policy – Defining the Need, **2012**, Brussels, Belgium; https://www.sto.nato.int/publications/STO%20Educational%20Notes/STO-EN-AVT-214/EN-AVT-214-01.pdf

IMX

The acronym IMX (*intermediate explosives*) has been introduced for the first time in the US back in the 1980s. Then, nitrate- and perchlorate-based formulations showing low-velocity detonation velocity were investigated (Figure I.5).

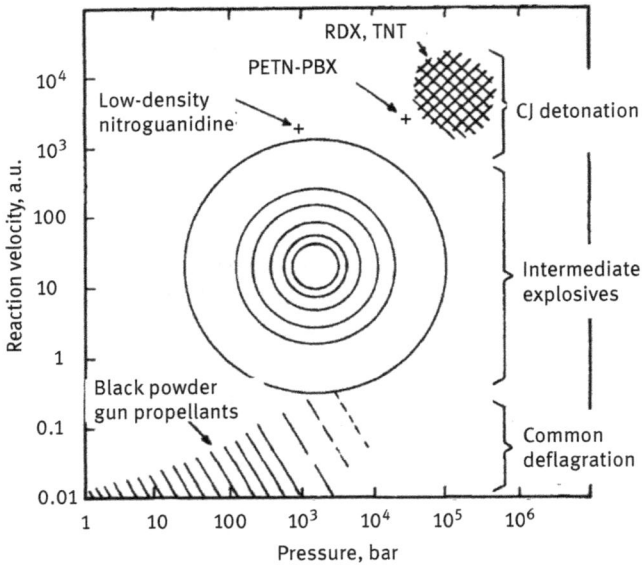

Figure I.5: Reaction dynamics of intermediate explosives after *Brown*.

Since the year 2000, IMX stands for insensitive explosives based on the insensitive melt-cast base DNAN developed conjointly at both Picatinny Arsenal and BAE. See *IMX 101* or *IMX 104*.

- J. A. Brown, M. Collins, Explosion Phenomena, *Intermediate Between Deflagration and Detonation*, Esso Research, Linden, October **1967**, 148 pp.

IMX-101

IMX-101 has been developed by ARDEC and BAE Ordnance Systems Inc. as a very insensitive melt-castable TNT replacement for use with large-caliber munitions like artillery and heavy mortar shells:
- 36.8 wt.-% nitroguanidine
- 19.7 wt.-% nitrotriazolone
- 43.5 wt.-% 2,4-dinitroanisole

A formulation containing about 80 wt.-% IMX101 and 20 wt.-% aluminum is called ALIMX-101 and is currently qualified as a replacement for the vulnerable H6 in large general-purpose bombs (e.g. Mk 82/BLU-111/B) (Table I.4).

Table I.4: Performance and
sensitiveness of IMX-101.

Parameter	
V_D (m s^{-1})	6,885
ρ (g cm^{-3})	1.64
Dp (°C)	213
$\varnothing_{cr}$ (mm)	64–66
Friction (N)	250
ELSGT 50% (GPa)	5.9
Impact NOL (J)	32

– A. Provatas, C. Wall, Evaluation of IMX Explosives IMX-101 & IMX-104 for the ADF, *PARARI*, **2013**, Canberra.
– E.-C. Koch, Insensitive High Explosives IV: Initiation & Detonation of Nitroguanidine, *Def. Tech.* **2019**, *15*, 467–487.

IMX-104

IMX-104 has been conjointly developed by ARDEC and BAE Ordnance Systems Inc. as insensitive melt-castable component B replacement for use with mortar ammunition (Table I.5):
– 15.3 wt.-% hexogen
– 53.0 wt.-% nitrotriazolone
– 31.7 wt.-% 2,4-dinitroanisole

Table I.5: Performance and
sensitiveness of IMX-104.

Parameter	
V_D (m s^{-1})	7400
ρ (g cm^{-3})	1.75
Dp (°C)	206
$\varnothing_{cr}$ (mm)	17–21
LSGT 50% (GPa)	3.25
Friction (N)	288
Impact NOL (J)	28

– Provatas, C. Wall, Evaluation of IMX Explosives IMX-101 & IMX-104 for the ADF, *PARARI*, **2013**, Canberra.

Incendiaries
Brandstoffe

Incendiaries are warfare agents that exploit the lethal and destructive effects of fire. In the *Protocol on Prohibitions or Restrictions on the use of Incendiary Weapons* entered into force on December 2, 1983, it is described under which circumstances the military use of incendiaries is prohibited. It is generally prohibited to use incendiaries of all kinds against civilians. The use of airborne incendiaries against military targets surrounded by civilians is prohibited. The use of obscurants that have incendiary effects as side effects is explicitly excluded from the ban. Typical incendiaries are napalm, white phosphorus, and pyrotechnic mixtures based on red phosphorus, magnesium (e.g. B299), thermite, and many others.

While the general military use of incendiaries has become ostracized (115 nations have so far (date 2020) ratified the above-mentioned UN protocol), there are still scenarios and missions which require the use of incendiaries in small-caliber ammunition to fight semihard and aerial targets or its bulk use in the emergency destruction of ordnance and IT equipment (denial of access) in command posts to become abandoned. Especially, insensitive munitions cannot be easily blown apart with some C4 but would require equivalent amounts of NEM for emergency decommission. Hence, using incendiaries to induce deflagration is much more effective for that purpose.

Regarding the destruction of biological and chemical warfare agents, see *agent defeat.*

- H. W. Koch, H. H. Licht, Brandstoffe, Brandmunition, Brandwirkung, *CO 34/74*, Bundesakademie für Wehrverwaltung, Mannheim, 30 September **1974**, 19 + X pages.
- Ministerrat Der Deutschen Demokratischen Republik Ministerium für Nationale Verteidigung, *Brandwaffen, A053/1/003*, Berlin, **1988**, 62 pp.
- Protocol on Prohibition or restrictions on the use of incendiary weapons, Geneva, Siwtzerland, 1980, accessed on October 20 2020 at http://disarmament.un.org/treaties/t/ccwc_p3/text

Initiators
Zündmittel

Initiators almost exclusively contain primary explosives (see for an exception: *EFI*) and serve the initiation, transfer, and enhancement of a detonation. It is distinguished between primary and secondary initiators. Primary initiators are often termed detonators. There are stab, impact, flame, and electrically triggered detonators. Booster and detonation cords belong to the secondary initiators

Insensitive high explosives, IHE

Insensitive high explosives, IHE

IHE are high explosives that undergo mass detonation. However, IHE are so insensitive that the probability for an accidental initiation is negligible.

Tests to classify high explosives as IHE are as follows:

- UN Test Series 7 (→ *EI(D)S = Extremely insensitive (detonating) substances*) which requires eight tests with a candidate substance and four tests with an item/munition.
- Department of Energy (DOE)-IHE Test, which requires 11 tests with a candidate substance.
- *Recommendations on the Transport of Dangerous Goods – Manual of Tests and Criteria* – 5th rev. Ed. United Nations, **2009**, pp. 157–175.
- *Explosives Safety Manual, MN471011*, Sandia National Laboratories, Albuquerque, **2007**.

Insensitive munitions

Insensitive munition

Insensitive munitions are munitions that fully comply with performance, availability, and other operational requirements. However, in comparison to standard ammunition, insensitive munitions do not lead to grave consequences when subjected to accidental stimuli (e.g. bullet impact or fire), and hence significantly minimize the probability of subsequent hazard for personel, and damage to the weapon system and/or logistic equipment. The worldwide first ever military specification to detail the investigation of the response of an ammunition to fast and slow heating (*FCO = fast cook-off* and SCO = slow cook-off) and small arms fire attack (*BI = bullet impact)* was the *DOD-STD-2105* filed by US Navy in 1982. The international standard in the field of ammunition safety are nowadays the NATO documents AOP-39, STANAG 4439, and the AOPs/STANAGS referenced therein. Many countries on a national level have developed even more strict regulations. STANAG 4439 defines stimuli (derived from actual operational *threats*) for munitions that are likely to occur in the life cycle of an ammunition.

The corresponding stimuli (and the underlying operational threat) and the maximum admissible reaction (response) of the ammunition are given in Tables I.6 and I.7.

Table I.6: Stimulus, definition, and maximum admissible response.

Stimulus	Requirement	Description	Operational threat	Letter
Fast heating, FH (fast cook-off, FCO)	≥Type V (burn)	Mean temperature T = 550–850 °C. T = 550 °C must be obtained within 30 s after start of the test	Fuel fire in storage, on a vehicle or airplane	A
Slow heating, SH (slow cook-off, SCO)	≥Type V (burn)	Heating rate 1–30 °C per hour starting from ambient	Fire in adjacent storage, vehicle or airplane	B
Bullet impact, BI	≥Type V (burn)	One to three projectiles 12.7 mm (0.50 cal) (AP = armour piercing) with a velocity of v = 400–850 m s^{-1}	Small arms attack	C
Fragment impact, FI	≥Type V (burn)	Steel fragment with mass of m = 15–65 g and a velocity of v = 2600–2200 m s^{-1}	Fragments from munition	D
Shaped charge jet impact, SJCI	≥Type III (explosion)	Shaped charge up to 85 mm	Shaped charge attack	E
Sympathetic reaction, SR	≥Type III (explosion)	Detonation of a donor munition of the same type in a lifecycle arrangement	Fiercest reaction of the same ammunition type in storage, vehicle or airplane	F

Table I.7: NATO nations IM policies.

			NATO		DEU	FRA			GBR	ITA			USA
Threat	**Test procedures**		**STANAG 4439**		**KMS**	**Instruction 211893**			**JSP 520**	**DG AT IM Guidelines**			**MIL-STD 2105D**
	STANAG	**Stimulus**	**IM**	**AASTP-1 SsD 1.2.3**		★	★★	★★★		F	FF	FFF	
A	4240	FH	V	V	V	IVa	V	Vb,c	V	V	V	V	V
B	4382	SH	V	V	V	III	V	Vc	V	V	V	V	V
C	4241	BI	V	V	V	III	V	Vc	V	V	V	V	V
D	4396	SR	III	III	III	III	III	IIIc	III	III	III	III	III
E	4496	FI	V		V		V	V	V		Id	V	V
		Hvy FI					III	IIIe,c			Id	V	
F	4526	SCJI	III		III		III	IIIc	III		Id	III	III

a No propulsion admissible; b only after 5 min; c energetic materials required to meet substance criteria specified in Test Series 7 in the UN Orange Handbook; d type I or better as defined by threat hazard assessment; e French National Standard NF T70-512.

Explosive reaction level – response types in accordance with AOP-39
detonation (type I)

The most violent type of munition reaction where the energetic material is consumed in a supersonic decomposition.

Primary evidence of a type I reaction is the observation or measurement of a shock wave with the magnitude and timescale of a purposely detonated calibration test or calculated value and the rapid plastic deformation of the metal casing contacting the energetic material with extensive high shear rate fragmentation.

Secondary evidence may include the perforation, fragmentation, and/or plastic deformation of a witness plate and ground craters of a size corresponding to the amount of energetic material in the munition.

Partial detonation (type II)

The second most violent type of munition reaction where some of the energetic material is consumed in a supersonic decomposition.

Primary evidence of a type II reaction is the observation or measurement of a shock wave with magnitude less than that of a purposely detonated calibration test or calculated value and the rapid plastic deformation of some, but not all of the metal casing contacting the energetic material with extensive high shear rate fragmentation.

Secondary evidence may include scattering of burned or unburned energetic material; the perforation, fragmentation, and/or plastic deformation of a witness plate and ground craters.

Explosion (type III)

The third most violent type of munition reaction with sub-sonic decomposition of energetic material and extensive fragmentation.

Primary evidence of a type III reaction is the rapid combustion of some or all of the energetic material once the munition reaction starts and the extensive fracture of metal casings with no evidence of high shear deformation resulting in larger and fewer fragments than observed from purposely detonated calibration tests.

Secondary evidence may include significant long-distance scattering of burning or unburned energetic material; witness plate damage; the observation or measurement of overpressure throughout the test arena with a peak magnitude significantly less than and significantly longer duration than that of a purposely detonated calibration test; and ground craters.

Deflagration (type IV)

The fourth most violent type of munition reaction with ignition and burning of confined energetic materials which leads to a less violent pressure release.

Primary evidence of a type IV reaction is the combustion of some or all of the energetic material and the rupture of casings resulting in a few large pieces that might include enclosures and attachments. At least one piece (e.g. casing, packaging, or energetic material) travels (or would have been capable of travelling) beyond 15 m and with an energy level greater than 20 J based on the distance versus mass relationships in Figure B.1. A reaction is also classified as type IV if there is no primary evidence of a more severe reaction and there is evidence of thrust capable of propelling the munition beyond 15 m.

Secondary evidence may include a longer reaction time than would be expected in a type III reaction; significant scattering of burning or unburned energetic material, generally beyond 15 m; and some evidence of pressure in the test arena which may vary in time or space.

Burn (type V)

The fifth most violent type of munition reaction where the energetic material ignites and burns non-propulsively.

Primary evidence of a type V reaction is the low-pressure burn of some or all of the energetic material. The casing may rupture resulting in a few large pieces that might include enclosures and attachments. No piece (e.g. casing, packaging, or energetic material) travels (or would have been capable of travelling) beyond 15 m and with an energy level greater than 20 J based on the distance versus mass relationships in Figure B.1. There is no evidence of thrust capable of propelling the munition beyond 15 m. A small amount of burning or unburned energetic material relative to the total amount in the munition may be scattered, generally within 15 m but no more than 30 m.

Secondary evidence may include some evidence of insignificant pressure in the test arena and for a rocket motor a significantly longer reaction time than if initiated in its design mode.

No reaction (type VI)

The least violent type of munition response where any reaction is self-extinguished immediately upon removal of the external stimulus.

Primary evidence of a type VI reaction is no reaction of the energetic material without a continued external stimulus; the recovery of all or most of the energetic material with no indication of sustained combustion; and no fragmentation of the casing or packaging greater than from a comparable inert test item.

Secondary evidence – none.

– *Hazard Assessment Test for Non-Nuclear Munitions*, MIL-STD-2105-D, Department of Defense, Washington, **2011**.
– *Policy for Introduction and Assessment of Insensitive Munitions (IM)*, STANAG 4439, NATO Standardization Agency, Brussels, Ed. 3, **2010**.
– *Guidance on the Assessment and Development of Insensitive Munitions*, AOP -39, NATO Standardization Agency, Brussels, Ed. 3, **2009**.
– *Konzept Munitionstechnische Sicherheit*, BMVg FüS IV 3, Bonn, August **2008**, 30 pp.

Intumescence
Intumeszenz

Intumescence is the ability of a solid material to expand upon exposure to heat. The underlying mechanism leading to expansion is the endothermic release of a vapor phase (H_2O, CO_2, etc.) thereby performing pressure–volume work against a stratified structure. Typical materials showing intumescence are stratified silicates such as vermiculite (hydrous phyllosilicate), graphite (intercalation) compounds, and closed microstructures filled with a blowing agent). Expanded structures are characterized by both increased surface area and reduced thermal conductivity. This behavior makes intumescent materials, on the one hand, ideal protective coatings for heat sensitive materials (protection against fast heating) (Figure I.6) and, on the other hand, perfect ingredients to increase the burning surface area of an energetic material and boost the burn rate accordingly (200–300 %) (Table I.8).

(a)   (b)

Figure I.6: Ammunition box with intiumescent coating before (a) and after exposure (b) to bon fire test. (Photos by courtesy of Dr. Volker Gettwert, © Fraunhofer ICT, further information can be found in V. Gettwert, T. Fischer, Reduction of the thermal sensitivity of ammunition by fire protective packaging, ICT Jata, 29 June – 2 July 2010, Karlsruhe, P-46).

Table I.8: Composition and ambient pressure burn rate of a pyrotechnic flare (1, 2) and rocket propellant (3, 4) with and without intumescent burn rate modifiers.

Component	1	2	3	4
Intumescent graphite (wt.-%)		5.0		3.0
Magnesium (wt.-%)	60.0	60.0		
Polytetrafluoroethylene (wt.-%)	35.0	30.0		
Viton® A (wt.-%)	5.0	5.0		
Ammonium perchlorate (wt.-%)			85.5	84.5
HTPB (wt.-%)			13.47	11.63
IPDI (wt.-%)			1.01	0.87
Iron(III) acetylacetonate (wt.-%)			0.02	0.02
Density (g cm^{-3})	1.881	1.852	1.678	1.711
Burn rate at ambient pressure (mm s^{-1})	3	8	1.6	4.5

- R. Stanek, *Verfahren zur Herstellung von brandschützenden Verbundwerkstoffen*, EP 694574A1, **1995**, Germany.
- A. Hahma, *Pyrotechnic Active Composition with a combustion accelerant*, AU 2013206582 B2, **2013**, Germany.
- F. Chassagne, M. Gillet, F. Daguisé, J.J. Serra, Modelling of intumescent coatings growth: simulatiuons from lab-scale to the large one, *EUROPYRO 2011*, 17–19 May **2011**, Reims, France, 239–255.
- V. Gettwert, T. Fischer, Reduction of the thermal sensitivity of ammunition by fire protective packaging, ICT Jata, 29 June – 2 July **2010**, Karlsruhe, P-46

Iodine pentoxide

Iodpentoxid

Formula		I$_2$O$_5$
GHS		03, 05
H-phrases		272-314
P-phrases		210-221-303+361+353-305+351+338-405-501a
UN		2085
REACH		LPRS
EINECS		234-740-2
CAS		[12029-98-0]
m$_r$	g mol^{-1}	333.81
ρ	g cm^{-3}	5.08
Mp	°C	300 (dec)
c$_p$	J g^{-1} K^{-1}	0.578
Δ_fH	kJ mol^{-1}	−177
Λ	wt.-%	+28.76

Iodine pentoxide, I_2O_5, is an alternative oxidizer in pyrotechnic compositions for ERA or agent defeat applications. I_2O_5 readily reacts with ambient moisture to give iodic acid in accordance with

$$3\,I_2O_5 + 3\,H_2O \rightarrow 2\,HI_3O_8 + 2\,H_2O \rightarrow 6\,HIO_3$$

Hence, it must be protected accordingly or processed in a dry box.

- R. Russell, S. Bless, M. Pantoya, Impact-driven Thermite Reactions with Iodine Pentoxide and Silver Oxide, *J. Energ. Mater.* **2011**, *29*, 175–192.
- J. Feng, G. Jian, Q. Liu, M. R. Zachariah, Passivated Iodine Pentoxide Oxidizer for Potential Biocidal Nanoenergetic Applications, *ACS Appl. Mater. Interfaces* **2013**, *5*, 8875–8880.
- D. K. Smith, M. L. Pantoya, J. S. Parkey, M. Kesmez, The water–iodine oxide system: a revised mechanism for hydration and dehydration, *RSC Advances*, **2017**, *7*, 10183–10191

Iron

Eisen

Aspect		Silvery grey – black
Formula		Fe
GHS		02, 07
H-phrases		228-319-335
P-phrases		210-241-261-305+351+338-405-501a
UN		3089
EINECS		231-096-4
CAS		[7439-89-6]
m_r	g mol^{-1}	55.847
ρ	g cm^{-3}	7.874
Mp	°C	1535
Bp	°C	3070
$\Delta_m H$	kJ mol^{-1}	14.9
$\Delta_b H$	kJ mol^{-1}	340.2
c_p	J mol^{-1} K^{-1}	24.98
κ	W m^{-1} K^{-1}	174
$\Delta_c H$	kJ mol^{-1}	−412.1
	kJ g^{-1}	−7.379
	kJ cm^{-3}	−58.102
Λ	wt.-%	−42.97

Due to its inherent oxidation in moist air, it must be processed at relative humidity <50%. In dry air, iron starts to oxidize at T = 220 °C. Stainless steel filings with high carbon content are used in sparklers to give the characteristic terminal particle explosions. Pure iron is used as fuel in thermal battery-heating pellets (*heat*). Figure I.7 shows the iron particles, which are suitable for the consolidation, in heat pellets due to favorable interlocking effects of the dendritic shape.

(a)

(b)

(c)

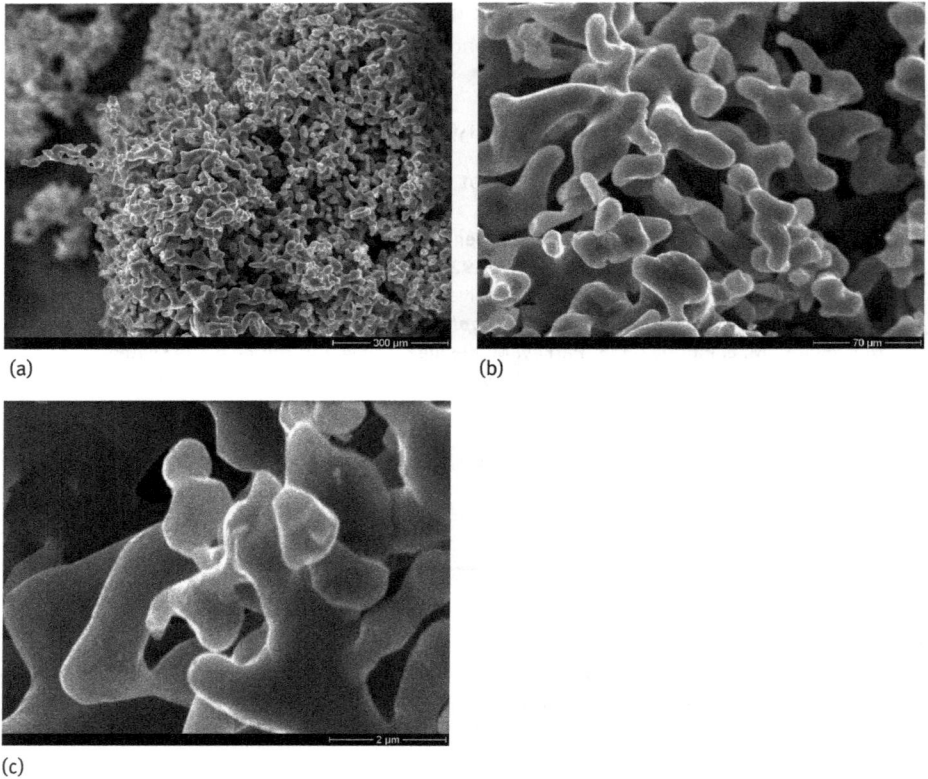

Figure I.7: Electron micrograph of dendritic iron powder.

Other applications for iron and steel filings are firework cones and oxygen-generating compositions. Elemental iron has also been suggested as color agent in HC smokes. In a side reaction, HC reacts with iron to yield the intense yellow colored, hygroscopic (DRH = 45%) but also extremely corrosive iron(III) chloride, which sublimes at a favorably low T = 320 °C:

$$2\,Fe_{(s)} + \; C_2Cl_{6(s)} \; \longrightarrow \; 2\,FeCl_{3(g)} + \; 2\,\{C\}$$

Iron has been used in gasless igniter and delay formulations of the type Fe/K[MnO$_4$] (*Ignit*).

– M. J. Tribelhorn, M. G. Blenkinsop, M. E. Brown, Combustion of some iron-fueled binary pyrotechnic systems, *Thermochim. Acta* **1995**, *256*, 291–307.
– H. Stoltzenberg, M. Leuschner, *Nebelsatz zur Herstellung von beständig gefärbten anorganischen Nebeln*, DE1188490, **1962**, Germany.

Iron(III) acetylacetonate, iron(III) pentandionate
Eisen(III)acetylacetonat

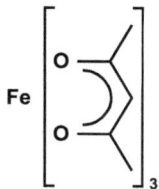

Fe(acac)$_3$		
Aspect		Orange-red crystals
Formula		C$_{15}$FeH$_{21}$O$_6$
GHS		07
H-phrases		302-315-319-335
P-phrases		261-302+352-305+351+338-321-405-501a
REACH		LPRS
EINECS		237-853-5
CAS		[14024-18-1]
m$_r$	g mol^{-1}	353.175
ρ	g cm^{-3}	1.348
Δ_fH	kJ mol^{-1}	−1268.92
Δ_cH	kJ mol^{-1}	−8047.2
	kJ g^{-1}	−22.786
	kJ cm^{-3}	−30.715
Λ	wt.-%	−160.82
Mp	°C	184

Iron(III) acetylacetonate is a catalyst for the polymerization of HTPB.

Iron oxides

Eisenoxide

		Ferric oxide	Black iron oxide
Formula		Fe_2O_3	$FeO \cdot Fe_2O_3$
Aspect		Orange-red	Black
alternative names		Hematite	Magnetite, hammer scale
CAS		[1309-37-1]	[1317-61-9]
m_r	g mol^{-1}	159.692	231.539
ρ	g cm^{-3}	5.24	5.18
$\Delta_f H$	kJ mol^{-1}	−824.25	−1,118.38
Λ	wt.-%	+10.02	+6.91
Mp	°C	1462 (dec)	1597
Dp	°C		1984

Iron(III) oxide, Fe_2O_3, and iron(II,III) oxide, $FeO \cdot Fe_2O_3$, are common oxidizers in first fires and delay compositions (Table E.7). Iron(II) oxide, FeO, (*ferrous oxide*) CAS [1345-25-1] does not find any use in pyrotechnics, however, it is utilized as non-stoichiometric compound, $Fe_{1-x}O$, with $0.05 \leq x \leq 0.12$ as a pyrophoric material in decoy flares.

Table I.9: Iron oxides in various pyrotechnic formulations.

	First fire	Igniter	Igniter	Thermite
Fe_2O_3 (wt.-%)		25	50	75
$FeO \cdot Fe_2O_3$ (wt.-%)	25			
$2PbO \cdot PbO_2$ (wt.-%)	25	25		
Al (wt.-%)				25
Si (wt.-%)	25	25		
Ti (wt.-%)	25	25	32.5	
Zr (wt.-%)			17.5	
NC (wt.-%)			+44	
Dp (°C)	762	865	456	800
Q (J g^{-1})	1506	1435	2635	3974

– A. K. Lay, *Decoy Countermeasures*, EP1948575B1, **2014**, UK.

ISL, Institut Saint Louis

The ISL is a binational defense research institute that has been founded conjointly in 1959 by the governments of France and Germany. The ISL is located in Saint

Louis, France in southern Alsace close to both German and Swiss border. The institute performs research in direct response to tasks defined by both French and German ministries of defence (German budget: 21 Mio. € in 2015). In addition, the ISL is conducting research for industrial third-party customer in the field of interior, external, and terminal ballistics; electromagnetic effects; energetic materials; lasers; and protection technologies. The institute is conjointly led by a French and German director which are appointed by their respective national ministries of defense, DGA and BMVg.

Many internationally renowned scientists have worked at ISL. Among these were the shock wave physicist *Prof. Herbert Oertel* (1918–2014), the former German directors and ballistics experts *Prof. Hubert Schardin*, Prof. *Richard Emil Kutterer* and *Rudi Schall* (1913–2002), and the French assistant director and detonics specialist *Claude Fauquignon* (1930–2009).

The current French director, since 2011, is *Christian de Villemagne* (*1963), the current German director, since 2014, is *Army General Dr. Ing. Thomas Czirwitzki* (*1958).

– Website: www.isl.eu/en/

Isophorone diisocyanate
Isophorondiisocyanat

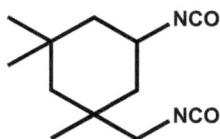

IPDI		
Aspect		Colorless to yellowish liquid with pungent odor
Formula		$C_{12}H_{18}N_2O_2$
GHS		06, 08, 09
H-phrases		331-334-315-319-317-335-411
P-phrases		273-285-302+352-305+351+338-309-310
UN		2290
REACH		LRS
EINECS		223-861-6
CAS		[4098-71-9]
m_r	g mol^{-1}	222.287
ρ	g cm^{-3}	1.061
$\Delta_f H$	kJ mol^{-1}	−372

$\Delta_c H$	kJ mol^{-1}	−6922.7
	kJ g^{-1}	−31.144
	kJ cm^{-3}	−33.043
Ω	wt.-%	−223.13
N	wt.-%	12.60
Mp	°C	−60
Bp	°C	158 bei P = 2 kPa

IPDI is used as a curing agent for hydroxyl-terminated polymers (e.g. *HTPB)* to produce polyurethanes.

– M. A. Daniel, Polyurethane Binder Systems for Polymer Bonded Explosives, *DSTO-GD-0492*, Weapons Systems Division Defence Science and Technology Organisation, Edinburgh South Australia, **2006**, 34 pp.

Isopropyl nitrate, IPN

Isopropylnitrat

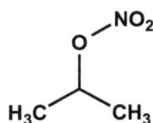

Formula		C$_3$H$_7$NO$_3$
REACH		LPRS
EINECS		216-983-6
CAS		[1712-64-7]
m_r	g mol^{-1}	105.093
ρ	g cm^{-3}	1.034
$\Delta_f H$	kJ mol^{-1}	−229.8
$\Delta_{ex} H$	kJ mol^{-1}	−607.3
	kJ g^{-1}	−5.779
	kJ cm^{-3}	−5.975
$\Delta_c H$	kJ mol^{-1}	−1951.2
	kJ g^{-1}	−18.566
	kJ cm^{-3}	−19.197
Ω	wt.-%	−98.96
N	wt.-%	13.32
Mp	°C	−82
Bp	°C	100
P_{vap}	kPa	3.42
V_D	m s^{-1}	5376 ms^{-1} at 1.34 g cm^{-3}
Impact	J	>50

IPN is frequently used as an explosive and fuel in slurry explosives in thermobaric charges (Table I.10).

Table I.10: Composition (wt.-%) and performance of thermobaric charges with IPN in comparison to PBXN-113.

Type	P_{CJ} (GPa)	E_{mech} (kJ cm^{-3})	V_D (km s^{-1})	P_{max} (kPa)	Impulse (kPa s)	T (°C)
30 IPN, 70 Al	4.40	3.83	3.22	268	586	567
30 IPN, 40 Al, 30 AN	11.81	10.21	5.55	468	620	927
30 IPN, 30 Al, 40 AP	15.84	11.46	6.06	496	710	488
30 IPN, 30 Al, 40 HMX	14.00	11.02	5.92	420	834	628
PBXN-113 (HMX/Al/HTPB: 35/45/20)	14.62	9.17	6.68	413	896	394

- S. Hall, G. Knowlton, Development, Characterisation and Testing of High Blast Thermobaric Compositions, *IPS*, Fort Collins, **2004**, pp. 663–678.
- F. Zhang, S. B. Murray, A. Yoshinaka, A. Higgins, Shock Initiation and detonabilty of Isopropyl Nitrate, *Det. Symp.*, **2002**, San Diego, pp. 781–790.

K

K-10

K-10, also known as Rowanite 8001, is a reddish limpid liquid. It is a blend from 2,4-dinitroethylbenzene and 2,4,6-trinitroethylbenzene in mass ratio 65/35. K-10 serves as plasticizer for propellants and high explosive formulations. K-10 does not fall under HD 1 but is HD 6.1 (toxic) material. Polymers treated with K-10 attain an intensive red-orange color. In greater concentrations (>20 wt.-%), K-10 impedes curing of hydroxyl-terminated polymers with isocyanates. K-10 is also incompatible with lead azide and other primary explosives.

- M. R. Andrews, S. E. Gaulter, J. Akhavan, P. Bolton, M. Till, FTIR Monitoring of Cure Rate and Effect of Plasticizer Content of a Series of Cross-linked Polynimmo -based systems, *ICT -Jata*, Karlsruhe, **2007**, P-112.
- A. Provatas, Energetic Polymers and Plasticisers for Explosive Formulations. A Review of Recent Advances, *DSTO-TR-0966*, Aeronautical and Maritime Research Laboratory, Melbourne, April **2000**, 51 pp.

Kast apparatus
Stauchungsapparat nach Kast

The Kast apparatus serves the comparative determination of the working capacity of solid high explosives. Therefore, a copper cylinder is compressed in a force-fit set up by the detonation of a cylindrical charge of the high explosive under investigation. The degree of compression correlates with the working capacity of the explosive. As different sized copper cylinders can be used in the test, the individual compression is translated by a procedure introduced by *Haid and Seile* into a standard compression number for a particular explosive and density. Figure K.1 shows a schematic depiction of the apparatus and Table K.1 displays the compression numbers for selected high explosives.

https://doi.org/10.1515/9783110660562-010

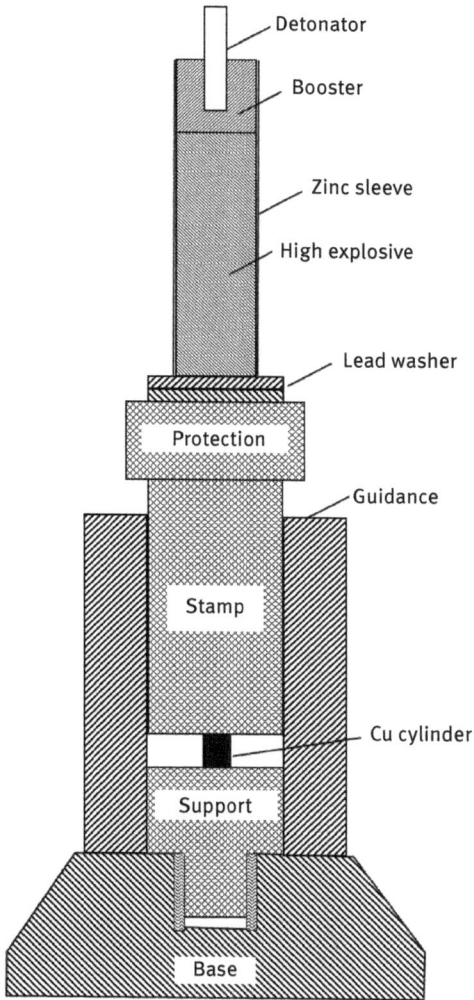

Figure K.1: Kast apparatus.

Table K.1: Compression values for various explosives.

Name	ρ (g cm^{-3})	Compression number (–)
1-Chloro-2,4,6-trinitrobenzene	1.66	7.53
Hexal (70/30)	1.81	15.5
Hexanitrodiphenylamine	1.32	7.38
Macarite (70/30)	2.75	5.80
Octal 70/30	1.81	20.76
Pentaerythritol tetranitrate	1.69	17.00

Table K.1 (continued)

Name	ρ (g cm^{-3})	Compression number (–)
Picric acid	1.68	12.20
Tetryl	1.53	9.72
1,3,5-Trinitrobenzene	1.60	7.60
Trinitrotoluene	1.59	10.1
Mercury fulminate	3.5	18

- P. Langen, P. Barth, Investigation of the Explosive Properties of HMX/Al 70/30, *Propellants Explos.* 1979, *4*, 129–131.
- H. Ahrens, International Study Group for the Standardization of the Methods of Testing Explosives, *Propellants Explosx.* **1977**, *2*, 7–20.
- A. Haid, H. Selle, Über die Sprengkraft und ihre Ermittlung, *Z. Sch. u. Spr.* **1934**, *29*, 11–14.

Keto-RDX
2,4,6-Trinitro-2,4,6-triazacyclohexanon

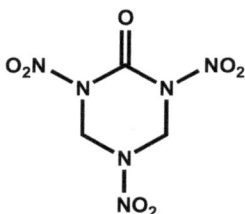

K-6, 2-Oxo-1,3,5-trinitro-1,3,5-triazacyclohexane		
Formula		$C_3H_4N_6O_7$
CAS		[115029-35-1]
m_r	g mol^{-1}	236.101
ρ	g cm^{-3}	1.932
$\Delta_f H$	kJ mol^{-1}	−41.84
$\Delta_{ex} H$	kJ mol^{-1}	−1360.9
	kJ g^{-1}	−5.764
	kJ cm^{-3}	−11.136
$\Delta_c H$	kJ mol^{-1}	−1710.2
	kJ g^{-1}	−7.244
	kJ cm^{-3}	−13.995
Λ	wt.-%	−6.78
N	wt.-%	35.60
Mp	°C	184
Dp	°C	205

Impact	J	3
Friction	N	42
V_D	m s^{-1}	8814 at 1.857 g cm^{-3}
P_{CJ}	GPa	37.98 at 1.857 g cm^{-3}

In the cylinder test, keto-RDX yields 4% more energy than HMX. While micrometric keto-RDX is more sensitive than RDX, nano-keto-RDX is about as sensitive as regular RDX, however, displaying a slightly reduced onset temperature for decomposition.

- A. R. Mitchel, P. F. Pagoria, C. L. Coon, E. S. Jessop, J. F. Poco, C. M. Tarver, R. D. Breithaupt, G. L. Moody, Nitroureas 1. Synthesis, Scale-up and Characterization of K-6, *Propellants Explos. Pyrotech.* **1994**, *19*, 232–239.
- Shokrolahi, A. Zali, A. M. Viazar, M. H. Keshavarz, H. Hajashemi, Preparation of Nano-K-6 (Nano-Keto RDX) and Determination of Its Characterization and Thermolysis, *J. Energ. Mater.* **2011**, *29*, 115–126.

KM smoke
KM-Nebel

The KM smoke (K = *Kalium* in German for potassium, M = magnesium) was developed in the mid-1980s by *Krone*. It is based on the work on fertilizer smokes which were initially intended as large area countermeasures against the acidification of forrestal soils and consecutive loss of nutrients causing then rapidly spreading forest decline in Germany (Table K.2).

Table K.2: Fertilizer smoke after *Krone*.

Magnesium, Mg (wt.-%)	20
Calcium carbonate, $CaCO_3$ (wt.-%)	50
Sodium chloride, NaCl (wt.-%)	15
Calcium hydroxide phosphate, $Ca_5(PO_4)_3(OH)$ (wt.-%)	5
Potassium perchlorate, $KClO_4$ (wt.-%)	10

The aerosol formed by this composition contains calcium carbonate, calcium hydroxide, various calcium phosphates, potassium chloride, magnesium oxide, and various magnesium phosphates, all of which show some extinction in the optical and near-infrared range. The actual KM smoke was first published about in 1988 (Table K.3). Due to the high relative deliquescence humidity (DRH) of its main aerosol constituents KCl (84%) and NaCl (75%), KM smoke shows a poor performance even compared to HC smoke (Table K.3).

Table K.3: Composition and performance of KM smoke.

Magnesium, Mg (wt.-%)	8
Potassium chloride, NaCl (wt.-%)	44
Potassium nitrate, KNO$_3$ (wt.-%)	27
Potassium perchlorate, KClO$_4$ (wt.-%)	5
Azodicarbonamide, C$_2$H$_4$N$_4$O$_2$ (wt.-%)	16
α_{VIS} (m^2 g^{-1}) at 50% RH	1.7
$\alpha_{1.5\ \mu m}$ (m^2 g^{-1})	0.2
Y$_f$ at 20% RH (–)	0.33

- U. Krone, W. Kühn, B. Georgi, A. Hüttermann, Pyrotechnically Generated Ca- and Mg-Aerosols, *IPS-Seminar*, **1984**, Colorado Springs, pp. 315–321.
- U. Krone, W. Kühn, B. Georgi, Entwicklung und Anwendung pyrotechnisch erzeugter Calcium – und Magnesiumaerosole, *IPS-Seminar & ICT Jata*, **1985**, Karlsruhe, Germany, pp. 36.
- U. Krone, *Pyrotechnische Mischung zur Erzeugung eines Tarnnebels und Anzündmischung hierfür*, DE-Patent 3728380, **1988**, Deutschland.
- U. Krone, A Non Toxic Pyrotechnic Screening Smoke for Training Purposes, *IPS-Seminar*, **1990**, Boulder Colorado, USA, pp. 581–586.

Koenen test, steel-sleeve test
Stahlhülsentest

The Koenen test was developed by *Koenen, Ide*, and *Däumler* in the 1950s to determine the explosiveness of a substance under investigation.

Table K.4: Possible reaction types in the Koenen test.

Typ	Effects	Result
O	Tube unchanged	No reaction
A	Bottom of tube bulged out	No reaction
B	Bottom and wall of the tube bulged out	No reaction
C	Bottom of tube split	No reaction
D	Wall of tube split	No reaction
E	Tube split into two fragments	No reaction
F	Tube fragmented into three or more mainly large pieces which in some cases may be connected with each other by a narrow strip	Explosion
G	Tube fragmented into mainly small pieces, closing device undamaged	Explosion
H	Tube fragmented into many very small pieces, closing device bulged out or fragmented	Explosion

The Koenen test is a mandatory test i.a.w. UN Test Series 1 (Test 1b) to determine whether a substance under investigation is an explosive or not. 10 g of the test substance are placed in the deep-drawn steel sleeve (25 $\varnothing$ × 0.5 × 75 mm) and the tube is closed with a nozzle plate that can have variable-sized orifice diameter ($\varnothing$ = 1–24 mm) and a closing nut. The tube then is positioned in a gimbal in a protective steel box and is heated by four propane burners from four sides until a reaction/explosion occurs (Tab. K. 4; Figure K.2). If no reaction occurs at 1 mm, the test is considered negative.

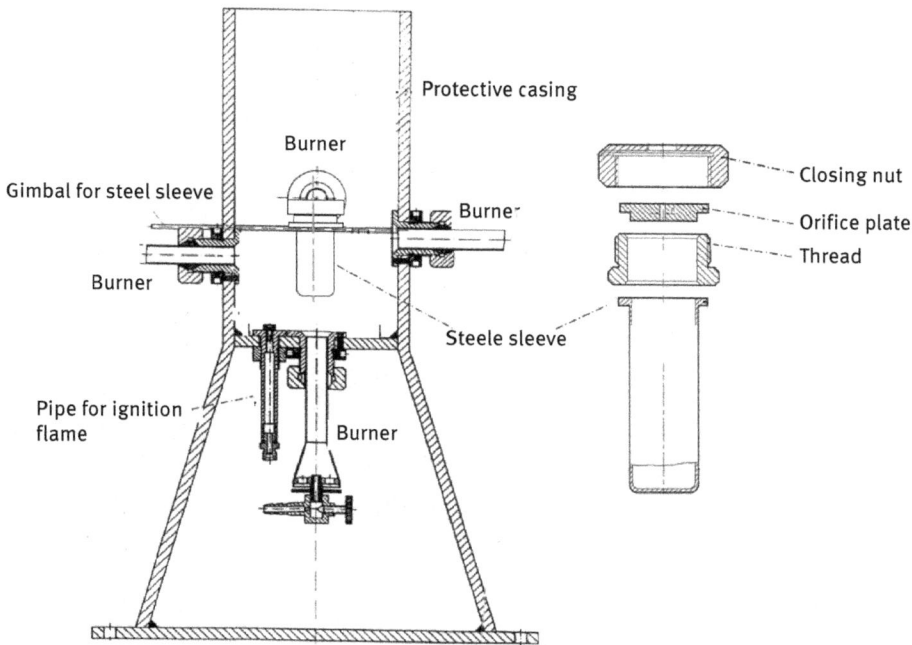

Figure K.2: (a) Koenen test setup (left); (b) steel sleeve for Koenen test (right).

- *Recommendations on the Transport of Dangerous Goods, Manual of Tests and Criteria,* 5th Edition, United Nations, Geneva, **2009**, pp. 33–38
- H. Koenen, K. H. Die, Über die Prüfung explosiver Stoffe, *Explosivstoffe*, **1956**, 4, 143–144.

Krone, Uwe (1938–2011)

Uwe Krone was a German chemist and pyrotechnic expert. After his chemistry studies at the Rheinische Friedrich-Wilhelms-Universität Bonn (1959–1967), he made his dissertation with J. Goerdeler in the field of organosulfur compounds. Having already had internships at NICO-Pyrotechnik in Trittau since 1960, he was hired as head of the chemical development department in 1967. In 1971, he developed the first pyrotechnic strobe flares and derived technical applications as distress signals. In the mid-1970s, he started working on VIS and IR obscurants and became internationally famous for the development of NT, KM, and the first real multispectral obscurant (VIS-IR-MMW) NG smoke.

- D. Cegiel, Uwe Krone 1938–2011, *Propellants Explos. Pyrotech.* **2012**, *37*, 7–8.
- U. Krone, Replacement of Toxic and Ecotoxic Components for Military Smokes for Screening in, P. C. Branco, H. Schubert, J. Campos, *Defense Industries, NATO -Science Series Volume 44*, **2001**, pp. 221–238.
- U. Krone, *Pyrotechnischer Satz zur Strahlungsemission*, DE2164437, **1971**, Germany.
- Entry in the German National Library: http://d-nb.info/482183659

Kutterer, Richard Emil (1904–2003)

Richard Emil Kutterer was a German physicist and ballistics expert. He made his dissertation with distinction under the guidance of *Carl Cranz* in 1935. Until 1945, he worked for the Heereswaffenamt Berlin (Ordnance Department of the Army) as an expert for small arms ballistics. In those years, he wrote a textbook on ballistics which saw two editions before 1945 and was significantly revised and enlarged as a third edition in 1959. After the war, he continued his scientific research at the *French-German Research Institute-ISL*. He was German director of ISL from 1965 until his retirement in 1969. Kutterer also taught aerospace technology at TU Karlsruhe. After his retirement at ISL, he became head of the ballistics research group at the *Ernst-Mach Institute* in Weil am Rhein.

- R. E. Kutterer, *Erdsatelliten- Ein Überblick*, Mittler, **1958**, 32 pp.
- R. E. Kutterer, *Ballistik*, 3. Aufl. Vieweg, **1959**, 304 pp.
- Entry in the German National Library: http://d-nb.info/369389905

L

Lactose
Milchzucker

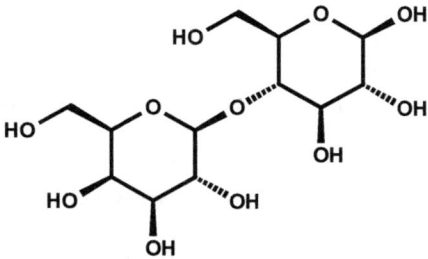

Formula		$C_{12}H_{22}O_{11}$
EINECS		200-559-2
CAS		[63-42-3]
m_r	g mol^{-1}	342,297
ρ	g cm^{-3}	1.59
$\Delta_f H$	kJ mol^{-1}	−2236.2
$\Delta_c H$	kJ mol^{-1}	−5630.4
	kJ g^{-1}	−16.449
	kJ cm^{-3}	−25.084
Mp	°C	202
T_p	°C	93.6 (α → β)
cp	J K^{-1} mol^{-1}	440
Λ	wt.-%	−112.18

The monohydrate $C_{12}H_{22}O_{11} \cdot H_2O$ CAS-No.: [64044-51-5] loses crystal water between 100 °C and 150 °C and strats to melt under decomposition at 202 °C. Lactose is an important fuel in colored-smoke compositions.

Laser initiation of explosives
Laserinitiierung von Explosivstoffen

In general, all energetic materials can be initiated or ignited by electromagnetic radiation. Östmark found the following relationships with the ignition energy. With short laser pulses, the ignitability is limited by the ignition energy density, ε_{crit}, with long duration pulses, the laser power, P_{crit}, is the limiting quantity

https://doi.org/10.1515/9783110660562-011

$$\varepsilon_{crit} = \rho \cdot c_p \cdot \alpha^{-1} \cdot (T_i - T_\infty)$$

$$P_{crit} = k \cdot \beta \cdot \sqrt{\pi} \cdot (T_i - T_\infty)$$

with ρ, density, c_p, specific heat, α, the optical absorption coefficient at the wavelength of the laser, T_i, ignition temperature, T_∞, ambient temperature, k, the heat conductivity, and b, spot radius.

As the earlier equations indicate, the thermal conductivity and optical absorption influence the ignition behavior of energetic materials. Laser ignition and initiation of energetic materials is used in optical detonators in emergency escape systems and *advanced reactive armour*.

- H. Östmark, Laser as a Tool in Sensitivity Testing of Explosives, *Detonation Symposium*, Albuquerque, **1985**, pp. 473–484.
- H. Östmark, Laser Ignition of Explosives Ignition Energy dependence of particle site, *GTPS*, Juan les Pins, **1987**, pp. 241–245.
- T. J. Blachowski, Status of Laser Initiation Efforts For Various Aircrew Escape System Applications, *IPS-Seminar*, Adelaide, **2001**, pp. 87–97.
- S. R. Ahmad, M. Cartwright, *Laser Ignition of Energetic Materials*, Wiley, **2015**, 283 pp.
- H. Scholles, R. Schirra, H. Zöllner, A Fast Low-Energy Optical Detonator, *IPS-Seminar*, Grand Junction, **2016**, pp. 422–428.

Applicable standards for laser initiation of explosives

NATO
- STANAG 4368 Electric and Laser ignition System for Rocket and Guided Missile Motors; Safety Design Requirements.
- STANAG 4560 Electro-Explosive Device, Assessment and Test Methods for Characterization.
- AOP 43 Electro-Explosive Devices Test Methods for Characterization Guidelines for STANAG 4560.

USA
- MIL-STD-1512 Electro-Explosive Subsystems, Electrically Initiated, Design Requirements and Test Methods.
- MIL-STD-1576 Electro-Explosive Subsystem Safety Requirements and Test Methods for space system.
- MIL-I-23659 Initiators, Electric, General Design Specification (MIL-DTL-23659).

LAX

LAX is the standard acronym preceding all explosive formulations that have been developed in the responsibility of Los Alamos National Laboratory.

Lead(II,IV) oxide
Mennige, Bleiorthoplumbat

Red lead		
Aspect		Orange-red powder
Formula		$[2PbO \cdot PbO_2] \equiv Pb_3O_4$
GHS		03, 08, 09, 07
H-phrases		272-360-373-400-410-302-332
P-phrases		260-261-281-304+340-405-501a
UN		1479
REACH		LRS, SVHC
EINECS		215-236-6
CAS		[1314-41-6]
m_r	g mol^{-1}	685.57
ρ	g cm^{-3}	9.05
Dp	°C	576
$\Delta_f H$	kJ mol^{-1}	−730.7
c_p	J K^{-1} mol^{-1}	147.2
Λ	wt.-%	+2.33

Pb_3O_4 is an oxidizer that was used frequently in gasless delay formulations. The thermal decomposition starting at 576 °C yields oxygen and PbO.

Lead azide
Bleiazid

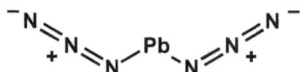

Aspect		Colorless to yellowish crystals
Formula		$Pb(N_3)_2$
REACH		LRS, SVHC
EINECS		236-542-1
CAS		[13424-46-9]
m_r	g mol^{-1}	291.24
ρ	g cm^{-3}	4.716
$\Delta_f H$	kJ mol^{-1}	+477
$\Delta_{ex} H$	kJ mol^{-1}	−477
	kJ g^{-1}	−1.638
	kJ cm^{-3}	−7.724
$\Delta_c H$	kJ mol^{-1}	−767

Λ	wt.-%	5.49
N	wt.-%	28.86
Dp	°C	350
Friction	N	0.3–0.5
Impact	J	0.5–4
V_D	m s^{-1}	3880 at 2.00 g cm^{-3}
P_{CJ}	GPa	1.7 at 2.00 g cm^{-3}
Trauzl	cm^3	110

$Pb(N_3)_2$ occurs in four distinct polymorphs, α, β, γ, and δ It is only the orthorhombic α-phase which has a favorable high density in the range between ρ = 4.68 and 4.716 g cm^{-3}. Additives of carbohydrates, for example, dextrin, favor the crystallization of the α-phase. Lead azide is practically insoluble in cold water and most common organic solvents. However, it dissolves in acetic acid and ethanolamine. As lead compounds in general are highly objectionable due to their toxicity, carcinogenity, and teratogenity, they have to be replaced in the foreseeable future, making lead azide obsolete.

- R. Matyas, J. Pachman, *Primary Explosives*, Springer, **2013**, pp. 72–88.

Lead block test, Trauzl test
Bleiblocktest

Lead block test was invented in 1884 by the Austrian chemist and businessman *Dr. Isidor Trauzl* (1840–1929). The Trauzl test serves the comparative determination of the working capacity of a detonating substance. The solid lead block for the test weights about 70 kg. It is cylindrical in shape (∅ = 20 cm, h = 20 cm) and bears a cylindrical cavity (∅ = 2.5 cm and 12.5 cm depth) machined concentrically into one face of the cylinder (Figure L.1). The volume of cavity before the test is 61.3 cm^3.

10 g of the explosive under investigation are wrapped in a trapezoidal piece of tin foil (150 × 70 × 130 mm) together with a standard No. 8 electric detonator. With the aid of a wooden slat, the sample is mildly pushed to fit at the bottom of the cavity. The remainder volume is filled with silica sand as stemming. In the case of liquid explosives, glycerin is used as stemming. The cavity after the test (*a*) is filled with a liquid and the determined volume minus the pre-test volume is the net expansion produced by the explosive. This determination is referred to a standard temperature of 15 °C and is compared with other explosives (Table L.1). The lead block test nowadays is only used with less brisant explosives. The reason for this is that with highly brisant explosives, in addition to the central expansion rarefactions, yield additional but closed cavities along the diagonal (*b*) which cannot be filled with liquid and hence lead to unrealistic low values (compare e.g. RDX and *HMX*).

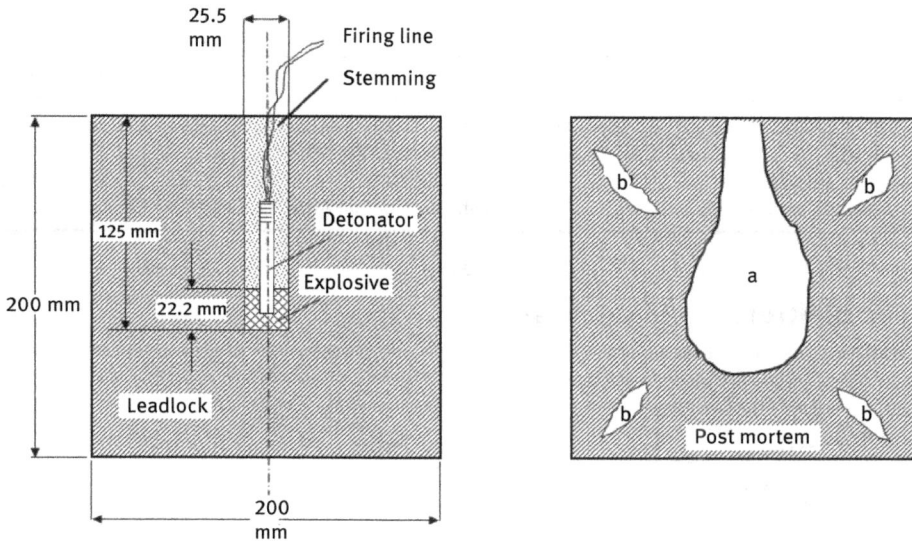

Figure L.1: Lead block test after *Trauzl*.

- A. Schmidt, Betrachtungen über die üblichen Methoden zur experimentellen Prüfung der Sprengwirkung von Explosivstoffen und ihre Bedeutung für die Beurteilung des Verhaltens der Explosivstoffe bei ihrer praktischen Anwendung, *Explosivstoffe*, **1959**, *7*, 225–231.
- H. Koenen, K. H. Die, K.-H. Swart, Sicherheitstechnische Kenndaten explosionsfähiger Stoffe, I. Mitteilung, *Explosivstoffe*, **1961**, *9*, 30–42.

Table L.1: Experimental net volume of various explosives.

Explosive	Expansion (mL)	Explosive	Expansion (mL)
Amatol 80/20	385	Guanylurea dinitramide	345–375
Ammonium nitrate	178	Urea nitrate	272
Ammonium perchlorate	194	Hexanitrostilbene	301
Ammonium picrate	280	Hexogen RDX	480
Lead azide	110	Methyl nitrate	610
Lead styphnate	130	Nitroguanidine	302
Composition A3	432	Nitroglycerine	520
Composition B	390	Nitrourea	310
Cyanuric triazide	415	Nitromethane	345
Diethylenglycol dinitrate	620	Nitropenta	520
Dinitrobenzene	242	Octogen HMX	428
Dinitrotoluene	240	PH-Salz	345–375
Dinitroxyethylnitramine	445	Picric acid	315

Table L.1 (continued)

Explosive	Expansion (mL)	Explosive	Expansion (mL)
Ethylendinitramine	366–410	R-Salz	370
Ethyltetryl	327	Triaminotrinitrobenzene	175
Guanidinium nitrate	240	Trinitrotoluene	300
Guandinium perchlorate	400	Flash composition Al/KClO$_4$ (20/80)	115

Lead dinitroresorcinate, basic

Bleidinitroresorcinat, basisch

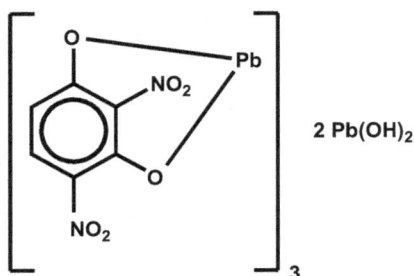

Tris[4,6-dinitro-1,3-benzenediolato(2-)]tetrahydroxypenta lead(II)

Aspect		Orange red
Formula		C$_{18}$H$_{10}$N$_6$O$_{22}$Pb$_5$
CAS		[55012-91-4]
m$_r$	g mol^{-1}	1638.26
ρ	g cm^{-3}	3.65
Mp		213
DSC onset	°C	185
Λ	wt.-%	−22.61
N	wt.-%	5.12
Impact	J	6

Lead dinitroresorcinate is used for detonative and pyrotechnic priming formulations and fast-burning delays.

- G. Taylor, A. Thomas, R. Williams, *Lead Compounds of 4,6-Dinitroresorcinole*, US3803190A, **1974**, UK.

Lead dioxide

Blei(IV)oxid

Aspect		Dark brown powder
Formula		PbO_2
GHS		08, 09, 07
H-phrases		360Df-373-400-410-302-332
P-phrases		260-261-281-304+340-405-501a
UN		1872
REACH		LRS, SVHC
EINECS		215-174-5
CAS		[1309-60-0]
m_r	g mol^{-1}	239.19
ρ	g cm^{-3}	9.643
Mp	°C	290 (Z)
$\Delta_f H$	kJ mol^{-1}	−282.8
c_p	J K^{-1} mol^{-1}	64.5
Λ	wt.-%	+6.69

PbO_2 has been used intensively as oxidizer in first fires and was a minor component in safety match head formulations. Its thermal decomposition starts at 290 °C and yields $2PbO \cdot PbO_2$ and oxygen. The mixed oxide decomposes at 576 °C to give PbO and more oxygen.

Lead nitrate

Bleinitrat

Formula		$Pb(NO_3)_2$
GHS		03, 08, 09, 07
H-phrases		272-360-373-400-410-302-332
P-phrases		210-220-221-260-405-501a
UN		1469
REACH		LRS, SVHC
EINECS		233-245-9
CAS		[10099-74-8]
m_r	g mol^{-1}	331.21
ρ	g cm^{-3}	4.535
Mp	°C	470 (dec)
$\Delta_f H$	kJ mol^{-1}	−451.87
Λ	wt.-%	+24.15
N	wt.-%	8.46

Lead nitrate decomposes at 350 °C to give lead(II) oxide, oxygen, and nitrogen dioxide in accordance with

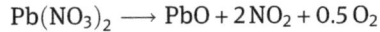

$$Pb(NO_3)_2 \longrightarrow PbO + 2NO_2 + 0.5O_2$$

Lead nitrate was used in igniter formulations. It has been suggested as the only oxidizer in a high-density propellant with aluminum. Allegedly lead nitrate is also used as a component in high-density explosives for explosive lenses in nuclear weapons (Table L.2).

Table L.2: High explosive formulations containing $Pb(NO_3)_2$.

Components (wt.-%)	Plumbatol, macarite	B-2174	B-2191	B-2192
Lead nitrate	70	11	11	11
TNT	30	–	–	–
Octogen	–	47	37	27
Ammonium perchlorate	–	30	40	50
Polyurethane	–	12	12	12
ρ (g cm^{-3})	2.89	1.83	1.83	1.836
V_D (m s^{-1})	4860	8540	–	–
Impact (J)	6.6	–	–	–

- J. Souletis, J. Groux, Continuous Observation of Mach Bridge and Mach Phenomena, *8th Detonation Symposium*, **1985**, NSWC White Oak, pp. 431–438.
- F. A. Marion, H. J. McSpadden, *Rocket containing lead oxidizer salt-high density propellant*, US3945202, **1973**, USA.
- H. H. M. Pike, R. E. Weir, The Passage of a Detonation Wave across the Interface between two Explosives, Report 22/50, Armament Research Establishment, Fort Halstead, UK, **1950**, 33 pp, declassified in 2008.

Lead styphnate

Bleistyphnat, Bleitricinat

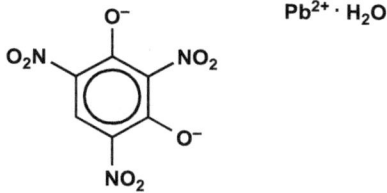

$$Pb^{2+} \cdot H_2O$$

Aspect		Gold orange – reddish brown
Formula		$C_6H_3O_9N_3Pb$
REACH		SVHC, LRS
EINECS		239-290-0
CAS		[15245-44-0]
m_r	g mol^{-1}	468.305
ρ	g cm^{-3}	3.02
$\Delta_f H$	kJ mol^{-1}	−837
$\Delta_c H$	kJ mol^{-1}	−2176
	kJ g^{-1}	−4.647
	kJ cm^{-3}	−14.033
Λ	wt.-%	−18.79
N	wt.-%	8.97
c_p	J K^{-1} g^{-1}	0.677
Dp	°C	252
V_D	m s^{-1}	5600 at 3.10 g cm
P_{CJ}	GPa	8.31 at 3.14 g cm^{-3}
Trauzl	cm^3	130

Lead styphnate is very sensitive to initiation by flames and hence is often used as thermal sensitizer in primary formulations used directly on bridge wires.

Leiber, Carl Otto (1934–2016)

Carl Otto Leiber was a German physicist and detonics expert. After his dissertation at the Technical University Stuttgart in 1963 and some years in the industry, he started working on detonics and explosives with the newly established CTI in 1971 (since 1976 called BICT and since 1997 called WIWEB) in Swisttal-Heimerzheim. His theory of reactive multiphase flow explaining both common HVD but also LVD gained him international recognition and made him a frequently sought-after expert on explosive accidents and insensitive munitions.

- C. O. Leiber, Sensibilisierung und Phlegmatisierung von Sprengstoffen, *Rheol. Acta*, **1975**, *14*, 85–91.
- C. O. Leiber, Die Detonation als Mehrphasenströmung, *Rheol. Acta*, **1975**, *14*, 92–100.
- C. O. Leiber, Aspects of the Mesoscale of Crystalline Explosives, *Propellants Explos. Pyrotech.* **2000**, *25*, 288–301.
- C. O. Leiber, *Assessment of Safety and Risk with a Microscopic Model of Detonation*, Elsevier, **2004**, 594 pp.
- A. S. Cumming, R. Wild, Carl-Otto Leiber, *Propellants Explos. Pyrotech.* **2016**, *41*, 603–604.
- Entry in the German National Library: http://d-nb.info/481964231

Liquid explosives
Flüssige Sprengstoffe

Liquid explosives can be either chemically uniform substances (the extremely impact sensitive organic nitrates: methyl nitrate, isopropyl nitrate, nitroglycerine, diethylene glycol, etc. or nitroaliphatics as the very insensitive nitromethane which needs to be sensitized) or mixtures of liquid oxidizers with nitrated and non-nitrated fuels, which is a virtually unlimited group of combinations with examples as given in Table L.3.

Table L.3: Liquid high explosives.

Name	Formulation	ρ (g cm^{-3})	V_D (m s^{-1})
Astrolite A-1-5	N_2O_4/hydrazine derivates	1.60	8600
		1.41	7500
Hellofit	HNO_3/ dinitrotoluene	1.45	7350
Nisalit	HNO_3/acetonitrile	1.24	6230
	hno_3/nitrobenzene	1.41	7500
Panclastit	N_2O_4/benzine	1.29	7200
	Tetranitromethane/toluene (86.5/13.5 wt.-%)	1.45	7100
	Tetranitromethane/borazine (92/8 wt.-%)	1.53	6700

Emulsion explosives actually can be also considered liquid explosives. They consist of mineral oil as the continuous phase, dispersed therein an aqueous ammonium nitrate solution and microballoons for sensitization. See also *WS-6D*.

Lithium

Aspect		Silvery
Formula		Li
GHS		02, 05
H-phrases		260-314-014
P-phrases		231-260-303+361+353-305+351+338-405a-501a+232
UN		1415
EINECS		231-102-5
CAS		[7439-93-2]
m_r	g mol^{-1}	6.941
ρ	g cm^{-3}	0.534
Mp	°C	180.54
Bp	°C	1347
c_p	J g^{-1} K^{-1}	2.99
$\Delta_m H$	J g^{-1}	4.93
$\Delta_v H$	kJ g^{-1}	147.7
c_L	m s^{-1}	6030
$\Delta_c H$	kJ mol^{-1}	298.95
	kJ g^{-1}	−43.070
	kJ cm^{-3}	−23.000
Λ	wt.-%	−39.92

Silvery white lithium in dry air quickly forms a golden oxide layer protecting it from further oxidation. However, in moist air, nitridation is facilitated and a combined porous coating from lithium nitride (Li_3N), lithium hydroxide monohydrate (LiOH · H_2O), and lithium carbonate (Li_2CO_3) forms, which allows further attack and degradation. Powdered lithium ignites depending on the particle size between T = 400–680 °C and burns with the formation of a dense white aerosol. Under dry nitrogen, ignition is observed above T > 800 °C. However, the presence of moisture can lower the ignition temperature as it catalyzes nitride formation. Since the 1950s, lithium is investigated as a possible energetic fuel in solid rocket propellants and hybrid propulsion systems. In addition, lithium is a fuel in *stored chemical energy propulsion system* (SCEPS) for naval applications with gaseous fluorine but more conveniently with inert fluorine compounds such as SF_6, C_4F_8, and $C_{11}F_{20}$.

- E.-C. Koch, Special Materials in Pyrotechnics III. Application of Lithium and its Compounds in Energetic Systems, *Propellants Explos. Pyrotech.* **2004** *29*, 67–80.

Lithium aluminum hydride
Lithiumaluminumhydrid

Lithiumalanat		
Formula		$LiAlH_4$
GHS		02, 05
H-phrases		224-260-314-019
P-phrases		210-231+232-280-305+351+338-403+233-501a
UN		3399
EINECS		240-877-9
CAS		[16853-85-3]
m_r	g mol^{-1}	37.954
ρ	g cm^{-3}	0.917
Mp	°C	125 (dec)
$\Delta_f H$	kJ mol^{-1}	−116.32
$\Delta_c H$	kJ mol^{-1}	−1592.1
	kJ g^{-1}	−41.948
	kJ cm^{-3}	−38.467
Λ	wt.-%	−168.62

Dry stable powder when handled in dry air. $LiAlH_4$ reacts hypergolically with water and decomposes in dry air above 150 °C in accordance with

$$3\,LiAlH_4 \longrightarrow Li_3AlH_6 + 2\,Al + 3\,H_2$$

$LiAlH_4$ is good soluble in diethyl ether S = 300 g·kg^{-1}. $LiAlH_4$ has been suggested as fuel in hybrid propulsion engines.

Lithium borohydride
Lithiumborhydrid

Lithium boranate		
Formula		$LiBH_4$
GHS		02, 06, 05
H-phrases		260-301-311-331-314-014
P-phrases		280-303+361+353-305+351+338-310-370+378i-402+404
UN		1413
EINECS		241-021-7
CAS		[16949-15-8]
m_r	g mol^{-1}	21.783
ρ	g cm^{-3}	0.66
Mp	°C	284
Dp	°C	380
$\Delta_f H$	kJ mol^{-1}	−190.5
$\Delta_c H$	kJ mol^{-1}	−1316.6

	kJ g^{-1}	−60.441
	kJ cm^{-3}	−39.891
Λ	wt.-%	−293.80

Lithium boranate reacts with water. In contrary to LiAlH$_4$, it is only slightly soluble in diethylether 25 g L^{-1}. Its decomposition proceeds at T > 380 °C in accordance with

$$2\,LiBH_4 \longrightarrow 2\,LiH + B_2H_6 \uparrow$$

It has been suggested as an energetic fuel in solid rocket propellants. Being prone to hydrolysis, it has been proposed to coat LiBH$_4$ with nickel or aluminum as a protection against moisture. Thus, coated LiBH$_4$ can be blended with AP and binder to yield stable propellants.

– W. C. Jenkin, *Method of encapsulation of lithium borohydride*, US 3.070.469, **1962**, USA.

Lithium chlorate
Lithiumchlorat

Formula		LiClO$_3$
REACH		LPRS
EINECS		236-632-0
CAS		[13453-71-9]
m$_r$	g mol^{-1}	90.390
ρ	g cm^{-3}	1.119
Mp	°C	129
Dp	°C	367
T$_p$	°C	41
T$_p$	°C	99
Δ$_f$H	kJ mol^{-1}	−293
Λ	wt.-%	+53.1

LiClO$_3$ is strongly hygroscopic and forms long, fine felt-type needles. It is a possible ingredient for oxygen generation in accordance with

$$LiClO_3 \longrightarrow LiCl + 3/2\,O_2$$

However, the alkaline reaction of the thermal decomposition products indicates a decomposition in accordance with

$$4\,LiClO_3 \longrightarrow 2\,Li_2O + 2\,Cl_2 + 5\,O_2$$

– C. Cohrt, *Combustion independent from ambient air*, US 4663933, **1985**, Germany.

Lithium dinitramide

Lithiumdinitramid

Formula		$LiN(NO_2)_2$
CAS		[154962-43-3]
m_r	g mol^{-1}	112.959
ρ	g cm^{-3}	2.175 g$\cdot$cm^{-3}
Dp	°C	398
$\Delta_f H$	kJ mol^{-1}	−220
Λ	wt.-%	+49.57

Lithium dinitramide (LiDN) is a strongly hygroscopic salt which rapidly forms the monohydrate $LiN(NO_2)_2 \cdot H_2O$, losing its crystal water at T = 63 °C. At T = 125 °C, LiDN decomposes into N_2O and $LiNO_3$. Hot stage microscopy reveals a darkening of LiDN at 125 °C indicating that the melting observed at 157 °C is surely not pure LiDN but a mixture of LiDN and some decomposition product or a eutectic with $LiNO_3$:

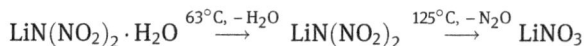

$$LiN(NO_2)_2 \cdot H_2O \xrightarrow{63°C, -H_2O} LiN(NO_2)_2 \xrightarrow{125°C, -N_2O} LiNO_3$$

So far no technical uses for LiDN have been suggested.

Lithium hydride

Lithiumhydrid

Formula		LiH
CAS		[7580-67-8]
GHS		02, 06, 05
H-phrases		260-331-314-014
P-phrases		280-303+361+353-305+351+338-321-405-501a
UN		1414
EINECS		231-484-3
m_r	g mol^{-1}	7.95
ρ	g cm^{-3}	0.820
Mp	°C	686.5
Dp	°C	972
$\Delta_f H$	kJ mol^{-1}	−91.230
$\Delta_c H$	kJ mol^{-1}	−350.9
	kJ g^{-1}	−44.140
	kJ cm^{-3}	−36.195
Λ	wt.-%	−201.28

Usually white but due to the presence of elemental Li grayish powder. Stable in dry air but hydrolyzing in contact with moisture according to

$$LiH + H_2O \longrightarrow LiOH + H_2 \uparrow$$

Lithium hydride is nicely soluble in many organic solvents, for example, diethyl ether. As molten LiH has a significant H_2-partial pressure ($p(H_2)_{686\ °C} = 27$ mbar), it is not possible to process it by melt casting but it needs to be consolidated mechanically. LiH has been suggested as fuel for hybrid propulsion systems (*lithergolic* systems) for *SCEPS* and for use as acoustic underwater countermeasures.

Lithium nitrate
Lithiumnitrat

Formula		$LiNO_3$
CAS		[7790-69-4]
GHS		03, 07
H-phrases		272-315-319-335
P-phrases		210-221-302+352-305+351+338-405-501a
UN		2722
EINECS		232-218-9
m_r	g mol^{-1}	68.946
ρ	g cm^{-3}	2.38
Mp	°C	264
Bp	°C	600
T_p	°C	−4.5
Dp	°C	474 °C
$\Delta_f H$	kJ mol^{-1}	−483.13
Λ	wt.-%	+58.01%
DRH	% RH	13 at 20 °C

Lithium nitrate ($LiNO_3$) forms limpid, strongly hygroscopic rhomboedric crystals which upon contact with moisture yield the trihydrate ($LiNO_3 \cdot 3H_2O$) [13453-76-4] ($-2.5\ H_2O$ 30 °C, $- 3\ H_2O$ 61 °C, $\rho = 1.55$ g$\cdot$cm^{-3}, $\Delta Hf_{298.15} = -1375$ kJ$\cdot$mol^{-1}). Its thermal decomposition proceeds as follows:

$$2\,LiNO_3 \longrightarrow 2\,LiNO_2 + O_2$$

$$2\,LiNO_3 \longrightarrow Li_2O + N_2 + 3/2\,O_2$$

The melting points of various binary and ternary eutectics of LiNO3 with the nitrates of Na, K, Rb, Cs, NH_4, Ag, Tl, urea, and Ca were given by *Ellern*. It has been observed that the impact sensitiveness of gun propellants based on RDX/HTPB can de reduced by addition of small amounts (0.3–5 wt.-%) $LiNO_3$. Eutectics from $LiNO_3$

with other alkaline and alkaline earth metal nitrates have been suggested as melt-cast bases for illuminants. $LiNO_3$ can be used for low-temperature autoignition compositions. Due to its unmatched high gravimetric oxygen content, lithium nitrate has often been considered as an oxidizer in solid rocket propellants. However, its low DRH complicates processing. *Douda* investigated the emission behavior of a $Mg/LiNO_3$ flare composition. From his work, we know that $LiNO_3$ does not work as a drop-in replacement for strontium nitrate in red flare formulations, though other books (*Meyer, Koehler, Homburg* and *Greenwood & Earnshaw*) assert its usefulness. In hot flames (Mg combustion), Li yields an overall orange flame due to excitation of Li minor series ($\lambda = 460$ nm), only in relatively cool flames with low or no Mg content at all and in the presence of high nitrogen fuels such as hexamine or 5-AT saturated red flames with long dominant wavelength can be obtained, however, at the significant expense of light intensity (*Klapötke*).

- H. Ellern, *Military and Civilian Pyrotechnics*, Chemical Publishing Company, New York, **1968**, 271.
- B. E. Douda, R. M. Blunt, E. J. Bair, Visible Radiation from Illuminating-Flare Flames: Strong Emission features, *J. Opt. Soc. Am.* **1970**, *60*, 1116–1119.
- H. W. Kruse, J. J. Bujak, Investigation of Eutectic Oxidizers for Use in Illumination Devices, *NWC TP 5395*, NWC, China Lake, CA, USA, **1973**, 22 pp.
- B. Martin, A. Lefumeux, *Ignition-sensitive low-vulnerability propellant powder*, USP 5.468.312, **1995**, France.
- E.-C. Koch, Special Materials in Pyrotechnics III. Application of Lithium and its Compounds in Energetic Systems, *Propellants Explos. Pyrotech.* **2004** *29*, 67–80.
- E.-C. Koch, C.-J. White, Is it possible to obtain a deep Red Pyrotechnic Flame Based on Lithium? *IPS-Seminar*, Rotterdam, **2009**, pp. 105–110.
- R. Meyer, J. Köhler, A. Homburg, *Explosives*, 7th Ed. Wiley, Weinheim, **2016**, pp. 210.
- N. N. Greenwood, A. Earnshaw, *Chemistry of the Elements*, Elsevier, New York, **1997**, pp. 89
- J. Glück, T. M. Klapötke, J. J. Sabatini, Flare or strobe: A tunable chlorine-free pyrotechnic system based on lithium nitrate, *Chemical Communications*, **2018**, *54*, 821–824.

Lithium perchlorate
Lithiumperchlorat

Formula		$LiClO_4$
GHS		03, 07
H-phrases		270-302-315-319-335
P-phrases		220-302+352-305+351+338-321-404-501a
UN		1481
EINECS		232-237-2
CAS		[7791-03-9]
m_r	g mol^{-1}	106.392

ρ	g cm^{-3}	2.428
Mp	°C	236
Dp	°C	438
T_p	°C	−19
Λ	wt.-%	+60.15
$\Delta_f H$	kJ mol^{-1}	381.00
DRH	% RH	70

$LiClO_4$ is a colorless, moderately deliquescent substance. Contrary to all other alkali metal perchlorates, it does not show any phase transitions before melting, underlining its similarity with the alkaline earth metal perchlorates. It crystallizes from a saturated solution in water as trihydrate $LiClO_4 \cdot 3\,H_2O$ [13453-78-6] (ρ: 1.841 g·cm^{-3}, −H_2O 98 °C, −H_2O 130−150 °C, $\Delta_f H$ = −1298 kJ·mol^{-1}). Its decomposition starts at 438 °C, and reaches a maximum at 470 °C in accordance with

$$LiClO_4 \longrightarrow LiCl + 2\,O_2$$

There are applications for $LiClO_4$ in fireworks: *Jennings–White* suggests both a pink lance (ξ($LiClO_4$) ~ 0.75) and strobe (ξ($LiClO_4$) ~ 0.5) from $LiClO_4$/hexamine($C_6H_{12}N_4$). $LiClO_4$/zirconium acetylacetonate yields pink flame *(Koch)*. Due to its hygroscopicity, $LiClO_4$ is not frequently used in technical applications. Together with LiO_2, $LiClO_4$ finds use in oxygen-generating compositions (→ *SCOG*).

– C. Jennings-White, Some Esoteric Firework Materials, *Pyrotechnica* **1990**, *XIII*, 26–32 + Plate VIII.
– E.-C. Koch, Evaluation of Lithium Compounds as Color Agents for Pyrotechnic Flames, *J. Pyrotech.* **2001**, *13*, 1–8.
– E.-C. Koch, Special Materials in Pyrotechnics III. Application of Lithium and its Compounds in Energetic Systems, *Propellants Explos. Pyrotech.* **2004** *29*, 67–80.

Lithium peroxide
Lithiumperoxid

Formula	Li_2O_2
GHS	03, 05
H-phrases	272-314
P-phrases	210-221-303+361+353-305+351+338-405-501a
UN	1472
EINECS	234-758-0
CAS	[12031-80-0]

m_r	g mol^{-1}	45.881
ρ	g cm^{-3}	2.30
Dp	°C	195
$\Delta_f H$	kJ mol^{-1}	−634.29
Λ	wt.-%	+34.87

Lithium peroxide is a colorless, crystalline non-hygroscopic powder, soluble in water but insoluble in ethanol. Li_2O_2 is used as a catalyst for the decomposition of sodium chlorate in SCOGs.

LLM-116, 4-amino-3,5-dinitropyrazole

4-Amino-3,5-dinitropyrazol

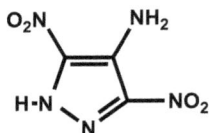

ADNP		
Aspect		Orange-yellow crystals
Formula		$C_3H_3N_5O_4$
CAS		[152678-73-4]
m_r	g mol^{-1}	173.088
ρ	g cm^{-3}	1.90
$\Delta_f H$	kJ mol^{-1}	−0.84
$\Delta_{ex}H$	kJ mol^{-1}	−820.3
	kJ g^{-1}	−4.739
	kJ cm^{-3}	−9.004
$\Delta_c H$	kJ mol^{-1}	−1609.3
	kJ g^{-1}	−9.298
	kJ cm^{-3}	−17,666
Λ	wt.-%	−32.35
N	wt.-%	40.46
Mp	°C	173.4 (dec)
Friction	N	>360
Impact	J	33.6 (50%)
V_D	m s^{-1}	8130 at 1.722 g cm^{-3} (LLM-116 /PIB 95/5)
P_{CJ}	GPa	31.4

LLM-116 is a very insensitive high explosive. Its adiabatic self-heating starts at T > 170 °C.

– R. D. Schmidt, G. S. Lee, P. F. Pagoria, A. R. Mitchell, R. Gilardi, Synthesis and Properties of a New Explosive, 4- Amino-3,5-dinitro-1HPyrazole, *UCRL-ID-148510*, LLNL, USA, **2001**.
– N. V. Muravyev, A. A. Bragin, K. A. Monogarov, A. S. Nikiforova, A. A. Korlyukov, I. V. Fomenkov, N. I. Shishov, A. N. Pivkina, 5-Amino-3,4-dinitropyrazole as a Promising Energetic Material, *Propellants Explos. Pyrotech.* **2016**, *41*, 999–1005.

LLM-175, 4-amino-4′-nitro-[3,3′,4′,3″]terfurazan

4″-Nitro[3,3′4′,3″-ter-1,2,5-oxadiazol]-4-amin, ANFF-1

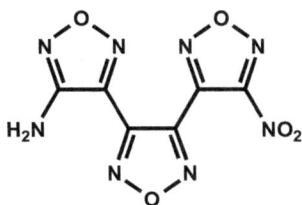

Formula		$C_6H_2N_8O_5$
CAS		[1613036-12-6]
m_r	g mol^{-1}	266.13
ρ	g cm^{-3}	1.782
Δ_fH	kJ mol^{-1}	667
$\Delta_{ex}H$	kJ mol^{-1}	−1490.3
	kJ g^{-1}	−5.600
	kJ cm^{-3}	−9.979
Δ_cH	kJ mol^{-1}	−3314.0
	kJ g^{-1}	−12.452
	kJ cm^{-3}	−22.190
Λ	wt.-%	−48.09
N	wt.-%	42.11
Mp	°C	99.4
Dp	°C	234
Friction	N	240
Impact	J	>35
V_D	m s^{-1}	7729 at 1.65 g cm^{-3} (LLM-175/PIB 95/5)
P_{CJ}	GPa	31.4

– A. DeHope, M. Zhang, K. T. Lorenz, E. Lee, D. Parrish, P. F. Pagoria, Synthesis and characterization of multicyclic oxadiazoles and 1-hydroxytetrazoles as energetic materials, *Chem. Heterocycl. Compd.* **2017**, *53*, 760–778.

LLM-201, 3-(4-nitro-1,2,5-oxadiazol-3-yl)-1,2,4-oxadiazol-5-amine

3-(4-Nitro-1,2,5-oxadiazol-3-yl)-1,2,4-oxadiazol-5-amin

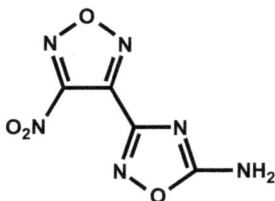

Formula		$C_4H_2N_6O_4$
CAS		[2130854-34-9]
m_r	g mol^{-1}	198.096
ρ	g cm^{-3}	1.736
$\Delta_f H$	kJ mol^{-1}	193.72
$\Delta_{ex}H$	kJ mol^{-1}	−928.7
	kJ g^{-1}	−4.688
	kJ cm^{-3}	−8.139
$\Delta_c H$	kJ mol^{-1}	−2053.9
	kJ g^{-1}	−10.368
	kJ cm^{-3}	−17.999
Λ	wt.-%	−40.38
N	wt.-%	42.42
Mp	°C	100.53
Dp	°C	261.2
Friction	N	>360
Impact	J	>35
V_D	m s^{-1}	7757 at 1.64 g cm^{-3} (LLM-201/Estane 95/5)

LLM-201 is an experimental high explosive that is currently under consideration as an insensitive melt-cast base in formulations like LH-55 (LLM-201/HMX: 50/50) (Table L.4).

- A. DeHope, M. Zhang, K. T. Lorenz, E. Lee, D. Parrish, P. F. Pagoria, Synthesis and characterization of multicyclic oxadiazoles and 1-hydroxytetrazoles as energetic materials, *Chem. Heterocycl. Compd.* **2017**, *53*, 760–778.
- P. Leonard, E.-G. Francois, *Final Report for SERDP WP-2209 Replacement melt-castable formulations for Composition B, LA-UR-12-24143*, Los Alamos National Laboratory, 19 May **2017**.

Table L.4: Performance and sensitiveness of LH-55.

Parameter	LH-55
V_D (m s^{-1})	8250
P_{CJ} (GPa)	29
ρ_{exp} (g cm^{-3})	1.75
$\varnothing_{cr}$ (mm)	<12
Friction (N)	>355
Impact (J)	12.5

LX

LX is the acronym for explosive formulations developed in the responsibility of Lawrence Livermore National Laboratory. Selected formulations are given in Appendix A.

M

Magnesium

Aspect		Silvery
Formula		Mg
GHS		02
H-phrases		228-251-261
P-phrases		210-231+232-241-280-420-501a
UN		1418
EINECS		231-104-6
CAS		[7439-95-4]
m_r	g mol^{-1}	24.305
ρ	g cm^{-3}	1.738
Mp	°C	648.8
Bp	°C	1107
$\Delta_m H$	kJ mol^{-1}	9.04
$\Delta_v H$	kJ mol^{-1}	127.6
c_p	J g^{-1} K^{-1}	0.855
c_L	m s^{-1}	5700
κ	W m^{-1} K^{-1}	171
T_i	K	1080
$\Delta_c H$	kJ mol^{-1}	−602
	kJ g^{-1}	−24.769
	kJ cm^{-3}	−43.048
Λ	wt.-%	−65.83

In moist air, magnesium quickly tarnishes to develop a complex coating as is schematically depicted in Figure M.1.

Mg can be ignited, with cerium sparks, safety matches, or propane torch, both as powder or band. Its first use in pyrotechnics was reported in 1865. With the large-scale use of magnesium as a light-weight alloy component in aircraft industry Mg became available at reasonable cost in the 1930s.

The combustion of magnesium is dependent on the particle size (surface area) and proceeds mainly in the gas phase for large surface area material. However, Mg band and coarse particles also show condensed phase combustion. Its superior thermochemical and physical properties (low heat of vaporization and high heat of oxidation/fluorination) make it the ideal energetic fuel in all kinds of pyrotechnic flare formulations, obscurants, and RAM solid propellants. Mg is not compatible with ammonium compounds:

$$Mg + 2\,NH_4^+ \longrightarrow Mg^{2+} + 2\,NH_3 + H_2$$

https://doi.org/10.1515/9783110660562-012

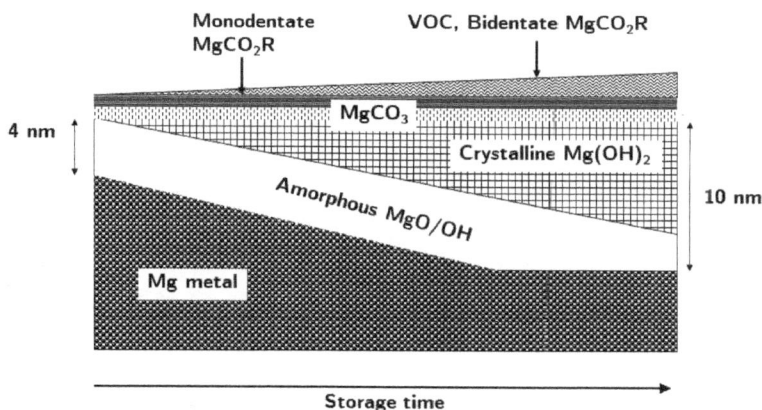

Figure M.1: Corrosion of Mg in air after *Fotea, Callaway & Alexander.*

In addition, ionic compounds of metals which are more noble than Mg (e.g. Cu^{2+}) yield the corresponding Mg salts leading to quick and dangerous deterioration. Hence, for certain applications chromated or generally passivated Mg (wolframate, phosphate coating) is applied. More recently, more corrosion-resistant Mg-rich compounds and alloys (e.g. AZ91E or Mg_3Al_4) are in use.

- P. Alenfelt, Corrosion Protection of Magnesium without the use of Chromates, *Pyrotechnica* **1995**, *XVI*, 44–49.
- E. L. Dreizin, Phase changes in metal combustion, *Frog. Energy Combust. Sci.* **2000** *26*, 57–78.
- C. Fotea, J. Callaway, M. R. Alexander, Characterisation of the surface chemistry of magnesium exposed to the ambient atmosphere, *Surf. Interface Anal.* **2006**, *38*, 1363–1371.
- C. Fotea, M. Alexander, P. Smith, J. Callaway, Surface Modification of Magnesium Particulates with Silanes Presented as Vapor Inhibition of ATmospheric Corrosion, *The Journal of Adhesion* **2008**, *84*, 389–400.
- N. Davies, P. Smith, J. Callaway, Coated Magnesium Powder for Pyrotechnic Decoy Flares for the Protection of Aircraft, *Nano-Scale Energetic Materials*, Strasbourg, **2009**.
- C. K. Wilharm, Combustion Performance of Coated Magnesium, *IPS-Seminar*, Fort Collins, **2008**, 31–38.
- T. J. Gudgel, F. Chapman, S. Sambasivan, T. Gillard, C. Wilharm, B. Douda, Inorganic Barrier Coating for the Protection against Humidity-Based Aging, *IPS-Seminar*, Fort Collins, **2008**, 25–29.
- P. Smith, *Surface-modified magnesium powders for use in pyrotechnic compositions*, GB-Patent 2450750B, **2008**, United Kingdom.
- D. L. Hastings, M. Schoenitz, K. M. Ryan, E. L. Dreizin, J. W. Krumpfer, Stability and Ignition of a Siloxane-Coated Magnesium Powder, *Propellants Explos. Pyrotech.* **2020**, *45*, 621–627.

Magnesium–aluminum alloy

Magnesium–Aluminum Legierung

Magnalium		
Aspect		Silvery
Formula		$Al_{12}Mg_{17}$ bis Al_3Mg_2
m_r	g mol^{-1}	178.16
ρ	g cm^{-3}	2.06
Mp	°C	450–462
$\Delta_f H$	kJ mol^{-1}	–205 (Al_3Mg_4)
$\Delta_c H$	kJ mol^{-1}	–4,715 (Al_3Mg_4)
	kJ g^{-1}	–26.465
	kJ cm^{-3}	–54.518

In the Mg–Al phase diagram (Fig. M.2), there are at least four compounds that can be unambiguously identified, Al_3Mg_2 (62.46% Al), AlMg (52.59% Al), Al_3Mg_4 (45.42% Al), and $Al_2Mg_3 \cong Al_{12}\,Mg_{17}$ (42.51% Al).

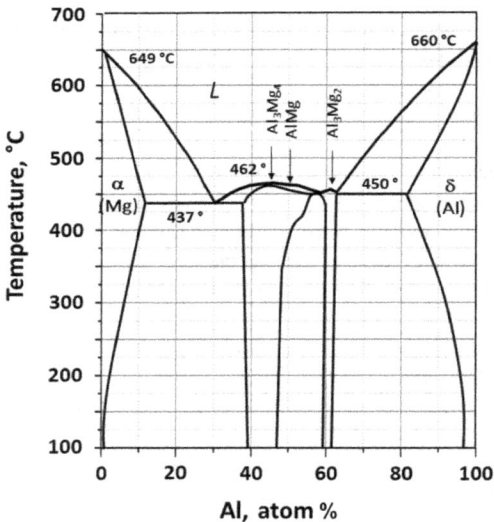

Figure M.2: Mg–Al phase diagram after Kurnakov and Mikheeva.

In comparison to its constituent metals (Al and Mg), magnalium is 10 times as hard (Brinell hardness: Al = 166 MPa, Mg = 254, Al_3Mg_2 = 1657 MPa) and hence can be grinded easily to give a powder. In contrary to Mg, it does not corrode when in contact with acidic components. Likewise, it is not as reactive with ammonium compounds as is magnesium. However, it is advised to coat magnalium when it is used with ammonium compounds. The flash point of magnalium powder is lower than any of its constituent metals (Figure M.3).

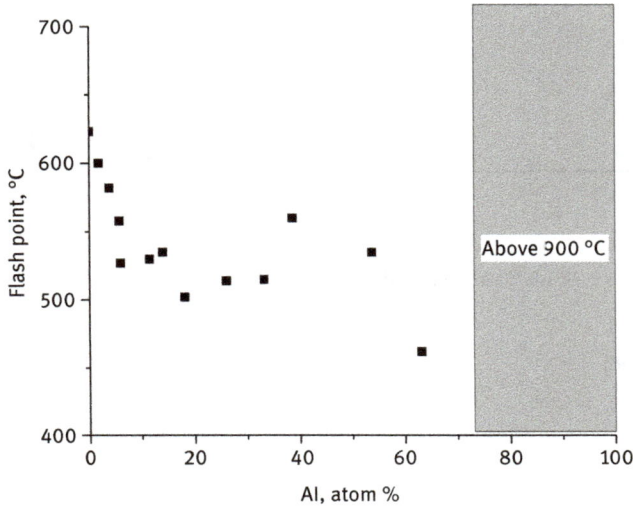

Figure M.3: Flash point of magnalium as a function of Al content.

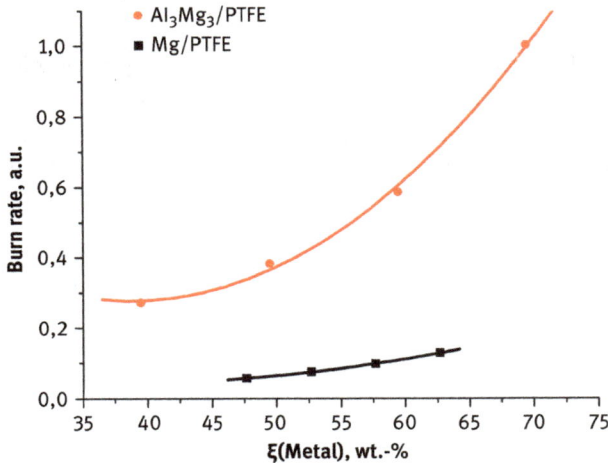

Figure M.4: Burn rate of Mg/PTFE and MgAl/PTFE.

Magnalium is used as fuel in crackling formulations, fountains, cone formulations, and in infrared decoy flare first fires and main compositions. Figure M.4 shows the burn rate for both Mg/PTFE and Mg_3Al_4/PTFE after *Cudzilo*.

- A. E. Vol, *Handbook of Binary Metallic Systems Structure and Properties*, *Vol.I*, Israel Program for Scientific Translations, Jerusalem, **1966**, pp. 160–196.
- S. Cudziło, W. A. Trzcinski, Study of High Energy Composites Containing Polytetrafluoroethylene, *ICT -Jata*, Karlsruhe, **1998**, P-151.

Magnesium hydride

Magnesiumhydrid

Formula		MgH$_2$
GHS		02
H-phrases		260
P-phrases		231-232-233-280-370+378i-402-501a
UN		2010
EINECS		231-705-3
CAS		[7693-27-8]
m$_r$	g mol^{-1}	26.321
ρ	g cm^{-3}	1.45
Δ$_f$H	kJ mol^{-1}	−75.31
Δ$_c$H	kJ mol^{-1}	−812.5
	kJ g^{-1}	−30.870
	kJ cm^{-3}	−44.761
Λ	wt.-%	−121.57
Dec	°C	284–300

Depending on its synthesis, MgH$_2$ can be either stable in air or be pyrophoric. MgH$_2$ reacts violently with water-producing hydrogen. Mg decomposes at 284 °C to give hydrogen and activated that is to say pyrophoric magnesium. Due to the significantly reduced thermal conductivity compared to metallic Mg compositions of MgH$_2$ and Sr(NO$_3$)$_2$, they have been advised for slow burning tracers.

- E.-C. Koch, V. Weiser, E. Roth, Combustion Behavior of Binary Pyrolants based on Mg, MgH$_2$, MgB$_2$, Mg$_3$N$_2$, Mg$_2$Si and Polytetrafluoroethylene, *IPS-Seminar*, Reims, **2011**, pp. 25–34.
- J. R. Ward, *Pyrotechnic composition*, US4302259, **1979**, USA.

Magnesium nitrate
Magnesiumnitrat

Formula		$Mg(NO_3)_2 \cdot 6\,H_2O$
GHS		03, 07
H-phrases		272-315-319-335
P-phrases		210-221-302+352-305+351+338-405-501a
UN		1474
EINECS		233-826-7
CAS		[13446-18-9]
m_r	g mol^{-1}	256.41
ρ	g cm^{-3}	1.636
$\Delta_f H$	kJ mol^{-1}	−2613.33
Λ	wt.-%	+31.20
Mp	°C	89
Bp	°C	330 (dec)

Magnesium nitrate hexahydrate is a deliquescent salt (DRH = 56% RH), which forms by hydration of the anhydrous compound $Mg(NO_3)_2$ [10377-60-3] ($\Delta_f H$ = −790.65 kJ mol^{-1}) via the dihydrate, $Mg(NO_3)_2 \cdot 2H_2O$ [15750-45-5] (ρ = 2.026 g·cm^{-3}, Mp = 130 °C). Magnesium nitrate hexahydrate has been tested in melt-castable high-explosives formulations and together with other nitrates forming eutectic mixtures as oxidizer in pyrotechnic illuminating flare compositions yielding spectral energies (cd s g^{-1}) comparable to $NaNO_3$.

–   H. W. Kruse, J. J. Bujak, Investigation of Eutectic Oxidizers For Use in Illumination Devices, *NWC-TP 5395*, NWC-China Lake, **1973**, 22 pp.

Magnesium peroxide

Magnesiumperoxid

Formula		MgO_2
GHS		03, 05
H-phrases		272-314
P-phrases		220-280-305+351+338-310
UN		1476
EINECS		238-438-1
CAS		[14452-57-4]
m_r	g mol^{-1}	56.305
ρ	g cm^{-3}	3.3
$\Delta_f H$	kJ mol^{-1}	−623
Dp	°C	300
Λ	wt.-%	+28.42

MgO_2 is used as co-oxidizer in tracer formulations.

- S. R. Lingampalli, K. Dileep, R. Datta, U. K. Gautam, Tuning the Oxygen Release Temperature of Metal Peroxides over a Wide Range by Formation of Solid Solutions, *Chem. Mater.* **2014**, *26*, 2720–2725

Magnesium silicide

Magnesiumsilicid

Formula		Mg_2Si
GHS		02
H-phrases		261
P-phrases		231+232-233-280-370+378a-402+404-501a
UN		2624
EINECS		245-254-5
CAS		[22831-39-6]
m_r	g mol^{-1}	76.710
ρ	g cm^{-3}	1.950
$\Delta_f H$	kJ mol^{-1}	−77.8
$\Delta_c H$	kJ mol^{-1}	−2034.9
	kJ g^{-1}	−26.532
	kJ cm^{-3}	−51.737
Dp	°C	280–300
κ	W m^{-1} K^{-1}	7.91
Λ	wt.-%	−83.43

Magnesium silicide (Mg_2Si) is a high energy fuel used both in high-blast explosives and fast burning IR-decoy flare formulations – yielding triple the pointance, L_λ (W sr^{-1} cm^{-2}) of comparable magnesium/Teflon®/Viton® (MTV) compositions.

- E.-C. Koch, V. Weiser, E. Roth, Combustion Behavior of Binary Pyrolants based on Mg, MgH_2, MgB_2, Mg_3N_2, Mg_2Si and Polytetrafluoroethylene, *IPS-Seminar*, Reims, **2011**, pp. 25–34.
- E.-C. Koch, A. Hahma, V. Weiser, E. Roth, S. Knapp, Metal-Fluorocarbon Pyrolants. XIII High Performance Infrared Decoy Flare Compositions Based on MgB_2 and Mg_2Si and Polytetrafluoroethylene/Viton A, *Propellants Explos. Pyrotech.* **2012**, *37*, 432–438.

Manganese
Mangan

Aspect		Grey-pink powder
Formula		Mn
GHS		02, 09
H-phrases		228-319
P-phrases		280-210-264-305-351-338-337-313
UN		3089
EINECS		231-105-1
CAS		[7439-96-5]
m_r	g mol^{-1}	54.94
ρ	g cm^{-3}	7.200
Mp	°C	1244
Bp	°C	2059
$\Delta_m H$	kJ mol^{-1}	12.06
$\Delta_v H$	kJ mol^{-1}	226.7
c_p	J g^{-1} K^{-1}	0.378
c_L	m s^{-1}	5560
κ	W m^{-1} K^{-1}	29.7
T_i	°C	223 (dust layer), 460 (dust cloud) both 325 mesh
$\Delta_c H$	kJ mol^{-1}	−520
	kJ g^{-1}	−9.465
	kJ cm^{-3}	−68.147
Λ	wt.-%	−58.243

Manganese is used as an auxiliary fuel in oxygen candles and in delay formulations.

- M. M. Markowitz, D. A. Boryta, H. Stewart, Jr., Lithium perchlorate Oxygen candle, *Chemical Industrial Engineering* **1964**, *3*, 321–330.
- Y. C. Montgomery, W. W. Focke, M. Atanasova, O. Del Fabbro, C. Kelly, Mn + Sb2O3 Thermite/Intermetallic Delay Compositions, *Propellants Explos. Pyrotech.* **2016**, *41*, 919–925.

Manganese dioxide

Mangandioxid, Braunstein

Aspect		Dark brown crystals
Formula		MnO_2
GHS		07
H-phrases		332
P-phrases		261-271-304+340-312
EINECS		215-202-6
CAS		[1313-13-9]
m_r	$g\ mol^{-1}$	86.937
ρ	$g\ cm^{-3}$	5.026
$\Delta_f H$	$kJ\ mol^{-1}$	−520
Dp	°C	533–570
Λ	wt.-%	+18.4
c_p	$J\ K^{-1}\ mol^{-1}$	54.4

Manganese dioxide (MnO_2) is used as oxidizer in delay formulations, red phosphorus-based obscurants and float and smoke signals (SR414) and also serves as combustion catalyst in solid rocket propellants, gun propellants (gps), and in the decomposition of chlorates and perchlorates in SCOGS.

- T. A. Vine, W. Fletcher, An Investigation of Failures to function of a red phosphorus marine marker, *IPS*, Westminster, **2002**, 477–489.

McLain, Joseph Howard (1917–1981)

McLain was an American solid-state chemist and pyrotechnics expert (Figure M.5). He studied at Washington College and received a BS in 1937. During World War II, he served as second Lieutenant at Edgewood Arsenal doing research on HC obscurants and delay formulations. After the war, he completed his dissertation at Johns Hopkins University in Maryland and returned to Washington College in 1946. There he became dean of the chemistry department in 1955. In 1980, he published a seminal textbook on pyrotechnics that, due to its unique background in solid state chemistry, even today is still an authoritative reading in the field.

Figure M.5: Joseph Mc Lain (1968).

– J. H. McLain, *Pyrotechnics – From the Viewpoint of Solid-State Chemistry*, The Franklin Institute Press, Philadelphia, **1980**, 243 pp.

Mercury fulminate
Knallquecksilber

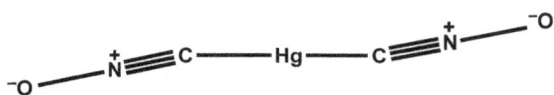

Aspect		Grey crystals
Formula		$C_2N_2O_2Hg$
REACH		LPRS
EINECS		211-057-8
CAS		[628-86-4]
m_r	g mol^{-1}	284.624
ρ	g cm^{-3}	4.467
Δ_fH	kJ mol^{-1}	−267.99
$\Delta_{ex}H$	kJ mol^{-1}	−1117

	kJ g^{-1}	−3.925
	kJ cm^{-3}	−17.533
Λ	wt.-%	−16.86
N	wt.-%	9.84
c_p	J K^{-1} mol^{-1}	141
Dp	°C	136
Friction	N	7 (BAM)
Impact	J	1 (BAM)
V_D	m s^{-1}	5200 at ρ = 4.2 g cm^{-3}
Trauzl	cm^3	315

Mercury fulminate, (MF) was the main primary explosive until the end of World War II. Then due to its poor thermal stability and high toxicity, it has been replaced first in the NATO countries mainly by lead azide and then after the fall of the iron curtain also became decomissioned in the former Warsaw pact countries. Technical qualities with embedded impurities decompose relatively quickly and lose their initiating capability. MF has its highest initiation capability at a density of about ρ = 3.2 g cm^{-3}. Starting at densities greater than ρ = 3.6 g cm^{-3} MF loses its initiation behavior and it is described as dead pressed.

- R. Matyáš, J. Pachman, *Primary Explosives*, Springer, **2013**, 39–59.
- W. Beck, J. Evers, M. Göbel, G. Oehlinger, T. M. Klapötke, The Crystal and Molecular Structure of Mercury Fulminate (Knallquecksilber), *Z. Anorg. Allg. Chem.* **2007**, *633*, 1417–1422.
- H. G. Otto, Notiz zum Kristallsystem des Knallquecksilbers, *Z. Schiess. Spreng.* **1943**, *38*, 85–86.

Methyl nitrate
Methylnitrat

H_3C ⁀ O ⁀ NO_2

Formula		CH$_3$NO$_3$
REACH		LPRS
EINECS		209-941-3
CAS		[598-58-3]
m_r	g mol^{-1}	77.04
ρ	g cm^{-3}	1.21 at 15 °C
$\Delta_f H$	kJ mol^{-1}	−155.90
$\Delta_{ex} H$	kJ mol^{-1}	−491.0
	kJ g^{-1}	−6.113
	kJ cm^{-3}	−7.385

$\Delta_c H$	kJ mol^{-1}	−978.2
	kJ g^{-1}	−12.697
	kJ cm^{-3}	−15.363
Λ	wt.-%	−10.38
N	wt.-%	18.18
Mp	°C	−82.3
Bp	°C	66.5
Friction	N	>360
Impact	J	0.2
V_D	m s^{-1}	1500–8000 (diameter dependent)

Methyl nitrate has an odor resembling chloroform. It has been used briefly as gelling aid in NC-based propellants and was tested as a ballistic modifier in propellants and gas-generating compositions but was skipped due to very high volatility. It is used as methylating agent in organic synthesis. Due to its extreme impact sensitivity, it has been often blended with fuels. Mixtures of methyl nitrate with methanol (*Myrol*) were used in World War II in German rocket planes and as volumetric explosive to clear bunker and foxholes. The viscosity of methyl nitrate is lower than water and explains for the extreme impact sensitvity (adiabatic pore collapse).

– H. Walter, B. Walter, B. H. Wilcox, *German Developments in High Explosives, FIAT Final Report No. 1035*, 9 April **1947**, Field Information Agency, Technical, 38 pp, formerly restricted.

Methyl violet test

Methylviolett Test

According to this test, nitrocellulose and single-based gps are heated to 134.5 °C whereas double, triple and other gps are heated to 120 °C with filter papers impregnated with methyl violet solution and fixed in the vapor phase above the samples. With NC no color change must occur in the first 30 min, whereas single-based gps must remain unchanged for at least 40 min and double- and triple-based gp may show a color change earliest after 60 min of heating. The color change is from violet over to blue green to salmon pink.

Metriol trinitrate

Metrioltrinitrat

TMETN, MTN		
Formula		$C_5H_9N_3O_9$
CAS		[3032-55-1]
m_r	g mol^{-1}	255.141
ρ	g cm^{-3}	1.488
$\Delta_f H$	kJ mol^{-1}	−425
$\Delta_{ex} H$	kJ mol^{-1}	−1311.0
	kJ g^{-1}	−5.138
	kJ cm^{-3}	−7.646
$\Delta_c H$	kJ mol^{-1}	−2828.8
	kJ g^{-1}	−11.087
	kJ cm^{-3}	−16.498
Λ	wt.-%	−34.49
N	wt.-%	16.47
Mp	°C	15.7
Bp	°C	182
T_{5ex}	°C	235
Friction	N	>360
Impact	J	0.8
Trauzl	cm^3	400

TMETN is a good energetic plasticizer for GAP-based formulations and also a gelling aid for NC. As TMETN gels at T = 110 °C only, it is blended with trimethylolethan triacetate (92/8), to facilitate gelling at T = 80 °C.

– K. Menke, S. Eisele, Rocket Propellants with Reduced Smoke and High Burning Rates, *Propellants Explos. Pyrotech.* **1994**, *22*, 112–119.

Mine-clearing torch

Minenräumfackel

Mine clearing torches are pyrotechnic flares to burn all kinds of UXO (Fig. M.6). They have a nozzle to produce a narrow jet of hot gaseous and condensed combustion products. The jet of a single torch can burn through a 5 mm mild steel plate. By this *cook-off* of the explosive fill in the UXO can occur. Commercial systems are based on Al/Ba(NO$_3$)$_2$ (e.g. *Fire-Ant*®/Chemring) or Al/CaSO$_4$·2 H$_2$O (z.B. *Dragon*®/Disarmco). A similar device however with different purpose is the *pyronol* torch, using a formulation based on Ni/Al/Fe$_2$O$_3$/PTFE used for underwater cutting and perforating operations. The advantage of pyrotechnic mine-clearing torches over detonating EOD devices is the logistical advantage of handling a H.D. 1.3 store over a H.D. 1.1.

Figure M.6: Comparison of the sizes of the ignition charge and (thermite) flares used from top to bottom: DM29 ignition charge, fire, EOD, PT with igniter, electric, hyperheat mine flare with electric match, Fire-Ant A210 (from *Kannenberger*).

–   G. Kannenberger, *Test and evaluation of pyrotechnical mine neutralisation means, ITEP Work Plan Project Nr. 6.2.4*, WTD 91, July **2005**, https//www.gichd.org/fileadmin/pdf/LIMA/GermanyITEP6.2.4.pdf

Mitigation device
Frühzündeinrichtung

A mitigation device typically is a safety feature in munitions and civilian stores (e.g. automobile air bags) that minimizes the explosive reaction level (ERL) upon slow and fast heating.

In automobile air bag gas generators the autoignition composition serves as mild thermal ignition to prevent the main gas-generating composition from deflagrating and creating fragments and blast effects. Typical autoignition compositions trigger ignition at T < 220 °C. The example formulation given below ignites at 170 °C:
–   13.1 wt.-% tetramethylammonium nitrate
–   9.1 wt.-% 5-aminotetrazole

- 52.1 wt.-% potassium perchlorate
- 25.7 wt.-% molybdenum

In munitions, both slow and fast heating can likewise effect catastrophic reaction. A popular concept in the thermal mitigation of munitions are venting devices to avoid pressure built-up. Venting can be achieved by structural elements that lose their strength due to thermal input (thermoplastics, low melting alloys, and glassy metals) and hence open up a confined energetic material. An energetic fill, on top of reacting mildly to fast heating, can expand and thereby push out a fuse (PAX 21) (Figure M.7).

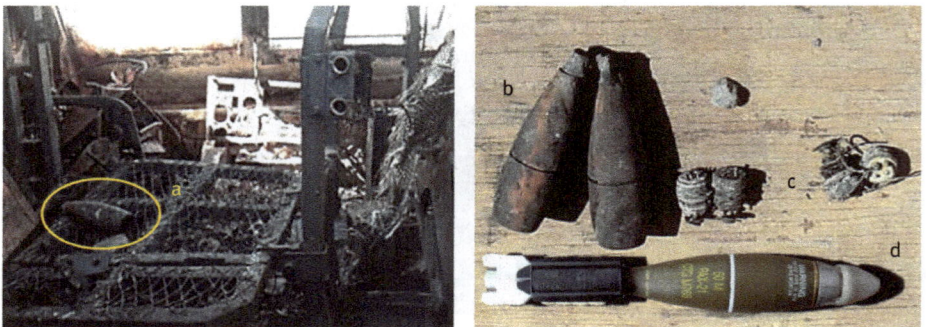

Figure M.7: Burnt-out M768 60 mm-mortar shell hull in destroyed MRAP vehicle (a); collected shell (b) and M783 fuzes (c); original shell with fuse cap and propellant (d) from *Smith* (2010).

Any container can be cut or split in part by detonating cords or linear-shaped charges. The concept of mitigation can also be understood to include protection in logistic configurations such as anti-fratricide bars in artillery shell storage to avert sympathetic reaction or intumescent coatings on ammunition boxes and bomb shells to increase the evacuation and/or firefighting response times.

- http://citeseerx.ist.psu.edu/viewdoc/download?doi=10.1.1.489.1533&rep=rep1&type=pdf accessed on April 5, 2020.
- P.-A. Prevot, C. Collet, E. Baker, *Mitigation Technologies for Warheads, L-226*, NATO-Munitions Safety Information Analysis Center (MSIAC), Brussels, February **2019**, 43 pp.
- J. Smith, R. Wong, P. Ferlazzo, W. Kuhnle, J. Niles, P. Ng, J. Stenkamp, IM in the Field – Experience of Reduced Sensitivity Mortar Cartridges to Actual Combat Threat Stimuli, *IMEMTS 2010*, 11–14. October **2010**, Munich, Germany.
- B. Eigenmann, K. Rudolf, M. Schildknecht, *System, das gegen eine unbeabsichtigte detonative Umsetzung unempfindlich ist*, DE10 2004005064, **2006**, Germany.
- G. D. Knowlton, C. P. Ludwig, *Low Temperature Autoignition Composition*, US6749702, **2004**, USA.

Molybdenum

Molybdaen

Aspect		Silvery
Formula		Mo
GHS		02, 08
H-phrases		250-252-315-319-335
P-phrases		210-222-302+352-305+351+338-405-501a
UN		1383
EINECS		231-107-2
CAS		[7439-98-7]
m_r	g mol^{-1}	95.94
ρ	g cm^{-3}	10.22
Mp	°C	2623
Bp	°C	4639
$\Delta_m H$	kJ mol^{-1}	39.10
$\Delta_v H$	kJ mol^{-1}	582.2
c_p	J g^{-1} K^{-1}	0.217
κ	W m^{-1} K^{-1}	142
c_L	m s^{-1}	6650
$\Delta_c H$	kJ mol^{-1}	−745
	kJ g^{-1}	−7.765
	kJ cm^{-3}	−79.361
Λ	wt.-%	−50.03

Molybdenum (Mo) serves as fuel in delay formulations and autoignition composi-tions. Mo compounds yield a citron coloration of flames due to its atomic lines at λ = 550.6, 553.3, 557.0, and 603.0 nm.

- J.-L. Dumont, *Compositions pyrotechniques pour la production d'artifices colores*, EP 0252803B1, France, **1992**.

Molybdenum trioxide

Molybdaentrioxid

Aspect		Colorless with green tinge, yellow as nanometric
Formula		MoO_3
GHS		08, 07
H-phrases		351-319-335
P-phrases		260-280h
EINECS		215-204-7
CAS		[1313-27-5]
m_r	g mol^{-1}	143.94
ρ	g cm^{-3}	4.632
$\Delta_f H$	kJ mol^{-1}	−745.6
Dp	°C	646 incipient, more pronounced after Mp
Mp	°C	795
Λ	wt.-%	11.12
c_p	J K^{-1} mol^{-1}	75.0

Molybdenum trioxide (MoO_3) is a popular oxidizer in thermites and particularly in nanothermites. MoO_3 partly sublimes before melting at 795 °C explaining its high reactivity in thermites.

– E. Lafontaine, M. Comet, *Nanothermites*, Wiley, **2016**, pp. 183.

Monoergol

Monoergols are homogeneous liquid propellants that do not require another component to deflagrate. They are either single-chemical compounds (e.g. hydrazine or propyl nitrate) or stable mixtures of compounds (e.g. *myrol*). Those monoergols decomposing when in contact with a catalyst are called catergols (e.g. 80% H_2O_2 at Ca $(MnO_4)_2$ catalyst).

– A. Dadieu, R. Damm E. W. Schmidt, *Raketentreibstoffe*, Springer, **1968**, pp. 99.

MSIAC

The NATO **M**unitions **S**afety **I**nformation **A**nalysis **C**enter (MSIAC) is a project funded by its member nations (as of 2020) Australia, Belgium, Canada, Finland, France, Germany, Italy, Netherlands, Norway, Poland, Republic of Korea, Sweden, Spain, the United Kingdom, and the United States. MSIAC's mission is to support its member nations in all technical and scientific issues related to safety of ammunition in the complete life cycle of an ammunition (*"from cradle to grave"*). Therefore, MSIAC maintains a team of technical specialist officers in the field of

- Energetic materials (high explosives, propellants, pyrotechnics)
- Warhead technology (detonics, terminal ballistics)
- Propulsion technology (internal ballistics, gun and rocket)
- Logistics (munitions transport and storage safety)
- Materials science
- Munition systems

led by a project manager and supported by information specialists. MSIAC neither runs test and research facilities nor does it finance third-party studies. The annual budget of MSIAC is between 1.3 and 1.8 Mio. EUR.

- https://www.nato.int/issues/iban/financial_audits/2017-msiac-2015-2016-2017-en.pdf
- Weblink www.msiac.nato.int

MTV, magnesium/Teflon®/Viton®

MTV stands for magnesium, Teflon®, and Viton®A, a common decoy flare and an igniter material. Teflon is a trade name for *polytetrafluoroethylene*, (PTFE), $(C_2F_4)_n$, and Viton® A is a trade name for the copolymer from hexafluoropropene and vinylidene fluoride. The predecessor of MTV, a mixture from magnesium, PTFE, and Kel-F wax (*polychlorotrifluoroethylene*) in the about mass proportions 5:3:2 was invented in the early 1950s at Naval Ordnance Test Station (NOTS), Inyokern (today: NAVSEA China Lake) in research for new igniter formulations for solid propellants. At about the same time, the research at NOTS into infrared-guided missiles (*Sidewinder*) necessitated the development of infrared target flares for drones. It was then recognized that Mg/PTFE-based formulations yield about an order of magnitude more infrared energy than common illuminating flares based on $Mg/NaNO_3$ (Table M.1).

Though the heat of combustion of both formulation is about the same ball park, it is for the particulate carbon released in the combustion zone of the Mg/PTFE which dramatically alters the emissivity ($\varepsilon_{\lambda,T} \sim 0.8$) of the flare flame mainly in the infrared range ($\lambda \sim 1$–5 µm):

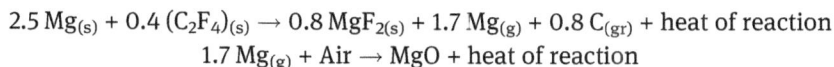

$$2.5\, Mg_{(s)} + 0.4\, (C_2F_4)_{(s)} \rightarrow 0.8\, MgF_{2(s)} + 1.7\, Mg_{(g)} + 0.8\, C_{(gr)} + \text{heat of reaction}$$
$$1.7\, Mg_{(g)} + Air \rightarrow MgO + \text{heat of reaction}$$

The burn rate and radiant intensity correlate with the magnesium content (Figure M.8). MTV is extremely sensitive towards electrostatic discharge and may react very violent in dry state (low velocity detonation).

Table M.1: Comparison of SR 580 illuminating composition versus Mg/PTFE IR-flare composition.

	SR 580	NOTS 702
Magnesium	60	54
Sodium nitrate	36	
Polytetrafluorethylene	–	30
Acaroid resin	4	
Kel-F-Wax	–	16
Q_{comb} (kJ g^{-1})	13.0	9.2
E_{Pbs} (J g^{-1} sr^{-1})	21	125

Derivates of MTV typically use halogen-free binders and correspondingly are termed:

MTE	– Elvax® (ethyl vinylacetate)
MTH	– Hytemp4454® (polyacrylate)
MTH	– Hycar4051® (polybutadiene)
MTK	– Kraton (styrene–butadiene copolymer)
MTTP	– Thermoplastic (polystyrol /plasticizer)
MTR	– Resin (styrene resin)

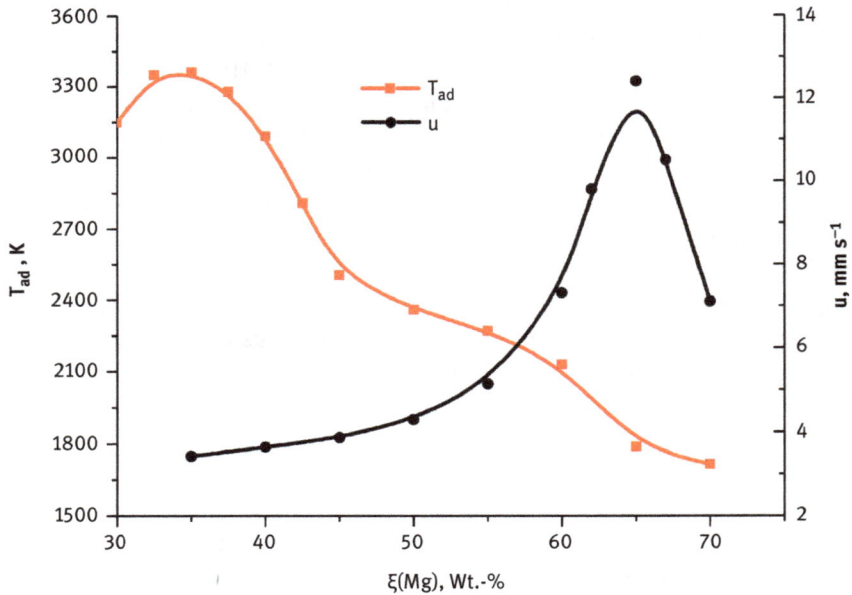

Figure M.8: Burn rate and adiabatic flame temperature of MTV.

- E.-C. Koch, *Metal-Fluorocarbon Based Energetic Materials*, Wiley-VCH, **2012**, 342 pp.

Muzzle flash
Mündungsfeuer

Upon firing of a gun, there are three independent luminous effects observable after the projectile has left the barrel.

Right in front of the muzzle, there is a small glow extending hemispherically into the air (*muzzle glow*). Separated from the muzzle glow by a dark area (of a few calibers length) is a zone of high luminance that is about 10–20 calibers in width and about double in length with a conical peak (*primary flash*). Depending on the oxygen balance of the propellant, the primary flash zone may transition directly into the secondary flash typically with a frayed conical top.

The hot propellant gases (T > 1300 °C) are visible at the muzzle as muzzle glow. Expansion of the gases cools them below T < 1300 °C and hence cause a dark zone. Adiabatic recompression of those gases explains the primary flash. In an inert atmosphere, the primary muzzle flash has a sharp conical boundary and reminds of a laminar burning premixed flame. In air however, the unburnt components of the propellant gases (CO, H_2, and soot) can undergo afterburn (provided their temperature is sufficiently high, T > 750 °C) and cause the fraying secondary flash.

From a tactical point of view, the muzzle flash is highly undesirable (blinding the gun crew, aids in localization of the gun position). With small-caliber weapons, it is hence tried to suppress the primary flash by shaping of the nozzle (e.g. longitudinal vents in the barrel) to minimize recompression. This also aids in suppressing the secondary flash as the combustion products cool down more quickly below their ignition temperature. However, with large-caliber weapons, those measures affecting the primary flash are not as quite effective. Only the secondary flash can be effectively countered by chemical suppressants that stop the radical chain mechanism of hydrocarbon combustion. Typically, chemical suppressants added to gun propellants are potassium salts (e.g. K_3AlF_6).

- O. K. Heiney, *Ballistics Applied to Rapid-Fire Guns*, in *Interior Ballistics of Guns*, H. Krier, M. Summerfield (Eds.), Progress in Astronautics and Aeronautics, Vol. 66, **1979**, pp. 87–112.
- N. N., *Interior Ballistics of Guns*, AMC Pamphlet, AMCP 706-150, US ARMY Materiel Command February **1965**.

N,N′,N″-Tris(2,4,6-trinitrophenyl)-2,4,6-pyrimidinetriamine

N,N′,N″-Tris(2,4,6-trinitrophenyl)-2,4,6-pyrimidintriamin

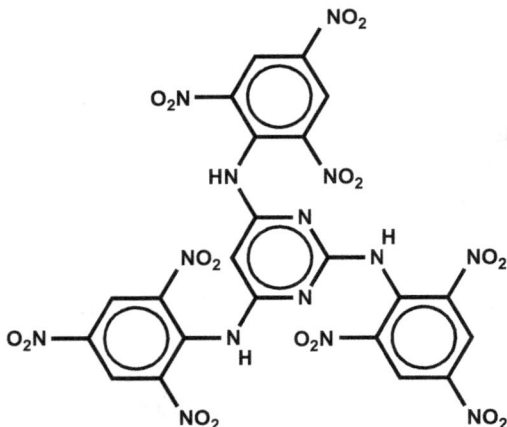

TPP		
Formula		$C_{22}H_{10}N_{14}O_{18}$
CAS		[41230-77-7]
m_r	g mol^{-1}	758.408
ρ	g cm^{-3}	1.90
$\Delta_f H$	kJ mol^{-1}	300
$\Delta_{ex}H$	kJ mol^{-1}	−3676.3
	kJ g^{-1}	−4.847
	kJ cm^{-3}	−9.210
$\Delta_c H$	kJ mol^{-1}	−10386.6
	kJ g^{-1}	−13.695
	kJ cm^{-3}	−26.021
Λ	wt.-%	−65.4
N	wt.-%	25.86
Mp	°C	301 (dec)

- B. W. Harris, J. L. Singleton, M. D. Coburn, Picrylamino-substituted heterocycles. VI. Pyrimidines, *J. Heterocy. Chem.* **1973**, *10*, 167–171.

https://doi.org/10.1515/9783110660562-013

N,N'-Dipicrylurea, DPU

Dipikrylharnstoff

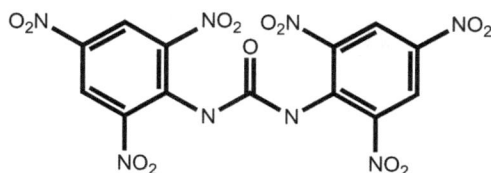

2,2',4,4',6,6'-Hexanitrocarbanilide

Aspect		Yellow rosette crystals
Formula		$C_{13}H_6N_8O_{13}$
CAS		[6305-08-4]
m_r	g mol^{-1}	482.2325
ρ	g cm^{-3}	1.885 (estimated in accordance with Ammon)
$\Delta_f H$	kJ mol^{-1}	−500 (estimated in accordance with Benson)
$\Delta_{ex}H$	kJ mol^{-1}	−2117.2
	kJ g^{-1}	−4.390
	kJ cm^{-3}	−8.276
$\Delta_c H$	kJ mol^{-1}	−5473.3
	kJ g^{-1}	−11.35
	kJ cm^{-3}	−21.394
Λ	wt.-%	−53.08
N	wt.-%	23.24
Mp	°C	140 (dec)
Dp	°C	203−209
T_{5ex}	°C	345
P_{CJ}	GPa	25.3 at 1.885 g cm^{-3}
V_D	m s^{-1}	7678 at 1.885 g cm^{-3}
Impact	J	~3

This high explosive has been reported by *Davis* to be suitable for booster applications.

– Davis, **1943**, pp. 188–189.

Nanodiamond

Nanodiamant

Upon detonation of ideal high explosives, the pressures prevailing in the CJ plane can shift the *Boudouard* equilibrium to the right side

$$2\,CO \rightarrow CO_2 + C$$

and can facilitate a phase change of amorphous carbon to diamond. Under these conditions, droplets of liquid carbon having nanometric dimensions pass through the P/T range (P = 10–16 GPa and T = 3400–2900 K) and crystallize as diamonds. It is hence that in the detonation products of oxygen-deficient high explosives under suitable conditions (isochoric detonation, P_{CJ} > 22 GPa, under inert gas (*dry*) or confined in water (*wet*) or in dry ice (*ice*)) up to 40 wt.-% of the carbon contained in the high explosive can be transformed into nanodiamonds. Nanodiamonds further form upon cryogenic quenching of oxygen-deficient flames. Apart from their mass use as abrasives nanodiamonds due to the extreme high thermal conductivity (k(diamond) = 2300 W m^{-1} K^{-1} compared to k (Ag) = 429), they have become increasingly popular as ballistic modifiers in pyrotechnics,

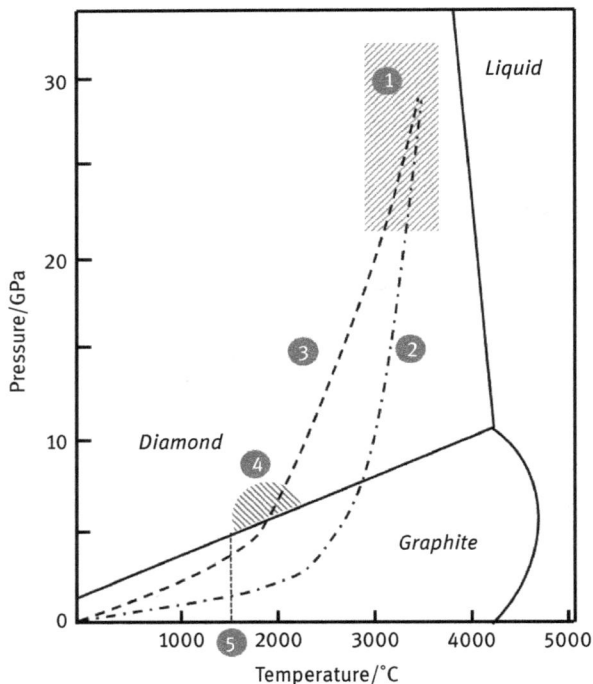

Figure N.1: Simplified phase diagram of carbon (explanations in the text) after *Baidakova & Vul'*.

propellants, and also to lower the shock sensitivity of high explosive formulations. Figure N.1 shows the simplified p–T phase diagram of carbon, with the indices: 1 = area of formation of nanodiamonds, p–T slope for 2 = *dry*, 3 = *wet* or *ice*, 4 = area of formation of diamond from graphite, 5 = Debye temperature of diamond (T_δ = 1577 °C).

- N. R. Greiner, D. S. Phillips, J. D. Johnson, F. Volk, Diamonds in Detonation Soot, *Nature* **1988**, *333*, 440–442.
- R. E. Clausing, L. L. Horton, J. C. Angus, P. Koidl (Eds.), *Diamond and Diamond-like Films and Coatings*, NATO ASI Series, Vol 266, Kluwer, **1991**, 911 pp.
- V. V. Danilenko, Specific features of Synthesis of Detonation Nanodiamonds. *Combust. Explos. Shock* **2005**, *41*, 577–588.
- M. Baidakova A. Vul', New prospects and frontiers of nanodiamond clusters *J. Phys. D. Appl. Phys.* **2007**, *40*, 6300–6311.
- M. Comet, V. Pichot, B. Siegert, D. Spitzer, J.-P. Moeglir, Y. Boehrer, Use of Nanodiamonds as a Reducing Agent in a Chlorate-based Energetic Composition, *Propellants Explos. Pyrotech.* **2009**, *34*, 166–173.
- E. B. Watkins, K. A. Velizhanin, D. M. Dattelbaum, R. L. Gustavsen, T. D. Aslam, D. W. Podlesak, R. C. Huber, M. A. Firestone, B. S. Ringstrand, T. M. Willey, M. Bagge-Hansen, R. Hodgin, L. Lauderbach, T. van Buuren, N. Sinclair, P. A. Rigg, S. Seifert, T. Gog, Evolution of Carbon Clusters in the Detonation Products of the Triaminotrinitrobenzene (TATB)-Based Explosive PBX 9502, *J. Phys. Chem. C* **2017** *121*, 23129–23140.
- E.-C. Koch, J. Licha, *unpublished research on shock sensitivity of PBXN-11 modified with nanodiamonds*, **2007**.

Nanolaminates

Nanolaminate

Energetic nanolaminates are reactive multilayers of metals capable to undergo exothermic *alloying reactions*. Typical metal combinations are Ni/Al or Zr/Al. Energy densities of up to $\Delta_{ex}H$ = −22 kJ cm^{-3} can be obtained.

Typical reaction temperatures are up to T = 3000 °C. By adjusting the thickness of the layers, the combustion rate and sensitiveness can be tailored. A commercial product is Nanofoil® (Ni/Al) produced by Indium Corporation. Finely chopped pieces of Nanofoil have been suggested as sensitive fillers for stab initiators. Large foils have been suggested as ballistic sleeves for infrared decoy flare payloads.

- T. W. Barbee Jr., T. Weihs, *Ignitable heterogeneous stratified structure for the propagation of an internal exothermic chemical reaction along an expanding wavefront and method of making same*, US5538795 A, **1996**, USA.
- A. E. Gash, T. W. Barbee Jr., O. Cervantes, Stab Sensitivity of Energetic Nanolaminates, *IPS*, **2006**, Fort Collins, 59–70.
- C. Dilg, D. B. Nielson, R. L. Tanner, *Flares including reactive foil for igniting a combustible grain thereof and methods of fabricating and igniting such flares*, US7469640B2, **2008**, USA.
- D. P. Adams, *Reactive Multilayers Fabricated by Vapor Deposition*, accessed at https://www.osti.gov/servlets/purl/1146570 on April 8, 2020.

Nanometric high explosive particles
Sprengstoffnanopartikel

Already before nanoparticles became commonly popular, *Moulard* in 1985 found that the shock wave sensitiveness of RDX decreases with decreasing particle size (4 µm versus 100 µm). In 1990, *Armstrong* also showed that the impact sensitiveness of RDX is inversely proportional to the crystallite size (d = 22 µm versus 100 and 1000 µm). Hence, RDX crystals with particle sizes in the two-figure nanometer range can be expected to be less sensitive than corresponding micrometric crystals. Table N.2 shows the impact and shock sensitiveness of different RDX types in formulations with chloroparaffins (Table N.1).

Table N.1: Composition of high explosives with common (µm) and nanometric RDX.

RDX type	RDX (wt.-%)	HMX (wt.-%)	Binder* (wt.-%)	Plasticizer# (wt.-%)	Density (g cm^{-3})	TMD (%)
Nano-A$	85.8	1.6	8.8	3.8	1.57	87.7
Nano-B&	85.0	1.3	9.6	4.1	1.58	88.2
Class 1	–	–	–	–	1.69	94.9
Class 5	79.3	8.4	8.6	3.7	1.57	87.7

*Chlorez700, #Paroil 170 T, $15–20 m^2 g^{-1}, &5–6 m^2 g^{-1}.

Table N.2: Sensitiveness of high explosives with common (µm) and nanometric RDX.

RDX type	NOL-SSGT-50% (GPa)	Impact-E-50% (J)
Nano-A$	2.49	14.3
Nano-B&	3.21	18.4
Class 1	1.96	–
Class 5	2.09	8.1

Nanometric RDX can be produced with the techniques depicted in Table N.3. Nanometric HMX (with reduced sensitiveness) can be prepared in principle in the same way, however, HMX tends to form the *y*-polymorph which has an inherently higher mechanical sensitiveness.

Table N.3: Techniques to produce nano-HE particles.

Method	Diameter (nm)	Reference	Notes
RESS rapid expansion of supercritical solutions	50–200	*Stepanov*	Expensive and energy intense
Flash crystallization	50–600	*Spitzer*	Continuous process, moderate yields 3 g/h
Electrospray	200–600	*Radacsi*	Low yields, highly flammable solvents
Ultrasonic	30–100	*Spitzer*	Continuous process, moderate yields 3 g/h
Vacuum condensation	50–200	*Frolov*	Low yields, however, at the advantage of no solvent inclusions
Wet milling	300	*Redner*	High yields, however, particle growth due to *Ostwald* ripening

- N. Radacsi, A. I. Stankiewicz, Y. L. M. Creyghton, A. E. D. M. van der Heijden, J. H. ter Horst, Electrospray Crystallization for High-Quality Submicron-Sized Crystals, *Chem. Eng. Technol.* **2011**, *34*, 624–630.
- B. Risse, D. Spitzer, D. Hassler, F. Schnell, M. Comet, V. Pichot, H. Muhr, Continuous formation of submicron energetic particles by the flash-evaporation technique, *Chem. Eng.* **2012**, *103*, 158–165.
- R. Patel, P. Cook, C. Crane, P. Redner, D. Kapoor, H. Grau, A. Gandzelko, Production and Coating of Nano-RDX using Wet Milling, *NDIA*, **2007**.
- H. Moulard, J. W. Kury, A. Delclos, The Effect of RDX Particle Size on the Shock Sensitivity of Cast PBX Formulations, *Detonation Symposium*, **1985**, Albuquerque, pp. 902–913.
- H. Moulard, Effects non lineaires de la granulometrie sur L'amorcabilite et al detonabilite d'un explosif coule a liant plastique, *EUROYPRO*, **1989**, La Grande Motte, pp. 17–21.
- R. W. Armstrong, C. S. Coffey, V. F. DeVost, W. L. Elban, Crystal Size Dependence for Impact Initiation of Cyclotrimethylenetrinitramine explosive, *J. Appl. Phys.* **1990**, *68*, 979–984.
- V. Stepanov, V. Anglade, W. Balas, A. Bezmelnitsyn, L. N. Krasnoperov, Processing and Characterization of Nanocrystalline RDX, *ICT -Jata*, **2008**, Karlsruhe, P-54.
- K.-Y. Lee, D. S. Moore, B. W. Asay, A. Llobet, Submicron-Sized Gamma-HMX I. Preparation and initial Characterization, *J. Energ. Mater.* **2007**, *25*, 161.
- D. S. Moore, K.-Y. Lee, S. I. Hagelberg, Submicron-Sized Gamma-HMX II Effect of Pressing on Phase Transition, *J. Energ. Mater.* **2008**, *26*, 70.
- E.-C. Koch, D. Schaffner, *Sensitivity of Nanoscale Energetic Materials I. High Explosives, L-158*, June **2009**, NATO -MSIAC, Brussels, 26 pp. https://www.msiac.nato.int/products-services/publications/l-158-sensitivity-of-nanoscale-energetic-materials-i-high-explosives, accessed on September 20 2020.

Nanothermites

Nanothermite

Nanothermites are thermites with at least one of the components being present as nanometric particles. Due to the high specific surface area of nanometric particles, the number of contact points between fuel and oxidizer is greatly increased and hence a more rapid reaction is encountered. However, not only do nanothermites burn fast, they are also much more sensitive towards accidental ignition, especially towards electrostatic discharge. Nanothermites have been suggested for actuators and even as replacements for primary explosives in green detonator applications.

– M. E. Brown, S. J. Taylor, M. J. Tribelhorn, Fuel-Oxidant Particle Contact in Binary Pyrotechnic Reactions, *Propellants Explos. Pyrotech.* **1998**, *23*, 320–327.
– R. J. Jouet, A. J. Schuman, MIC/Al Incidents at Indian Head 1998–2005, *MIC Safety Meeting*, Los Alamos National Laboratory, 13 Feb **2006**.
– E.-C. Koch, D. Schaffner, *Sensitivity of Nanoscale Energetic Materials II. Fuels, Pyrolants and Propellants, L-162*, December **2009**, NATO -MSIAC, Brüssel, 29 pp.
– A. Gromov, U. Teipel, *Metal Nanopowders- Production, Characterization and Energetic Applications*, Wiley-VCH, **2014**, 417 pp.
– C. Rossi, *Al-based Energetic Nanomaterials*, Wiley, **2015**, 154 pp.
– V. E. Zarko, A. A. Gromov, *Energetic Nanomaterials, Synthesis, Characterization and Application*, Elsevier, **2016**, 374 pp.
– E. Lafontaine, M. Comet, *Nanothermites*, Wiley, **2017**, 327 pp.
– B. Khasainov, M. Comet, B. Veyssiere, D. Spitzer, Comparison of Performance of Fast-Reacting Nanothermites and Primary Explosives, *Propellants Explos. Pyrotech.* **2017**, *42*, 754–772.

Napalm

Liquid aliphatic hydrocarbons form stable gels when mixed with 6–13 wt.-% of the aluminum salts of **na**phthenic acid (CAS-Nr. [1338-24-5]) and **palm**itic acid (CAS-No. [57-10-3]). The gels have high tackiness, are highly flammable, and burn stable, but do not drain away on the soil or evaporate. The general term **napalm** is used to describe a large number of different incendiary gels based on gasoline or kerosene. Table N.4 describes the characteristic properties of napalm.

Table N.4: Composition and heat of combustion of napalm.

Composition	
JP-4	94.9
Al-salts of naphthenic and palmitic acid (wt.-%)	4.3
Silica (wt.-%)	0.2
m-Cresol	0.6
$\Delta_c H$ (kJ g^{-1})	44.35

The burn time, t_B (min), of napalm is a function of mass, m (kg), and can be approximated with the following expression:

$$t_b = 9 \cdot m^{0.8}$$

The surface burn rate is about 25–40 g m^{-2} min^{-1}. A standard fire (1 kg incendiary gel on a mild steel plate (25 × 25 cm) over the total burn time yields the following maximum temperatures (Table N.5).

Table N.5: Maximum temperatures.

Location	T (°C)
Inside the incendiary gel	600
Below the mild steel plate	400
5 cm above the steel plate	650
25 cm above the steel plate	800

-   H. W. Koch, H. H. Licht, Brandstoffe, Brandmunition, Brandwirkung, *Bericht CO 34/74*, ISL, Saint-Louis, 12.12.**1974**, 19 + X pp.

Naphthalene

Naphthalin

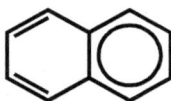

Formula		C$_{10}$H$_8$
GHS		02, 08, 09, 07
H-phrases		228-351-400-410-302
P-phrases		210-241-280-281-405-501a
UN		1334
EINECS		202-049-5
CAS		[91-20-3]
m_r	g mol^{-1}	128.17
$\rho_{25\ °C}$	g cm^{-3}	1.1536
$\Delta_f H$	kJ mol^{-1}	77.95
$\Delta_c H$	kJ mol^{-1}	−5156.5
	kJ g^{-1}	−40.232
	kJ cm^{-3}	−46.411
Mp	°C	80.5
Bp	°C	218
$\Delta_m H$	kJ mol^{-1}	18.81
$\Delta_v H$	kJ mol^{-1}	70.85
c_p	J K^{-1} mol^{-1}	196.06
Λ	wt.-%	−299.58

Naphthalene is a non-hygroscopic fuel with a characteristic smell. It has been suggested as fuel in incendiaries and obscurants and is used in the LK-6-TM gas-generator composition (Mg/NaNO₃/naphthalene) of the SA-6 missile.

- E. Nourdin, *Composition incendiaire et projectile incendiaire dispersant une telle composition*, EP663376A1, **1994**, France.
- L. A. Klyachko, L. Y. Kashporov, N. A. Silin, Combustion of Magnesium – Sodium Nitrate Mixtures. III Concentration Limits of the Combustion of Magnesium – Sodium Nitrate – Organic Fuel Mixtures, *Fisika Goreniya I Vzryva* **1993**, 79–83.
- Espagnac, G. D. Sauvestre, *Method for opaquing visible and infrared radiance and smoke ammunition which implements this method*, US Patent 4 697 521, **1983**, France.

N-Butyl-*N*-(2-nitroxyethyl)nitramine, Bu-NENA

N-Butyl-N-(2-nitroxyethyl)nitramin, Butyl-NENA

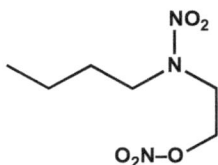

Aspect		Liquid
Formula		$C_6H_{13}N_3O_5$
REACH		LPRS
EINECS		279-976-7
CAS		[82486-82-6]
m_r	g mol^{-1}	207.186
ρ	g cm^{-3}	1.22
$\Delta_f H$	kJ mol^{-1}	−192
$\Delta_{ex}H$	kJ mol^{-1}	−985.2
	kJ g^{-1}	−4.755
	kJ cm^{-3}	−5.801
$\Delta_c H$	kJ mol^{-1}	−4027.0
	kJ g^{-1}	−19.437
	kJ cm^{-3}	−23.713
Ω	wt.-%	−104.25
N	wt.-%	20.28
Mp	°C	−9
Bp	°C	205 (dec)
Friction	N	<353
Impact	J	1
STANAG		4583

Bu-Nena was prepared during World War II in the USA for the first time. Bu-Nena is used as energetic plasticizer for insensitive gun propellants and solid rocket propellants with low combustion temperature. For stabilization, it is typically treated with low amounts (0.1 wt.-%) of MNA. In the Abel test, at 82.2 °C, there is a coloration after 5–6 min.

NEAK

NEAK is the acronym for a eutectic melting at 98–99 °C and recrystallizing at 82 °C It has the following composition:

57–57.1 wt.-% ammonium nitrate
25–25.3 wt.-% ethylenediammine dinitrate
10–10.1 wt.-% potassium nitrate
8–7.5 wt.-% nitroguanidine

The performance of NEAK modified with additional NGu is given in Table N.6.

Table N.6: Composition and performance of NEAK.

Composition	
Nitroguanidine (wt.-%)	49.10
Ethylenediammine dinitrate (wt.-%)	25.00
Ammonium nitrate (wt.-%)	21.15
Potassium nitrate (wt.-%)	3.75
Microballoons (wt.-%)	0.90
ρ (g cm^{-3})	1.64
Parameter	
V_D (m s^{-1})	7030 (ø = 30 mm)
	7420 (ø = 36 mm)
$P_{CJ\ (calc)}$ (GPa)	29.3
Impact$_{50}$ (J)	26
Gap test$_{50}$ (GPa)	5.7–6.6

- W. E. Voreck Jr., *Castable High Explosive Compositions of Low Sensitivity*, US Patent 4421578, **1983**, USA.
- W. Spencer, H. H. Cady, The Ammonium Nitrate with 15 wt% Potassium Nitrate-Ethylenediamine Dinitrate-Nitroguanidine System, *Propellants Explos.* **1981**, 6, 99–103.

NENA

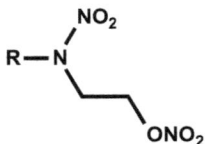

NENA compounds are *N*-alkyl-substituted nitratoethylnitramines, with R = methyl, ethyl, and n-butyl. Nenas were developed in the 1940s in the USA and have since been used as energetic plasticizer for gun propellants. An important NENA derivative is n-butyl-NENA (Bu-Nena).

– R. V. Cartwright, Volatility of NENA and other Energetic Plasticizers Determined by Thermogravimetric Analysis, *Propellants Explos. Pyrotech.* **1995** *20*, 51–57.

NG smoke
NG-Nebel

The NG smoke (NG = **N**ico-**G**raphit) was developed in the early 1990s by *Krone*. It then was the first real multispectral (triband) screening smoke (VIS, IR, and MMW). The pyrotechnic composition contains a graphite intercalation compounds (e.g. graphite hydrogensulphate), which expands as the compositions burns. The resulting expanded graphite particles are very effective dipoles for millimetric radiation and show significant scattering in the mid on long IR. A typical composition featuring burn rates between u = 1–2 mm/s is given thereafter:

– 48 wt.-% graphite hydrogensulphate
– 23 wt.-% potassium perchlorate
– 16 wt.-% magnesium
– 6 wt.-% expanded graphite
– 4 wt.-% azodicarbonamide
– 3 wt.-% phenolic resin

The IR (8–14 μm) mass extinction coefficient for this obscurant is $\alpha_\lambda \sim 0.5\ m^2\ g^{-1}$. Both at 35 and 95 GHz, two-path attenuation of up to 30 dB is achieved.

– U. Krone, E. Schulz, K. Möller, *Pyrotechnischer Nebelsatz für Tarnzwecke und dessen Verwendung in einem Nebelkörper*, DE4337071, **1995**, Germany.
– D. Cegiel, *personal communication*, January **2020**.

Nickel

Aspect		Silvery white
Formula		Ni
GHS		08, 02, 07
H-phrases		372-351-317
P-phrases		260-261-302+352-321-405-501a
UN		3089
EINECS		231-111-4
CAS		[7440-02-0]
m_r	g mol^{-1}	58.693
ρ	g cm^{-3}	8.908
Mp	°C	1455
Bp	°C	2730
$\Delta_m H$	kJ mol^{-1}	17.47
$\Delta_v H$	kJ mol^{-1}	369
c_p	J g^{-1} K^{-1}	0.398
c_L	m s^{-1}	5810
κ	W m^{-1} K^{-1}	83
$\Delta_c H$	kJ mol^{-1}	−239.7
	kJ g^{-1}	−4.084
	kJ cm^{-3}	−36.38
Λ	wt.-%	−27.26

Nickel serves as a fuel in various delay formulations and is also a popular component in dystectic mixtures with other metals such as *Pyronol* or *Pyrofuze*.

Nipolit, nipolite

Nipolit is a solid thermoplastic explosive that was developed by German chemist *Dr. Erich von Holt* at WASAG in World War II as an alternative use for non-conformant solvent-free gun powders. Gun powder sheets (Ger.: *Pulverfelle*) were mixed with organic **ni**trates (such as PETN) and blended by rolling on heated cylinders and extruded into the desired shape. After cooling Nipolit attained high mechanical strength and would even allow regular machining, drilling, and cutting operations. Nipolit is waterproof and was often used without any additional containment particularly in hand grenades and all kinds of engineering charges. Thin Nipolit charges that could be pushed through the observation slits of tanks (Ger.: *Briefschlitzbombe*) definitely exaggerated its reputation as a tank killer and wonder weapon. After the war, Holt patented his formulation and renamed it Holtex (Tab. N.7). A Me-too derivative is "Detasheet."

Table N.7: Composition and performance of Nipolit and postwar Holtex.

Composition	Nipolit	Holtex
NC (12.6–12.7% N) (wt.-%)	34.10	50
DEGDN (wt.-%)	30.00	17
PETN (wt.-%)	35.00	33
Stabilizer (wt.-%)	0.75	
MgO (wt.-%)	0.05	
Graphite (wt.-%)	0.10	
ρ (g cm^{-3})	1.61	1.63
V_D (m s^{-1})	7452	7510
P_{CJ} (GPa)	22.0	22.7
$V/V_0 = 7.2$ (kJ cm^{-3})	−6.40	−6.47

- B. T. Fedoroff, *Dictionary of Explosives, Ammunition and Weapons*, Picatinny Arsenal, Technical report 2510, Dover New Jersey, **1958**, Ger-117.
- K. Böhlein, HOLTEX – Ein neuer Sprengstoff?, *Explosivstoffe*, **1962**, *10*, 156–157.
- E. von Holt, *Nitrocellulose Containing Explosive Compositions and Methods of preparing Same*, US-Patent 3186882, **1965**, Germany.

Nitric acid

Salpetersäure, Ignol, Salbei

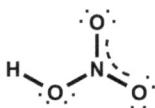

Aspect		Colorless yellowish liquid
Formula		HNO_3
GHS		03, 05
H-phrases		272-314
P-phrases		210-221-303+361+353-305+351+338-405-501a
UN		2031
EINECS		231-714-2
CAS		[7697-37-2]
m_r	g mol^{-1}	63.013
ρ	g cm^{-3}	1.503
$\Delta_f H$	kJ mol^{-1}	−173
Λ	wt.-%	+63.48
N	wt.-%	22.23
c_p	J K^{-1} mol^{-1}	109.8
Mp	°C	−41.6
Bp	°C	82.6

Pure nitric acid (100%) is a colorless liquid fuming in air. Exposure to air yields water uptake. At temperatures above 0 °C, the acid slowly decomposes under the evolution of brown NO_2 leading to a slight yellowing of the acid:

$$4\,HNO_3 \Leftrightarrow 4\,NO_2 + 2\,H_2O + O_2$$

This solution is called white fuming nitric acid (WFNA). Further, NO_2 dissolution yields an intense red coloration: red fuming nitric acid (RFNA). The abbreviations often bear attached figures to describe the content of water and NO_2:

WFNA-7 (7 wt.-% water)
RFNA-3-14 (3% water, 14% NO_2)

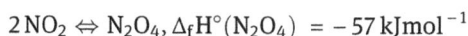

$$2\,NO_2 \Leftrightarrow N_2O_4, \Delta_f H°(N_2O_4) = -57\,kJmol^{-1}$$

RFNA is used as a hypergolic oxidizer together with liquid fuels such as amines (aniline, hydrazine, UDMH, etc.), vinyl alcohol, furfuryl alcohol, and even with solid fuels like polybutadiene in hybrid propulsion. Heating aqueous nitric acid yields loss of water. Upon reaching T = 121.8 °C, the composition of both liquid and vapor phases is identical to 69.2 wt.-% HNO_3 (molar ratio HNO_3:H_2O = 1:1.65) (azeotrope) and the acid cannot be further concentrated by evaporation at ambient pressure. The azeotropic solution is called *concentrated nitric acid* and freezes at −38 °C forming an ionic solid of oxonium nitrate $[H_3O^+][NO_3^-]$. Nitric acid is the most important starting material for the production of inorganic nitrates, organic nitro-, nitramine, nitrimine, and nitrate compounds. A mixture of 88 wt.-% WFNA and 12 wt.-% *fuming sulfuric acid* (a 100% H_2SO_4 containing additional 65 wt.-% SO_3 which is also called *oleum*) is called mixed acid and is frequently used as a nitration agent.

RFNA with up to 5 wt.-% ammonium bifluoride ($NH_4F \cdot HF$) yields much less corrosion in stainless steel containers and is called inhibited RFNA: IRFNA.

– A. Dadieu, R. Damm, E. W. Schmidt, *Raketentreibstoffe*, Springer, Vienna, **1968**, 427–451.
– R. Steudel, *Chemistry of the Non-Metals*, 2nd Edition, De Gruyter, Berlin, **2020**, 417–419.

Nitrimino-5-nitro-hexahydro-1,3,5-triazine, NNHT

2-Nitroimino-5-nitrohexahydro-1,3,5-triazin

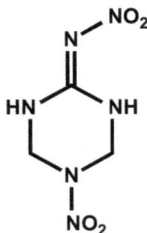

Aspect		Yellow crystals
Formula		$C_3H_6N_6O_4$
CAS		[130400-13-4]
m_r	g mol^{-1}	190.118
ρ	g cm^{-3}	1.730
$\Delta_f H$	kJ mol^{-1}	68
$\Delta_{ex}H$	kJ mol^{-1}	−953.8
	kJ g^{-1}	−5.017
	kJ cm^{-3}	−8.720
$\Delta_c H$	kJ mol^{-1}	−2106.1
	kJ g^{-1}	−11.078
	kJ cm^{-3}	−19.164
Λ	wt.-%	−42.08
N	wt.-%	44.20
Mp	°C	207 (dec)
T_{5ex}	°C	240
Impact	J	2.6
Friction	N	>355

NNHT was regarded for quite some time as a future insensitive high explosive. This was based on the fact that NNHT shares structural similarities with both insensitive nitroguanidine and powerful RDX. However, it turned out that NNHT is more sensitive to impact than RDX and less powerful than NGu.

- I. J. Dagley, M. Kony, G. Walker, Properties and Impact Sensitiveness of cyclic Nitramine Explosives Containing Nitroguanidine Groups, *J. Energ. Mater.* **1995**, *13*, 35–56.
- A. J. Bracuti, Crystal structure of 2-nitrimino-5-nitro-hexahydro-1,3,5-triazine, *J. Chem. Cryst.* **2004**, *34*, 135–140.
- A.M. Astachov, A. D. Vasiliev, M. S. Molokeev, A. A. Nefedov, L. A. Kruglyakova, V. A. Revenko, E. S. Buka, S-Nitrimino-5-Nitrohexahydro-1,3,5-triazine, Structure and Properties, *NTREM-2005*, Pardubice, **2005**, 443–456.
- A. Hahma, Ignition and Combustion of Aluminum in High Explosives, *J. Pyrotech.* **2007**, *26*, 24–46.

Nitro-1,2,4-triazol-5-one, NTO, ONTA

5-Nitro-2,4-dihydro-1,2,4-triazol-3-on

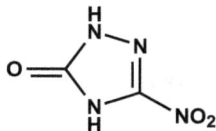

Formula		$C_2H_2N_4O_3$
REACH		LRS
EINECS		213-254-4
CAS		[932-64-9]
m_r	g mol^{-1}	130.063
ρ	g cm^{-3}	1.93
$\Delta_f H$	kJ mol^{-1}	−97
$\Delta_{ex}H$	kJ mol^{-1}	−528.3
	kJ g^{-1}	−4.062
	kJ cm^{-3}	−7.840
$\Delta_c H$	kJ mol^{-1}	−975.9
	kJ g^{-1}	−7.503
	kJ cm^{-3}	−14.481
Λ	wt.-%	−24.6
N	wt.-%	43.08
Dp	°C	264
Friction	N	296
Impact	J	>50 (BAM)
V_D	m s^{-1}	7860 at 1.80 g cm^{-3}
		8500 at 1.91 g cm^{-3}
P_{CJ}	GPa	29.4 at 1.80 g cm^{-3}
		27.8 at 1.781 g cm^{-3}
$\varnothing_{cr}$	mm	<3

NTO is an important powerful insensitive high explosive. It is used in a number of melt-castable and cure-castable IHE such as *IMX-101*, *IMX-104*, XF13333, and B-2248 (Table N.8). NTO is soluble in water (S ~ 10 g l^{-1}). Bearing acidic protons, NTO forms stable and mostly insensitive salts with a number of nitrogen bases (hydrazine, hydroxylamine, guanidine, ethylenediamine, etc.) which are of interest for use as high nitrogen sources in gas generators and gun propellants. The electrochemical oxidation of NTO in acidified solutions via comproportionation yields the very insensitive high explosive *AZTO*.

Table N.8: Compositions containing NTO and their properties.

Composition	XF13333	B2248	Properties	XF13333	B2248
NTO (wt.-%)	48	46	ρ	1.727	1.685
HMX (wt.-%)	–	42	V_D (m s^{-1})	7143	8,050
TNT (wt.-%)	31	–	P_{CJ} (GPa)	22.4	–
Al, spherical (wt.-%)	14	–	Friction (N)	296	>355
HTPB (wt.-%)	–	12	Impact (J)	>50	
D2-wax (wt.-%)	7	–	$\varnothing_{cr}$ (mm)	60	13

- K.-Y. Lee, M. M. Stinecipher, Synthesis and Initial Characterization of Amine Salts of 3-Nitro -1,2,4-triazol-5-one, *Propellants Explos. Pyrotech.* **1989**, *14*, 241–244.
- A. Becuwe, A. Delclos, Low-Sensitivity Explosive Compounds for Low Vulnerability Warheads, *Propellants Explos. Pyrotech.* **1993**, *18*, 1–10.
- E. F. Rothgery, F. W. Migliaro Jr., *Hydroxylammonium Salts of 5-Nitro-1,2,4-triazol-3-one*, US Patent 5,274105, USA, **1993**.

4-Nitro-2-(1-nitro-*1H*-1,2,4-triazol-3-yl)*2H*-1,2,3-triazole

4-Nitro-2-(1-nitro-1H-1,2,4-triazol-3-yl)-2H-1,2,3-triazol

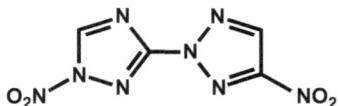

Aspect		Light yellow crystals
Formula		$C_4H_2N_8O_4$
CAS		[210708-11-5]
m_r	g mol^{-1}	226.111
ρ	g cm^{-3}	1.82
$\Delta_f H$	kJ mol^{-1}	515.47
$\Delta_{ex}H$	kJ mol^{-1}	−914.9
	kJ g^{-1}	−4.046
	kJ cm^{-3}	−7.364
$\Delta_c H$	kJ mol^{-1}	−2375.4
	kJ g^{-1}	−10.505
	kJ cm^{-3}	−19.120
Λ	wt.-%	−35.38
N	wt.-%	49.56
Dp	°C	160

Friction	N	240
Impact	J	1.5
T_i	°C	256
V_D	m s^{-1}	8300 at 1.82 g cm^{-3}
		7760 at 1.57 g cm^{-3*}
P_{CJ}	GPa	30.8 at ρ = 1.82 g cm^{-3}
$\sqrt{2E}$	m s^{-1}	2750 at ρ = 1.82 g cm^{-3}
		2550 at ρ = 1.57 g cm^{-3*}

*Determined in a formulation with 7 wt.-% wax.

4-Nitro-2-(1-nitro-1H-1,2,4-triazol-3-yl)-2H-1,2,3-triazole is significantly more sensitive than isomeric DNBTR.

- H. H. Licht, H. Ritter, Synthesis and Explosive Properties of Dinitrobitriazole, *Propellants Explos. Pyrotech.* **1997**, *22*, 333–336.

4-Nitro-2-(5-nitro-*1H*-1,2,4-triazol-3-yl)*2H*-1,2,3-triazole

4-Nitro-2-(5-nitro-1H-1,2,4-triazol-3-yl)2H-1,2,3-triazol

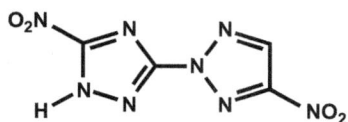

DNBTR		
Formula		$C_4H_2N_8O_4$
CAS		[159536-71-7]
m_r	g mol^{-1}	226.111
ρ	g cm^{-3}	1.89
Δ_fH	kJ mol^{-1}	189.90
$\Delta_{ex}H$	kJ mol^{-1}	−986.3
	kJ g^{-1}	−4.362
	kJ cm^{-3}	−8.244
Δ_cH	kJ mol^{-1}	−2049.8
	kJ g^{-1}	−9.066
	kJ cm^{-3}	−17.134
Λ	wt.-%	−35.38
N	wt.-%	49.56
Mp	°C	178
Dp	°C	247
Friction	N	>360
Impact	J	>15
V_D	m s^{-1}	7985 at 1.89 g cm^{-3}
$\sqrt{2E}$	m s^{-1}	2610 at ρ = 1.89 g cm^{-3}

DNBTR is extremely insensitive when compared with isomeric 4-nitro-2-(5-nitro-1*H*-1,2,4-triazol-1-yl)2*H*-1,2,3-triazole.

– H. H. Licht, H. Ritter, New Energetic Materials from Triazoles and Tetrazines, *J. Energ. Mater.* **1994**, *12*, 223–235.

Nitro-2,4,6-triaminopyrimidine-1,3-dioxide, NTAPDO

2,4,6-Triamino-5-nitro-pyrimidin-1,3-dioxid

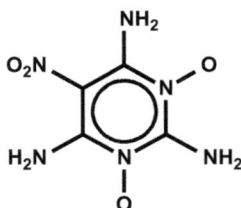

TX 1033		
Aspect		Yellow crystals
Formula		$C_4H_6N_6O_4$
CAS		[19867-41-5]
m_r	g mol^{-1}	202.128
ρ	g cm^{-3}	1.81
Δ_fH	kJ mol^{-1}	20
$\Delta_{ex}H$	kJ mol^{-1}	−925.2
	kJ g^{-1}	−4.577
	kJ cm^{-3}	−8.285
Δ_cH	kJ mol^{-1}	−2451.6
	kJ g^{-1}	−12.129
	kJ cm^{-3}	−21.953
Λ	wt.-%	−55.41
N	wt.-%	41.58
Impact	cm	91
T_i	°C	256
V_D	m s^{-1}	8025 at 1.81 g cm^{-3}
P_{CJ}	GPa	28.01 at ρ = 1.81 g cm^{-3}

NTAPDO is easily accessible and less sensitive than RDX.

– R. W. Millar, S. P. Philbin, R. P. Claridge, J. Hamid, Novel Insensitive High Explosive Compounds Based on Heterocyclic Nuclei Pyridines, Pyrimidines, Pyrazines and Their Benzo Analogues, *ICT-Jata* **2002**, V4.

Nitro-4,6-bis(5-amino-3-nitro, 1*H*-1,4-triazol-1-yl)pyrimidine

1,1'-(5-Nitro-4,6-pyrimidindiyl)bis[3-nitro-1H-1,2,4-triazol-5-amin]

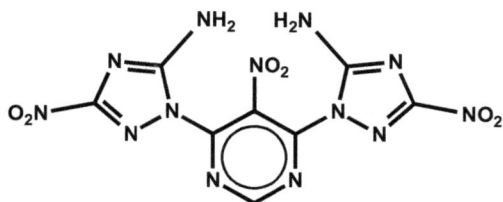

DANTNP, ANTAPM		
Aspect		Yellow crystals
Formula		$C_8H_5N_{13}O_6$
CAS		[141227-99-8]
m_r	g mol^{-1}	379.211
ρ	g cm^{-3}	1.865
$\Delta_f H$	kJ mol^{-1}	431
$\Delta_{ex}H$	kJ mol^{-1}	−1668.5
	kJ g^{-1}	−4.400
	kJ cm^{-3}	−8.206
$\Delta_c H$	kJ mol^{-1}	−4293.7
	kJ g^{-1}	−11.323
	kJ cm^{-3}	−21.117
Λ	wt.-%	−52.74
N	wt.-%	48.02
Impact	cm	91
Dp	°C	328
T_i	°C	256
V_D	m s^{-1}	8200 at 1.81 g cm^{-3}
P_{CJ}	GPa	23.4 at 1.81 g cm^{-3}
$\sqrt{2E}$	m s^{-1}	2090

- C. Wartenberg, P. Charrue, F. Laval, Conception, synthese et characterization d'un nouvel Explosif insensible et energetique: le DANTNP, *Propellants Explos. Pyrotech.* **1995**, *20*, 23–26.

Nitroacetylene

Nitroethin

H———≡———NO$_2$

Aspect		Yellowish oil
Formula		C$_2$HNO$_2$
CAS		[32038-80-5]
m$_r$	g mol^{-1}	71.04
ρ	g cm^{-3}	1.222 (calculated)
Δ$_f$H	kJ mol^{-1}	187.6
Δ$_{ex}$H	kJ mol^{-1}	−450.1
	kJ g^{-1}	−6.336
	kJ cm^{-3}	−7.743
Δ$_c$H	kJ mol^{-1}	−1117.6
	kJ g^{-1}	−15.732
	kJ cm^{-3}	−19.225
Λ	wt.-%	−56.30
N	wt.-%	19.72
Bp	°C	121.4

Nitroacetylene forms through nitrodesilylation of trimethylsilylacetylene. It has not been isolated in substance yet but only as solution. It has been used in the Khand reaction to give substituted benzenes.

- M.-X. Zhang, P. E. Eaton, I. Steele, R. Gilardi, Nitroacetylene HC≡CNO$_2$, *Synthesis* **2002**, 2013–2018.
- G. K. Windler, M.-X. Zhang, R. Zitterbart, P. F. Pagoria, K. P. C. Vollhardt, En route to Dinitroacetylene Nitro(trimethylsilyl)acetylene and Nitroacetylene Harnessed by Dicobalt Hexacarbonyl, *Chem. Eur. J.* **2012**, *18*, 6588–6603.
- G. K. Windler, P. F. Pagoria, K. P. C. Vollhardt, Nitroalkynes A Unique Class of Energetic Materials, *Synthesis* **2014**, 2383–2412.

Nitrocellulose, NC

Nitrocellulose, Cellulosenitrat

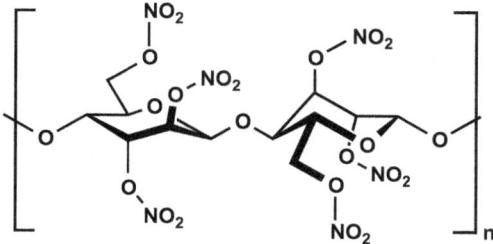

NC, CAS-No. [9004-70-0] is obtained through nitration of cellulose with mixed acid. The raw material for cellulose is either cotton-linters or purified cellulose from pine wood. Today, the security of supply with constant quality cellulose is possibly one of the major greatest challenges for NC manufacturers.

Depending on the degree of nitration of cellulose nitrogen contents between 6.76 up to 14.15 wt.-% are feasible (Tables N.9 – N.11). If the terminal glucopyranose-rings are taken into account (which bear four OH groups each) then with short polymer chains nitrogen contents up to 14.17 wt.-% N are accessible. The above formula depicts a maximum number x = six nitrate groups.

Table N.9: Properties and nitrogen content.

x, Number O-NO$_2$ groups	N content (wt.-%)	NC type (use)
6	14.15	Not accessible on large technical scale
5.5	13.45	Guncotton
5	12.75	Pyrocellulose
4.5	11.96	Collodion*
4	11.11	Collodion
3.5	10.16	Collodion

*Also called pyroxyllin (mostly used in lacquers).

The raw NC is purged consecutively with 0.5% sulfuric acid (to destroy unstable products) and thereafter with boiling sodium carbonate solution. Finally, the fibers are milled down in a multistep process and are boiled repeatedly with water. To protect NC against the autocatalytic decomposition by split nitrogen oxide, NO and nitrous acid, HNO$_2$, urea derivatives (Akardite II = N'-methyl-N,N-diphenylurea), and aromatic amines (diphenylamine) are added as stabilizers. The stabilizers scavenge both NO and HNO$_2$.

Figure N.2 depicts the mass loss of pure NC without stabilization and a NC stabilized with Akardite II at 90 °C. While non-stabilized NC quickly decomposes following

Table N.10: Composition and performance of different nitrocellulose types.

Nitrogen content (wt.-%)	11.1	12.62	13.00	13.15	13.20	13.45	14.14
ρ (g cm^{-3})	1.653	1.655	–	1.656	–	1.657	1.659
Λ (wt.-% O)	−44.49	−34.36	−31.38	−30.83	−30.50	−28.84	−24.24
$\Delta_f H$ (kJ mol^{-1})	−753.4	−707.2	−694.4	−689.2	−687.5	−678.6	−652.7
Q_{ex} (kJ g^{-1})-$H_2O_{(l)}$	3.07	4.07	4.29	4.38	4.41	4.59	4.85
Q_{ex} (kJ g^{-1})-$H_2O_{(g)}$	2.70	3.62	3.68	3.96	4.00	4.10	4.40
V_{ex} (cm^3 g^{-1})-$H_2O_{(g)}$	983	900	880	874	868	857	829
T_{ex} (°C)	2100	2840	3025	3095	3130	3245	3580
Gas composition at T_{ex}							
CO_2	9.5	10.5	13.9	14.3	14.4	16.0	18.4
CO	45.0	43.8	40.8	40.5	40.2	38.6	36.8
H_2	18.6	9.5	9.6	9.2	9.0	7.0	6.6
N_2	9.1	11.1	11.7	11.9	12.1	12.5	9.0
H_2O	17.7	25.1	24.0	24.1	24.3	25.0	23.2

Table N.11: Sensitiveness and performancve of various nitrocellulose types.

Nitrogen content (wt.-%)	13.3	14
Impact (J)	3 (BAM)	1.5
Friction (N)	>353	
DSC onset (°C)	180–220	
c_p (J g^{-1} K^{-1})	0.996	
Koenen (mm)	20	
Trauzl (cm^3)	370	

an initial period of low mass loss of 4 days, stabilized NC only undergoes slow reaction with the stabilizer until it is consumed which at the tested level is at 92 days.

Hot-stage microscopy reveals that NC melts at T = 200 °C before decomposing just a few degrees higher than 220 °C. NC is hygroscopic with the degree of free hydroxyl groups being proportional to the water uptake. The solubility of NC in ether alcohol and acetone is depicted in Figure N.3. Acceptance testing of nitrocellulose includes determination of:

- nitrogen content
- ether alcohol solubles
- acetone insoluble
- stability

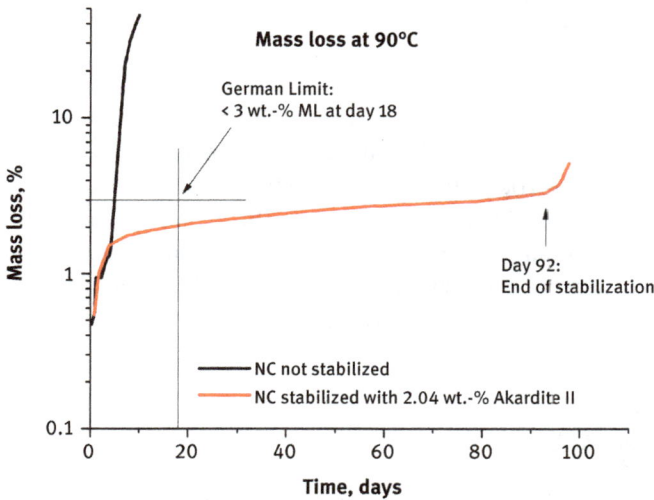

Figure N.2: Mass loss of NC at 90 °C with and without stabilizer after *Bohn*.

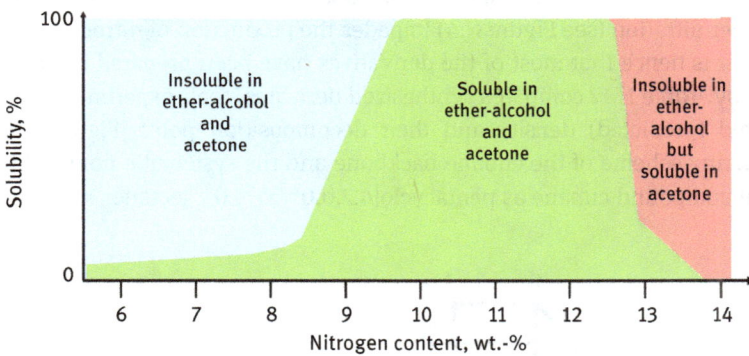

Figure N.3: Solubility of NC as a function of nitrogen content.

Nitrocellulose is the most important component in gun propellants, energetic binders (e.g. D2-wax), and few high explosives (Nipolit, Detasheet). In more recent years, NC, with lower nitrogen content (pyrocellulose e.g Xylocoll®), is used extensively as flame expander in illuminants, spectrally matched flares, and smoke-free indoor pyrotechnics.

– B. Vogelsanger, R. Sopranetti, P. Folly, *Stanag 4178 Ed. 2 Testing of Nitrocellulose*, **2010**.
– M. A. Bohn, Prediction of In-Service Time Period of Three Differently Stabilized Single-Base Propellants, *Propellants Explos. Pyrotech.* **2009**, *34*, 252–266.
– M. A. Bohn, NC-based energetic materials – stability, decomposition ans ageing, *Nitrocellulose- Supply, Ageing and Characterization*, Aldermaston, UK, 24–25 April **2007**.

Nitrocubanes

Nitrocubane

Nitrocubanes belong to the group of polynitrated polyhedral. Due to the high density, high energy content, and low sensitiveness, they are interesting energetic materials. While the unsubstituted cubane, $(CH)_8$, deflagrates when stroke with a hammer, the nitrated derivatives are insensitive to impact! This is due to the strong electron withdrawing effect of the nitrogroups strengthening the C–C bonds in the cage. Vicinal donor–acceptor substituted cubanes eventually decompose by stabilization through bond scission and formation of iminium nitronates (Figure N.4).

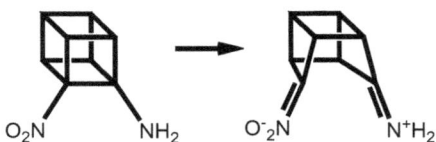

O_2N NH_2 O^-_2N N^+H_2

Figure N.4: Rearrangement and ring cleavage of vicinal amino-nitro cubanes.

The complicated synthesis of the starting materials cubane or cubane-1,4-dicarboxylic acid and the further nitration (see Figure N.4) impedes the production of nitrocubanes on a larger level. It is hence that most of the derivatives have been prepared in milligram amounts only. Table N.12 compares synthesized derivates, their experimental as well as calculated (predicted) density and their decomposition point. Figure N.5 shows the numbering scheme of the cubane backbone and the systematic nomenclature of the parent compound cubane as pentacyclo[$4.2.0.0^{2,5}.0^{3,8}.0^{4,7}$]octane.

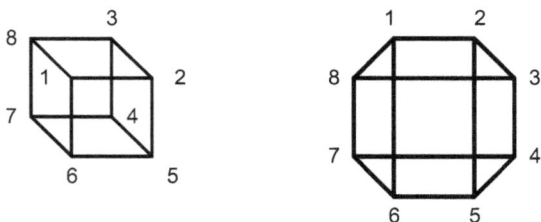

Figure N.5: Numbering of the cubane cage.

Despite the initial expectations, the experimental density of octanitrocubane (ONC) is only $\rho = 1.98$ g cm^{-3} and therefore significantly lower (−8%) than predicted. This is accounted to the strong electrostatic repulsion of the eight nitro groups impeding a denser mode of packing. With heptanitrocubane (HpNC), however, the density is 4%

higher than with ONC. Calculations indicate that HpNC is hence more powerful than ONC (Table N.13) and even has a higher metal acceleration capability ($V/V_0 = 7.2$) than CL-20.

Table N.12: Density and decomposition of cubane and various nitrocubanes.

Compound	Density, calc (g cm^{-3})	Density, exp. (g cm^{-3})	Dec (°C)
Cubane	1.290	1.30	>200
Nitrocubane	1.474	1.453	>200
1,4-Dinitrocubane	1.660	1.660	257
1,3,5-Trinitrocubane	1.77	1.76	267
1,3,5,7-Tetranitrocubane	1.86	1.81	277
1,2,3,5,7-Pentanitrocubane	1.93	1.96	>200
1,2,3,4,5,7-Hexanitrocubane	2.02	1.931	>200
1,2,3,4,5,6,8-Heptanitrocubane	2.07	2.028	>200
1,2,3,4,5,6,7,8-Octanitrocubane	2.13	1.979	275

Table N.13: Calculated performance of ONC, HpNC, and reference explosives CL-20 and HMX.

Compound	ONC	HpNC	CL-20	HMX
CAS-No	[99393-63-2]	[99393-62-1]	[135285-90-4]	[2691-41-0]
m_r (g mol^{-1})	464.132	419.136	438.20	296.156
ρ (g cm^{-3})	1.979	2.028	2.044	1.905
$\Delta_f H$ (kJ mol^{-1})	435	356	403	84
Λ (wt.-%)	0	−9.54	−10.95	−21.61
V_D (m s^{-1})	9492	9618	10,065	9,329
$V/V_0 = 7.2$ (kJ cm^{-3})	−11.31	−11.75	−11.31	−9.76
P_{CJ} (GPa)	40.32	46.48	48.23	39.75

- P. E. Eaton, M. Zhang, R. Gilardi, N. Gelber, S. Iyer, R. Surapaneni, Octanitrocubane: A New Nitrocarbon, *Propellants Explos. Pyrotech.* **2002**, *27*, 1–6.
- M. Zhang, P. E. Eaton, R. Gilardi, Hepta- und Octanitrocubane, *Anwt. Chem.* **2000**, *112*, 422–426.
- K. A. Lukin, J. Li, P. E. Eaton, N. Kanomata, J. Hain, E. Punzalan, R. Gilardi, Synthesis and Chemistry of 1,3,5,7-Tetranitrocubane Including Measurements of its Acidity, Formation of o-Nitro Anions, and the First Preparations of Pentanitrocubane and Hexanitrocubane, *J. Am. Chem. Soc.* **1997**, *119*, 9591–9602.

Nitrodiphenylamine, 2NDPA

2-Nitrodiphenylamin

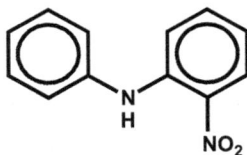

Formula		$C_{12}H_{10}N_2O_2$
GHS		07
H-phrases		315-319-335
P-phrases		261-302+352-305+351+338-321-405-501a
EINECS		204-348-6
CAS		[119-75-5]
m_r	g mol^{-1}	214.224
ρ	g cm^{-3}	1.366
Mp	°C	75.5
Bp	°C	215
$\Delta_f H$	kJ mol^{-1}	78.99
$\Delta_c H$	kJ mol^{-1}	−6230.4
	kJ g^{-1}	−29.084
	kJ cm^{-3}	−39.729
Λ	wt.-%	−201.65
N	wt.-%	13.08

2-NDPA is a stabilizer for gun propellants and *OTTO fuel*. Its reaction products with nitrogen oxides are carcinogenic.

– A. J. Bellamy, M. H. Sammour, Stabilizer Reactions in Cast Double Base Rocket Propellants. Part III: Evidence for Stabilizer Interaction during Extraction of Propellant for HPLC Quantitative Analysis, *Propellants Explos. Pyrotech.* **1993**, *18*, 46–50.

Nitroglycerine, NGl

Nitroglycerin

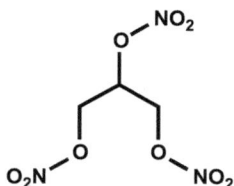

Propane-1,2,3-triyl-1,2,3-trinitrate, glycerine trinitrate, NG		
Formula		$C_3H_5N_3O_9$
REACH		LRS
EINECS		200-240-8
CAS		[55-63-0]
m_r	g mol^{-1}	227.087
ρ	g cm^{-3}	1.593
$\Delta_f H$	kJ mol^{-1}	−369.87
$\Delta_{ex}H$	kJ mol^{-1}	−1424.2
	kJ g^{-1}	−6.272
	kJ cm^{-3}	−9.991
$\Delta_c H$	kJ mol^{-1}	−1525.1
	kJ g^{-1}	−6.716
	kJ cm^{-3}	−10.699
Λ	wt.-%	+3.52
N	wt.-%	18.50
c_p	J g^{-1}	1.3
Dp	°C	223-225
P_{vap}	Pa	3.3 10^{-2}
Friction	N	>355
Impact	J	0.2 (BAM)
V_D	m s^{-1}	8980 at 1.59 g cm^{-3} in 51 mm steel pipe with 3 mm wall strength
Trauzl	cm^3	520
Koenen	mm	24, type H
$\varnothing_{cr}$	mm	<2 mm at 20 °C

Nitroglycerine was the first homogeneous high explosive manufactured on an industrial scale. It is still of paramount importance as an energetic ingredient in double- and triple-base propellants and also find occasional use in certain civilian-type dynamites

Regarding HVD, *LVD*, and *SVD* of NGl, see *Detonation*. NGl is a vasodilatant and hence used as medication against *angina pectoris*.

– E. Contini, N. Flood, D. McAteer, N. Mai, J. Akhavan, Low hazard small-scale synthesis and chemical analysis of high purity nitroglycerine (NG), *RSC Adv.* **2015**, *5*, 87228–87232.

Nitroguanidine, NGu, NQ
Nitroguanidin

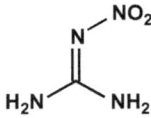

Picrite, guanite		
Formula		$CH_4N_4O_2$
REACH		LRS
EINECS		209-143-5
CAS		[556-88-7]
m_r	g mol^{-1}	104.068
ρ	g cm^{-3}	1.77
$\Delta_f H$	kJ mol^{-1}	−86
$\Delta_{ex}H$	kJ mol^{-1}	−405.7
	kJ g^{-1}	−3.898
	kJ cm^{-3}	−6.900
$\Delta_c H$	kJ mol^{-1}	−879.2
	kJ g^{-1}	−8.448
	kJ cm^{-3}	−14.953
Λ	wt.-%	−30.75
N	wt.-%	53.83
Dp	°C	257
Friction	N	>355
Impact	J	>50 (BAM)
V_D	m s^{-1}	8546 at 1.77 g cm^{-3}
P_{CJ}	GPa	26.8 at 1.704 g cm^{-3} (NQ/Estane 95/5)
$\varnothing_{cr}$	mm	<14 mm at 1.52 g cm^{-3}
Koenen	mm	<1, type C

Nitroguanidine is a very insensitive high explosive. It was initially prepared in 1877 by Josselin. However, it was only in the 1930s following the work by *Smith* that NGU became accessible on an industrial scale (Figure N.6).

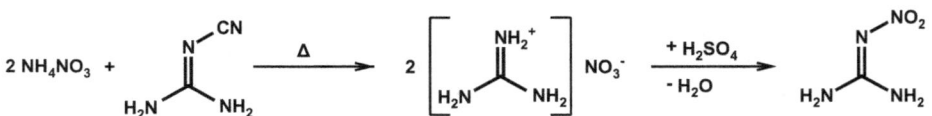

Figure N.6: Synthesis of NGu.

The first large-scale use occurred in wartime Germany in the so-called Gudol powders that featured low combustion temperatures and hence were less erosive and led to longer barrel lifetime.

Typical NGU-crystal qualities are low bulk density (LBD; $\rho \sim 0.3$ g cm^{-3}), high bulk density (HBD; $\rho \sim 0.9$ g cm^{-3}), and spherical high bulk density (SHBD; $\rho \sim 1.0$ g cm^{-3}). NGu finds use in triple- and quadruple-base gun propellants such as MTLS powder and in insensitive high explosives such as *IMX 101* or AFX-770. NGu is more powerful but also less sensitive than FOX-12 and thanks to its simple synthesis also by far cheaper than FOX-12.

Very often, the molecular structure of NGu has been erroneously depicted as that of a nitroamine (Figure N.7). However, reactivity and spectroscopic investigations in the solid state and in solution clearly show that NGu is a nitroimine. Table N.14 depicts common particle types, their properties, and use.

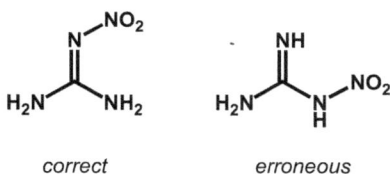

correct   erroneous

Figure N.7: Correct nitroimine and erroneous nitroamine structure of NGu.

Table N.14: Properties and use of different NGu morphologies.

Parameter	Unit	UF	LBD	HBD	SHBD
Crystal density	g cm^{-3}		1.72–1.73	1.75	1.76
Bulk density	g cm^{-3}		0.3	0.9	1.15
Surface area	m^2 g^{-1}	>100	4–25	4	<0.5
Crystallite size	nm		~120	~120	
Morphology	–	Spherical	Needle	Prismatic	Spheroidal
Diameter	µm	0.300	3–15	60–70	50–1,000
Use	–	Gun prop.	Gun prop.	Rocket prop., HE	HE

- C. S. Choi, Refinement of 2-Nitroguanidine by Neutron Powder Diffraction, *Acta Cryst.* **1981**, *B37*, 1955–1957.
- S. Bulusu, R. L. Dudley, J. R. Autera, Structure of Nitroguanidine: Nitroamine or Nitroimine? New NMR Evidence from ^{15}N-Labeled Sample and ^{15}N Spin Coupling Constants, *Magnet. Res. Chem.* **1987**, *25*, 234–238.
- J. Bracuti, Crystal Structure Refinement of nitroguanidine, *J. Chem. Crystallog.* **1999**, *29*, 671–676.

– E.-C. Koch, Insensitive High Explosives III. Nitroguanidine, Synthesis – Structure – Spectroscopy – Sensitiveness, *Propellants Explos. Pyrotech.* **2019**, 44, 267–292.
– E.-C. Koch, Insensitive High Explosives IV. Nitroguanidine, Initiation & Detonation, *Defence Technology*, **2019**, *15*, 467–487
– E.-C. Koch, Insensitive High Explosives V. Ballistic Properties and Vulnerability of Nitroguanidine-Based Propellants, *Propellants Explos. Pyrotech.* **2020**, *45*, accepted for publication.

Nitroguanylurea, NGUU

Nitroguanylharnstoff

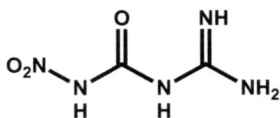

Formula		$C_2H_5N_5O_3$
CAS		[28787-21-5]
m_r	g mol^{-1}	147.093
ρ	g cm^{-3}	1.82
$\Delta_f H$	kJ mol^{-1}	−308
$\Delta_{ex} H$	kJ mol^{-1}	−407.0
	kJ g^{-1}	−2.767
	kJ cm^{-3}	−5.036
$\Delta_c H$	kJ mol^{-1}	−1193.6
	kJ g^{-1}	−8.115
	kJ cm^{-3}	−14.769
Λ	wt.-%	−38.07
N	wt.-%	47.61
Mp	°C	150 (dec)

NGUU has been suggested as an additive to suppress the muzzle flash in double-base propellants.

– R. Boyer, *Propellant*, CA 635781, Canada, **1962**.

Nitromethane
Nitromethan

Formula		CH_3NO_2
GHS		02, 07
H-Phrases		226-302
P-Phrases		260-370+380+375
UN		1261
EINECS		200-876-6
CAS		[75-52-5]
m_r	g mol^{-1}	61.04
ρ	g cm^{-3}	1.139
$\Delta_f H$	kJ mol^{-1}	−112.55
$\Delta_{ex}H$	kJ mol^{-1}	−290.7
	kJ g^{-1}	−4.762
	kJ cm^{-3}	−5.424
$\Delta_c H$	kJ mol^{-1}	−709.7
	kJ g^{-1}	−11.627
	kJ cm^{-3}	−13.243
Ω	wt.-%	−39.32
N	wt.-%	22.95
Mp	°C	−29.2
Bp	°C	101.15
P_{vap}	kPa	3.2
V_D	m s^{-1}	6210 at 1.139 g cm^{-3}
Impact	J	30
Koenen	mm	<1, type O
$\varnothing_{cr}$	mm	18

Nitromethane (NM) was made for the first time in 1872. However, due to its extreme insensitiveness, it was not until 1938 that it could be shown that NM is able to undergo detonation. NM is used frequently as a model compound to investigate the behavior of nitro-compounds in ageing, combustion, and detonation. Nitromethane is also investigated as possible monopropellant in gel propulsion.

– U. Teipel, U. Förter-Barth, Rheological Behavior of Nitromethane Gelled with Nanoparticles, *J. Propul. Power* **2005**, *21*, 40–43.
– S. Kelzenberg, N. Eisenreich, W. Eckl, Modelling Nitromethane Combustion, *Propellants Explos. Pyrotech.* **1999**, *24*, 189–194.

Nobel, Alfred (1833–1896)

Alfred Nobel studied chemistry with the Russian chemist *Nikolai N. Sinin* (1812–1880) in St. Petersburg. There he learned about the difficulties and hazards to work with nitroglycerine as an explosive. As nitroglycerine does not initiate reliably when ignited with a blow from a safety fuse, Nobel invented the concept of an initiating charge (1863). He discovered that NGl could be absorbed by diatomaceous earth (Kieselguhr) to form the plastic but safe gurdynamite (1864). He later found that nitrocellulose would act as a gelling aid on nitroglycerine and invented blasting gelatine in 1875. Finally in 1887, coincidentally with Abel, Nobel developed the first diphenylamine-stabilized double-base propellant *Ballistite*.

Figure N.8: Alfred Nobel http://sok.riksarkivet.se/sbl/Presentation.aspx?id=8143.

- U. Larsson, *Alfred Nobel – Networks of Innovation*, Nobel Museum, **2008**, 216 pp.
- Website of the Nobel Museum in Stockholm http://www.nobelmuseum.se/en

Novichok

Nowischok

Novichok is the overall term for a series of chemical nerve agents developed in the late 1980s in the former Soviet Union. In the account of Russian dissident, Vil S Mirzayanov, a former chemical scientist involved in the program, the molecular structures of several agents codenamed A230, A232, and A234 were disclosed which are given in Figure N.9. Novichok agents have been allegedly used in the deliberate poisoning of GRU defector Sergei Viktorovich Skripal (2018) as well as Russian politician Alexej Nawalny (2020).

The particular danger associated with the Novichok agents is their extreme persistence as is evidenced by their slow hydrolysis.

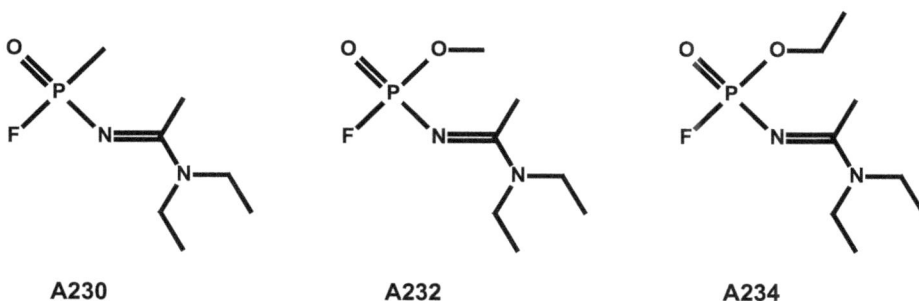

A230   **A232**   **A234**

Figure N.9: Alleged structures of Novichok agents.

- https://www.opcw.org/media-centre/news/2020/09/opcw-provides-technical-assistance-germany-regarding-allegations-chemical accessed on September 18 **2020**.

Table N.9: Properties of Novichok agents compared to *G* and *V* nerve agents.

Property		A230	A232	A234	VX	GB
Alternative designations		Substance 84			EA1701	Sarin
CAS-No		[2442944-33-8] [2387496-12-8]	[2387496-04-8] [2308498-31-7]	[2387496-06-0] [2422944-37-2]	[50782-69-9]	[107-44-8]
Formula		$C_7H_{16}FN_2PO$	$C_7H_{16}FN_2PO_2$	$C_8H_{18}FN_2PO_2$	$C_{11}H_{26}NPO_2S$	$C_4H_{10}FPO_2$
Molecular mass	g mol^{-1}	194.19	210.19	224.22	267.37	140.09
Boiling point	°C	61–62	70–71	73–74	256	147
Density	g cm^{-3}	1.612	1.515	1.414	1.062	1.09
Rate of hydrolysis*	µM min^{-1}	0.17	0.061	0.0032	0.246	6.68
Hydrolysis in water, $t_{1/2}$	H				1000	39
Δ_fH (gas) (calculated)#	kJ mol^{-1}	−667	−885	−903	−643	−895
Δ_{vap}H (calculated)	kJ mol^{-1}			280		105
Dipole moment (calculated)	Debye	3.307	5.083	5.087	0.981	2.368

*Hydrolysis by organophosphorus acid anhydrolase (OPAA) enzyme (*Harvey et al.*);
#Semiempiric PM3.

- S. P. Harvey, L. R. McMahon, F. J. Berg, Hydrolysis and enzymatic degradation of Novichok nerve agents, *Heliyon*, **2020**, *6*, e03153.
- H. Bhakhoa, L. Rhyman, P. Ramasami, Theoretical study of the molecular aspect of the suspected Novichok agent A234 of the Skripal poisoning, *Roy. Soc Open Sci.* **2019**, *6*, rsos.181831.
- E. Nepovimova, K. Kuca, Chemical warfare agent NOVICHOK – mini-review of available data, *Food Chem. Tox.* **2018**, *121*, 343–350.
- V. S. Mirzayanov, *State Secrets, An Insider's Chronicle of the Russian Chemical Weapons Program*, Outskirts Press Inc., Denver, **2009**, pp. 142–168.
- A. E. Smithson, V. S. Mirzayanov, R. Lajoie, M. Krepon, *Chemical Weapons Disarmament in Russia: Problems and Prospects, Report No. 17*, October **1995**, Henry L Stimson Center, 73 pp.

NT smoke
NT-Nebel

The aerosol from common HC-type smokes contains $Zn(H_2O)_nCl_2$ (with n = 4,5,6) which are very strong acids that are both corrosive and yield severe lesions to organic tissue. Inhalation of HC aerosol as well as dermal and ocular contact has hence led to countless grave chemical burns and numerous fatalities. In view of this *Krone* investigated, the possibility to impede the formation of aquo complexes by introducing NH_3 as an alternative and possibly stronger binding σ-donor ligand. He succeeded in doing so with developing the NT obscurants (Table N.15).

Table N.15: NT obscurants.

Component	I	II	III	IV
Ammonium perchlorate (wt.-%)	30.0	31.3	34.0	40.5
Pergut®S20 (wt.-%)	15.0	15.0	15.0	15.0
Ammonium chloride (wt.-%)	14.3	13.0	10.3	3.8
Zinc oxide (wt.-%)	31.3	31.3	31.3	31.3
Dioctyl phthalate (wt.-%)	9.4	9.4	9.4	9.4
Burn rate (mm s^{-1})	0.7	0.9	1.0	1.4
pH of aerosol (–)	6.2	6.0	5.9	5.0

Upon combustion of NT, the formal reaction yields

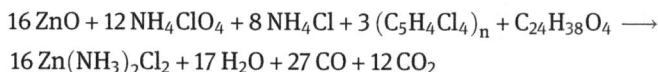

$$16\,ZnO + 12\,NH_4ClO_4 + 8\,NH_4Cl + 3\,(C_5H_4Cl_4)_n + C_{24}H_{38}O_4 \longrightarrow$$
$$16\,Zn(NH_3)_2Cl_2 + 17\,H_2O + 27\,CO + 12\,CO_2$$

The ammines of the type $[Zn(NH_3)_2]Cl_2$ are indeed only mildly acidic (see pH values in Table N.16) proving Krones idea. However, a severe shortfall is that those complexes rapidly hydrolyze when in contact with say mucous membranes. This happens upon inhalation and hence the corrosive effects occur without initial warning symptoms. In addition, NT is less effective an obscurant when compared with HC in view of its low yield factor $Y_F < 1$.

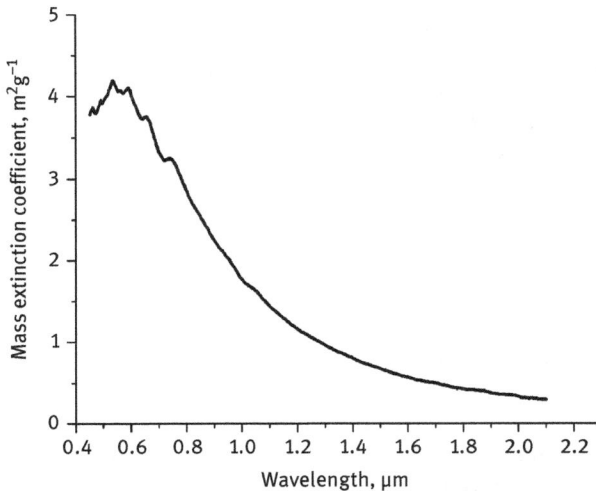

Figure N.10: Mass extinction coefficient of NT smoke at 13% RH and 22 °C.

Table N.16: NT obscurants, performance and sensitiveness.

Component	NT(III)	Obscurant		50% RH	70% RH
Density (g cm^{-3})	1.9–2.1	NT(III)	$\alpha_{630\ nm}$ (m^2 g^{-1})	4.8	5.1
Heat of explosion (kJ g^{-1})	2.283	HC	$\alpha_{630\ nm}$ (m^2 g^{-1})	4.6	5.1
Ignition temperature (°C)	263				
DSC onset (°C)	198	NT(III)	Y_F (g g^{-1})	0.6	0.7
Koenen test (mm)	10, type E	HC	Y_F (g g^{-1})	1.1	1.3
Friction (N)	360				
Impact (J)	10				

- U. Krone, K. Moeller, *Smoke Composition*, US-Patent 4,376,001, Germany, **1983**.
- U. Krone, Herstellung und Bearbeitung eines Neuen Pyrotechnischen Nebelsatzes NT-Nebel, *ICT-Jata*, Karlsruhe, **1981**, pp. 211–234.
- U. Krone, Pyrotechnisch erzeugte Tarnnebel – Vom HC zum NG Nebel. *Sprengstoffgespräch*, Lünen, **1996**, 161–167.
- *ARAED-TR-95014*, Picatinny Arsenal, October **1995**, 31 pp.

O

Obscurant compositions

Nebelsätze

Pyrotechnic obscurants upon combustion yield a primary aerosol of high hygroscopicity = low deliquescent relative humidity. The aerosol then attracts water from the atmosphere thereby producing droplets.

The figures of merit of pyrotechnic obscurants are the aerosol yield, the mass consumption rate, and the mass extinction coefficient in a defined spectral range.

The aerosol yield depends on the chemical composition of the primary aerosol, the relative humidity, and temperature. It is written as the ratio of the aerosol weight (m_A) by the mass of the pyrotechnic composition burnt (m_P):

$$Y_A = \frac{m_A}{m_P}$$

Like the aerosol yield, the mass extinction coefficient of the aerosol, α_λ ($m^2\ g^{-1}$), also depends on the relative humidity and consequently temperature. Figure O.1 depicts the use of screening smoke for the self-protection of vehicles.

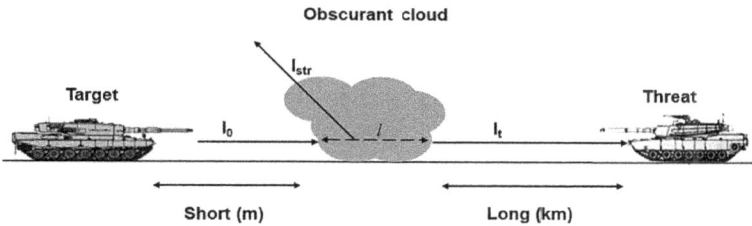

Figure O.1: Application of screening smoke for vehicle self-protection.

With I_0 being the radiant intensity of the target, I_t the transmitted intensity by an aerosol cloud of length l, $T_{obsc} = \frac{I_0}{I_t}$ the transmission, and $\alpha_\lambda = -\frac{\ln T_{obsc}}{cl}$ the wavelength-dependent mass extinction coefficient.

The mass extinction coefficient α_λ is the sum of the scattering and absorption properties of the aerosol particles

$$a(\lambda) = a(\lambda)(sca) + a(\lambda)(abs)$$

While the chemical composition affects absorption, size and shape of the particles determine the scattering behavior.

https://doi.org/10.1515/9783110660562-014

Typical pyrotechnic compositions produce a primary aerosol by any of the following three reactions:

- Halogenation of a metal, for example, Mg or Zn

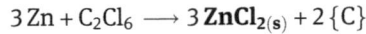

$$3\,Zn + C_2Cl_6 \longrightarrow 3\,\textbf{ZnCl}_{2(s)} + 2\,\{C\}$$

- Vaporization of a hygroscopic halide, for example, KCl or NH$_4$Cl

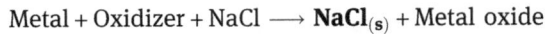

$$Metal + Oxidizer + NaCl \longrightarrow \textbf{NaCl}_{(s)} + Metal\ oxide$$

- Sublimation of phosphorus

$$Metal + Oxidizer + P_R \longrightarrow {}^1/_4\textbf{P}_{4(g)} + Metal\ oxide$$

While the metal halides start attracting moisture, the phosphorus vapor first has to react with the atmospheric oxygen to give phosphorus pentoxide, which then consecutively reacts with moisture in a strongly exothermic reaction to give phosphoric acid. The latter then starts absorbing more water to become hydrated. Table O.1 shows the typical composition of pyrotechnic obscurants and Table O.2 shows the performance of the aerosols in the visible range as function of relative humidity. Figure O.2 depicts the mass extinction coefficient for both obscurant types in the infrared spectral range.

Table O.1: Composition of selected pyrotechnic obscurants.

	HC smoke	RP smoke
Aluminum (wt.-%)	3.6	
Zinc oxide (wt.-%)	48.7	
Hexachloroethane (wt.-%)	47.6	
Red phosphorus (wt.-%)		79.8
Sodium nitrate (wt.-%)		14
Epoxy binder (wt.-%)		4.2
Silica (wt.-%)		2.0

Table O.2: Performance of selected obscurants as a function of relative humidity.

	HC	RP
Y_F at 20 °C (g g^{-1})		
20% RH	1.25	2.99
50% RH	1.58	3.45
80% RH	2.77	4.61
α_{VIS} at 20 °C (m^2 g^{-1})		
20% RH	3.5	3.5
50% RH	3.9	4.2
80% RH	2.9	3.8

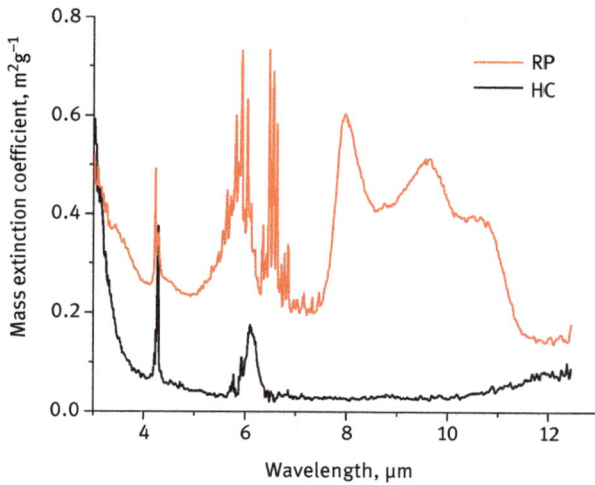

Figure O.2: Mass extinction coefficient of typical HC and RP aerosols in the IR after *Scheunemann*.

- W. Scheunemann, Über die optischen Eigenschaften von Nebelpartikeln im Infraroten, *ICT -Jata*, Karlsruhe **1981**, pp. 235–251.
- E.-C. Koch, Military Applications of Phosphorus and its Compounds, *Propellants Explos. Pyrotech.* **2008**, *33*, 165–176.
- D. W. Hoock Jr., R. A. Sutherland, Obscuration Countermeasures, in J. S. Accetta, D. L. Shumaker (Eds.) *The Infrared and Electro-Optical Systems Handbook, Vol. 7.*, **1996**, pp. 359–493.

Octal

Oktal

Analogous to *hexal*, octal is a blend from octogen (high melting explosive (HMX)) phlegmatized with 5 wt.-% with aluminum in weight ratio 70/30 (Table O.3).

Table O.3: Composition and properties of octal.

Composition	
Octogen (wt.-%)	66.70
Aluminum (wt.-%)	30.00
Montan wax (wt.-%)	3.30
V_D (m s^{-1})	7810 at 1.81 g cm^{-3}
Trauzl (cm^3)	553 at 1.23 g cm^{-3}
Koenen (mm)	3 (F)
Friction (N)	240
Impact (J)	7.5
c_p (J g^{-1} K^{-1})	1.073
Dp (°C)	160

The thermal decomposition of octal starts at T = 160 °C by the HMX dissolved in wax.

– P. Langen, P. Barth, Investigation of the Explosive Properties of HMX /Al 70/30, *Propellants Explos.* **1979**, *4*, 129–131.

Octogen, HMX

Oktogen

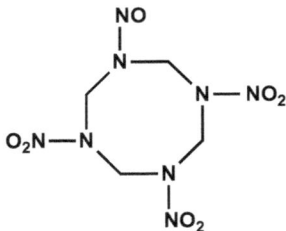

Cyclotetramethylenetetranitramine

Formula		$C_4H_8N_8O_8$
REACH		LRS
EINECS		220-260-0
CAS		[2691-41-0]
m_r	g mol^{-1}	296.156
ρ	g cm^{-3}	1.906 (β-polymorph)
$\Delta_f H$	kJ mol^{-1}	84.01
$\Delta_{ex}H$	kJ mol^{-1}	−1738.6
	kJ g^{-1}	−5.870
	kJ cm^{-3}	−11.189
$\Delta_c H$	kJ mol^{-1}	−2801.4
	kJ g^{-1}	−9.459
	kJ cm^{-3}	−18.029
Ω	wt.-%	−21.61
N	wt.-%	37.83
Dp	°C	280
Koenen	mm	8, H
Trauzl	cm^3	480

Polymorph		α	β	γ	δ	ε
ρ	g cm^{-3}	1.839	1.902	1.780	1.759	1.919
Unit cell	–	Rhombohedric	Monoclinic	Monoclinic	Hexagonal	Monoclinic
Friction	N	108	120	144	108	
Impact	J	2	7.5	2	1	–
c_p	J g^{-1} K^{-1}	1.038	1.017	1.109	1.310	–

V_D	m s^{-1}	9100 at 1.90 g cm^{-3}
P_{CJ}	GPa	39.5 at 1.90 g cm^{-3}
TL		1376-820
STANAG		4284
MIL-DTL		45444C

Octogen, abbreviated HMX (High Melting Explosive, compared to RDX which melts at much lower temperature), is the most powerful high explosive (next to ε-CL-20) that is manufactured on an industrial scale. It is used in high-performance applications such as shaped charges (SC), explosive formed penetrators (EFP) and in booster charges. HMX was discovered and characterized independently in World War II by *Whitmore* in the USA in 1942 and *Fischer* in Germany in 1943. In analogy to "Hexogen," Fischer devised the name Oktogen while Whitmore used the above explained term HMX. Both in the USA and Germany, explosives were evaluated with regards to their output in, then common, Trauzl test. However, yielding a lower expansion volume than RDX, HMX was disregarded both in the USA and Germany. It was only in the early 1950s that density as a determining factor in detonative performance was recognized which led to its introduction. In military applications, typical particle size distributions are utilized to achieve high filling degrees in charges. Those particle size classes are called classes 1 to 5 (see Table O.4).

Table O.4: Octogen-particle sizes.

Mesh No. Grain fraction	Web width (μm)	Passing sieve (%)					
		1	2	3	4	5	6
8	2,360				100		
12	1,700			99–100	85–100		99–100
35	500				10–40		
50	300	84–96	100	25–55			90–100
100	150	40–60		10–30	0–15		50–80
120	125		98–100				
200	75	14–26		0–20			15–45
325	45	3–13	75–100			98–100	5–25

- H. Fischer, Notiz über die Darstellung von Oktogen, *Chem. Ber.* **1949**, *82*, 192–193.
- H. H. Licht, HMX (Octogen) and its Polymorphic Forms, *Chem. Stab. of Expl.*, Tyringe, Sweden, **1970**, 168–179.
- M. Herrmann, W. Engel, N. Eisenreich, Thermal Expansion, Transitions, Sensitivities and Burning Rates of HMX, *Propellants Explos. Pyrotech.* **1992**, *17*, 190–195.
- B. L. Korsunski, S. M. Aldoshin, S. A. Vozchikova, N. I. Golovina, N. V. Chukanova, G. V. Sjilov, A new crystalline HMX polymorph: ε-HMX, *Russ. J. Phys. Chem. B* **2010**, 4, 934–941.

Octol
Oktol

Octol is the term describing melt-castable formulations based on TNT with high octogen content (Table O.5). Due to the high difference in density casting, octol yields a significant density gradient with an accumulation of HMX at the bottom of the casting. Octol is used frequently in SC and explosively formed penetrators.

Table O.5: Composition and properties of octol.

	Type I octol 75/25	Type II octol 70/30
Octogen (wt.-%)	75	70
TNT (wt.-%)	25	30
ρ cast (g cm^{-3})	1.800	1.790
ρ cast under vacuum (g cm^{-3})	1.810–1.825	1.805–1.810
TMD (g cm^{-3})	1.835	1.822
ρ (g cm^{-3})	1.81	1.81
V_D (m s^{-1})	8643	8319
P_{CJ} (GPa)	31.4	
$\Delta_{ex}H$ (kJ g^{-1})		4.732
Impact (J)	–	20
c_p (J K^{-1} g^{-1})	–	0.84
NOL-SLGT (GPa)	1.86	
BICT water-gap test (GPa)		2
SCO (reaction type/°C)	I/201	
$\sqrt{2E_G}$ (m s^{-1})	2910	2790

Modifications containing additional aluminum are called *octonal (see HTA-3)*.

ODTX, one-dimensional time to explosion

The ODTX test serves the determination of thermal sensitiveness and thermal explosion phenomena (cook-off) of energetic materials. With ODTX, the threshold temperatures to explosion, time to explosion, as well as the response of the energetic material under investigation are determined. From this information, the activation energy and frequency factors of the thermal explosion can de derived to deduce conditions for safe transport and storage of an energetic material. The ODTX can be conducted either with cylindrical or spherical samples of different sizes. Very often, the set-up shown in Figure O.3 is used. It shows a spherical sample of the explosive ($\varnothing$ = 12.7 mm) which form fits between two electrically heated aluminum cylinders with hemispherical cavities ($\varnothing$ = 75 mm, l = 50 mm).

Both calottes are sealed off by a copper ring and two circumferential blades pressing into it. This set-up is held force fit in a hydraulic press and is able to withstand gas pressures up to p = 150 MPa. At given temperature, the time to explosion is measured with a microphone indicating the event.

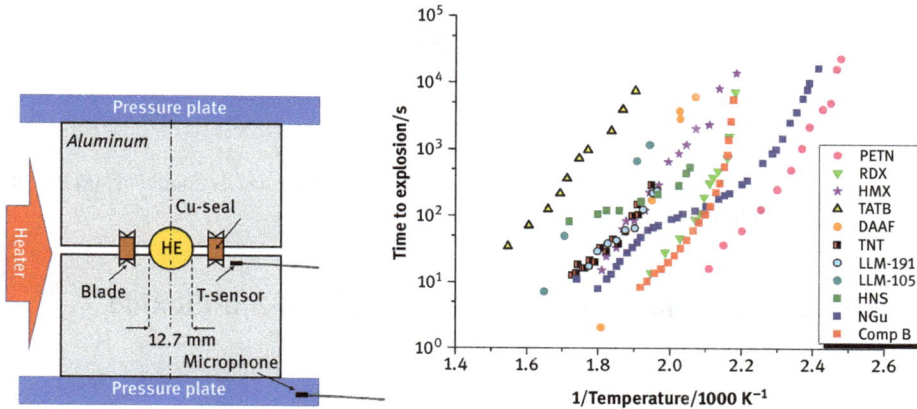

Figure O.3: Setup for small ODTX test (left) and typical values (right).

– Michael L. Hobbs, Michael J. Kaneshige, and William W. Erikson, Predicting Large-scale Effects During Cookoff of PBXs and Melt-castable Explosives, *26th ICDERS*, July 30th – August 4th, **2017**, Boston, MA, USA, 6 pp.

OTTO-II

OTTO-II (Tab. O.6) is a stabilized monoergol used in naval torpedoes invented by and named after *Dr. Otto Reitlinger* (1891–1971) in 1963. It is insensitive to weapons' attack and, thanks to its high flash point, is not prone to accidental ignition when spilled:

– 75 wt.-% propylene glycol dinitrate
– 22.5 wt.-% dibutyl sebacate
– 1.5 wt.-% nitrodiphenylamine

Table O.6: Properties of OTTO fuel.

Parameter	Value
Color	Reddish orange
CAS-No.	106602-80-6
Density (g cm^{-3})	1.2314 at 25 °C
Freezing point (°C)	−32
Vapor pressure (Pa)	11.69
Flash point (s)	120–122
I_{sp} (s) (0.1:6.89 MPa expansion)	207.8

- O. Reitlinger, *Liquid monopropellants of reduced shock sensitivity and explodability*, US Patent 4,026,739, USA, **1977**.

1*H*,4*H*-[1,2,5]Oxadiazolo[3,4-c][1,2,5]oxadiazol-3,6-dioxide

1H,4H-[1,2,5]Oxadiazolo[3,4-c][1,2,5]oxadiazol 3,6-dioxid

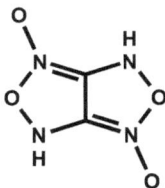

ODOD		
Formula		$C_2H_2N_4O_4$
CAS		[928170-96-1]
m_r	g mol^{-1}	146.06
ρ	g cm^{-3}	2.00±0.1
Δ_fH	kJ mol^{-1}	264
$\Delta_{ex}H$	kJ mol^{-1}	−1024.9
	kJ g^{-1}	−7.017
	kJ cm^{-3}	−14.034
Δ_cH	kJ mol^{-1}	−1336.9
	kJ g^{-1}	−9.153
	kJ cm^{-3}	−18.306
Ω	wt.-%	−10.95
N	wt.-%	38.36
V_D	m s^{-1}	7975 at ρ = 1.67 g cm^{-3}
P_{CJ}	GPa	31 at ρ = 1.67 g cm^{-3}
$\sqrt{2E_G}$	% HMX	130 at V/V$_0$ = 2.20

Wallin et al. suggested ODOD in 2007 as a prospective high explosive.

– H. Östmark, S. Wallin, P. Goede, High Energy Density Materials (HEDM) Overview, Theory and Synthetic Efforts at FOI, *Cent. Eur. J. Energ. Mater.* **2007**, 4, 83–108.

Oxite

Oxite is an extrudable igniter mixture for gun propellants. The name oxite refers to the positive oxygen balance (Λ = +13.90 wt.-%) which facilitates ignition of an under-balanced substrate. Oxite yields only little hot particles mainly due to condensation of KCl:

30.79 wt.-% nitrocellulose, 13.5% N
20.69 wt.-% nitroglycerine (wt.-%)
47.49 wt.-% potassium perchlorate (wt.-%)
1.00 wt.-% centralite I
0.03 wt.-% soot

– C. Roller, B. Strauss, D. S. Downs, "Development of Oxite – A Strand Igniter Material for LOVA Propellant," CPIA Publication 432, Vol. II, pp. 377–390, October **1985**.

Oxygen balance
Sauerstoffbilanz

The oxygen balance, Λ (wt.-%), of a substance gives its surplus or deficiency of oxygen by weight. For a substance with the formula $C_a\text{-}H_b\text{-}F_c\text{-}N\text{-}P_d\text{-}O_e\text{-}S_f$, the oxygen balance can be determined with the below equation.

It has to be taken into account that the number of hydrogen atoms reduces by the exact number of fluorine atoms, as the formation of HF is both thermodynamically as well as kinetically favored over the formation of H_2O:

$$\Lambda = -\frac{15.9994 \cdot 100}{m_r}\left(2a + \frac{(b-c)}{2} + 2.5d + 2f - e\right)$$

For a composite energetic material consisting of a fuel A, forming an oxidation product with a mol oxygen and an oxidizer B, delivering b mol oxygen with the corresponding molecular masses, m_r, and the molar amount x and y of fuel and oxidizer respectively Λ is given by

$$\Lambda = \frac{15.9994 \cdot 100}{x \cdot m_r(A) + y \cdot m_r(B)}(b \cdot y - a \cdot x)$$

Regarding the influence of Λ on both heat of explosion and heat of combustion, *see* Figure E.6, p. 318.

Oxyliquit

Oxyliquit describes blasting explosives based on porous carbonaceous fuels soaked *on-site* with liquid oxygen (Table O.7). Typical fuels are powdered cork, acetylene soot, or wood meal. In the lead block test, oxyliquites yield volumina between 500 and 700 cm^3.

Table O.7: Properties of oxyliquit explosives.

Property	Porous fuel				
	Acetylene soot	Soot	Cork powder	Wood meal	Peat
Density (g cm^{-3})	1.04	0.72	0.63	0.82	0.53
$\Delta_{ex}H$ (kJ g^{-1})	−9.121	−8.347	−6.945	−6.422	−6.987
Trauzl (cm^3)	535	530	510	450	485
V_D (m s^{-1})	4760	4680	3300	3610	3275

- T. Urbański, *Chemie und Technologie der Explosivstoffe, Band III*, VEB Deutscher Verlag für Grundstoffindustrie, Leipzig, **1964**, pp. 326–329

P

PAX

PAX is an acronym for Picatinny Arsenal explosive. It precedes a two or three figure code and a letter indicating P pressable and M melt-castable formulations.

PBX

PBX is a general acronym for plastic bonded explosives. In addition, it is an intro-
duced nomenclature with high explosives (HEs) developed in the authority of the
US Navy and the US Department of Energy (DOE).
  Within the US Navy, the code reads:

PBXN-Number   for qualified high explosives

The numbers characterize
  – 1–99   pressable explosives
  – 100–199 castable explosives
  – 200–299 extrudable explosives
  – 300–399 injectable explosives

Experimental formulations (not yet qualified) of the US Navy start with following
codes:
Place of development
  – PBXC China Lake
  – PBXW White Oak
  – PBXIH Indian Head
  – PBXAF Air Force

Followed by a number. See Tables on pages 709–714 in the Appendix.
  Formulations developed in the authority of the DOE start with the code PBX followed by a four-digit number.

https://doi.org/10.1515/9783110660562-015

PELE (Penetrator with enhanced lateral effects)

PELE is a non-explosive projectile to fight hard and semi-hard targets with enhanced lateral fragmentation. The projectile consists of a cylinder jacket of high-density metal (which can have high brittleness) and a filler of low density (e.g. PE or Al). Upon impact (Fig. P.1), the metal jacket penetrates the target material while the low-density filler is compressed. With increasing penetration, the pressure inside the cylinder increases thereby affecting an increasing expansion of the jacket. With successive penetration of the target, a plug of material accumulates in front of the cylindrical jacket which is finely sheared off as the jacket perforates the target. Evolving behind the target the jacket is now no longer confined to the side and explosively expands and undergoes fragmentation.

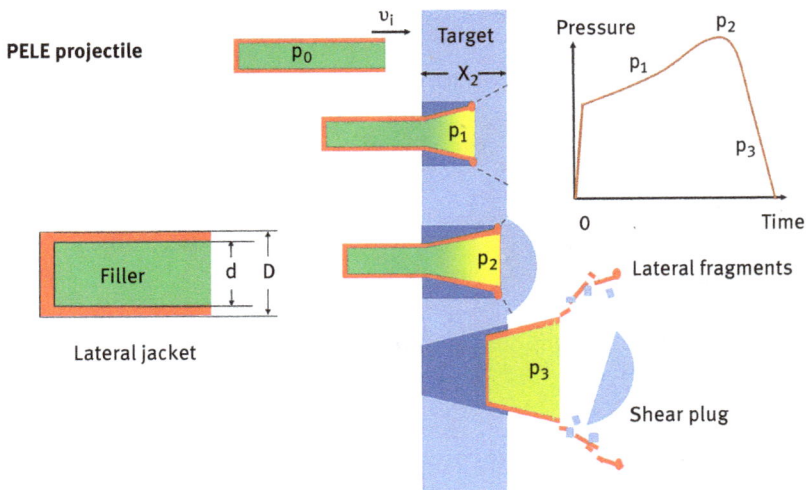

Figure P.1: Schematic mode of action of a PELE projectile (after *Paulus et al.*).

The fragmentation (Figure P.2) increases with
- decreasing wall thickness of the jacket, D-d
- decreasing ductility of the jacket
- with increasing impact velocity, v_i
- with increasing thickness of the target, x_z
- with increasing density of the filler, ρ_F

The pressure in the filler p_i depends linearly on the impact velocity v_0.

As only weak shock waves are generated at typical projectile velocities, the pressure can be described utilizing the acoustic impedance of the filler and target material:

$$p_i = v_i \cdot \frac{a_F \cdot a_T}{a_F \cdot a_T} \text{ mit } a_i = \rho_i \cdot c_L$$

- J. Verreault, Analytical and numerical description of the PELE fragmentation upon impact with thin target plates, *Int. J. Impact Eng.* **2015**, *76*, 196–206.
- G. Paulus, V. Schirm, Impact behavior of PELE projectiles perforating thin target plates, *Int. J. Impact Eng.* **2006**, *33*, 566–579.
- G. Paulus, B. Wellige, G. Gütter, G. Koerber, *Möglichkeiten zur Variation der Lateralwirkung bei PELE*, Schwerpunktaufgabe 2.11, Deutsch-Französisches Forschungsinstitut ISL, Saint Louis, March 2 **2006**.
- G. Paulus, B. Wellige, G. Gütter, *Zum Durchschlag dünner Platten mit PELE-Geschossen*, Schwerpunktaufgabe 2.11, Deutsch-Französisches Forschungsinstitut ISL Saint Louis, March **2005**.

Pentaerythrite tetranitrate, PETN

Pentaerythritoltetranitrat, Nitropenta, Pentrit, Pentastit

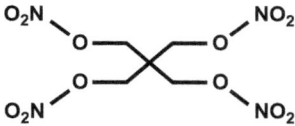

Formula		$C_5H_8N_4O_{12}$
REACH		LRS
EINECS		201-084-3
CAS		[78-11-5]
m_r	$g\ mol^{-1}$	316.138
ρ	$g\ cm^{-3}$	1.778
$\Delta_f H$	$kJ\ mol^{-1}$	−462
$\Delta_{ex}H$	$kJ\ mol^{-1}$	−1875.1
	$kJ\ g^{-1}$	−5.931
	$kJ\ cm^{-3}$	−10.546
$\Delta_c H$	$kJ\ mol^{-1}$	−2648.9
	$kJ\ g^{-1}$	−8.379
	$kJ\ cm^{-3}$	−14.898
Λ	wt.-%	−10.12
N	wt.-%	17.72
c_p	$J\ K^{-1}\ mol^{-1}$	790
$\Delta_{sub}H$	$kJ\ mol^{-1}$	122
Dp	°C	192

Friction	N	60 (BAM)
Impact	J	3 (BAM)
	cm	12 (tool 12)
V_D	m s^{-1}	7975 at $\rho = 1.67$ g cm^{-3}
P_{CJ}	GPa	31 at $\rho = 1.67$ g cm^{-3}
Trauzl	cm^3	523
Koenen	mm	6

PETN is used as a main filler in hand grenades (phlegmatized with wax) as well as in melt-cast formulations based on TNT (wt.-% 50/50) as Pentolite. Plastic explosives based on PETN have become known as *Semtex*.

- T. R. Gibbs, A. Popolato, LASL Explosive Property Data, University of California Press, Berkeley, **1980**, pp. 130–151.

Pentazenium salts
Pentazeniumsalze

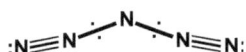

Increased valence bond formula of pentazenium cation with proportionate electron pair bonds (thin lines) to describe the geometry of the pentazenium cation observed in crystallographic investigation.

Pentazenium compounds contain the double-tilted N_5^+ cation, CAS-Nr. [236412-64-9]. The first pentazenium salts have been prepared and characterized by *Christe* in 1999 by reaction at low temperatures. With non-energetic anions (e.g. SbF_6^-), pentazenium salts are stable up to 70 °C before they decompose. However, with energetic anions such as the nitrogen rich $P(N_3)_6^-$ or $B(N_3)_4^-$ explosive decomposition occurs upon reaching T = −20 °C. The long sought-after hypothetical pentazenium azide, $[N_5][N_3]$, and pentazenium pentazolate, $[N_5][N_5]$, cannot be formed as the lattice energy for those ionic compounds is too small. In addition, the ionization potential of N_5^- and N_3^- would need to be higher than the electron affinity of N_5^+ (which is not) to avoid immediate electron transfer and decomposition to N_2.

- Christe, K. O., Wilson, W. W., Sheehy, J. A. and Boatz, J. A. N5+: A Novel Homoleptic Polynitrogen Ion as a High Energy Density Material, *Angew. Chem. Int. Ed.* **1999**, 38, 2004–2009.
- Haiges, R., Schneider, S., Schroer, T. and Christe, K. O., High-Energy-Density Materials Synthesis and Characterization of N5 +[P(N3)6] −, N5 + [B(N3)4] −, N5 + [HF2] − · n HF, N5 + [BF4] −, N5 + [PF6] −, and N5 + [SO3 F] −. *Angew. Chem.* **2004**, 116, 5027–5032.
- M. Rahm, Chalmers University, Sweden, *personal communication*, September **2020**.

Pentazole and pentazolates

Pentazol und Pentazolate

Pentazole, HN_5 CAS-Nr.: [289-19-0], is a promising building block in energetic materials due to its high nitrogen content, N = 98,58 wt.-%, and its high positive heat of formation ($\Delta_f H$ = +443 kJ mol^{-1}). Its anion, pentazolate, *cyclo*-N_5^- (isoelectronic with cyclopentadienyl, $C_5H_5^-$, see, e.g. *Ferrocen*) has been identified in 2004 and has been tamed in a temperature stable salt for the first time in 2017 [*cyclo*-$N_5]_6[H_3O]_3[NH_4]_4$ [Cl] (decomposition at T = 160 °C). Since then, quite a number of stable pentazolates have been prepared in an attempt to determine its performance as an energetic material. Figure P.2 shows various nitrogen containing cations tested so far which all yield pentazolates with decomposition points not exceeding 120 °C. Table P.1 displays the ambient temperature density and sensitiveness as well as the calculated heat of formation and estimated detonative properties with CHEETAH 7.0.

Table P.1: Sensitiveness and performance of various nitrogen rich pentazolates.

Nr.	m_r	ρ (g cm^{-3})	Dp (°C)	$\Delta_f H$ (kJ mol^{-1})	N (wt.-%)	V_D (m s^{-1})	P_{CJ} (GPa)	Impact (J)	Friction (N)
1	88.072	1.317	102	+112.1	95.42	8810	21.70	>50	???
2	144.119	1.483	88	+312.3	86.12	7960	20.14	24	360
3	173.137	1.567	110	+203.4	72.81	6920	18.9	14	160
4	103.087	1.686	85	+471.3	95.11	10400	37.00	6	100
5	104.071	1.601	104	+371.7	80.75	9930	35.80	6	60
6	144.179	1.218	81	+297.2	58.29	5880	10.08	35	>360

- S. Wallin, H. Östmark, N. Wingborg, P. Goede, E. Bemm, M. Norrefeldt, A. Pettersson, J. Pettersson, T. Brinck, R. Tryman, High Energy Density Materials (HEDM) – A Literature Survey, *FOI -R-1418-SE*, December **2004**, Tumba, Sweden.
- C. Zhang, C. Sun, B. Hu, C. Yu, M. Lu, Synthesis and characterization of the pentazolate anion cyclo-N_5^- in $(N_5)_6(H_3O)_3(NH_4)_4Cl$, *Science*, **2017**, *355*, 374–376
- P. Wang, Y. Xu, Q. Lin, M. Lu, Recent advances in the syntheses and properties of polynitroigen pentazolate anion cyclo-N_5^- and its derivatives, *Chem. Soc. Rev.* **2018**, *47*, 7522–7538.
- C. Yang, C. Zhang, Z. Zheng, C. Jiang, J. Luo, Y. Du, B. Hu, C. Sun, K. O. Christe, Synthesis and Characterization of cyclo-Pentazolate Salts of NH_4^+, NH_3OH^+, $N_2H_5^+$, $C(NH_2)_3^+$ and $N(CH_3)_4^+$, *J. Am. Chem. Soc.* **2018**, *140*, 16488–16494.

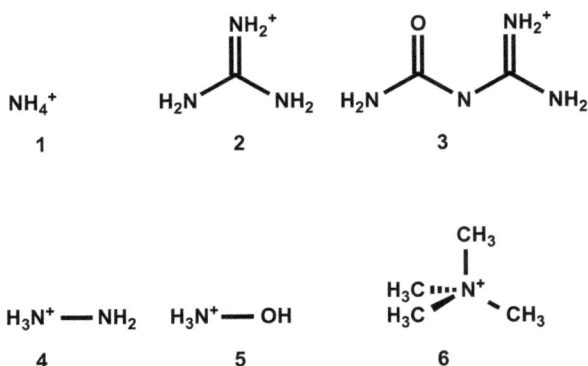

Figure P.2: Cations of prepared and characterized pentazolates.

Pentolite
Pentolit

Pentolite is a melt-castable explosive based on TNT/PETN (50/50 wt.-%) (Table P.2). Pentolit 50/50 with $\rho = 1.56$ g cm^{-3} serves as a standard donor in the NOL-LSGT.

Table P.2: Composition and performance of pentolite.

	Pentolite 50/50
PETN (wt.-%)	50
TNT (wt.-%)	50
TMD (g cm^{-3})	1.71
ρ (g cm^{-3})	1.66
V_D (m s^{-1})	7620
P_{CJ} (GPa)	23.7
$\Delta_{ex}H$ (kJ g^{-1})	5.146
Impact (cm)	30.48
$\varnothing_{cr}$ (mm)	6.7
Dp (°C)	142
c_p (J K^{-1} g^{-1})	1.09
Trauzl (cm^3)	366

Perchloric acid

Perchlorsäure

Formula		$HClO_4$
GHS		03, 05, 07
H-phrases		271-314-302
P-phrases		221-283-303+361+353-305+351+338-405-501a
UN		1873
CAS		[7601-90-3]
m_r	g mol^{-1}	100.459
ρ	g cm^{-3}	1.761
$\Delta_f H$	kJ mol^{-1}	−40.58
Mp	°C	−112
Bp	°C	130
Λ	wt.-%	+63.71

Pure perchloric acid is a limpid liquid with low viscosity fuming slightly at ambient temperature. The concentrated aqueous solutions are viscous. The azeotrope (72 wt.-% $HClO_4$/28 wt.-% H_2O) boils at 203 °C and ambient pressure. The pure acid is obtained by distilling a mixture of the azeotrope and a threefold surplus of highly concentrated sulfuric acid at 90–160 °C and reduced pressures between p = 26 and 40 kPa.

- A. F. Holleman, E. Wiberg, N. Wiberg, *Anorganische Chemie, Band 1 Grundlagen und Hauptgruppenelemente*, 103. Auflage, De Gruyter, Berlin, **2017**, pp. 521–522.
- G. Brauer, *Handbuch der präparativen Anorganischen Chemie*, Band 1, Ferdinand Enke, Stuttgart, **1975**, pp. 327–329.

Perfluoroisobutene, 1,1,3,3,3-pentafluoro-2-(trifluoromethyl) prop-1-ene

Perfluorisobuten, Octafluorisobuten

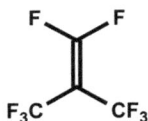

PFIB		
Formula		C_4F_8
CAS		[382-21-8]
EINECS		609-533-9
m_r	g mol^{-1}	200.03
ρ	g cm^{-3}	1.59 (0 °C)
ρ	g l^{-1}	8.2 (25 °C)
$\Delta_f H$	kJ mol^{-1}	−1600 (estimated)
c_p	J mol^{-1} K^{-1}	158.9
T_{crit}	°C	96.25
Mp	°C	−130
Bp	°C	7
Fp	°C	−36.4
Λ	wt.-%	−31.99
LCt$_{50}$	ppm	<1

PFIB is a highly toxic colorless and odorless gas that may form from the thermolysis of PTFE under anaerobic conditions (such as thermal treatment (pyrolysis) of plastic waste). It is not absorbed by activated charcoal and hence can break through protective masks. Its inhalation yields lung edema and further targets the liver. It is assumed that its unusual high toxicity is due to its strong electrophily which causes rapid depletion of the intracellular active antioxidative nucleophiles.

- H. Arito, R. Soda, Pyrolysis products of polytetrafluorethylene and polyfluoroethylenepropylene with reference to inhalation toxicity, *Ann. Occup. Hyg.* **1977**, *20*, 247–255.
- J. Patocka, J. Baigar, Toxicology of perfluoroisobutene, *ASA Newsl.* **1998**, *5*, 16–18. Accessed at https://www.researchgate.net/publication/315767483_Toxicology_of_Perfluoroisobutene

Phosgene, CG

Phosgen, Grünkreuz, D-Stoff, Carbonyldichlorid

Formula		COCl$_2$
CAS		[75-44-5]
EINECS		200-870-3
m_r	g mol^{-1}	98.916
ρ	g cm^{-3}	1.434 (0 °C)
ρ	g l^{-1}	4.12 (25 °C)
$\Delta_f H$	kJ mol^{-1}	−220.08
$\Delta_v H$	kJ mol^{-1}	24.9
c_p	J mol^{-1} K^{-1}	60.7
Mp	°C	−127.8
Bp	°C	7.56
T_{crit}	°C	183
P_{crit}	MPa	5.6
Λ	wt.-%	0
LCt$_{50}$	ppm	3200

The sweet odor of phosgene at low concentration resembles that of rotten apples or moist hay. Phosgene is only toxic upon inhalation. However, it is extremely malicious in that its effects become noticeable only after a delay of 3–12 h and hence lethal doses can be inhaled unnoticed.

– S. Franke, *Lehrbuch der Militärchemie Band 1*, Militärverlag der Deutschen Demokratischen Republik, Berlin **1977**, pp. 338–343.

Phosphorus nitride

Phosphor(V)nitrid

Formula		P_3N_5
REACH		LPRS
EINECS		235-233-9
CAS		[12136-91-3]
m_r	g mol^{-1}	162.955
ρ	g cm^{-3}	2.77
Dp	°C	>700
$\Delta_f H$	kJ mol^{-1}	−962
$\Delta_c H$	kJ mol^{-1}	−1295.4
	kJ g^{-1}	−7.949
	kJ cm^{-3}	−22.019
Λ	wt.-%	−73.64

P_3N_5 forms three phases α, γ, and δ, with α-phase being commercially available. The γ-phase has a density of $\rho = 3.65$ g cm^{-3} and for the δ-phase a density of $\rho = 4.75$ g cm^{-3} is prognosticated, which exceeds the volumetric phosphorus content of RP by nearly 15 % (Table P.3).

Table P.3: P-content of RP, WP, and P_3N_5 allotropes.

Parameter	WP	RP	α-P_3N_5	γ-P_3N_5	δ-P_3N_5
ρ (g cm^{-3})	1.82	2.36	2.77	3.65	4.75
Mol P cm^{-3}	0.0588	0.0762	0.0510	0.0672	0.0874
% RP	77.2	100	66.9	88.2	114.7

P_3N_5 serves as a safe and maintenance-free replacement for RP. P_3N_5 stored in moist air does not form any acids nor does it release any phosphanes. Binary formulations of P_3N_5 with nitrates and even chlorates (compare that with Armstrong's mixture!) are friction insensitive and only mildly impact sensitivity. Table P.4 compares sensitiveness of oxygen balanced RP/KNO$_3$ and selected P_3N_5 oxidizer formulations.

Table P.4: Comparison of oxygen balanced sensitiveness of RP/nitrate and P_3N_5/oxidizer formulations.

	RP/KNO$_3$ 27/73	P_3N_5/KNO$_3$ 35/65	P_3N_5/KClO$_3$ 35/65	P_3N_5/KClO$_4$ 38/62
Friction (N)	40	>360	>360	360
Impact (J)	5	20	7.5	>30

- E.-C. Koch, S. Cudziło, Safer pyrotechnic Obscurants Based on Phosphorus(V) Nitride, *Angew. Chem. Int. Ed.* **2016**, 55, 15439–15442.
- E.-C. Koch, *Pyrotechnischer Nebelsatz zum Erzeugen eines Tarnnebels*, DE-Patent 102016103810B3, **2017**, Germany

Phosphorus, red (amorphous)

Roter Phosphor, RP

Formula		P_R
GHS		02
H-phrases		228-412
P-phrases		210-240-241-273-280-501a
UN		1338
CAS		[7723-14-0]
m_r	g mol^{-1}	30.974
ρ	g cm^{-3}	2.36
Mp	°C	600–615 (under inert atmosphere)
$T_i(O_2)$	°C	~400-440
$\Delta_{sub}H$	kJ mol^{-1}	−128.1
Δ_cH	kJ mol^{-1}	−735.5
	kJ g^{-1}	−23.746
	kJ cm^{-3}	−56.040
Λ	wt.-%	−129.136
Specification		STANAG 4679

Coarse-grained red phosphorus (RP) appears violet brown, while smaller particle sizes appear adobe red or even orange red. RP always bears a characteristic fishy to garlic-type smell which is due to the constant formation of phosphane (PH_3). RP is obtained through monotropic transformation of white phosphorus (WP) under high temperature and pressure (technically realized by treatment of WP with overheated steam at 270–350 °C in a ball mill).

RP is extremely prone to ignition by any given stimuli (impact, friction, shock, temperature, electric discharge, etc.) and readily burns in air.

Though amorphous with regard to a macroscopic dimensions, RP is best described as a distorted version of Hittorf's phosphorus. Figure P.3 shows the three-bonded phosphorus and a section of the polymeric network

Due to its free electron pair, the three-bonded phosphorus is generally very prone to electrophilic attack and oxidation. That is why phosphorus in air and in contact with moisture undergoes the following reactions:

$$2\,P_R + 3\,H_2O_{(g)} \rightarrow PH_{3(g)} + H_3PO_{3(l)} \quad \Delta_RH \;-240\;kJ\,mol^{-1}$$

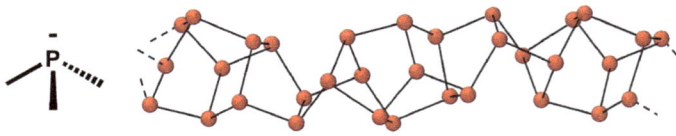

Figure P.3: Three-bonded phosphorus (left), and section of polymeric network of red phosphorus with three-bonded phosphorus (right).

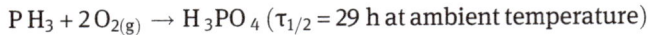

$$PH_3 + 2O_{2(g)} \rightarrow H_3PO_4 \ (\tau_{1/2} = 29 \text{ h at ambient temperature})$$

As a consequence, unstabilized RP is always heavily contaminated with all kinds of phosphoric acids which in turn affect the stability and safety of RP and the formulations based thereon. In addition, following Ostwald's rule, RP undergoes phase change into self-inflammable and toxic WP when heated rapidly (e.g. adiabatic compression of RP-fines) or subjected to shock. It is hence that over the last decades technical qualities of RP have been developed which show increased safety and longevity compared to non-stabilized RP. These qualities contain basic and amphoteric metal oxides, hydroxides, and hydrates (Al(OH$_3$, Mg(OH)$_2$, SnO·H$_2$O, AgO·H$_2$O) that both serve to buffer acid traces as well as to suppress the formation of phosphane. In addition, treatment with long chain paraffins or polymers with decreased permeability for both oxygen and water vapor reduce the amounts of fines (d_p < 1 μm) that are susceptible to rapid oxidation and ignition by ESD or adiabatic compression. However, even state of the art RP (e.g. HB 600 from Clariant) still develop PH$_3$ as depicted in Figure P.4 and given in Tab P.5 which makes venting (release of PH$_3$) and general maintenance of RP-based obscurant ammunition inevitable.

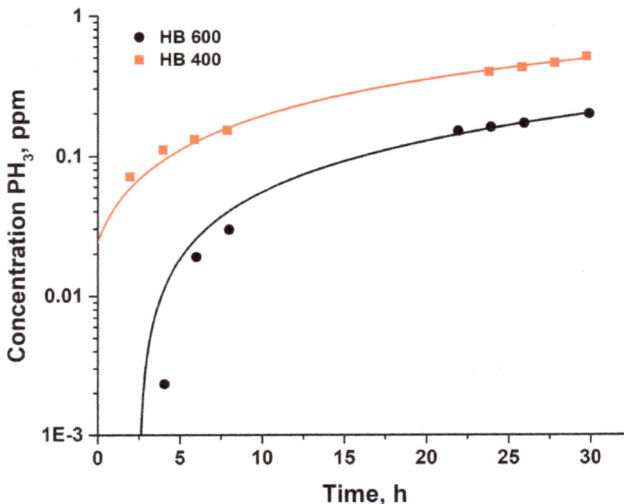

Figure P.4: Evolution of PH$_3$ from PR at 20 °C and 65% RH taking into account the oxidation of PH$_3$ to phosphoric acid after *Lissel*.

Table P.5: Comparison of HB 400 and HB 600 phosphorus.

Parameter	HB 400	HB 600	NATO-STANAG 4679
RP content (wt.-%)	88–92	>91	>90
WP content (wt.-%)	<0.02	<0.02	<0.02
pH of suspension in water (-)	≥6	≥6	
<10 μm (% of g/g)		≤45	
<400 μm (% of g/g)		≥99.5	
Phosphine split rate (ppm/24 h) at 25 °C and 75% RH			≤10
Phosphine split rate (ppm/24 h) at 25 °C and 65% RH	0.39	0.16	

A major disadvantage of RP is the extreme mechanical sensitiveness of its mixtures with oxidizers which has caused a great many number of fires and explosions in the manufacturing industry. In addition, the extreme explosiveness of its reaction with chlorates (*Armstrong's mixture*) and perchlorates requires hermetically sealed-off separate production lines to avoid any cross contamination and explosions resulting from it in industrial manufacture. Finally, plastic bonded obscurants based on RP/metal/nitrates are so impact and shock sensitive that experimental large caliber ammunitions filled with those payloads have undergone in-bore prematures leading to destruction of howitzers. Consequently, the internal design of those shells has to account for the immense shock sensitivity. The main use for RP is in *safety matches*. Second next is the use of RP in pyrotechnic obscurant formulations. In these compositions, the fierce exothermic reaction between an energetic fuel and an oxidizer delivers the necessary heat (eq. (1)) to volatilize the phosphorus (eq. (2)). The vaporized phosphorus spontaneously reacts with the atmospheric oxygen to give phosphorus pentoxide (eq. (3)). Fuel + Oxidizer → Products + Heat

$$\text{Fuel} + \text{Oxidizer} \rightarrow \text{Products} + \text{Heat} \tag{1}$$

$$4\,P_R + \text{Heat} \rightarrow P_{4(gas)} \tag{2}$$

$$P_{4(gas)} + 5\,O_2 \rightarrow P_4O_{10} + \text{Heat} \tag{3}$$

The latter then hydrolyses with atmospheric moisture to give orthophosphoric acid (eq. (4)), which depending on the relative humidity and temperature can hydratize up to stages of n = 13–14 (eq. (5))

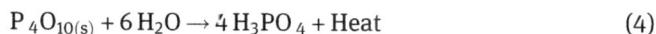

$$P_4O_{10(s)} + 6\,H_2O \rightarrow 4\,H_3PO_4 + \text{Heat} \tag{4}$$

$$H_3PO_4 + n\,H_2O_{(g)} \rightarrow H_3PO_4 \cdot n \cdot H_2O + n \cdot \text{Heat} \tag{5}$$

The consecutive release of heat makes these aerosols particularly effective for screening in the far (thermal) infrared range (λ = 8–14 µm). Another advantage is the yield of aerosol, Y_F, which can reach values of up to 10 at about 80% RH and 20 °C which is unmatched by any other pyrotechnic obscurant. Finally, orthophosphoric acid as the main aerosol constituent is relative benign both with respect to occupational health and environmental concerns when compared to aerosols from metal-halocarbon reactions (HC smoke) or other obscurant materials (brass powder, $TiCl_4$, etc.). Other uses for RP are in spectrally matched decoy flares, marine location markers which have been in use for many decades, and most recently in blast-enhanced explosives where phosphorus plays out its proverbial role as an oxophilic element thereby interacting with any oxygeneous product behind the CJ plane.

- E. Lissel, Elektrochemische Detektion gasförmiger Zersetzungsprodukte zur Beurteilung der Stabilität pyrotechnischer Phrases – Aufbau einer Prüfapparatur und erste Erfahrungen, *2. Internationaler Workshop des WIWEB*, Swisttal-Heimerzheim, **1999**.
- E.-C. Koch, Special Materials in Pyrotechnics IV The Chemistry of Phosphorus and its Compounds, *J. Pyrotech.* **2005**, *21*, 39–50.
- E.-C. Koch, D. Clement. Red Phosphorus-Based Blast Enhanced Explosives, *ICT-JATA*, **2007**, V-30.
- E.-C. Koch, Blast Effect Charge, US-Patent Application 20080006167A, Germany, **2007**.
- E.-C. Koch, Special Materials in Pyrotechnics V Military Applications of Phosphorus and its Compounds, *Propellants Explos. Pyrotech.* **2008**, *33*, 165–176.
- G. O. Rubel, Predicting the Droplet Size and Yield Factors of a Phosphorus Smoke as a Function of Droplet Composition and Ambient Relative Humidity Under Tactical Conditions, Report *ARCSL-TR-78057*, Aberdeen, **1978**.
- K. Raupp, W. Brand, H. Lang, W. Kukla, *Munition zur Erzeugung eines Nebels*, DE10065816A1, **2000**, Germany.
- E.-C. Koch, A. Dochnahl, *Pyrotechnische Wirkmasse zur Erzeugung eines im Infraroten stark emissiven und im Visuellen undurchdringlichen Aerosols*, EP000001173394B9, **1999**, Germany.
- M. Weber, *Pyrotechnische NebelPhrases*, DE3238444C2, **1986**, Germany.
- NATO -Standardization Agency, Brussels, Belgium STANAG Specification for Red Phosphorus, amorphous, microencapsulated (for use in pyrotechnics), No. 4679, March **2013**.
- G. Manton, R. M. Endsor, M. Hammond, An Effective Mitigation for Phosphine Present in Ammunition Container Assemblies and in Munitions Containing Red Phosphorus, *Propellants Explos. Pyrotech.* **2014**, *39*, 299–308.
- a) E.-C. Koch, *IPS*, Fort Collins, **2008**, pp. 531–537; http//fudder.de/phosphor-granate-explo diert-behandlung-in-spezialzelt–118693541.html; http//www.badische-zeitung.de/neuenburg/erneut-unfall-bei-rheinmetall–12489221.html;
- https://www.badische-zeitung.de/brand-in-ruestungsfirma-rheinmetall-80-000-euro-schaden
- https://www.clariant.com/de/Solutions/Products/2014/03/18/16/34/Red-Phosphorus-HB-600 accessed on April 22 2020.

Picramic acid

Pikraminsäure, 2-Amino-4,6-dinitrophenol

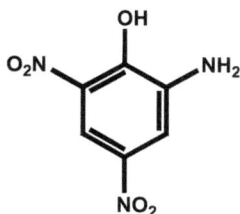

Aspect		Red crystals
Formula		$C_6H_5N_3O_5$
REACH		LRS
EINECS		202-544-6
CAS		[96-91-3]
m_r	g mol^{-1}	199.123
ρ	g cm^{-3}	1.69
$\Delta_f H$	kJ mol^{-1}	−248.53
$\Delta_{ex} H$	kJ mol^{-1}	−788.0
	kJ g^{-1}	−3.957
	kJ cm^{-3}	−6.688
$\Delta_c H$	kJ mol^{-1}	−2827.2
	kJ g^{-1}	−14.198
	kJ cm^{-3}	−23.995
Λ	wt.-%	−76.33
N	wt.-%	21.10
Mp	°C	180
Dp	°C	217
Friction	N	>355 (BAM)
Impact	J	>40 (BAM)

Picramic acid is a very insensitive but also weak explosive. It has been suggested as a component in detonators.

- K. Budik, R. Cunek, J. Kulhanek, *Pyrotechnical charge*, CS-Patent 206828B1, Czechoslovakia, **1981**.
- H. H. H. Wurzenberger, J. T. Lechner, M. Lommel, T. M. Klapötke, J. Stierstorfer, Salts of Picramic Acid – Nearly Forgotten Temperature-Resistant Energetic Materials, *Propellants Explos. Pyrotech.* **2020**, https://doi.org/10.1002/prep.201900402.

Picratol

Pikratol

Picratol is an insensitive melt-castable HE based on *TNT* and *ammonium picrate* (Table P.6).

Table P.6: Composition and properties of picratol.

Ammonium picrate (wt.-%)	52
TNT (wt.-%)	48
ρ (g cm^{-3})	1.67
V_D (m s^{-1})	6970
P_{CJ} (GPa)	20.3
T_{5ex} (°C)	285
$\varnothing_{cr}$ (mm)	13.9

Picric acid, PA

Pikrinsäure, 2,4,6-Trinitrophenol, Ekrasit, Füllung 88

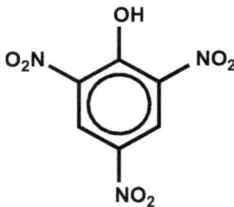

Aspect		Yellow crystals
Formula		$C_6H_3N_3O_7$
REACH		LPRS
EINECS		201-865-9
CAS		[88-89-1]
m_r	g mol^{-1}	229.106
ρ	g cm^{-3}	1.767
$\Delta_f H$	kJ mol^{-1}	−249
$\Delta_{ex} H$	kJ mol^{-1}	−1040.6
	kJ g^{-1}	−4.542
	kJ cm^{-3}	−8.026
$\Delta_c H$	kJ mol^{-1}	−2540.9
	kJ g^{-1}	−11.090
	kJ cm^{-3}	19.597

Λ	wt.-%	−45.39	
N	wt.-%	18.34	
c_p	$J K^{-1} mol^{-1}$	240	
$\Delta_m H$	$kJ mol^{-1}$	20	
$\Delta_v H$	$kJ mol^{-1}$	88	
Mp	°C	122.5	
Dp	°C	183	
Friction	N	>355 (BAM)	(dry substance)
Impact	J	7 (BAM)	
	cm	12 (Tool 12)	
V_D	$m s^{-1}$	7680 at $\rho = 1.76$ g cm^{-3}	
P_{CJ}	GPa	26.5 at $\rho = 1.76$ g cm^{-3}	
Trauzl	cm^3	315	
Koenen	mm	4	

Picric acid was used as an explosive filler in World war I. However, due to its corrosive behavior and the inherent formation of mechanically extremely sensitive metal picrates in metal shells in contact with usual construction materials such as brass, copper, iron, lead, and zinc led to its replacement by more powerful and less offensive explosives.

- A. Stettbacher, *Schiess- und Sprengstoffe*, 2. Aufl. Verlag von Johannn Ambrosius Barth, Leipzig, **1933**, pp. 281–286.

Picrylamino-1,2,4-triazole, PATO

N-(2,4,6-Trinitrophenyl)-1H-1,2,4-triazol-5-amin

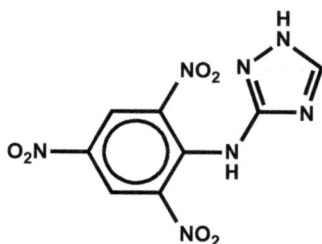

Aspect		Yellow crystals
Formula		$C_8H_5N_7O_6$
CAS		[18212-12-9]
m_r	g mol^{-1}	295.171
ρ	g cm^{-3}	1.936
$\Delta_f H$	kJ mol^{-1}	151

$\Delta_{ex}H$	kJ mol^{-1}	-1373.1
	kJ g^{-1}	-4.652
	kJ cm^{-3}	-9.024
$\Delta_c H$	kJ mol^{-1}	-4013.7
	kJ g^{-1}	-13.598
	kJ cm^{-3}	-26.353
Λ	wt.-%	-67.75
N	wt.-%	33.22
Mp	°C	310 (dec
Friction	N	>360
Impact	J	>50
V_D	m s^{-1}	7490
P_{CJ}	GPa	24.3

PATO is a temperature-resistant and -insensitive HE.

- M. D. Coburn, *3-Picrylamino-1,2,4-triazole and its preparation*, US-patent 3,483,211, USA, **1969**.

Plate dent test

The plate dent test serves the assessment of the detonation pressure of a HE. (Table P.7). A cylindrical charge (ø = 40 mm, l = 200 mm) (~ 400 g) of the HE under investigation is centered on a witness plate (152 × 152 × 51 mm) from mild steel (S235JR + AR) resting on two other identical plates. The HE charge is initiated with a booster charge (ø = 40 mm, l = 51 mm) (PBX9205) placed on top. The dent depth, T, correlates with the detonation pressure:

$$P_{CJ} = 0.02642 + T \cdot 3.3373 [GPa]$$

Licht and *Schwab* could show that the crater volume much better correlates with the detonation pressure.

Table P.7: Plate dent results and P_{CJ} values for selected high explosives.

HE	Density (g cm^{-3})	Dent, T (mm)	P_{CJ} (GPa)
BTF	1.838	3.05	10.2
HMX	1.730	10.07	33.5
NM	1.133	4.15	13.8
PETN	1.665	9.75	32.4
PYX	1.63	1.96	6.6

Table P.7 (continued)

HE	Density (g cm⁻³)	Dent, T (mm)	P_CJ (GPa)
RDX	1.754	10.35	34.5
TATB	1.87	8.31	27.8
Tetryl	1.681	8.10	27.0
Comp B	1.71	8.47	28.2
Octol 75/25	1.802	9.99	33.3
Pentolit 50/50	1.655	7.84	26.3
HMX/NGu/Kel-F 65.7/26.4/7.79	1.814	9.8	32.7
HMX/NGu/Estane 29.7/64.9/5.4	1.709	7.4	24.7
HMX/Viton 88/12	1.852	10.211	34.1

- H. H. Licht, J. Schwab, Der Plate-Dent-Test, *Bericht R-112/91*, ISL, Saint Louis, 28. 10. **1991**, 20 pp.
- T. R. Gibbs, A. Popolato (Eds.), *LASL Explosive Property Data*, University of California Press, **1980**, pp. 280–287.

Politzer, Peter (1937*)

Peter Politzer is a native Czech immigrated American theoretical chemist and international authority in the prediction and description of energetic materials sensitiveness and performance. He studied chemistry and completed his PhD at Western Reserve University in Cleveland and after a two-year postdoctoral phase at Indiana State University in 1966, accepted a professorship at University of New Orleans. He retired from UNO in 2006 but remains active in the field with a private research company.

- T. Clark, M. Hennemann, J. Murray, P. Politzer, Halogen bonding: the σ-hole, *J. Molec. Model.* **2007**, *13*, 291–296.
- P. Politzer, J. Martinez, J. S. Murray, M. C. Concha, A. Toro-Labbe, An electrostatic interaction correction for improved crystal density prediction, *Mol. Phys.* **2009**, *107*, 2095–2101.
- J. S. Murray, M. C. Concha, P. Politzer, Links between surface electrostatic potentials of energetic molecules, impact sensitivities and $C-NO_2/N-NO_2$ bond dissociation energies, *Mol. Phys.* **2009**, *107*, 89–97.
- P. Politzer, J. S. Murray, High Performance, Low Sensitivity: Conflicting or Compatible?, *Propellants Explos. Pyrotech.* **2016**, *41*, 414–425.
- http://www.clevetheocomp.org/research_EnergeticMaterials.html

Poly(azido)-methyloxetane, poly-AMMO

Polyazidomethylmethyloxetan

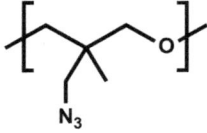

Aspect		Brown amorphous material
Formula		$(C_5H_9N_3O)_n$
CAS		[89883-49-8]
m_r	g mol^{-1}	127.146
ρ	g cm^{-3}	1.17
Δ_fH	kJ mol^{-1}	43.89
$\Delta_{ex}H$	kJ mol^{-1}	−422.3
	kJ g^{-1}	−3.321
	kJ cm^{-3}	−3.886
Δ_cH	kJ mol^{-1}	−3297.8
	kJ g^{-1}	−25.938
	kJ cm^{-3}	−30.347
Λ	wt.-%	−169.88
N	wt.-%	33.05
Dp	°C	180
Friction	N	160
Impact	J	7.5

Poly-AMMO is still an experimental energetic binder in gun propellants.

– T. Keicher, A. Kawamoto, Neue energetische Materialien und Komponenten Energetische Thermoplastische Elastomere (E-TPE), *ICT -Symposium*, **2005**.

Poly(bisazido)-methyloxetan, poly-BAMO

Polybisazidomethyloxetan

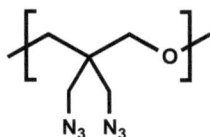

Formula		$(C_5H_8N_6O)_n$
CAS		[59595-53-8]
m_r	g mol^{-1}	168.158
ρ	g cm^{-3}	1.2
$\Delta_f H$	kJ mol^{-1}	371.50
$\Delta_{ex}H$	kJ mol^{-1}	−756.7
	kJ g^{-1}	−4.500
	kJ cm^{-3}	−5.401
$\Delta_c H$	kJ mol^{-1}	−3530.9
	kJ g^{-1}	−20.998
	kJ cm^{-3}	−25.197
Λ	wt.-%	−123.69
N	wt.-%	49.98
Mp	°C	65.8
Dp	°C	175
Friction	N	192
Impact	J	10

Poly(bisazido)-methyloxetan is investigated as an experimental energetic binder in gun propellants.

– T. Keicher, A. Kawamoto, M. Kaiser, Synthese und Eigenschaften von GAP -Poly-BAMO Copolymer, *ICT -Symposium*, **2007**.

Poly(carbon monofluoride), PMF

Graphitfluorid

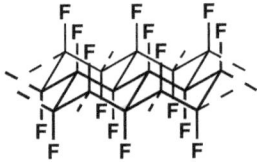

Graphite fluoride, Carbofl4or®		
Formula		$(-CF_x-)_n$, with x 0.33-1.2
CAS		[11113-63-6]
GHS		07
H-phrases		315-319-335
P-phrases		261-302+352-305+351+338-321-405-501a
EINECS		234-345-5
$m_{r(monomer)}$	g mol^{-1}	31.009 for x = 1.00
ρ	g cm^{-3}	2.65-2.69
c_p	J K^{-1} g^{-1}	0.89
Dp	°C	610
$\Delta_f H$	kJ mol^{-1}	−196
Λ	wt.-%	−103.2
Φ	wt.-%	61.3 for x = 1.00

Graphite fluoride was first prepared by the German chemist Otto Ruff (1871–1939) in 1934. Depending on its fluorine content, poly(carbon monofluoride) (PMF) is a creme to colorless microcrystalline superhydrophobic material. PMF is a superior oxidizer in high performance blackbody flares yielding 10 times higher radiant intensity than PTFE-based formulations due to formation of expanded graphite. PMF is advantageously used in multispectral (VIS/IR/MMW) obscurants as expanded graphite is an effective multispectral scattering material. PMF has been used as a modifier for boron-based rocket propellant. PMF is also utilized as high temperature lubricant and has been used to prepare nano-allotropes of carbon (nanotubes, nano-carpet rolls, graphene, etc.) by reductive elimination of fluorine will all kinds of halophilic fuels.

- E.-C. Koch, *Metal-Fluorocarbon Based Energetic Materials*, Wiley-VCH, Weinheim, **2012**, pp. 200–201.
- S. Cudziło, M. Szala, A. Huczko, M. Bystrzejewski, Combustion Reactions of Poly(Carbon Monofluoride), (CF)n,, with Different Reductants and Characterization of the Products, *Propellants Explos. Pyrotech.* **2007**, *32*, 149–154.

– E.-C. Koch, Metal/Fluorocarbon Pyrolants: VI. Combustion Behavior and Radiation Properties of Magnesium /Poly(Carbon Monofluoride) Pyrolant, *Propellants Explos. Pyrotech.* **2005** *30*, 209–215.
– E.-C. Koch, *Pyrotechnischer Satz zur Erzeugung von IR-Strahlung*, EP 1090895B1, **2003**, Germany.
– E.-C. Koch, Defluorination of Graphite Fluoride Applying Magnesium, *Z. Naturforsch.* **2001**, *56b*, 512–516; accessed at http://zfn.mpdl.mpg.de/data/Reihe_B/56/ZNB-2001-56b-0512.pdf
– T. Liu, I. Shyu, Y. Hsia, Effect of fluorinated graphite on combustion of boron and boron-based fuel-rich propellants, *J. Propul. Power* **1996**, *12*, 26–33.
– O. Ruff, O. Bretschneider, Die Reaktionsprodukte der verschiedenen Kohlenstoffformen mit Fluor II (Kohlenstoff-monofluorid), *Z. anorg. u. allg. Chem.* **1934**, *217*, 1.

Poly[2,2,2-trifluoroethoxy-5′,6′dinitratohexanoxy]phosphazene

Poly[2,2,2-trifluorethoxy-5′,6′dinitratohexanoxy]phosphazen

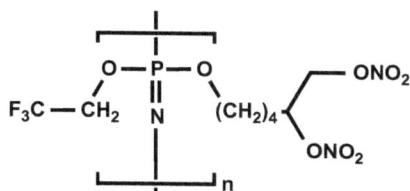

PPZ-E		
Aspect		Light brown polymer
Formula		$[PN(OC_2H_2F_3)(O(C_6H_{11}(ONO_2)_2)]_n$
CAS		not assigned
m_r	$g\ mol^{-1}$	367.17
ρ	$g\ cm^{-3}$	1.52
Δ_fH	$kJ\ mol^{-1}$	−1516 ± 66
$\Delta_{ex}H$	$kJ\ mol^{-1}$	−386.3
	$kJ\ g^{-1}$	−1.052
	$kJ\ cm^{-3}$	−1.599
Δ_cH	$kJ\ mol^{-1}$	−5338.0
	$kJ\ g^{-1}$	−14.538
	$kJ\ cm^{-3}$	−22.098
Λ	wt.-%	−67.54
N	wt.-%	11.44
T_g	°C	−55
Dp	°C	186
Friction	N	>360
Impact	J	>20
Koenen	mm	1, type O

PPZ-E is the best investigated energetic polyphosphazene to date. At small diameters (8 mm, ρ = 1.3 g cm^{-3}) it does not propagate a detonation. Despite its energetic character, it is not an HD 1 material.

- A. J. Bellamy, A. E. Contini, P. Golding, Bomb Calorimetric Correlation Study between Chemical Structure and Enthalpy of Formation for a Linear Energetic Polyphosphazene, *Centr. Eur. J. Energ. Mater.* **2013**, *10*, 3–15.
- A. J. Bellamy, P. Bolton, J. D. Callaway, A. E. Contini, N. Davies, P. Golding, M. K. Till, J. N. Towning, S. J. Trussell, Energetic Polyphosphazenes – A New Category of Binders for Pyrotechnic Formultions, *32th International Pyrotechnics Seminar*, Karlsruhe, Germany, **2005**.
- P. Golding, S. J. Trussell, Energetic Polyphosphazenes – a new category of binders for energetic Formulations, *NDIA*, 15–17 November **2004**, San Francisco, USA.

Polychlorotrifluoroethylene, PCTFE

Polychlortrifluorethylen

Kel-F® (3 M)		
Formula		(-CFClCF$_2$-)$_n$
CAS		[9002-83-9]
$m_{r(monomer)}$	g mol^{-1}	116.470
ρ	g cm^{-3}	2.10 – 2.13
Mp	°C	211
$\Delta_m H$	kJ mol^{-1}	4.659
c_p	J K^{-1} g^{-1}	0.835
Dp	°C	>400
$\Delta_f H$	kJ mol^{-1}	−600
$\Delta_c H$	kJ mol^{-1}	−187.0
	kJ g^{-1}	−1.606
	kJ cm^{-3}	−3.421
Λ	wt.-%	−27.47
Φ	wt.-%	48.9
LOI	Vol.%	99.5

PCTFE was invented in 1934 by Scherer and Schloffer in Germany and was the first commercially produced fluorinated polymer to enter the market. PCTFE and copolymers with vinylidene fluoride (*Kel-F 800*) are frequently used as binders for insensitive HEs such as TATB or DAAF. Furthermore, PCTFE is also used as binder and fluorine source in blackbody decoy flare formulations. PCTFE can be produced from polymerization of gaseous chlorotrifluoroethylene by UV radiation or efficiently by radiative curing with gamma rays from a ^{60}Co source.

- F. Schloffer, O. Scherer, *Verfahren zur Darstellung von Polymerisationsprodukten*, DE-Patent 677071, **1939**, Germany.
- C. F. Parrish, J. E. Short, W. T. Biggs, *Radiation-Polymerization Binder for Mk 48 Decoy Flares, RDTR No. 232*, Naval Ammunition Depot Crane, Indiana, February **1972**, 28 pp.

Polyisobutylene, PIB

Polyisobuten, Oppanol (BASF)

mineral jelly		
Formula		$(-C_4H_8-)_n$
CAS		[9003-27-4]
EINECS		618-360-8
$m_{r(monomer)}$	g mol^{-1}	56.106
ρ	g cm^{-3}	0.92
$\Delta_f H$	kJ mol^{-1}	−87
$\Delta_c H$	kJ mol^{-1}	−2629.4
	kJ g^{-1}	−46.865
	kJ cm^{-3}	−43.115
Λ	wt.-%	−342.19

Depending on the degree of polymerization a tacky plastic (~ 3000 g mol^{-1}) to rubber like elastic colorless material (40,000–120,000 g mol^{-1}). Polyisobutylene is the binder in C4 plastic explosive.

Polynitrated polyhedra

Polynitropolyeder

Nitrated hydrocarbons with polyhedral carbon structures have been suggested in 1981 for the first time as high-performance energetic materials. Polyhedra like tetrahedron and cubane possess significantly higher density than their open chain isomers. On top, the high strain energy resulting from the deviation of typical valence bond angles yields an extraordinary high energy content in comparison to unstrained compounds. Figure P.5 depicts the influence of ring size and structure on strain energy and density of selected cyclic and derived cage hydrocarbon compounds. This explains the attractivity of the carbon frameworks 1, 2, 5, 6, and 7 as building blocks for polynitrated polyhedra (see also *nitrocubanes*, *eicosanitrododecahedrane* and *1,3,3-trinitroazetidine*).

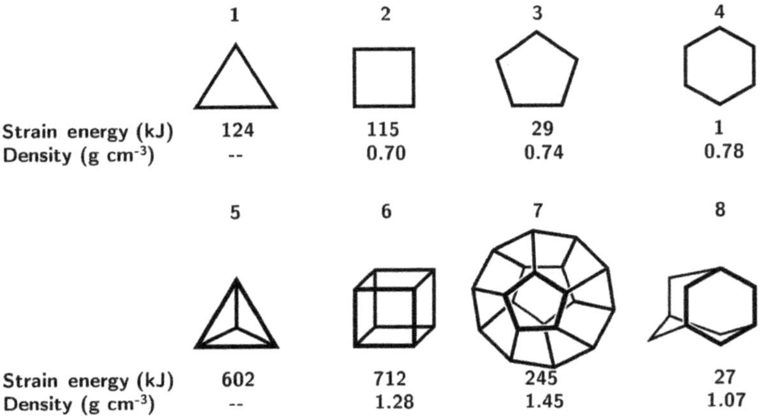

	1	2	3	4
Strain energy (kJ)	124	115	29	1
Density (g cm^{-3})	--	0.70	0.74	0.78

	5	6	7	8
Strain energy (kJ)	602	712	245	27
Density (g cm^{-3})	--	1.28	1.45	1.07

Figure P.5: Density and strain energy of hydrocarbons cyclopropane (**1**), cyclobutane (**2**), cyclopentane (**3**), cyclohexane (**4**), tetrahedrane (**5**), cubane (**6**), dodecahedrane (**7**), and adamantane (**8**).

- G. P. Sollott, J. Alster, E. E. Gilbert, Research Towards Novel Energetic Materials, *J. Energ. Mater.* **1986**, 4, 5–28.
- O. Sandus, Detonation Performance Calculations on Novel Explosives, *Sixth Annual Working Group Meeting on Synthesis of High Energy Density Materials*, Kiamesha Lake, NY, **1987**, 150–167.
- J. Alster, S. Iyer, O. Sandus, *Molecular Architecture Versus Chemistry and Physics of Energetic Materials*, in S. N. Bulusu, eds., *Chemistry and Physics of Energetic Materials*, Kluwer **1990**, 641–652.
- H. Hopf, *Classics in Hydrocarbon Chemistry*, Wiley-VCH, **2000**, 547 pp.

Polynitratomethylmethyloxetane, poly-NIMMO
Polynitratomethylmethyloxetan

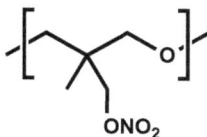

Aspect		Brown amorphous material
Formula		$(C_5H_9NO_4)_n$
CAS		[107760-30-5]
m_r	g mol^{-1}	147.131
ρ	g cm^{-3}	1.26
$\Delta_f H$	kJ mol^{-1}	−309
$\Delta_{ex}H$	kJ mol^{-1}	−628.5
	kJ g^{-1}	−4.271
	kJ cm^{-3}	−5.382
$\Delta_c H$	kJ mol^{-1}	−2944.8
	kJ g^{-1}	−20.015
	kJ cm^{-3}	−25.219
Λ	wt.-%	−114.18
N	wt.-%	9.52
T_g	°C	−30
Dp	°C	187
Friction	N	>360
Impact	J	>50

– T. Keicher, J. Böhnlein-Mauss, Auf dem Weg zu NC -freien Treibladungspulvern mit Energetischen Thermoplastischen Elastomeren (E-TPE), *ICT -Symposium*, **2010**.

Poly-nitrogen compounds
Polystickstoffverbindungen

Homoleptic nitrogen compounds are potential future energetic materials. Their application (*but not necessarily their synthesis and production!*) would yield environmentally neutral dinitrogen, N_2, as exhaust which, in addition, leaves no visual infrared or millimetric wavelength signature. On top homoleptic nitrogen compounds would allow to make a big step in performance far beyond the possibilities of CHNO compounds.

Table P.8: Synopsis of HMX and CL-20 with potential polynitrogen compounds.

Compound	$\Delta_f H$ (kJ mol^{-1})	ρ (g cm^{-3})	V_D (km s^{-1})	P_{CJ} (GPa)	$V/V_0 = 2.2$ HMX = 100	I_{sp} (s)
HMX	74.75	1.901	9.3	39,3	100	266
CL-20	393	2.04	10.0	47,8	119	273
N$_4$ [(1)]	760	~2.3	15.5	122	310	424
N$_{60}$ [(2)]	6780	1.97	12.3	65	161	331
Poly-N [(3)]	290	3.9	10.8	133	265	288
Poly-CO-N$_2$	1640	3.983	7.3	764	1379	640

(1) tetraazatetrahedrane, (2) aza-fullerene-[60], (3) polymeric nitrogen p = 125 GPa.

- C. Yoo, M. Kim, J. Lim, Y. J. Ryu, I. G. Batyrev, Copolymerization of CO and N$_2$ to Extended CON$_2$ Framework Solid at High Pressures, *J. Phys. Chem. C.* **2018**, *122*, 13054–13060.
- E. Bemm H. Östmark C. Eldsäter, P. Goede, N. Roman, S. Wallin, N. Wingborg, Development of energetic materials over the time span 2004–2025. Special forcast, *FOI -R–1992 – SE*, FOI, Tumba, **2004**, 61 pp.
- M. I. Eremets, A. G. Gavriliuk, I. A. Trojan, D. A. Dzivenko, R. Boehler, single-bonded cubic form of nitrogen, *Nature Materials*, **2004**, *3*, 558–563.

Polynitropolyphenylene, PNP
Polynitrophenylen

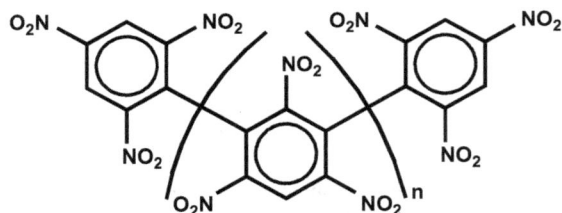

Aspect		Brown amorphous material
Formula		(C$_6$HN$_3$O$_6$)$_n$(C$_6$H$_2$N$_3$O$_6$)$_2$, n ~ 6–11
CAS		[70977-31-0]
m$_r$	g mol^{-1}	211.09
ρ	g cm^{-3}	1.8–2.2
$\Delta_f H$	kJ mol^{-1}	−711
$\Delta_{ex} H$	kJ mol^{-1}	−493.5
	kJ g^{-1}	−2.338
	kJ cm^{-3}	−4.208

$\Delta_c H$	kJ mol^{-1}	−1793.0
	kJ g^{-1}	−8.494
	kJ cm^{-3}	−15.290
Λ	wt.-%	−49.72
N	wt.-%	20.10
Dp	°C	298
Friction	N	240 (kR)
Impact	J	4
Koenen	mm	20, type F, 24 type B

Polynitropolyphenylene (PNP) is an amorphous polymer with molecular weights between 1500 and 2600 g mol^{-1}. It nicely dissolves in acetone or esters. PNP is much more temperature resistant than NC. The caseless ammunition 4.73 × 33 mm (DM 18) for the projected German assault rifle G11 initially used NC as binder and energetic filler. However, it was prone to cook-off and was replaced by a NC-free formulation containing HMX as the main energetic filler and PNP as the temperature resistant and hence cook-off-proof binder.

A representative caseless gun propellant formulation using PNP and tested in the G11 rifle is given thereafter:

DM 18-TLP

8 wt.-%	Polyvinylbutyral
86 wt.-%	HMX
6 wt.-%	PNP
+1−4 wt.-%	Nylon fibres

– K. H. Redecker, R. Hagel, Polynitropolyphenylene, a High-Temperature Resistant Non-Crystalline Explosive, *Propellants Explos. Pyrotech.* **1987**, *12*, 196–201.
– K. Redecker, *Hülsenlose Treibmittelkörper*, DE2843477C2, **1987**, Germany.
– B. Berger, B. Haas, G. Reinhard, Einfluss des Bindergehaltes auf das Reaktionsverhalten pyrotechnischer Mischungen, *ICT -Jata*, Karlsruhe, **1995**, V-2.

Polyphosphazenes

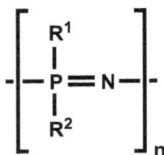

Polyphosphazene

$$\left[\begin{array}{c} R^1 \\ | \\ -P = N- \\ | \\ R^2 \end{array} \right]_n$$

Polyphosphazenes belong to the group of inorganic polymers and contain the essential monomeric unit $-N = P(R)_2-$. Polymers with $R^1 = O\text{-}CH_2\text{-}CF_3$ and $R^2 = $ alkoxy bearing explosophoric substituents are investigated mainly in the UK since the 1990s as potential energetic binders for HEs, propellants, and pyrotechnics. So far explosophores with nitrato and azido groups as well as polymers with ionic groups have been investigated (Figure P.6).

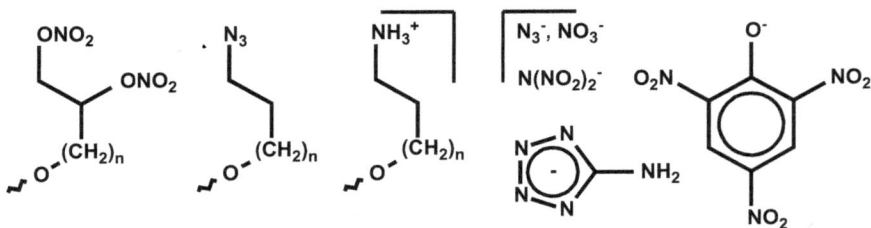

Figure P.6: Various phosphazene monomers and building blocks.

- P. R. Bolton, P. Golding, C. B. Murray, M. K. Till, S. J. Trussell, Enhanced Energetic Polyphosphazenes, *IMEMTS 2006*, Bristol, UK, **2006**.
- J. E. Mark, H. R. Allcock, R. West, *Inorganic Polymers*, Oxford University Press, Oxford, **2005**, 62–153.

Polytetrafluorethylene
Polytetrafluoroethylen

PTFE (acr.), Teflon ® HOSTAFLON®		
Formula		$(-C_2F_4-)_n$
REACH		LPRS
EINECS		618-337-2
CAS		[9002-84-0]
$m_{r(monomer)}$	$g\ mol^{-1}$	100.016
ρ	$g\ cm^{-3}$	2.20–2.31
Mp	°C	328 resp. 340 (upon renewed fusion of a solidified PTFE melt)
$\Delta_m H$	$kJ\ mol^{-1}$	3.6
c_p	$J\ K^{-1}\ mol^{-1}$	102.02
κ	$W\ K^{-1}\ m^{-1}$	0.024
Dp	°C	(605)
$\Delta_f H$	$kJ\ mol^{-1}$	−809.60
$\Delta_c H$	$kJ\ mol^{-1}$	−666.9
	$kJ\ g^{-1}$	−6.68
	$kJ\ cm^{-3}$	−14.696
Λ	wt.-%	−31.99
Φ	wt.-%	75.97
LOI	vol.%	99.5

PTFE is a white, microcrystalline hydrophobic powder or flakes. It is the main oxidizer in blackbody decoy flare compositions together with fuels like magnesium and/ or magnalium and Viton ® as binder (magnesium/Teflon ®/Viton® = MTV). Solid bulk PTFE is usually not attacked by anything at ambient temperature. However, it is attacked by molten metals or solutions of alkali and alkaline metals in liquid ammonia and then undergoes exothermic defluorination yielding amorphous carbon and the metal fluoride. In a similar way PTFE serves as oxidizer and carbon source in infrared screening smokes, reactive fragments, and delay formulations. As PTFE has a very low friction coefficient (PTFE/PTFE = 0.04), it is also often used as a pressing aid in consolidating HEs (e.g. PBXW-7).

– E.-C. Koch, *Metal-Fluorocarbon Based Energetic Materials*, Wiley-VCH, **2012**, 342.
– R. N. Walters, S. M. Hackett, R. E. Lyon, Heats of Combustion of High Temperature Polymers, *Fire Mater.* **2000**, *24*, 245–252.

Polyvinyl chloride, PVC
Polyvinylchlorid

Formula		$[C_2H_3Cl]_n$
CAS		[9002-86-2]
m_r	g mol^{-1}	62.499
ρ	g cm^{-3}	1.3–1.45
$\Delta_f H$	kJ mol^{-1}	−94
$\Delta_c H$	kJ mol^{-1}	−1123.8
	kJ g^{-1}	−17.981
		−23.375
Λ	wt.-%	−128.00
Cl-content	wt.-%	56.73
Dp	°C	218
LOI	vol.%	>45

PVC was first suggested by *Pastor* as a color intensifier for green illuminating flare compositions allowing to replace dangerous barium chlorate hydrate. Pyrolysis and combustion yields a number of problematic chloro-organic compounds.

- K. Smit, M. Morgan, R. Pietrobon, Pyrotechnic Films based on Thermites Covered with PVC, *Propellants Explos. Pyrotech.* **2019**, *44*, 37–40.
- C. Huggett, B. C. Levin, Toxicity of the Pyrolysis and Combustion Products of Poly (Vinyl Chlorides): A Literature Assessment, *Fire Mater.* **1987**, *2*, 131–142.
- G. Pastor, *Chloratfreier Leuchtsatz für Leuchtsterne*, DE-Patent 727865, **1942**, Germany.

Polyvinyl nitrate, PVN
Polyvinylnitrat

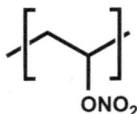

ONO$_2$

Aspect		Colorless to brown material
Formula		$[C_2H_3NO_3]_n$
CAS		[26355-31-7]
m_r	g mol^{-1}	89.051
ρ	g cm^{-3}	1.2–1.6
$\Delta_f H$	kJ mol^{-1}	−102
$\Delta_{ex} H$	kJ mol^{-1}	−169,820.3
	kJ g^{-1}	−1907 at 11.76% N
		−4129 at 15.71% N

$\Delta_c H$	kJ mol^{-1}	−1141
	kJ g^{-1}	−15664 at 11.76% N
		−12648 at 15.71% N
Λ	wt.-%	−44.92
N	wt.-%	15.73
Dp	°C	134
Friction	N	>200
Impact	J	9
V_D	m s^{-1}	6500 at $\rho = 1.5$ g cm^{-3}
P_{CJ}	GPa	26.5 at $\rho = 1.76$ g cm^{-3}
Trauzl	cm^3	330
Koenen	mm	6–8

PVN is an amorphous thermoplastic polymer with molecular weight ~ 200,000 g mol^{-1}. It dissolves in acetone or esters. Originally targeted as an NC replacement, PVN does not meet the requirements due to its too low softening temperature T = 35–50 °C and inherent lack of thermal stability. Stabilizers tested with PVN are DPA and 2NDPA. PVN has been qualified in Germany as a crystallization aid for composition B.

– E. Backof, Polyvinylnitrat – Eine Komponente für Treib- und Explosivstoffe, *ICT-Jata*, **1981**, 67–84.

Polyvinylidene fluoride, PVDF
Polyvinylidendifluorid

Formula		[CH$_2$CF$_2$]$_n$
CAS		[24937-79-9]
m_r	g mol^{-1}	64.035
ρ	g cm^{-3}	1,75–1,80
$\Delta_f H$	kJ mol^{-1}	−500
$\Delta_c H$	kJ mol^{-1}	−572.9
	kJ g^{-1}	−8.946
	kJ cm^{-3}	−16.103
Tg	°C	−40
Mp	°C	154–184
Λ	wt.-%	−99.94
Φ	wt.-%	59.4
Dp	°C	218
LOI	vol.%	43.6

PVDF is a piezoelectric polymer appearing in a number of polymorphs. Its copolymer with hexafluoropropene is *Viton*® *A*. PVDF is utilized as binder in pyrotechnics and HE formulations. PVDF-based pyrotechnics show adjustable impact sensitivity (by electric voltage).

- S. L. Row, L. J. Groven, Smart Energetics: Sensitization of the Aluminum-Fluoropolymer Reactive System, *Adv. Eng. Mater.* **2018**, *20*, 1700409.
- R. S. Janesheski, L. J. Groven, S. F. Son, Fluoropolymer and aluminum piezoelectric reactives, *17th Biennial International Conference of the APS Topical Group on Shock Compression of Condensed Matter*, **2011**, Chicago.

Potassium benzoate

Kaliumbenzoat

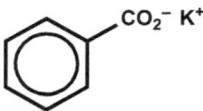

Formula		$C_7H_5O_2K$
REACH		LPRS
EINECS		209-481-3
CAS		[582-25-2]
m_r	g mol^{-1}	160.219
$\rho_{25\,°C}$	g cm^{-3}	1.558
Mp	°C	~ 350
Dp	°C	373
$\Delta_f H$	kJ mol^{-1}	−610
$\Delta_c H$	kJ mol^{-1}	−3040.0
	kJ g^{-1}	−18.975
	kJ cm^{-3}	−29.563
Λ	wt.-% O_2	−149.80
c_p	J mol^{-1} K^{-1}	172.8
S	g L^{-1}	70.7 (in H_2O at 20 °C)

Potassium benzoate is mildly hygroscopic and therefore at relative humidity levels >80% RH forms the trihydrate ($C_7H_5O_2K \cdot 3\,H_2O$) [6100-02-3]. Potassium benzoate is used as a fuel in whistling compositions together with potassium perchlorate as oxidizer. Figure P.7 displays the burn rate, exothermicity, and calculated adiabatic temperature for $KClO_4/C_7H_5O_2K$. Potassium benzoate has been proposed as a fuel firework rocket motors with diphenyl being the melt cast base: diphenyl/$KClO_4$/KNO_3/$C_7H_5O_2K$ {5/45/15/40}. Potassium benzoate containing whistling compositions (SR136) have been suggested as well as payloads for spectrally adapted decoy flares.

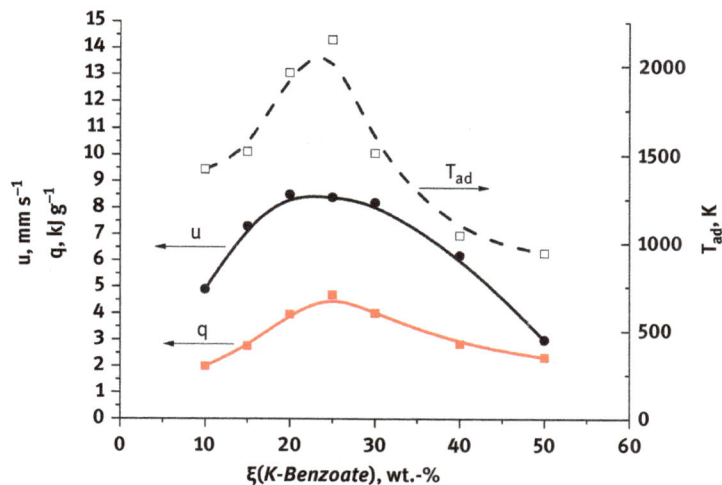

Figure P.7: Heat of explosion, adiabatic temperature, and burn rate of K-benzoate/KClO$_4$.

- Wei-Wei Yang, You-Ying Di, Zhen-Fen Yin, Yu-Xia Kong, Zhi-Cheng Tan, Low-temperature heat capacities and standard molar enthalpy of formation of potassium benzoate C$_7$H$_5$O$_2$K$_{(s)}$, *Int. J. Thermophys.* **2009**, *30*(2), 542–554.
- J. Callaway, T. D. Sutlief, *Infra-red emitting decoy flare*, U.S. Patent Application 2004/0011235 A1, **2004**, GB.
- E.-L. Charsley, J. J. Rooney, S. B. Warrington, T. T. Griffiths, T. A. Vine, A Study of the Potassium Benzoate-Potassium Perchlorate Pyrotechic System, *IPS*, Grand Junction, **2000**, 381–392.
- B. Cook, Novel Powder Fuel for Firework Display Rocket Motors, *J. Pyrotech.* **1998**, *7*, 59–68.

Potassium bromate
Kaliumbromat

Formula		KBrO$_3$
GHS		03, 06
H-phrases		271-301
P-phrases		210-221-283-301+310-405-501a
UN		1484
EINECS		231-829-8
CAS		[7758-01-2]
m$_r$	g mol^{-1}	167.00
ρ	g cm^{-3}	3.25
Mp	°C	434
Δ$_f$H	kJ mol^{-1}	−332.1
c$_p$	J mol^{-1} K^{-1}	104.9
Λ	wt.-%	+28.74

KBrO$_3$ has been suggested as an experimental oxidizer in deep orange red (*Jennings-White*) and deep blue signal flare compositions (*Koch*). As a precursor for KBr, it has been suggested too for use in *fire extinguishing* compositions.

- C. J. White, Some Esoteric Firework Materials, *Pyrotechnica*, **1990**, *XIII*, 26–32 + Plate VIII.
- P. L. Posson, M. L. Clark, *Flame suppressant aerosol generant*, US 8,182,711, **2012**, USA.
- E.-C. Koch, Spectral Investigation and Color Properties of Copper(I) Halides CuX (X = F, Cl, Br, I) in Pyrotechnic Combustion Flames, *Propellants Explos. Pyrotech.* **2015**, *40*, 799–802.

Potassium chlorate
Kaliumchlorat

Formula		KClO$_3$
GHS		03,07,09
H-phrases		271-302-332-411
P-phrases		210-221-283-306+360-371+380+375-501a
UN		1485
CAS		[3811-04-9]
m$_r$	g mol^{-1}	122.55
ρ	g cm^{-3}	2.320
Mp	°C	370
Dp	°C	400
c$_p$	J mol^{-1} K^{-1}	100
Δ$_f$H	kJ mol^{-1}	−397.73
Λ	wt.-%	+39.17

For many decades, potassium chlorate was the most important oxidizer in pyrotechnics next to potassium nitrate. This dominant position is due to its exothermic decomposition making it facile to burn any kind of fuel with it:

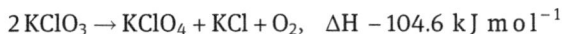

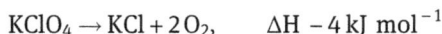

$$2\,KClO_3 \rightarrow KClO_4 + KCl + O_2, \quad \Delta H \; -104.6 \; kJ\,mol^{-1}$$

$$KClO_4 \rightarrow KCl + 2\,O_2, \quad \Delta H \; -4\,kJ\,mol^{-1}$$

With fuels, potassium chlorate forms mechanically sensitive systems also explaining its role as an oxidizer in safety matches.

Potassium chlorate reacts with strong acids (e.g. H$_2$SO$_4$) to give the very sensitive chloric acid, HClO$_3$, which disproportionates into perchloric acid and the highly reactive chlorine dioxide:

$$KClO_3 + H_2SO_4 \rightarrow HClO_3 + KHSO_4$$

$$3\,HClO_3 \rightarrow HClO_4 + ClO_2 + H_2O$$

Chlorine(IV) dioxide even in a pure state is a highly unstable/explosive substance which also shows hypergolic reaction with a great variety of oxidizable substances. The reactivity of $KClO_3$ with acids also determines the stability and compatibility of $KClO_3$ with other components in formulations. Sublimed sulfur (flower of sulfur) due to its innate traces of H_2SO_4 can undergo self-ignition based on the earlier reactions. Even natural resins (e.g. *gum arabic*) containing acids (*D*-glucuronic acid) may induce self-ignition when mixed with potassium chlorate *(Krone, 1997)*. *Leiber* showed that the exothermic decomposition of pure $KClO_3$ makes this prone to undergo low-velocity detonation when initiated sufficiently strong.

The high sensitiveness and hazards have led to replacement of $KClO_3$ by $KClO_4$ in many formulations except colored smoke formulations and few Bengal compositions. Metal oxide additives may lower the decomposition temperature of $KClO_3$ as depicted in Table P.9.

Table P.9: Influence of various metal oxides on the onset for $KClO_3$ decomposition.

Additive	T_d, °C
None	582 ± 5
Cr_2O_3	280 ± 6
CoO	348
Co_3O_4	350 ± 7
Fe_2O_3	373 ± 10
Cu_2O	393 ± 11
MnO_2	395
CuO	430
NiO	480
Ag_2O	508 ± 31
ZnO	523 ± 25
MgO	544 ± 6
TiO_2	545
Al_2O_3	568 ± 8

- C. O. Leiber, IHE – 2000, Wunsch und Wirklichkeit, *11. Sprengstoffgespräch*, Nonnweiler, **1987**.
- U. Krone, Grundlagen der Pyrotechnik, *Carl Cranz Seminar*, Weil am Rhein, **1997**.
- G. Krien, Thermoanalytische Ergebnisse von pyrotechnischen Ausgangsstoffen, *Az. 3.0-3/3712/75*, BICT, Swisttal-Heimerzheim, **1975**, 235 pp.
- D. Chapman, R. K. Wharton, G. E. Williamson, Studies of the Thermal Stability and Sensitiveness of Sulfur/Chlorate Mixtures, Part 1. Introduction, *J. Pyrotech.* **1997**, *6*, 30–35.
- D. Chapman, R. K. Wharton, J. E. Fletcher, G. E. Williamson, Studies of the Thermal Stability and Sensitiveness of Sulfur/Chlorate Mixtures, Part 2. Stoichiometric Mixtures, *J. Pyrotech.* **1998**, *7*, 51–57.

— D. Chapman, R. K. Wharton, J. E. Fletcher, A. E. Webb, Studies of the Thermal Stability and Sensitiveness of Sulfur/Chlorate Mixtures, Part 3. The Effects of Stoichiometry, Particle Size and Added Materials, J. Pyrotech. 2000, 11, 16–24.

Potassium dichromate

Kaliumdichromat

Formula		K_2CrO_7
GHS		03, 08, 06, 05, 09
H-phrases		350-340-360-272-330-301-312-372-314-334-317-410
P-phrases		201-280-301+330+331+310-304+340+310-305+351+338-308+313
UN		3087
REACH		LRS
EINECS		231-906-6
CAS		[7778-50-9]
m_r	g mol^{-1}	294.184
ρ	g cm^{-3}	2.69
Tp	°C	241
Dp	°C	398
$\Delta_f H$	kJ mol^{-1}	−2061
$\Delta_{pt} H$	kJ mol^{-1}	1.53
$\Delta_m H$	kJ mol^{-1}	36.7
c_p	J K^{-1} mol^{-1}	219.7
Λ	wt.-%	+16.32

Potassium dichromate has been used in the past in delay formulations and as a combustion catalyst in safety matches.

Potassium dinitramide, KDN

Kaliumdinitramid

KDN		
Formula		$KN(NO_2)_2$
CAS		[140456-79-7]
m_r	g mol^{-1}	145.116
ρ	g cm^{-3}	2.201
Mp	°C	128-139
Dp	°C	177
$\Delta_f H$	kJ mol^{-1}	−264.18
Λ	wt.-%	+38.59

In the DSC, potassium dinitramide (KDN) shows a mild exothermal reaction at 105 °C before undergoing fusion at 128 °C. *Dawe* tested the combustion behavior and recorded IR-emission spectra of KDN/boron – and KDN/boron/HTPB as potential IR decoy flare formulations. KDN has been suggested as a phase stabilizer for ammonium nitrate.

– J. R. Dawe, M. D. Cliff, Metal Dinitramides New Novel Oxidants for the Preparation of Boron Based Flare Compositions, *IPS-Seminar*, **1998**, Monterey, 789–810.

Potassium dinitrobenzofuroxan, KDNBF

Kaliumdinitrobenzofuroxan, Kaliumbenzanat

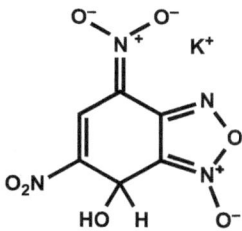

Aspect		Gold-orange platelets
Formula		$C_6H_3KN_4O_7$
REACH		LPRS
EINECS		249-543-7
CAS		[29267-75-2]
m_r	$g\ mol^{-1}$	282.209
ρ	$g\ cm^{-3}$	1.58
$\Delta_f H$	$kJ\ mol^{-1}$	−431
$\Delta_{ex} H$	$kJ\ mol^{-1}$	−731.8
	$kJ\ g^{-1}$	−2.593
	$kJ\ cm^{-3}$	−5.731
$\Delta_c H$	$kJ\ mol^{-1}$	−2539.6
	$kJ\ g^{-1}$	−8.999
	$kJ\ cm^{-3}$	−14.219
Ω	wt.-%	−39.69
N	wt.-%	19.85
Dp	°C	208
c_p	$J\ mol^{-1}\ K^{-1}$	241
Impact	J	0.3 – 0.4 (BAM)
Friction	N	2.5 (BAM)

KDBNF is a low energetic primary explosive and due to its mechanical sensitivity, it is used to sensitize other primary explosives corresponding formulations.

Potassium ferrate (VI)

Kaliumferrat(VI)

Aspect		Dark violet crystals
Formula		$K_2[FeO_4]$
CAS		[13718-66-6]
m_r	g mol^{-1}	198.05
ρ	g cm^{-3}	2.829
Dp	°C	200
Δ_fH	kJ mol^{-1}	−1027
$\Delta_{dec}H$	kJ mol^{-1}	−44.35
Λ	wt.-%	+16.16

Potassium ferrate (VI) is used as an environmentally friendly oxidizer in sewage treatment. It has also been considered as an alternative oxidizer in pyrotechnic delay and thermal battery-heater formulations. Working with ferrates(VI) requires the exclusion of moisture to avoid oxidative splitting of water in accordance with:

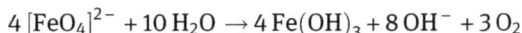

$$4\,[FeO_4]^{2-} + 10\,H_2O \rightarrow 4\,Fe(OH)_3 + 8\,OH^- + 3\,O_2$$

- C. K. Wilharm, A. Chin, S. K. Pliskin, Thermochemical Calculations for Potassium Ferrate(VI), K_2FeO_4, as a Green Oxidizer in Pyrotechnic Formulations, *Propellants Explos. Pyrotech.* **2014**, *39*, 173–179.
- A. Chin, S. Pliskin, C. K. Wilharm, *Ferrate Based Pyrotechnic Formulations*, US-Patent-Appl. 201414519874, **2014**, USA.

Potassium hexafluoroaluminate

Kaliumaluminumfluorid

Potassium cryolithe	
Formula	K_3AlF_6
GHS	08, 07, 09
H-phrases	332−372-411
P-phrases	260
WGK	1
UN	3288

REACH		LPRS
EINECS		237-409-0
CAS		[13775-52-5]
m_r	g mol^{-1}	258.267
ρ	g cm^{-3}	1.34
Pt1	°C	132
Pt2	°C	153
Pt3	°C	306
Mp	°C	995
c_p	J mol^{-1} K^{-1}	221.66
$\Delta_f H$	kJ mol^{-1}	−3326
Λ	wt.-%	0

Potassium fluoroaluminate is used as muzzle flash suppressing agents in gun propellants. With $KAlF_4$, it forms a eutectic melting at 558 °C.

- W. T. Thompson, D. G. W. Goad, Some thermodynamic properties of K_3AlF_6-$KAlF_4$ melts, *Can. J. Chem.* **1976**, *54*, 3342–3349.

Potassium nitrate

Kaliumnitrat

Formula		KNO_3
GHS		02
H-phrases		272
P-phrases		210-220-221-280-370+378-501a
UN		1486
EINECS		231-818-8
CAS		[7757-79-1]
m_r	g mol^{-1}	101.103
ρ	g cm^{-3}	2.109
$\Delta_f H$	kJ mol^{-1}	−494.63
Mp	°C	334
Bp	°C	400
$Pt_{(rh \rightarrow tg)}$	°C	129
Dp	°C	533
Λ	wt.-%	+39.56

KNO_3 is non hygroscopic but tends to agglomerate under its own weight. It is hence that anticaking agents (e.g. finely disperse silico acid) are frequently added to it (0.1–1.0 wt.-%). Potassium nitrate is the oldest oxidizer known to mankind. It was probably used in ancient China long before common era. Though with the advent of organic nitrates, nitro, and nitramine compounds in the last 150 years, the paramount position of potassium nitrate as ingredient in explosives has diminished; it is still an important ingredient in many civilian and military pyrotechnic formulations.

The thermal decomposition of KNO_3 starts at T = 533 °C in accordance with

$$2\,KNO_3 \longrightarrow K_2O + N_2 + 5/2\,O_2$$

Potassium perchlorate
Kaliumperchlorat

Formula		$KClO_4$
GHS		03, 07
H-phrases		271-302
P-phrases		210-221-283-306+360-371+380+375-501a
UN		1489
REACH		LRS
EINECS		231-912-9
CAS		[7778-74-7]
m_r	g mol^{-1}	138.549
ρ	g cm^{-3}	2.52
$\Delta_f H$	kJ mol^{-1}	−432.79
Mp	°C	590
Dp	°C	>561
Pt$_{(rh \leftrightarrow k)}$	°C	299
$\Delta_{pt}H$	kJ mol^{-1}	13.8
Λ	wt.-%	+46.19
c_p	J K^{-1} mol^{-1}	110.3

The non hygroscopic $KClO_4$ is one of the most important oxidizers in pyrotechnics. It has long been used as a co-oxidizer in military red and green flare compositions and is currently frequently found in many spectral flare formulations as well as flash charges for use in stun grenades and training ammunition. Its thermal decomposition commences at T = 561 °C and becomes rapid just above its melting point at T = 607 °C. Table P.10 shows the influence of various metal oxides on the decomposition temperature of $KClO_4$.

Table P.10: Influence of various metal oxides on the onset for $KClO_4$ decomposition.

Additive	T_d (°C)	Additive	T_d (°C)
None	607	MnO_2	532
Cr_2O_3	438	NiO	556 ± 13
Co_3O_4	463	MgO	567 ± 12
CoO	492	Ag_2O	570
Cu_2O	498 ± 9	Al_2O_3	582 ± 11
Fe_2O_3	500	TiO_2	787
CuO	517	ZnO	605

- K. H. Stern, *High Temperature Properties and Thermal Decomposition of Inorganic Salts with Oxyanions*, CRC Press, **2001**, pp. 202.

Potassium periodate
Kaliumperiodat

Formula		KIO_4
GHS		03, 07
H-phrases		272-315-319-335
P-phrases		210-221-302+352-305+351+338-405-501a
UN		1479
REACH		LPRS
EINECS		232-196-0
CAS		[7790-21-8]
m_r	g mol^{-1}	230.00
ρ	g cm^{-3}	3.618
Dp	°C	580
$\Delta_f H$	kJ mol^{-1}	−467.23
Λ	wt.-%	+27.82

KIO_4 has been suggested as an alternative oxidizer for delay compositions.

- J. S. Brusnahan, A. P. Shaw, J. D. Moretti, W. S. Eck, Periodates as Potential Replacements for Perchlorates in Pyrotechnic Compositions, *Propellants Explos. Pyrotech.* **2017**, *42*, 62–70.

Potassium permanganate

Kaliumpermanganat, Kaliumanganat(VII)

Formula		K[MnO$_4$]
GHS		03, 09, 07
H-phrases		272-400-410-302
P-phrases		210-220-221-280-301+312-501a
UN		1490
REACH		LRS
EINECS		231-760-3
CAS		[7722-64-7]
m$_r$	g mol^{-1}	158.034
ρ	g cm^{-3}	2.703
Dp	°C	224 (exo)
Δ$_f$H	kJ mol^{-1}	−837.22
Λ	wt.-%	+25.31

Potassium permanganate is used as oxidizer in delay formulations with antimony as fuel. With many organic substances (e.g. glycerole), KMnO$_4$ gives a hypergolic reaction. The thermal decomposition occurs in accordance with

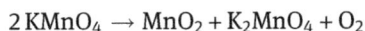

$$2\,KMnO_4 \rightarrow MnO_2 + K_2MnO_4 + O_2$$

With sulfuric acid potassium permanganate forms in accordance with

$$2\,H_2SO_4 + 2\,KMnO_4 \rightarrow Mn_2O_7 + 2\,KHSO_4 + H_2O$$

the extremely impact sensitive dimanganese heptoxide. Mn$_2$O$_7$ is an oily liquid with a metallic green luster (Mp 9.5 °C), decomposing at 55 °C and exploding at 95 °C. This reaction is also responsible for the high sensitiveness and poor stability of mixtures of KMnO$_4$ with flowers of sulfur which are known to contain traces of sulfuric acid.

- M. E. Brown, S. J. Taylor, M. J. Tribelhorn, Fuel – Oxidant Particle Contact in Binary Pyrotechnic Reactions, *Propellants Explos. Pyrotech.* **1998**, *23* 320–327.
- G. Brauer, *Handbuch der Präparativen Anorganischen Chemie, Band III*, Enke-Verlag, **1981**, pp. 1583.

Potassium peroxodisulfate, KPODS

Kaliumperoxodisulfat, Kaliumpersulfat

Formula		$K_2S_2O_8$
GHS		03, 08, 07
H-phrases		272-334-302-335-315-319-317
P-phrases		210-221-285-305+351+338-405-501a
UN		1492
REACH		LRS
EINECS		231-781-8
CAS		[7727-21-1]
m_r	g mol^{-1}	270.33
ρ	g cm^{-3}	2.450
Dp	°C	~ 100 (exo)
$\Delta_f H$	kJ mol^{-1}	−1916
Λ	wt.-%	+17.76

KPODS is a mild oxidizing ingredient (*Oxone™*) in the synthesis of, for example, azoxy compounds. In addition, it has been suggested as an oxidizer for shock-insensitive smoke formulations based on RP.

– W. Steinicke, G. Skorns, A. Schiessl, H. Büsel, W. Badura, *Hochbelastbarer Nebelformkörper mit Breitbandtarnwirkung*, DE3028933C1, **1989**, Germany.

Potassium picrate

Kaliumpikrat

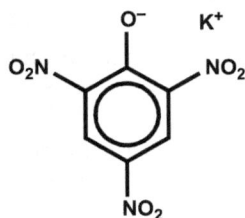

Potassium-2,4,6-trinitrophenolate	
Aspect	Red-yellow platelets
Formula	$C_6H_2KN_3O_7$
REACH	LPRS
EINECS	209-361-0

CAS		[573-83-1]
m_r	g mol^{-1}	267.196
ρ	g cm^{-3}	1.955
$\Delta_f H$	kJ mol^{-1}	−498
$\Delta_{ex} H$	kJ mol^{-1}	−948.0
	kJ g^{-1}	−3.548
	kJ cm^{-3}	−6.936
$\Delta_c H$	kJ mol^{-1}	−2328.7
	kJ g^{-1}	−8.715
	kJ cm^{-3}	−17.039
Λ	wt.-%	−38.92
N	wt.-%	15.73
Dp	°C	325
$T_{p(?)}$	°C	250

Potassium picrate has been in use with the so-called *picrate powders* by the end of the nineteenth century. Mixtures having equal mass fraction of potassium picrate and KNO_3 have been used in the past as whistling compositions.

- Z. G. Szabó, J. Száva, Factors which influence the thermal decomposition of potassium picrate, *8th International Combustion Symposium*, Pasadena, CA, 28 Aug – 3 Sept **1962**, 863–872.

Price, Donna (1913–1993)

Dr. Donna Price was an US American detonation specialist. She studied at Cornell University where she completed her dissertation in physics in 1937. Price in an oral witness statement from her colleagues at the former Naval Ordnance Laboratory (NOL), White Oaks, was considered the "mother of modern detonation physics" (the late *Sigmund J. Jacobs (1913–2006)*, Kamlet-Jacobs, JCZ, etc., bears the corresponding title "father of modern detonation physics"). Price spent the majority of her professional life at NOL, White Oaks to research the basics of initiation and propagation of shock waves in condensed HEs.

- S. J. Jacobs & D. Price, "Research Program Planning in Task Area of Physics and Chemistry of Detonation," NOL memo to XC, 28 January **1966**. 10. J. M. Majowicz and S. J. Jacobs, "Initiation to Detonation of High Explosives by Shocks," presented at the Lehigh Meeting of Fluid Dynamics division, APS, Nov. 1957 and Majowicz and Jacobs NAVORD report 5710, 1 March **1958** (CONF.) for the NOL wedge test
- D. Price, Contrasting Patterns in the Behavior of High Explosives, *11th Symposium (Int.) on Combustion*, Combustion Inst. Pittsburgh, **1967**, 693–702.
- D. Price, A. R. Clairmont Jr., Explosive Behavior of Simplified Propellant Models, *Combust. Flame*, **1977**, *29*, 87–93.
- D. Price, Critical Parameters for Detonation Propagation and Initiation of Solid Explosives, *NSWC-TR 80–339*, Naval Surface Weapons Center, Dahlgren, Virginia, **1981**, 94 pp.

Primary explosives
Initialsprengstoffe, Zündstoffe

Primary explosives, in contrary to common secondary explosives like TNT or RDX, after thermal ignition instantly undergo transition from deflagration to detonation. Primary explosives are also very sensitive towards friction and impact and mostly exhibit extreme sensitiveness towards ESD. In view of the charge accumulation on a human body (~ 0.3 J), this can be sufficient to initiate primary explosives.

Powerful primary explosives such as lead azide ($Pb(N_3)_2$) (LA) or silver azide (AgN_3)(SA) (in the past also the now obsolete mercury fulminate, $Hg(ONC)_2$) serve to initiate secondary HEs. To trigger LA or SA, less powerful but thermally or mechanically more sensitive materials such as tetracen ($C_2H_8N_{10}O$) or lead tricinate ($Pb(C_6HN_3O_8) \cdot H_2O$) (LT) are used in electric fuseheads, stab, friction, and percussion caps.

As lead compounds (LA and LT) are banned internationally for their toxicity and environmental contamination (REACH, EPA) green alternatives are currently developed.

Table P.11 shows important primary explosives and their specific properties.

Table P.11: Properties of important primary explosives.

Compound	$Pb(N_3)_2$	$Ag(N_3)_2$	Pb-tricinate	DDNP	Tetracen
Stability (–)	Good	Good	Good	Medium	Poor
Ignitability (–)	Medium	Medium	Very good	Very good	Poor
DTA decomposition (°C)	298	307	275	155	130
Impact (J)	5	2	0.2	0.2	0.1
Friction (N)	0.05	0.1	1–4	6–9	1–3
ESD (J)	10^{-10}	10^{-10}	10^{-4}		
Threshold mass* loose (mg)	170	110	1000	?	250
consolidated (mg)	50	5	No	?	No

*Necessary amount to initiate PETN.

- K. D. Oyler, *Green Primary Explosives* in T. Brinck (Ed.), *Green Energetic Materials*, Wiley, New York, **2014**, 103–177.
- R. Matyáš, J. Pachman, *Primary Explosives*, Springer, **2013**, 338 pp.
- P. Lechner, *Aufbau, Eigenschaften und Wirkungsweise von Zündstoffen und explosiven Komponenten*, BG-Chemie Kursunterlagen, RUAG Ammotec, **2004**.
- H. D. Fair, R. F. Walker (Eds.) *Energetic Materials Volume 1 – The Physics and Chemistry of the Inorganic Azides*, *Volume 2 – The Technology of the Inorganic Azides*, Plenum Press, New York, **1977**.
- A. D. Yoffe, *The Inorganic Azides*, in C. B. Colburn (Eds.), Developments in Inorganic Chemistry, Volume I, Elsevier, Amsterdam, **1966**, pp. 72–149.

Projectile

Geschoß, Kugel, Wurfkörper

Synonymous terms for projectile are bullet or missile.

In general, projectiles are bodies accelerated by expanding vapors (e.g. combustion products, metal vapor, and pressurized air) in a barrel or through a throat of the body itself (missile). Also, bodies accelerated by means of electromagnetic effects fall under the term projectile.

Depending on the diameter of the projectile in the military terminology, three groups of ammunition are determined: small caliber, medium caliber, and large caliber ammunition; whereas for the description of civilian ammunition different smaller reference diameters have been taken. Table P.12 shows the caliber designations for military and civilian ammunition.

Table P.12: Civilian and military caliber and weapon types.

	Small caliber	Medium caliber	Large caliber
Military	≤15.24 mm/0.6 cal	16–40 mm	>40 mm
Type of gun	Small arms*	Autocannon	Mortar
			Antitank
			Artillery
Civilian	≤5.59 mm/0.22 cal	–	>9 mm/0.38 cal
Type of gun	Hand guns#		Hand guns

*Automatic pistol, machine pistol, submachine gun, assault rifle, machine gun (also miniguns on board vehicles); #6-shot, automatic pistol, and rifle.

Calibers are often given as metric values (in mm or cm). Naval artillery is using multiples of inches (1 inch = 25.4 mm).

Hand gun calibers use both fractions of inches (e.g. *0.30 cal, 0.50 cal, cal 0.38*) or millimeter often extended by the length of the cartridge (e.g. *9 × 19 mm, 5.56 × 45 mm, 7.62 x 51 mm*). Special pyrotechnic weaponry (signal guns and signal projectors) bear other caliber designations:

- *caliber 4* = 25.4 mm
- *caliber 12* = 19 mm

Prussian blue

Berliner Blau, Vossenblau

Iron(III) hexacyanidoferrate(II/III), Ferric ferrocyanide		
Aspect		Darkblue powder
Formula		$Fe_4[Fe(CN)_6]_3$
REACH		LPRS
EINECS		247-875-5
CAS		[14038-43-8]
m_r	g mol^{-1}	859.242
ρ	g cm^{-3}	1.8
$\Delta_f H$	kJ mol^{-1}	+1184
Dp	°C	>250
Λ	wt.-%	−80.07
N	wt.-%	29.34

The water-insoluble Prussian blue is frequently used as burn rate modifier and due to its high positive heat of formation and its high nitrogen content, it is also an interesting energetic fuel in pyrotechnics.

– M. Weber, F. Hinzmann, *Pyrotechnische Ladung*, DE-Patent 3031369, **1982**, Germany.

PTX, 3,7,8-trinitro-pyrazolo[5,1-c][1,2,4]-triazin-4-amine

3,7,8-Trinitropyrazolo[5,1-c][1,2,4]-triazin-4-amin

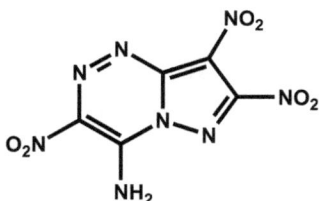

Aspect		Yellow crystalline material
Formula		$C_5H_2N_8O_6$
CAS		[1268011-02-4]
m_r	g mol^{-1}	270.12
ρ	g cm^{-3}	1.946 at 300 K
$\Delta_f H$	kJ mol^{-1}	370
$\Delta_{ex} H$	kJ mol^{-1}	−1469.41
	kJ g^{-1}	−5.4399
	kJ cm^{-3}	−10.586

$\Delta_c H$	$kJ\ mol^{-1}$	-2623.4
	$kJ\ g^{-1}$	-9.712
	$kJ\ cm^{-3}$	-18.900
Λ	wt.-%	-29.62
N	wt.-%	41.48
Dp	°C	246 (onset)
Friction	N	>324-360
Impact	cm	58.4 ± 13.5 (2.5 kg weight type 12 tool)
V_D	$m\ s^{-1}$	8998 at 1.946 g cm^{-3} (CHEETAH 5.0)
	$m\ s^{-1}$	9018 at 1.946 g cm^{-3} (CHEETAH 2.0)
P_{CJ}	GPa	36.04 at 1.946 g cm^{-3} (CHEETAH 5.0)
	GPa	38.98 at 1.946 g cm^{-3} (CHEETAh 2.0)

PTX is an insensitive HE with high predicted performance. It is currently investigated as a photoactive HE for optical initiation.

- J. A. Bjorgaard, A. E. Sifain, T. Nelson, T. W. Myers, J. M. Veauthier, D. Craig, R. J. Scharff, S. Tretiak, Two-Photon Absorption in Conjugated Energetic Molecule, *J. Phys. Chem. A.* **2016**, *120*, 4455–4464.
- M. C. Schulze, B. L. Scott, D. A. Chavez, A high density pyrazolo-triazine explosive (PTX), *J. Mater. Chem. A* **2015**, *3*, 17963–17965.
- I. L. Dalinger, C. Waltz, T. K. Shkineva, G. P. Popova, B. I. Ugrak, S. A. Shevlev, Nitropyrazoles, 18. Synthesis and transformations of 5-amino-3,4-dinitropyrazole, *Russ. Chem. Bull.* **2010**, *59*, 1631–1638.

Pyrodex™

Pyrodex is a pyrotechnic gun propellant. Its composition is basically a blend of about 60 % of a three-component black powder (KNO_3/charcoal/sulfur: 75/15/10) and 30 % of a whistling composition ($KClO_4$/Na-benzoate: 66/34) with an added binder (dextrin) and a high nitrogen compound (DCD):
- 45 wt.-% potassium nitrate
- 9 wt.-% charcoal
- 6 wt.-% sulfur
- 19 wt.-% potassium perchlorate
- 11 wt.-% sodium benzoate
- 6 wt.-% dicyandiamide
- 4 wt.-% dextrin
- 1 wt.-% water

It was invented by *Dan Pawlak* who perished in the explosion of his factory manufacturing the powder in 1977. Though lower in specific density than black powder Pyrodex, it is just more energetic to replace black powder volumetrically in a 1:1 ratio. Figure P.8 shows time pressure curve for firings of 5.184 g (80 grain) each black powder and Pyrodex in an instrumented gun producing about equal integrals (work) but showing a milder pressure built-up for Pyrodex which is beneficial for shooting.

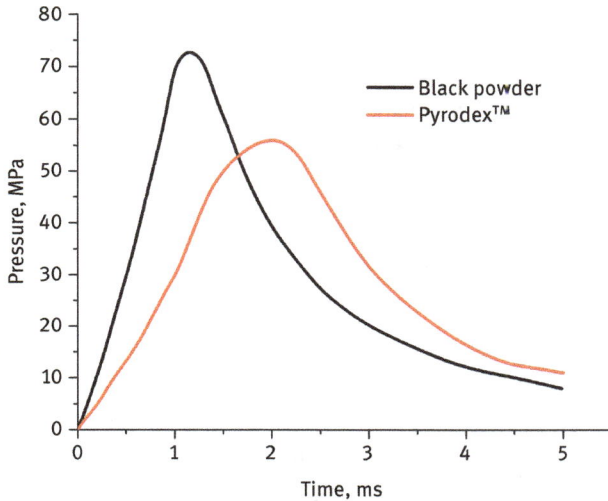

Figure P.8: Pyrodex versus black powder pressure–time curves.

– E. C. Bender, The Analysis of Dicyandiamide and Sodium Benzoate in Pyrodex by HPLC in *Crime Laboratory Digest*, U.S. Department of Justice, Federal Bureau of Investigation, **1989**, *16*, 76–77.
– D. E. Pawlak, M. Levenson, *Deflagrating Propellant Compositions*, US-Patent 4,128,443, USA, **1978**.

Pyrofuze®

Pyrofuze® is a woven thread from thin palladium and aluminum fibres. It burns fiercely forming an alloy releasing about q = 1.4 kJ g^{-1}.

– Y. Chozev, A.E. Fuhs and J. Kol, Burning Time and Size of Aluminum, Magnesium, Zirconium, Tantalum, and Pyrofuze Particles Burning in Steam, AIAA-86-1336, *AIAA/ASME Joint Thermophysics and Heat Transfer Conference*, **1986**, Boston.

Pyronol®

Pyronol is a formulation originally developed for use in underwater cutting torches. It contains

- 31.0 wt.-% nickel
- 27.4 wt.-% aluminum
- 34.4 wt.-% iron(III) oxide
- 7.0 wt.-% polytetrafluoroethylene

The formulation is pressed in a tube which is fitted with a refractory nozzle from graphite to focus the hot combustion products. The main reaction is the alloying reaction between Ni and Al. The addition of PTFE and Fe_2O_3 serves to lower the ignition temperature and to allow the formation of volatile products (e.g. AlF, AlF_2, and CO_2) to avoid clogging up the nozzle.

- H. H. Helms, A G. Rozner, *Pyrotechnic Composition*, US3695951, **1972**, USA.

Pyrophoricity

Pyrophorizität

Pyrophoricity is the property of a material (gas/liquid/solid) to spontaneously ignite when it comes into contact with oxygen and moisture. It is a special form of *hypergolicity* with oxygen and/or moisture being the triggering oxidizer. Many elements behave pyrophoric when present as finely dispersed material with high surface area (e.g. Ti, WP (P_4), U, and Zr). Many non-metal hydrides are pyrophoric (e.g. diphosphane, P_2H_6, *nido*-pentaborane, B_5H_9, tetrasilane, and Si_4H_{10}). Some liquid pyrophorics are used as starters in fuel air explosive warheads. Liquid pyrophorics (e.g. triethylaluminum, TAE) have also been used for some time as payloads for spectral decoy flares but are now obsolete. Currently, porous iron and finely dispersed iron oxides of the type FeO_x (x < 1), are used for aerial decoy flares with very little visible signature.

- D. B. Ebeoglu, C. W. Martin, *The Infrared Signature of Pyrophorics*, Air Force Armament Lab Eglin AFB, **1974**, 34 pp.
- E. G. Kayser, C. Boyars, Spontaneously Combustible Solids – A Literature Search, *NSWC/WOL/TR 75-159*, NSWC White Oak, May **1975**, 39 pp.
- J. D. Callaway, J. N. Towning, R. Cook, P. Smith, D. G. McCartney, A. J. Horlock, *Pyrophoric Material*, US 8430982B2, **2013**, UK.

Pyrotechnic compositions
Pyrotechnische Sätze (Formulierungen)

Pyrotechnic compositions serve the production of heat, electromagnetic radiation, noise, gas, solids, and aerosols and can be utilized to create other specific effects which as Table P.13 shows can be used either in the civilian field or in defense.

Table P.13: Pyrotechnic effects and their applications.

Effect	Application	
	Civilian	Defence
Heat	Destruction of data storage devices	Thermal batteries
Hot particles	Igniter, fireworks	Igniter, IR-decoys
Light	Fireworks, distress signals (SOLAS)	Illuminating and signaling ammunition, tracers
Electrons	Radar reflection (outer space)	Radar spoof
Infrared radiation	–	Decoy flares, NIR-illuminating
Report, crackle, strobe and whistle	Fireworks, bird banger, avalanche trigger	Training, simulation, stun grenades
Gas, unspecific	Airbags, actuators	Buoyancy body
Gas, specific	Hydrogen generator	Oxygen generator (SCOG)
Condensed matter	Thermite, material synthesis (SHS)	Thermite
Aerosols	Colored smoke, frost protection, hail suppression	Colored smoke, obscurants

Pyrotechnics must contain at least one oxidizer (oxidizing agent and electron acceptor) and one fuel (reducing agent and electron donor). In addition, pyrotechnics most often contain other components such as auxiliary oxidizers and fuels, processing aids, phlegmatizers, burn rate modifiers, and components to yield particular effects. Table P.14 depicts typical components in pyrotechnic formulations.

Table P.14: Components in pyrotechnic formulations.

Purpose	Example	About mass percentage
Fuel	Magnesium, titanium, zirconium, zinc, boron, aluminum, antimony, charcoal, silicon, phosphorus, sulfur, hydrocarbon polymers, carbohydrates	5–60 wt.-%
Oxidizer	Nitrates, chlorates, perchlorates, dinitramides, peroxides, sulfates, chlorine and fluorine compounds	40–75 wt.-%

Table P.14 (continued)

Purpose	Example	About mass percentage
Color agents	Compounds of sodium, the alkaline earth metals (Ca, Sr, Ba), Mo, Cu, Sm, Yb, Eu, Tm	0–15 wt.-%
Halogen source	PVC, chlorinated paraffin, hexachloroethane	0–15 wt.-%
Burn rate modifier	CuO, Fe_2O_3, soot, graphite	0–5 wt.-%
Burning surface enhancer	Expanded graphite, intumescent compounds	0–5 wt.-%
Aerosol source	Red phosphorus, ammonium chloride, chloroacetophenon (CN)	0–65 wt.-%
Friction aid	Glass powder, silicon carbide, antimony sulfide	0–5 wt.-%
Pressing aid and phlegmatizing agent	Zinc stearate, PTFE, boron nitride	0–5 wt.-%
Binder	Gum arabic, polyacrylates, polyvinyl acetate, polychloroprene, polyvinylbutyral, etc.	0–15 wt.-%
Thermal conductivity	Carbon fibers, copper powder, nanodiamonds	0–5 wt.-%

Pyrotechnics do not require atmospheric oxygen to burn, however, the ambient phase (high altitude–low altitude atmosphere, seawater, open space vacuum, etc.) always affects the combustion process, influences the rate and degree of energy release, and the quality and extent of the desired effect. The oxygen balance of pyrotechnic compositions can vary widely (Tab. P.15); even for one and the same system, a stable combustion may occur over a broad range of oxygen balance (see *barium chromate*).

Table P.15: Oxygen balance of selected pyrotechnic compositions.

Balance	Examples
Negative	–Magnesium-containing illuminating and signalling compositions, decoy compositions based on magnesium/Teflon®/Viton® (MTV) –Screening smokes based on red phosphorus –Colored smokes
Neutral	–Gas-generating compositions –Spectrally adapted flare compositions –Rocket propellants –Report charges –Whistling compositions –Black powder –HC obscurants –Thermite
Positive	–Oxygen generators –Strobe compositions

Pyrotechnic compositions need to fulfil certain criteria to be able to be manufactured and used.

- Both ingredients and reaction products must be safe with regard to occupational health as well as environmental toxicity.
- The components of pyrotechnic formulations must be compatible as it is needed to be the components of formulations in contact with one another.
- The components should be available on the market in quantity and sufficient quality to be readily used.

Pyrotechnic formulations

- should be able to be produced with minimum technical complexity to assure low manufacturing costs on the one side and avoid product defects resulting from wrong handling on the other side
- must be safe in production, storage, and use
- must withstand the environmental conditions in storage when used particularly in other climatic zones
- Must not become unserviceable due to storage under the earlier conditions and must not undergo alteration of its sensitiveness

Though plenty of progress in other technological areas has been made particularly in the last decades, there are not yet mature alternatives for many pyrotechnic compositions and the devices containing them. This is due to the generally favorable properties of pyrotechnics which comprise

- high gravimetric and volumetric energy density,
- long service life and high reliability,
- low triggering energy, and
- cheap manufacture.

- A. A. Shidlovsky, *Fundamentals of Pyrotechnics*, **1965**, 373 pp.
- J. C. Cackett, *Monograph on Pyrotechnic Compositions*, Ministry of Defence, **1965**, 131 pp.
- N. N., *Engineering Design Handbook Military Pyrotechnics Series Part One Theorie and Application*, US Army Materiel Command, **1967**, XVI + 216 pp.
- H. Ellern, *Military and Civilian Pyrotechnics*, CRC Press, **1968**, 464 pp.
- J. H. McLain, *Pyrotechnics*, The Franklin Institute Press, **1980**, 243 pp.
- A. P. Hardt, B. L. Bush, B. T. Neyer, *Pyrotechnics*, Pyrotechnica Publications, **2001**, 430 pp.
- K. & B. Kosanke et al. *Pyrotechnic Chemistry*, Journal of Pyrotechnics Inc. **2004**, 350 pp.
- T. M. Klapötke, G. Steinhauser, "Green" Pyrotechrcs: A Chemists' Challenge, *Angew. Chem. Int. Ed.* **2008**, *47*, 3330–3347.
- J. A. Conkling, C. J. Mocella, *Chemistry of Pyrotechnics*, 3rd Ed., CRC Press, **2019**, 225.
- E.-C. Koch, *Metal-Fluorocarbon Based Energetic Materials*, Wiley-VCH, **2012**, XVIII + 342 pp.
- J. J. Sabatini Advances toward the Development of „Green" Pyrotechnics, in T. Brinck (Eds.), *Green Energetic Materials*, Wiley-VCH, **2014**, pp. 63–101.

PYX, 2,6-bis(picrylamino)-3,5-dinitropyridine

2,6-Dipikrylamino-3,5-dinitropyridin

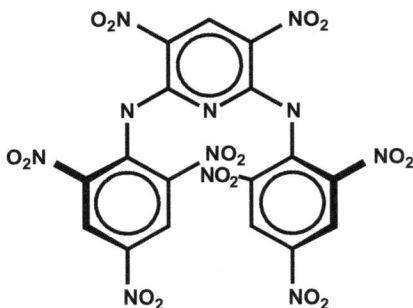

Aspect		Yellow crystals
Formula		$C_{17}H_7N_{11}O_{16}$
CAS		[38082-89-2]
m_r	g mol^{-1}	621.307
ρ	g cm^{-3}	1.757
Δ_fH	kJ mol^{-1}	80
$\Delta_{ex}H$	kJ mol^{-1}	−2930.8
	kJ g^{-1}	−4.7171
	kJ cm^{-3}	−8.288
Δ_cH	kJ mol^{-1}	−7770.2
	kJ g^{-1}	−12.506
	kJ cm^{-3}	−21.974
Λ	wt.-%	−55.37
N	wt.-%	24.80
Dp	°C	360
Friction	N	>360
Impact	cm	138 (type 12 tool)
V_D	m s^{-1}	7380 at 1.75 g cm^{-3}
P_{CJ}	GPa	23.72 at 1.75 g cm^{-3}

PYX is a high-temperature-resistant insensitive explosive mainly used in exploration of oil wells.

– T. M. Klapötke, J. Stierstorfer, M. Weyrauther, T. G. Witkowski, Synthesis and Investigation of 2,6-Bis(picrylamino)-3,5-dinitro-pyridine (PYX) and Its Salts, *Chem. Eur. J.* **2016**, *22*, 8619–8626.

Q

Quick match

Anzündlitze

Quick match is a fuse burning with a vigorous open flame (for a depiction, see Figure B.10, p. 117). It consists of a core of grained black powder in a thin sheet of textile fibers. An embedded copper wire in the core allows to bend the quick match into any desired shape and lends some mechanical strength as well. Color coding of the outer textile fibers indicates slow burning (yellow, $u = 23 \pm 5\,\mathrm{s\,m^{-1}}$) and fast burning (red, $u = 10 \pm 2\,\mathrm{s\,m^{-1}}$).

https://doi.org/10.1515/9783110660562-016

R

Reactive fragments
Reaktive Splitter

Reactive fragments are warhead casings that after fragmentation and collision with a target structure undergo a shock-induced deflagration. In addition to typical terminal ballistic effects, distinct thermal effects and blast effects result from the deflagration of reactive fragments (see also *vaporific effect*). Figure R.1 compares the effect of inert and reactive fragments on the structural units of a medium-range missile.

Inert fragments   Reactive fragments

Missile body

Missile electronics

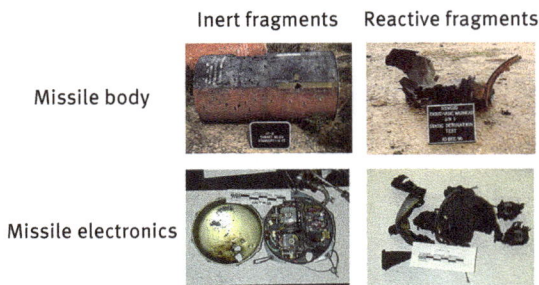

Figure R.1: Effect of inert and reactive fragments on identical structural units of a missile after *NRC*.

For quite some time, aluminum/PTFE was the standard reactive material in reactive fragments. The following is a typical formulation:
RM4
 – 26.5 wt.-%   Aluminum, H5
 – 73.5 wt.-%   Polytetrafluoroethylene, Teflon® 7A
                 ρ: 2.26 g cm^{-3}

Though a meticulous hot sintering and "forging" process for Al/PTFE has been developed over the years, the density and mechanical strength of RM4 are insufficient to meet current demands which have been defined by DARPA as follows:
 – Density > 7.8 g cm^{-3}
 – Energy content > 6.3 kJ/g
 – Compressive strengths minimum > 344 MPa, optimum > 689 MPa
 – Blast impulse minimum double (better quadruple) to common warhead based on steel casing with a mass/explosive ratio = 3

More advanced reactive fragment materials are based on low-melting alloys (based on bismuth or tin) to allow high density and high strength.

https://doi.org/10.1515/9783110660562-017

- G. Heilig, N. Durr, M. Sauer, Mesoscale Mechanics of Reactive Materials for Enhanced Target Effects, *Report I-69/12*, Fraunhofer Inst. f. Kurzzeitdynamik, Freiburg, **2013**, 85 pp.
- N. N., *Advanced Energetic Materials*, National Research Council, **2004**, p. 21.
- M. E. Grudza, W. J. Flis, H. L. Lam, D. C. Jann, R. D. Ciccarelli, *Reactive Material Structures*, DE Technologies Inc., **2014**, 348 pp.
- B. N. Ashcroft, D. B. Nielson, D. W. Doll, *Reactive Compositions Including Metal*, US Patent 8075715, **2011**, USA.
- G. D. Hugus, E. W. Sheridan, G. W. Brooks, *Structural Metallic Binders for Reactive Fragmentation Weapons*, US Patent 8746145B2, **2014**, USA.
- R. G. Ames, A Standardized Evaluation Technique for Reactive Warhead Fragments, *Ballistics Symp.*, Taragona, **2007**, 49–58.

Rim value, rim (burn) rate
Rinnenwert

The rim value is the burn time for a material obtained from a standardized metal rim. Typical rims have a measuring path of 50 cm. A stainless steel plate with dimensions 70 × 7.5 × 1 cm is furnished with a machined square rim (5 × 5 mm) placed on the middle line of the flat part of the plate extending over its full length. Indentations on both sides of the rim mark an exact 50 cm path.

The powdered substance under investigation is trickled into the rim, possible cavities are settled by tapping against the rim. With a soft wiper made from a conductive elastomer or carbon fibre brush, any supernatant material is scraped away. The material is ignited remotely at a distance of minimum 2 cm before the dent marks to allow for a stable burn to establish. The time between the dent marks is taken optically or could be taken with other instrumentation. The rim value is an important characteristic for not only fine metal powders like titanium, zirconium, and its alloys (ZrNi, ZrFe, etc.) but also slow-burning pyrotechnic compositions.

Rocket, missile
Rakete

A rocket is a streamlined body that is propelled by the directed expulsion of pressurized gases (combustion products or hot steam) through a nozzle (throat). In the absence of aerodynamic drag and gravitational forces, the maximum velocity of a rocket, v_R, is related to the exit velocity of the gases, w, and the mass ratio between the total weight (rocket + propellant) = m_0 and the net weight of the rocket (without propellant mass), m:

$$v_R = w \cdot \ln \frac{m_0}{m}$$

This equation was established by the Russian rocket mathematician and physicist *Konstantin Ziolkowski* (1857–1935) and is referred to as Ziolkowski's rocket equation. The equation was later independently established by *Hermann Oberth* (1894–1989) and *Robert Goddard* (1882–1945), explaining why Ziolkowski, Obert, and Goddard are today referred to as pioneers of modern rocketry.

The equation explains that high velocities afford low net weight of the rocket and the application of separable stages (to further reduce the net weight) supports this.

– A. Dadieu, R. Damm, E. W. Schmidt, *Raketentreibstoffe*, Springer-Verlag, Vienna, **1968**, pp. 4–6.

RS-RDX, reduced sensitivity RDX

RS-RDX, I-RDX® (trademark of EURENCO), and VI-RDX (used by ISL) describe RDX qualities which show reduced shock sensitivity (*gap test*) compared to common RDX. RS-RDX is obtained through a special crystallization procedure which yields crystals with regular morphology and very few internal defects and surface activities.

In 1989, *van der Steen* recognized that spheroidal and ellipsoidal RDX crystals with smooth surfaces possess lower shock sensitivity than polyhedral crystals with rough surface area and cavities located therein.

Relevant in terms of identifying particularly insensitive RDX crystals is the good correlation between shock sensitivity and the number and quality of optically discernible defects (*Hudson*), as well as the onset for adiabatic self-heating (ARC) (T > 198 °C) (*Bohn*). Other common methods (HFC, NQR, XRD, AFM, DSC, BAM-Fall-hammer, etc.), however, are not effective in telling the difference between common and reduced-sensitivity RDX (*Doherty & Watt*).

While the application of RS-RDX in plastic-bonded-pressable and cure-castable formulations greatly contributes to the reduction in sensitiveness, their use in melt-castable formulations based on TNT (e.g. Comp B) does not yield any improvements. This is because RDX is soluble in molten TNT and upon solidification RDX, in accordance with *Ostwald's step rule*, crystallizes out in the least stable polymorph yielding common sensitivity levels.

– M. A. Bohn, H. Pontius, Thermal Behavior of Energetic Materials in Adiabatic Selfheating Determined by ARC™, *43. ICT -JATA*, **2012**, P57.
– R. Hudson, Investigating the factors influencing RDX shock sensitivity, *PhD-Thesis*, Cranfield University, **2012**, 209 pp.
– R. M. Doherty, D. S. Watt, Relationship Between RDX Properties and Sensitivity, *Propellants Explos. Pyrotech.* **2008**, *33*, 4–13.
– L. Borne, F. Schesser, From RS-RDX to VI-RDX A New Step, *EUROPYRO 2007*, **2007**, 737–751.
– A.C. Van der Steen, H. J. Verbeek, J. J. Meulenbrugge, Influence of RDX Crystal Shape on the Shock Sensitivity of PBXs, *9. Int. Det. Symposium*, **1989**, 83–88.

Run-to-detonation distance

Anlaufstrecke

The run-to-detonation distance is the distance a high explosive at given density, temperature, confinement, and charge diameter attains a steady-state detonation velocity after subjected to a shock of variable pressure. The run-to-detonation distance information is typically displayed as the so-called *pop-plot* (P, x) after the US explosives engineer *Alphonse Popolato* (1921–2015) who investigated the correlation. Figure R.2 shows the run-to-detonation distance of various high explosives at a given density as a function of initiation pressure.

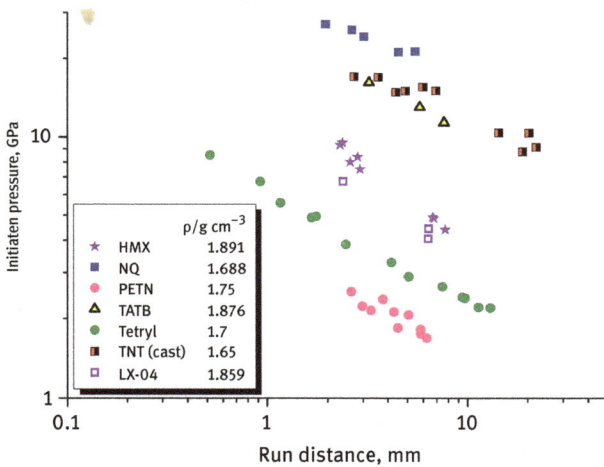

Figure R.2: Run-to-detonation distance after *Popolato et al.*

– T. R. Gibbs, A. Popolato (Eds.), *LASL Explosive Property Data*, University of California Press, **1980**, 471.

S

Safety match
Streichholz, Sicherheitszündholz

A safety match or just match is a soft wood splint coated at one end with a friction sensitive pyrotechnic composition. The match head mainly contains potassium chlorate, fuels, and inert fillers. Up to the 1990s, the match heads did contain small amounts of sulfur to aid the ignitability and dichromates and lead compounds as combustion catalysts (Table S.1). These materials are nowadays inacceptable in household applications and instead of the earlier additives, small amounts of red phosphorus (RP) as sensitizers are now used in the matchhead too.

Table S.1: Composition of safety match head formulations.

	"Old"	"New"
Potassium chlorate (wt.-%)	50	~50
Gelatine (wt.-%)	9.3	~10
Sulfur (wt.-%)	1.5	
Phosphorus, red (wt.-%)		<2
Potassium dichromate (wt.-%)	0.4	
Manganese dioxide (wt.-%)	4.3	
Glass powder	24	
Binder	2.5	
Buffer and additives	8	~38 (Kaolin)

European safety matches consist of poplar wood (lat.: *populus*) splint. This is impregnated first with an aqueous solution of ammonium phosphate, $NH_4H_2PO_4$, to prevent incandescence of the burnt match. To protect against moisture, the splint is later soaked in warm paraffin and dipped right after in the pasty match head formulation and dried thereafter.

To ignite the matchhead, this has to be scratched against a phosphorus containing striker face. While the match head formulations changed appreciably over time, the striker formulations have not changed very much and only antimony sulfide, which is both a friction aid and fuel, has been replaced due to its noxious combustion products. Table S.2 shows old and new striker formulations.

https://doi.org/10.1515/9783110660562-018

Table S.2: Composition of striker formulations for match boxes.

	"Old"	"New"
Phosphorus, red (wt.-%)	40	40
Antimony(III) sulfide (wt.-%)	28	
Manganese dioxide (wt.-%)		30
Polyvinylacetate (wt.-%)	30	25
Buffer and additives (wt.-%)	2	5 ($CaCO_3$)

- H. Hartig, *Zündwaren*, VEB Fachbuchverlag Leipzig, 2. Aufl. **1971**, 310 pp.
- http://www.i-m.de/gefahrstoffe/261313.pdf

Samarium

Formula		Sm
GHS		02
H-phrases		250-260-252
P-phrases		210-222-231+232-280-422a-501a
UN		1383
EINECS		231-128-7
CAS		[7440-19-9]
m_r	g mol^{-1}	150.36
ρ	g cm^{-3}	7.520
Mp	°C	1072
Bp	°C	1788
$\Delta_{melt}H$	kJ mol^{-1}	8.62
$\Delta_{vap}H$	kJ mol^{-1}	166.4
c_p	J K^{-1} mol^{-1}	29.54
c_L	m s^{-1}	2700
κ	W m^{-1} K^{-1}	13.3
Δ_cH	kJ mol^{-1}	−913.7
	kJ g^{-1}	−6.077
	kJ cm^{-3}	−45.697
Λ	wt.-%	−15.96

Samarium is a soft malleable silvery metal which tarnishes quickly in air thereby forming a voluminous white oxide layer. In terms of reactivity (EN(Sm) = 1.17), it is about in between magnesium (1.31) and calcium (1.0). Due to its low price and its beneficial thermochemical properties, it merits consideration in niche applications in pyrotechnics. Sm powder and pyrotechnic formulations containing it burn with a pink flame.

- F. Lederle, J. Koch, W. Schade, E. G. Hübner, Color-Changing Sparks from Rare Earth Metal Powders, *Z. Anorg. Allg. Chem.* **2020**, *646*, 37–46.
- A. Abraham, N. A. MacDonald, E. L. Dreizin, Reactive Materials for Evaporating Samarium, *Propellants Explos. Pyrotech.* **2016**, *41*, 926–935.
- E.-C. Koch, V. Weiser, E. Roth, S. Kelzenberg, Consideration of some 4 *f*-Metals as New Flare Fuels Europium, Samarium, Thulium and Ytterbium, *ICT*, **2011**, Karlsruhe, V1.

Sarin, GB

(RS)-Propan-2-yl methylphosphonofluoridat

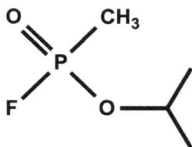

Isopropoxymethylphosphorylfluorid, Trilon 144		
Formula		$C_4H_{10}FO_2P$
CAS		[107-44-8]
m_r	g mol^{-1}	140.093
ρ	g cm^{-3}	1.0887
Mp	°C	−56
Bp	°C	152
Λ	wt.-%	−148.47
$\Delta_f H$	kJ mol^{-1}	−1006.3 (QBS-QB3)
$\Delta_c H$	kJ mol^{-1}	−2878.3
	kJ g^{-1}	−20.550
	kJ cm^{-3}	−22.368
c_p	J mol^{-1} K^{-1}	167.3
LCt$_{50}$	ppm	15 (at high-level physical stress)
	ppm	100 (at rest)
ICt$_{50}$	ppb min^{-1}	8 or 50

Sarin is a nerve agent that quickly exerts its toxic effects upon inhalation, dermal or ocular exposure. Near its boiling point, GB decomposes into propylene and methylphosphonic acid fluoride.

- S. Franke, *Lehrbuch der Militärchemie, Band 1*, Militärverlag der Deutschen Demokratischen Republik, 2. Auflage, Leipzig, **1977**, pp. 398–422.

Schardin, Hubert (1902–1965)

Figure S.1: Hubert Schardin, photograph by Sigrid Schardin (1963). https://commons.wikimedia.org/wiki/File:Schardin.jpg.

Hubert Schardin studied at the TH in Berlin-Charlottenburg and Munich and completed his dissertation with honors in 1934 under the supervision of Carl Cranz. In 1936 Schardin became head of the Institute for Technical Physics and Ballistics at the Technical Academy of the Luftwaffe in Berlin. In 1937, he became professor at the TH Berlin-Charlottenburg. After World War II together with the French General Robert Cassagnou, he cofounded a research institute in Saint-Louis in France specializing on ballistics which officially became the French-German Research Institute ISL in 1959. There Schardin was German director until his appointment as assistant secretary of state in the German Ministry of Defence in Bonn in 1964.

Schardin's scientific achievements have been plentiful mainly in field of high-speed photography (*Cranz-Schardin Camera*), ultra-short time measurement, and the Schlieren photography.

- H. Schardin, Die Schlierenverfahren und ihre Anwendungen, in F. Hund (Eds.) *Ergebnisse der Exakten Naturwissenschaften*, Bd. XX, Springer, **1942**, 303–439 + Farbtafel I + II.
- H. Schardin (Eds.) *Beiträge zur Ballistik und Technischen Physik*, Ambrosius/Barth, **1938**, 216 pp.
- C. Cranz, H. Schardin, Kinematographie auf ruhendem Film und mit extrem hoher Bilderfrequenz, *Z. Physik*, **1929**, *56*, 147–149.
- Entry in the German National Library: http://d-nb.info/gnd/118606506

Schießwolle

The first torpedo charges used by the German imperial navy consisted of water moist nitrocellulose which in German was also referred to as "Schießwolle." This literally translates to "guncotton." In tradition and also as a disguise future formulations in World War I and World War II, beared the same term "Schießwolle." These were the first insensitive naval munitions based on bullet proof melt-castable insensitive HEs based on hexadinitrophenylamine, TNT, and aluminum (Tab. S.3).

Table S.3: Different types of "Schießwolle."

Schießwolle	18		19	36	39	39a
Other names assigned	*Hexamit, TSMV1-101, S-1*			S-2	S-3	
Aluminum (wt.-%)	16		27	25	25	35
Ammonium nitrate (wt.-%)					30	5
Nitropenta (wt.-%)			25			
Hexanitrodiphenylamine (wt.-%)	24			8	5	10
Trinitrotoluene (wt.-%)	60		48	67	45	50

– E.-C. Koch, Insensitive Munitions, Propellants Explos. Pyrotech. **2016**, *41*, 407.

Schubert, Hiltmar (1927*)

Hiltmar Schubert is a German chemist. He studied chemistry at the University of Lübeck and at the TH Karlsruhe and completed his dissertation in 1959 with *Karl Meyer*. Schubert was director of the Fraunhofer Institut für Chemische Technologie (ICT) from 1972 to 1994. In 1976, he founded the topical journal *Propellants Explosives* (which became Propellants Explosives Pyrotechnics in 1982) and was its main editor until 2000. Schubert always stayed on top of the explosives field and was able to prepare his institute for upcoming challenges in the field. Examplary was his initiative in the field of *insensitive HEs* as a member of the prestigious group, Air Senior National Representatives, under which he initiated the development of nitroguanidine-based EIDS at ICT and his activities that helped establishing detection of HEs at ICT to support counter-terrorism, national security, and detente.

– H. Schubert, *Ein Leben für die Explosivstoff-Forschung im Nachkriegsdeutschland*, Fraunhofer ICT, Pfinztal Berghausen, **2007**, 61 pp.
– H. Schubert, F. Schedlbauer, "ASNR-Langzeit Technologieprogramm über unempfindliche Sprengstoffe (IHE), Ergebnisbericht der 1. und 2. Phase, *Sprengstoffgespräch*, Lünen, **1995**, 315–336.

- H. Schubert, A. Kuznetsov, *Detection and Disposal of Improvised Explosives, NATO Security through Science Series*, Springer, **2006**, 239 pp.
- Entry in the German National Library: http://d-nb.info/482374292

Self-contained oxygen generator (SCOG)
Sauerstoffgenerator

Self-contained oxygen generators, also known as Solid fuel oxygen generator (SFOG), serve the supply with oxygen under distress situations on boards of submarines at high altitude and even in space (e.g. *SFOG* on board the International Space Station (ISS). Compared to pressurized oxygen or cryogenic oxygen, SCOGs are more easy and safer to handle and serve as low pressure source of oxygen. Typical compositions used in SCOGS contain alkali chlorate (V or VII) and a metallic fuel such as steel wool or powdered manganese (Table S.4). Other additives (peroxides) serve as scavengers for chlorine which forms upon decomposition of perchlorates and chlorates in traces and for carbon dioxide which forms from the carbon present in steel. Typical SCOGs (mass: 8–12 kg) develop over their entire burn time (50–60 min) about 2000–3000 L oxygen. The functional safety of SCOGs is dealt with in a NASA report.

Table S.4: Compositions of SCOGs.

Component	A	B
$NaClO_3$ (wt.-%)	92	
$LiClO_4$ (wt.-%)		84.8
BaO_2 (wt.-%)	4	
Li_2O_2 (wt.-%)		4.2
Steel wool (wt.-%)	4	
Manganese (wt.-%)		10.9
O_2 content (wt.-%)	41	51

- J. Graf, *Chlorate Oxygen Generator (Oxygen Candle) Review of the History of Candle Development*, February **2017** Houston TX; for eventual publication in NESC Report accessible at: https://ntrs.nasa.gov/archive/nasa/casi.ntrs.nasa.gov/20170002051.pdf
- Investigation, Analysis, and Testing of Self-Contained Oxygen Generators, *WSTF-IR-1129-001-08*, NASA, White Sands, **2008**, 87 pp.

Semtex

Semtex is a trade name for a variety of plastic explosives manufactured by the Czech company *Synthesia*, Pardubice (see Table S.5). Semtex A and Semtex H were misused very often in the past in terrorist attacks (e.g. Lockerbie-bombing, 1988), building up international pressure which finally led to the decision by the manufacturer to add a taggant for detection of the explosives in safety locks in airports. Between 1990 and 1991, Semtex was tagged with EGDN. Nowadays, Semtex is tagged with 2,3-dimethyl-3,4-dinitrobutane or 4-nitrotoluene.

Table S.5: SEMTEX compositions.

Semtex	A	1A	10	H
RDX	41.72			4.65
PETN	41.38	83.5	85.0	76.94
Styrene-butadiene rubber	9.0	4.1	3.7	9.4
Styrene-butadiene rubber	Sudan I	0.002	0.002	Sudan IV
N-Phenyl-2-naphthylamine	x			x
N-Octyl phthalate/butyl citrate	7.9	12.4		9.0
Dibutylformamide			11.3	
Ethylcentralite		x	x	

- S. Moore, M. Schantz, W. MacCrehan, Characterization of Three Types of Semtex (H, 1A, and 10), *Propellants Explos. Pyrotech.* **2010**, *35*, 540–549.

Sensitiveness and sensitivity
Empfindlichkeit

The response of an explosive towards accidental stimuli (sensitiveness) determines its safe handling. The response of an explosive towards intentional stimuli (sensitivity) determines the possible range of its use. Typical stimuli to affect reactions in explosive materials are as follows:
- impact
- shock
- friction
- heat
- flames
- electromagnetic radiation
- electrostatic discharge

There are standardized methods to determine the sensitiveness of explosives (e.g. Orange Handbook or AOP -7 and the corresponding STANAGs).

To increase the safety of handling and processing of explosives, they are typically phlegmatized. A well-documented and most prominent historic example is the taming of nitroglycerine by Alfred Nobel with the subsequent development of gur-dynamite and finally the gelling of NGl with nitrocellulose to give the safe blasting gelatine.

Nowadays, many crystalline HEs are phlegmatized.

We have to differentiate between a temporal phlegmatization to meet logistic requirements such as those manifested in UN Test Series 3 or permanent phlegmatization.

Temporal phlegmatization is, for example, done with crystalline nitramines and nitroguanidine by addition of ~ 25 wt.-% water. While the friction and impact sensitiveness of RDX and HMX become low enough to allow transport as HD 1.1D, nitroguandine, upon addition of minimum 20 wt.-% water, loses its mass explosive potential and becomes HD 4.1 (inflammable solid). Another example though not an HE is red phosphorus (RP). As RP is prone to ignition by friction, impact and very much by electrostatic discharge, it is often stored and transported as a water moist (20 wt.-%) filter cake (see HB 600 under *RP*).

Permanent phlegmatization is necessary to meet operational requirements, such as gun launch safety (high set back forces and high spin), and increasingly important to meet insensitive munitions requirements (see *IM*). Therefore, crystalline explosives can be produced in an extraordinary quality that features very low shock sensitivity (*RS-RDX*), explosive crystals can be embedded in an insensitive melt cast or flexible cure castable binder, or consolidated with the aid of a thermoplastic binder. All of which have their pros and cons depending on the particular HE, formulation, and intended application.

While the earlier mentioned points intend to lower the sensitiveness of HEs to enable safe handling and meeting operational and IM-policy requirements, it is also frequently necessary to enhance the sensitivity to enable safe use of a HE by means of **sensitization**.

Nitromethane (NM) and 1,3,5-triamino-2,4,6-trinitrobenzene (TATB) are both extremely shock-insensitive HEs. NM needs sensitization with microballoons (tiny air-filled hollow glass spheres) to become detonable within certain limits. TATB can be prepared sonochemically from 1,3,5-trichloro-2,4,6-trinitrobenzene and gaseous ammonia thereby yielding a porous material (with pore size dimensions influenced by frequency and intensity of ultrasonication) which is more shock sensitive than bulk crystals. The underlying concept is that a passing shock wave (both in sensitized NM and porous TATB) yields pore collapse and consequently temperature increase to give hot spots to seed the start of a reaction.

- Recommendations on the Transport of Dangerous Goods – Manual of Tests and Criteria – 5th rev. Ed. United Nations, **2009**, 450 pp.

Sensitivity correlations
Empfindlichkeitskorrelationen

Though there is no overarching mathematically valid correlation between the sensitiveness of a chemically uniform covalent or ionic explosive and other parameters of a substance accessible either by experiment or theoretical method, there are still obvious trends which support the understanding of the existing materials and assist in the development of new materials. With respect to impact sensitiveness, the following parameters have been investigated and discussed so far.

Oxygen balance

- Kamlet & Adolph *Propellants Explos. Pyrotech.* **1979**, 4, 30

Chemical composition

- G. T. Afanas'Ev, T. S. Pivina, D. V. Sukhachev, *Propellants Explos. Pyrotech.* **1993**, *18*, 309–316
- D. E. Bliss, S. L. Christian, W. S. Wilson, Impact Sensitivity of Nitramines, *J. Energ. Mat.* **1991**, *9*, 319–348
- W. C. Lothrop, G. R. Handrick, The Relationship between Performance and Constitution of Pure Organic Explosive Compounds, *Chem. Rev.* **1949**, *44*, 419–445.
- J. R. Stine, On predicting properties of explosives - Detonation Velocity, *J Energ Mat* **1990**, *8*, 41–73.
- S. R. Jain, Energetics of propellants, Fuels and Explosives; a Chemical Valence Approach, *Propellants Explos. Pyrotech.* **1987**, *12*, 188–195.
- L. R. Rothstein, R. Petersen, Predicting High Explosive Detonation Velocities from their Composition and Structure, *Propellants Explos. Pyrotech.* **1979**, *4*,56–60 &86 (Erratum).
- L. R. Rothstein, Predicting High Explosive Detonation Velocities from their Composition and Structure (II), *Propellants Explos. Pyrotech.* **1981**, *6*, 91–93.
- A. R. Martin, H. J. Yallop, Some Aspects of detonation. Part 1. – Detonation velocity and chemical constitution, *Trans. Faraday Soc.* **1958**, *54*, 257–263.
- A. R. Martin, H. J. Yallop, Some Aspects of detonation. Part 2. – Some Aspects of detonation. Part 2. – Detonation velocity as a function of oxygen balance and heat of formation, *Trans. Faraday Soc.* **1958**, *54*, 257–263.
- A. Mustafa, A. A. Zahran, Tetryl, Pentyl, Hexyl, and Nonyl. Preparation and Explosive Properties, *J. Chem. Eng. Data* **1963**, *8*, 135–150.

Orbital energies

- M. M. Kuklja, E. V. Stefanovich, A. B. Kunz, An excitonic mechanism of detonation initiation in explosives, *J. Chem. Phys.* **2000**, *112*, 3417–3423.
- H. Zhang, F. Cheung, F. Zhao, X. Cheng, Band gaps and the possible effect on impact sensitivity for some nitro aromatic explosive materials. *Int. J. Quantum Chem.* **2009**, *109*, 1547–1552.
- S. V. Bondarchuk, Quantification of Impact Sensitivity Based on Solid-State Derived Criteria, *J. Phys. Chem. A* **2018**, *122*, 5455–5463.

Vibrational transitions

- J. Sharma, B. C. Beard, M. Chaykovsky, Correlation of impact sensitivity with electronic levels and structure of molecules, *J. Phys. Chem.* **1991**, *95*, 1209–1213.
- M. R. Manaa, L. E. Fried, Intersystem Crossings in Model Energetic Materials, *J. Phys. Chem. A* **1999**, *103*, 9349–9354

Chemical shift in NMR

- S. Zeman, New Aspects of the Impact Reactivity of Nitramines, *Propellants Explos. Pyrotech.* **2000**, *25*, 66–74.
- S. Zeman, M. Jungová, Sensitivity and Performance of Energetic Materials, *Propellants Explos. Pyrotech.* **2016**, *41*, 426–451.

Molecular surface electrostatic potential

- J. S. Murray, P. Lane, P. Politzer, Effects of strongly electron-attracting components on molecular surface electrostatic potentials application to predicting impact sensitivities of energetic molecules, *Mol. Phys.* **1998**, *93*, 187–194
- B. M. Rice, J. J. Hare, A Quantum Mechanical Investigation of the Relation between Impact Sensitivity and the Charge Distribution in Energetic Molecules, *J. Phys. Chem. A* **2002**, *106*, 1770–1783.

Electrostatic potential of the median of a C-NO$_2$ bond

- J. S. Murray, P. Lane, P. Politzer, A relationship between impact sensitivity and the electrostatic potentials at the midpoints of C-NO$_2$-bonds in nitroaromatics, *Chem. Phys. Lett.* **1990**, *168*, 135–139.

Heat of fusion

- S. Zeman, New Aspects of the Impact Reactivity of Nitramines, *Propellants Explos. Pyrotech.* **2000**, *25*, 66–74.

Activation energy

- I. Fukuyama, T. Ogawa, A. Miyake, Sensitivity and Evaluation of Explosive Substances, *Propellants Explos. Pyrotech.* **1986**, *11*, 140–143.
- H. Xiao, J. Fana, Z. Gua, H. Dong, Theoretical study on pyrolysis and sensitivity of energetic compounds (3) Nitro derivatives of aminobenzenes, *Chem. Phys.* **1998**, *226*, 15–24.
- C. Xu, X. Heming, Y. Shulin, Theoretical investigation on the impact sensitivity of tetrazole derivatives and their metal salts, Chem. Phys. 1999, 250, 243–248.
- S. Zeman, New Aspects of the Impact Reactivity of Nitramines, *Propellants Explos. Pyrotech.* **2000**, *25*, 66–74.

Rate of constant of decomposition

- S. Zeman, New Aspects of the Impact Reactivity of Nitramines, *Propellants Explos. Pyrotech.* **2000**, *25*, 66–74.

Bonding energy/total energy

- L. E Fried, M R. Manaa, P. F Pagoria, R. L. Simpson, Design and Synthesis of Energetic Materials, *Ann. Rev. Mater. Res.* **2001**, *31*, 291–321.

$C-NO_2$ and $N-NO_2$ bond energy

- F. J. Owens, Calculation of energy barriers for bond rupture in some energetic molecules *J. Mol. Struct. (Theochem)* **1996**, *370*, 11–16.
- B. M. Rice, S. Sahu, F. J. Owens, Density functional calculations of bond dissociation energies for NO_2 scission in some nitroaromatic molecules, *J. Mol. Struct. (Theochem)* **2002**, *583*, 69–72.
- X. Song, X. Cheng, X. Yang, D. Li, R. Linghu, Correlation between the bond dissociation energies and impact sensitivities in nitramine and polynitro benzoate molecules with polynitro alkyl groupings, *J. Hazard. Mater.* **2008**, *150*, 317–321.
- J. Li, A multivariate relationship for the impact sensitivities of energetic N-nitrocompounds-based on bond dissociation energy, *J. Hazard. Mater.* **2010**, *174*, 728–733.
- J. Li Relationships for the Impact Sensitivities of Energetic C-Nitro Compounds Based on Bond Dissociation Energy, *J. Phys. Chem. B* **2010**, *114*, 2198–2202.

Bond lengths

- P. Politzer, J. S. Murray, P. Lane, Shock-sensitivity relationships for nitramines and nitroaliphatics, *Chem. Phys. Lett.* **1991**, 181, 78–82.

Molecular electronegativity

- J. Mullay, A Relationship between Impact Sensitivity and Molecular Electronegativity, *Propellants Explos. Pyrotech.* **1987**, *12*, 60–63.
- E.-C. Koch, The Hard and Soft Acids and Bases (HSAB) Principle–Insights to Reactivity and Sensitivity of Energetic Materials, *Propellants Explos. Pyrotech.* **2005**, *30*, 5–16.

Molecular geometry

- V. Belik, V. A. Potemkin, N. S. Zefirov, Correlation between geometrical structure of molecules and impact sensitivity of explosives, *Dokl. Akad. Nauk. SSSR* **1989**, *308* 882–886.

Energy transfer rate to vibrations

- L. E. Fried, A. J. Ruggiero, Energy Transfer Rates in Primary, Secondary, and Insensitive Explosives, *J. Phys. Chem.* **1994**, *98*, 9786–9791
- K. L. McNesby, C. S. Coffee, Spectroscopic Determination of Impact Sensitivities of Explosives, *J. Phys. Chem. B* **1997**, *101*, 3097–3104.

Polarity of bonds

- A. Delpuech, J. Cherville, Relation entre la Structure Electronique et al Sensibilite au Choc des Explosifs Scondaires Nitres-Critere Moleculaire de Sensibilite, Propellants Explos. 1978, 3, 169–175.

Substitution patterns

- F. J. Owens, Relationship between impact induced reactivity of trinitroaromatic molecules and their molecular structure, *J. Mol. Struct. (Theochem)* **1985**, *121*, 213–220.

Bond order

- H. Xiao, J. Fana, Z. Gua, H. Dong, Theoretical study on pyrolysis and sensitivity of energetic compounds (3) Nitro derivatives of aminobenzenes, *Chem. Phys.* **1998**, *226*, 15–24.

Free volume in crystal lattice

- P. Politzer, J. S. Murray, Impact Sensitivity and Crystal Lattice Compressibility/Free Space, *J. Mol. Model.* **2014**, *20*, 222–231.
- P. Politzer, J. S. Murray, High Performance, Low Sensitivity Conflicting or Compatible? *Propellants Explos. Pyrotech.* **2016**, *41*, 414–425.

Shaped charge and explosively formed projectiles

Hohlladung, P(rojektilbildende)-Ladung

A shaped charge (SC) is an explosive charge which produces a focused, concentrated, and enhanced effect on a target structure. Therefore, the charge is covered with a convex concave liner made of dense malleable metal (e.g. copper) in the direction of the planned effect. Linear charges bear an angular symmetric cavity extending over the length of the charge while cylindrical charges bear an acute conical cavity ($\alpha = \frac{1}{2}$ cone angle) on one of the circular planes. Figure S.2 shows the principal construction of a cylindrical shaped charge. Charges with blunt angled liners ($2\alpha > 120°$) produce a slug and are called explosively formed projectile (EFP, see later).

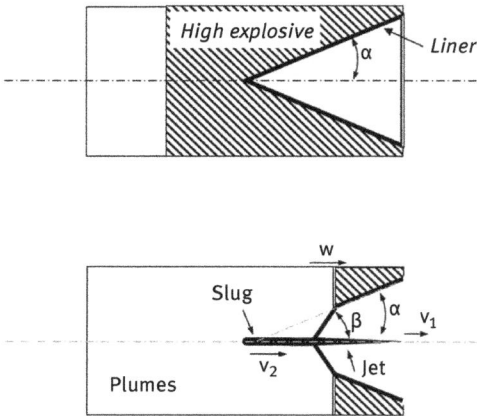

Figure S.2: Schematic depiction of the processes going on upon detonation of a cylindrical metal lined shaped charge.

Upon detonation of the HE (detonation velocity = w), the shock front accelerates the liner to assume an angle β, which is characteristic for the combination of HE and metal. The detonation upends the liner completely to give a front jet and back slug travelling at two discrete velocities $v_1 > v_2$.

Small particles in front of the jet may assume very high velocities approaching up to $v_1 = 25$ km s^{-1} (Figure S.3). The back side of the liner is converted into a thick slug travelling at the lower velocity v_2.

For the velocities the following equations apply:

$$v_1 = w \cdot \frac{\sin(\beta - \alpha)}{\cos\alpha} \cdot \left[\frac{1}{\sin\beta} + \cot\beta + \tan\frac{1}{2}(\beta - \alpha)\right]$$

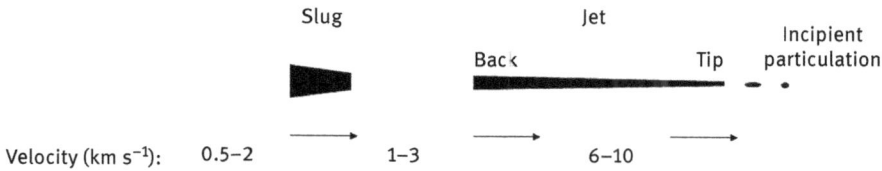

Figure S.3: Schematic depiction of the detonatively shaped liner.

and

$$v_2 = w \cdot \frac{\sin(\beta - \alpha)}{\cos\alpha} \cdot \left[\frac{1}{\sin\beta} - \cot\beta - \tan\frac{1}{2}(\beta - \alpha)\right]$$

With the masses of jet (1) and slug (2)

$$m_1 = \frac{1}{2}m(1 - \cos\beta) \text{ and } m_2 \frac{1}{2}m(1 + \cos\beta)$$

with m being the total mass of the liner.

The penetration of the jet of density (ρ_j) into a target of density (ρ_t) can be described by hydrodynamics. For the stagnation pressure at the impact point we can write:

$$\frac{1}{2}\rho_j \times (v_1 - u)^2 = \frac{1}{2}\rho_t \times u^2$$

Then the velocity of the jet tip reaching the crater bottom is given by

$$u = \frac{\sqrt{\frac{1}{2}\rho_j} \cdot v_1}{\sqrt{\frac{1}{2}\rho_j} - \sqrt{\frac{1}{2}\rho_t}}$$

In addition to hydrodynamics influencing the depth of penetration ("$\sqrt{\rho}$-law"), the penetration is further influenced by other parameters such as the type of liner, type of HE (influence on jet velocity), and distance between SC and target (*standoff*) (Figure S.4). That is, at too large distance, particulation of the jet occurs together with detrimental cross velocities. However, at too short distances maximum elongation of the jet has not yet been achieved. Both effects reduce penetration. Further the opening angle of the cone influences the rate at which the liner upends and thereby also limits the jet tip velocity (Figure S.5). Finally, the depth of penetration correlates with the reciprocal of the yield strength of the target material, as the crater diameter becomes narrow with increasing yield strength.

Figure S.6 depicts the influence of liner material on depth of penetration. For military shaped charges, mainly HMX-based formulations are in use (Octol, PBXN-5, -9, -11, and -110). In civilian applications (e.g. linear cutting charges and jet tapper charges for furnaces and oil rigs) temperature-resistant HEs such as HNS are in use.

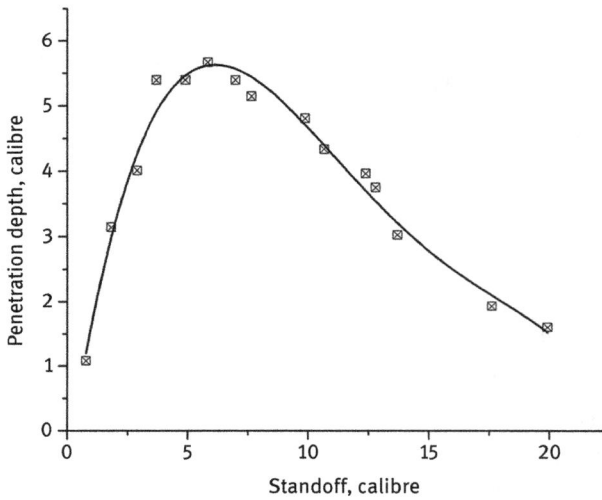

Figure S.4: Depth of penetration as a function of standoff for Cu liner.

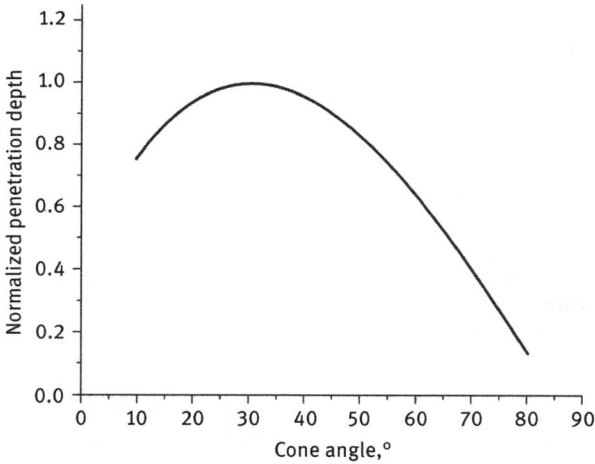

Figure S.5: Depth of penetration as a function of cone angle α for Cu liner.

Projectile forming charges, also called EFP charges (EFP or penatrator), belong to the family of the forming charges. Figure S.7 shows SC, hemispherical charge (HEMI), and EFP. The affiliation to the family results from the similar architecture of the different charge types. In every case, there is a HE filling and a metal liner in the charge. When the HE is initiated, the detonation front forges the metal liner into projectiles

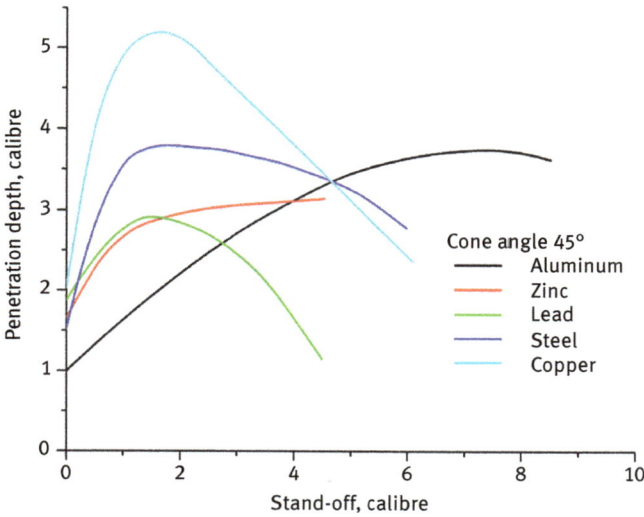

Figure S.6: Depth of penetration as a function of caliber of different liners.

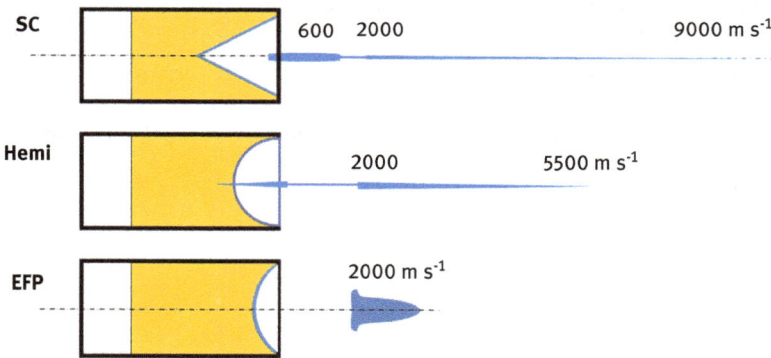

Figure S.7: Family of explosively formed projectile charges: shaped charge, hemispherical charge, and EF charge.

with high velocities. In the case of the SC and the HEMI, very fast jets are generated which particulate due their high velocity gradients as depicted in Figure S.7.

In contrary to these charges for the EFP charge, the angle (if a conical liner is used) or the curvature (if a dish-shaped liner is used, as depicted in Figure S.7) is selected very large leading to very small velocity gradients only. This gradient is now used to forge the liner to a projectile, which means that the kinetic energy is transformed into a plastic forming energy and heat.

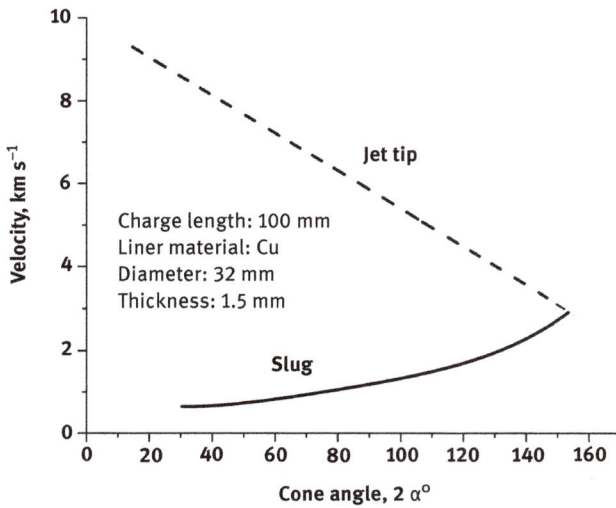

Figure S.8: Velocity of jet tip and slug as function of cone angle.

The missing particulation leads to the nearly independence of the perforation performance of the EFP projectile from the stand-off, besides a slight deceleration effect due to air-drag – compared to the quick reduction with SCJ and HEMI jets depicted in Figure S.9. With modern EFP charges a perforation performance of 1.1 to 1.3 charge diameter (caliber) can be expected in about distance of

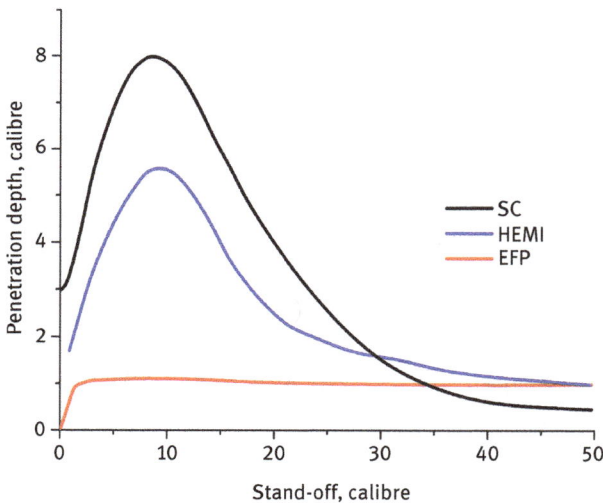

Figure S.9: Stand-off curves for SC, HEMI, and EFP.

1000 calibers stand-off. This is an important feature with regard to modern long-distance active defense systems (e.g. with future tank systems). Here top attacks play an important role to compensate the reduced perforation performance compared to SC and HEMI which typically attack the heavier protected lateral armor. Also, long distance attacks to the rear or side of a tank system is another potential application exploiting the feature of an EFP charge.

A successful explosive forging process of an EFP liner (modern liners are dish shaped) depends strongly on a number of key design parameters, which have to be harmonized carefully:

- liner contour (curvature or angle, wall thickness, and mass distribution)
- liner material (pure iron, copper, tantalum, etc.)
- casing of the charge (wall thickness, and material)
- charge dimensions (shape, length L, and diameter D, L/D ratio)
- detonation wave shaper
- type of high explosives (mass C and type)

As an example, the influence of the liner contour shall be demonstrated. The angle of a conical liner (or the curvature of a dish) influences strongly the velocity gradient of the self-forging projectile (depicted in Figure H.8). This velocity gradient on the other side is responsible for the result of the L/D ratio of the built projectile. And finally, this L/D ratio determines the perforation performance of the EFP-projectile. If the selected velocity gradient is too large, the projectile will particulate as it is the case with HEMI and SC charges. At very blunt angles ($2\alpha > 150°$) a significant part of the mass can be lost through radial particulation due to shear spall. Figure H.7 depicts the relative velocities of both jet tip and slug as a function of cone angle 2α.

- K. Weimann, *Ausgewählte Kapitel der Kurzzeitdynamik, Teil 4 – Projektilbildende Ladungen (in German)*. Ernst-Mach-Institude, Freiburg, Germany, **1998**, 7–27.
- R. A. Brimmer, Manual for Shaped-Charge Design, *NAVORD Report 1248*, China Lake, USA, **1950**.
- R. E. Kutterer, „Die Hohlladung " in *Ballistik*, 3. Aufl. Vieweg, **1959**, pp. 248–253.
- M. Held, Explosive Formed Projectiles, *3. Int Symposium on Ballistics*, Karlsruhe, Germany, **1977**.

Shellac
Schellack

Shellac is a resinous exudation of the insect *laccifer lacca* that feeds on the sap of host tree *schleichera trijuga*. The composition of shellac varies with the seasons and can be described as $C_6H_{9.6}O_{1.6}$; CAS-No. [9000-59-3]. The main components of shellac are aleuritic acid [533-87-9] and schellolic acid [4448-95-7]. The heat of formation of shellac is about $\Delta_f H = -440$ kJ mol^{-1}. Schellack serves as both fuel and binder in pyrotechnics. Between $T = 40$ and 94 °C, it gives of moisture and starting at T = 190 °C shellac undergoes decomposition which is completed at about 500 °C. Shellac has densities between 1.05 and 1.2 g cm^{-3}.

- W. Meyerriecks, Organic Fuels Composition and Formation Enthalpy Part II – Resins, Charcoal, Pitch, Gilsonite and Waxes, *J. Pyrotech.* **1999**, *9*, 1–19

Shock tube
Stoßrohr

Shock tubes are test setups to generate plane compression shocks in gases. They have been invented by *Vieille* in 1899. They serve an amazingly broad scope of physical and chemical investigations such as supersonic fluid dynamics of objects (missiles, projectiles), and optical properties of shocked gases.

The term *shock tube* is also frequently used to describe fast non-electric (and hence safe against electromagnetic interference) ignition means based on thin polymer tubings (OD = 3 mm, ID = ~ 1.5 mm) coated on the inside with a film of energetic materials such as Al/HMX (50/50) (EXEL®) or Al/KClO$_4$/Fe$_3$O$_4$(50/25/25) (Lightning thermo tube) (Figure S.10). After thermal ignition, the shock tubes speed up through DDT to LVD with velocities in the 1100 m s^{-1} (LTT) to 1900 m s^{-1} (EXEL) ballpark. The explosion is contained in the tube and only at the end of a tube or lateral opening a distinct spray of hot particles is obtained. The run-up distance to LVD with LTT is abot 200 mm (t = 100 µs), with Excel t = 400 µs.

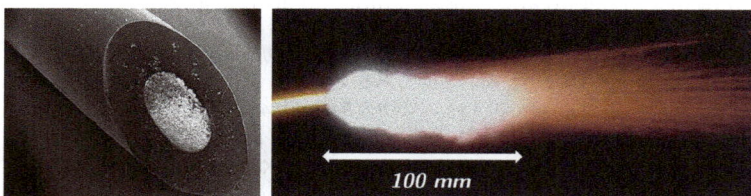

Figure S.10: Electron micrograph of LTT and flame picture.

- K. Kosanke, Evaluation of LTT as a Pyrotechnic Ignition System, *Workshop on Pyrotechnic Combustion Mechanisms*, Fort Collins, **2006**.
- H. Oertel, *Stossrohre*, Springer-Verlag, Vienna, **1966**, 1030 pp.
- A. G. Gaydon, I. R. Hurle, *The Shock Tube in High-Temperature Chemical Physics*, Chapman and Hall, London, **1963**, 307 pp.

SHS, self-propagating high-temperature synthesis
Autotherme Hochtemperatursynthese

Self-propagating high-temperature synthesis (SHS) is a collective term to describe exothermic reactions utilized to produce condensed material. While the classical thermite reaction and any other aluminothermic process are usually associated with SHS, the term is broader in scope and also encompasses all kinds of metathetic reactions, alloying reactions but also decomposition reactions as is depicted exemplary in the following:

Synthesis from the elements

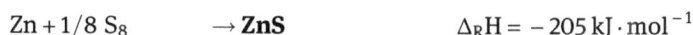

$$Ti + C \qquad \rightarrow \mathbf{TiC} \qquad \Delta_R H = -185 \, kJ \cdot mol^{-1}$$

$$Zn + 1/8 \, S_8 \qquad \rightarrow \mathbf{ZnS} \qquad \Delta_R H = -205 \, kJ \cdot mol^{-1}$$

Metathetic reactions

$$MoCl_5 + 2.5 Na_2S \quad \rightarrow \mathbf{MoS_2} + 5\,NaCl + 0.5S, \quad \Delta_R H = -891 \, kJ \cdot mol^{-1}$$

$$Ba_3N_2 + 3MnI_2 \quad \rightarrow \mathbf{Mn_3N_2} + 3\,BaI_2, \qquad \Delta_R H = -600 \, kJ \cdot mol^{-1}$$

Decompositions reactions

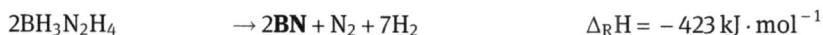

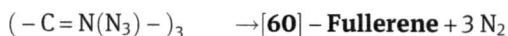

$$2BH_3N_2H_4 \qquad \rightarrow \mathbf{2BN} + N_2 + 7H_2 \qquad \Delta_R H = -423 \, kJ \cdot mol^{-1}$$

$$(-C=N(N_3)-)_3 \quad \rightarrow [\mathbf{60}] - \mathbf{Fullerene} + 3\,N_2$$

$$\Delta_R H = -657 \, kJ \cdot mol^{-1}$$

The prominent characteristics of SHS are the following:
- High reaction temperature, $T \gg 500 \,°C$. Hence removal of volatile impurities (vaporization, sublimation)
- thermodynamic product control
- destruction of the morphology of the starting materials
- transport path between 1 and 1000 µm
- diffusion coefficients between 10^{-7} and $10^{-14} \, cm^2 \, s^{-1}$

SHS reactions can be triggered by any means (thermal, microwave, shock waves; see *reactive fragments*).

- P. Hardt, S. L. McHugh, S. L. Weinland, Chemistry and Shock Initiation of Intermetallic Reactions, *IPS*, Vail, **1986**, pp. 255- 274.
- W. L. Frankhouser, K. W. Brendley, M. C. Kieszek, S. T. Sullivan, *Gasless Combustion Synthesis of Refractory Compounds*, Noyes Publications, **1985**, 152 pp.
- A.A. Borisov, L DeLuca, A. Merzhanov, Y. B. Scheck, *Self-Propagating High Temperature Synthesis of Materials,* Taylor 6 Francis, **2002**, 339 pp.
- Webbook on SHS http://www.ism.ac.ru/handbook/shsf.htm

Silicon

Silicium

Aspect		Dark grey powder with metallic luster
Formula		Si
GHS		02, 07
H-phrases		228-315-319-335
P-phrases		210-280 g-305+351+338
UN		3089
EINECS		231-130-8
CAS		[7440-21-3]
m_r	g mol^{-1}	28.08553
ρ	g cm^{-3}	2.329
Mp	°C	1412
Bp	°C	3217
$\Delta_m H$	kJ mol^{-1}	50.21
$\Delta_v H$	kJ mol^{-1}	383.3
κ	W K^{-1} m^{-1}	83.7
$\Delta_c H$	kJ mol^{-1}	−911
	kJ g^{-1}	−32.437
	kJ cm^{-3}	−75.545
Λ	wt.-%	−113.93

Contrary to boron which does not support combustion in its crystalline modification, silicon burns either as amorphous or crystalline material. Especially the latter is a highly energetic but also ageing and corrosion proof fuel. Silicon is used mainly in technical and defense applications. As *Schmied* could show silicon burns in the condensed phase only and hence is not suitable as a visible flare fuel which is just the reason why it is used as fuel in NIR flares and tracers. Both bulk and powdered crystalline silicon can be porosified by either wet chemical (HF-etching) or electrochemical processes to give mesoporous silicon which yields highly reactive (and sensitive!) composites with both oxygenous and fluorinated oxidizers.

Si wafers with porosified structures filled by oxidizers are hence investigated for new applications (detonator on chip).

The surface of HF-etched silicon is functionalized with hydrogen atoms, specifically with $\equiv$ Si-H, = SiH$_2$ and –SiH$_3$ groups. It is hence that the cavities of porosified silicon (Si$_p$) are extremely hydrophobic and thereby impede the intended infiltration with aqueous solution of common oxidizers explaining why the experimentally determined performance of many oxidizer at Si$_p$ composites with water-soluble salts (e.g. NaClO$_4$) do not live up to the theoretical predictions.

- Schmied, Licht- und Farbintensitäten pyrotechnischer Phrases durch Variation der Metallkomponente, *Arbeitstagung-Wehrtechnik, Chemie und Physik der Explosivstoffe III*, **1968**, Mannheim.
- E.-C. Koch, D. Clément, Special Materials in Pyrotechnics VI. Silicon – An Old Fuel with New Perspectives, *Propellants Explos. Pyrotech.* **2007**, *32*, 205–212.
- M. DuPlessis, A Decade of Porous Silicon as Nano-Explosive Material, *Propellants Explos. Pyrotech.*, **2014**, *39*, 348–364.
- R. Ramchandran, V. S. Vuppuluri, T. J. Fleck, J. F. Rhoads, I. E. Gunduz, S. F. Son, Influence of Stoichiometry on the Thrust and Heat Deposition of On-Chip Nanothermites, *Propellants Explos. Pyrotech.* **2018**, *43*, 258–266.

Silver azide
Silberazid

Aspect		Colorless – grey crystals
Formula		AgN$_3$
REACH		LPRS
EINECS		237-606-1
CAS		[13863-88-2]
m$_r$	g mol^{-1}	149.888
ρ	g cm^{-3}	5.1
Δ$_f$H	kJ mol^{-1}	312
Λ	wt.-%	0
N	wt.-%	28.03
c$_p$	J K^{-1} mol^{-1}	73.37 at 250 °C
P$_t$	°C	190
Mp	°C	289
Dp	°C	307
Friction	N	
Impact	J	1–4
V$_D$	m s^{-1}	3830 at 2.00 g cm^{-3}
P$_{CJ}$	GPa	*1.9* at 2.14 g cm^{-3}
Trauzl	cm^3	115

The light sensitivity of silver azide is much less pronounced than say with silver halides or silver nitrate. Silver azide has a higher specific initiation capability than lead azide. On top it cannot be dead pressed. However, silver azide is not compatible with tetracen and as such cannot fully replace lead azide.

– R. Matyáš, J. Pachman, *Primary Explosives*, Springer, **2013**, 39–59.

Silver fulminate

Silberfulminat, Knallsilber

Formula		AgCNO
CAS		[5610-59-3]
m_r	g mol^{-1}	149.885
ρ	g cm^{-3}	4.107
$\Delta_f H$	kJ mol^{-1}	180
Λ	wt.-%	−10.67
N	wt.-%	9.34
$\Delta_{ex}H$	kJ mol^{-1}	−1970
Dec	°C	236–241
T_{5ex}	°C	170
Impact	J	>1–2
V_D	m s^{-1}	1700 at 2.00 g cm^{-3}
P_{CJ}	GPa	*1.9* at 2.14 g cm^{-3}
Trauzl	cm^3	115

Crystalline silver fulminate is an extremely sensitive primary explosive. It is hence that is does not find technical application. However, silver fulminate is used in small amounts in toy snappers which may contain up to 2.5 mg silver azide per item. Silver fulminate was also a popular item used by alchemists trying to impress potential sponsors.

– C. Wentrup, Fulminating Gold and Silver, *Angew. Chem. Int. Ed.* **2019**, *58*, 14800–14808.
– R. Matyáš, J. Pachman, *Primary Explosives*, Springer, **2013**, 39–59.

Sinoxid®

Sinoxid® is the registered trademark by RUAG Ammotec, Fürth. Sinoxids are chlorate-free tetracene-containing primer mixtures leading to minimum corrosion. Table S.6 shows two general compositional ranges.

– H. Rathsburg, E. von Herz, *Verfahren zur Herstellung von ZündPhrasesn*, DE Patent 518885, **1931**, Germany.
– R. Hagel, K. Redecker, Sintox – A New, Non-Toxic Primer Composition by Dynamit Nobel AG, *Propellants Explos. Pyrotech.* **1986**, *11*, 184–187.

Table S.6: SINOXID® compositions.

	wt.-%	wt.-%
Lead styphnate	20–73	25–55
Tetracene	1–40	0.5–5
Barium nitrate	6–49	25–45
Lead dioxide	0–10	5–10
Calcium silicide	0–16	3–15
Antimony(III) sulfide	0–27	0–10
Glass powder	0–16	0–5

Sintox®

Sintox is a registered trademark by RUAG Ammotec, Fürth. Sintox are barium- and lead-free tetracene containing primer mixtures leading to minimum corrosion. Table S.7 shows an exemplary composition.

Table S.7: SINTOX® composition.

Tetracene	5
Diazodinitrophenole	20
Zinc peroxide	50
Titanium	5
Nitrocellulose	19.5
Stabilizer for NC	0.5

– R. Hagel, K. Redecker, *Blei- und Bariumfreie AnzündPhrases,* DE-OS-3321943A1, **1984,** Germany.

Sodium azide
Natriumazid

Formula	NaN_3
GHS	06, 09
H-phrases	300-400-410-032
P-phrases	273-280-302+352-309-310-501a
UN	1687
EINECS	247-852-1
CAS	[26628-22-8]

m_r	g mol^{-1}	65.0099
ρ	g cm^{-3}	1.846
Dp	°C	375
$\Delta_f H$	kJ mol^{-1}	−21.3
$\Delta_c H$	kJ mol^{-1}	−185.7
	kJ g^{-1}	−2.856
	kJ cm^{-3}	−5.273
c_p	J K^{-1} mol^{-1}	80.0
Λ	wt.-%	−12,31
N	wt.-%	64.64

Sodium azide is non hygroscopic. It has been used intensively in the early years of automobile airbag gas generators as a source for nitrogen. However, its toxicity and the sensitiveness of its formulations led to its replacement by less toxic high nitrogen compounds (see *airbag*). Sodium azide has been used as halogenophile in the solid-state synthesis of unusual carbon allotropes.

- S. Cudziło, A. Huczko, S. Gachot, M. Monthioux, W. A. Trziński, Synthesis of Ceramic and Carbon Nanostructures by self-sustaining combustion of mixtures of halogenated hydrocarbons with reducers, *NTREM-2003*, Pardubice **2003**, pp. 69–75.

Sodium chlorate
Natriumchlorat

Formula		NaClO$_3$
CAS		[7775-09-9]
GHS		03, 07, 09
H-phrases		271-302-411
P-phrases		210-221-283-306+360-371+380+375-501a
UN		1495
EINECS		231-887-4
m_r	g mol^{-1}	106.441
ρ	g cm^{-3}	2.49
Mp	°C	263
Dp	°C	465
$\Delta_f H$	kJ mol^{-1}	−365.77
Λ	wt.-%	+45.09
Koenen	mm	type <1, E

Due to its low price and good solubility, sodium chlorate has been used for decades as a non-selective herbicide (Rasikal®, Unkraut-ex®, Netosol®, etc.). Being once virtually available from any drugstore, it also made its way into countless home-made explosives (HME). Mixtures from sodium chlorate and fuels are extremely sensitive due to the exothermic decomposition of it in accordance with

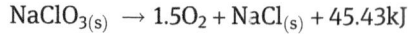

$$NaClO_{3(s)} \rightarrow 1.5O_2 + NaCl_{(s)} + 45.43kJ$$

Leiber found that the exothermic decomposition also makes pure $NaClO_3$ present in bulk quantities (>20 kg) an extremely dangerous and detonable material. $NaClO_3$ is used in *SCOGs*.

- C.-O. Leiber, IHE – 2000, Wunsch und Wirklichkeit, *11. Sprengstoffgespräch*, Nonnweiler, **1987**.
- Y. Zhang, J. C. Cannon, *Chemical Oxygen Generating Composition Containing Li₂O₂*, US Patent 5.279.761, **1994**, USA.
- C.-O. Leiber, *Assessment of Safety and Risk with a Microscopic Model of Detonation*, Elsevier, Amsterdam, **2003**, 258–260.

Sodium dinitramide
Natriumdinitramid

NaDN (acr.), SDN (acr.)		
Formula		$NaN(NO_2)_2$
CAS		[160150-82-3]
m_r	g mol⁻¹	129.007
ρ	g cm⁻³	2.09
Mp	°C	101–107
Dp	°C	156
$\Delta_f H$	kJ mol⁻¹	−229.87
Λ	wt.-%	+43.41
N	wt.-%	32.57

The thermal decomposition of NaDN starts with a weak exothermal reaction at 91 °C accompanied by a mass loss of ~ 6%. Additional exothermal reactions at 156 °C and 210 °C correspond to successive loss of N_2O. *Dawe* has tested the combustion properties and IR emission of NaDN/boron – and NaDN/boron/HTPB formulations.

- J. R. Dawe, M. D. Cliff, Metal Dinitramides New Novel Oxidants for the Preparation of Boron Based Flare Compositions, *IPS-Seminar*, **1998**, Monterey, 789–810.

Sodium nitrate

Natriumnitrat

Formula		$NaNO_3$
GHS		03, 07
H-phrases		272-302
P-phrases		210-220-221-280-301+312-501a
UN		1498
EINECS		231-554-3
CAS		[7631-99-4]
m_r	g mol^{-1}	84.995
ρ	g cm^{-3}	2.261
Mp	°C	307
Bp	°C	380
Dp	°C	521
Λ	wt.-%	+47.06
N	wt.-%	16.48
$\Delta_f H$	kJ mol^{-1}	−467.85
$T_{p(rh \leftrightarrow tg)}$	°C	273

$NaNO_3$ is a colorless hygroscopic powder (DRH = 75% RH). For protection against moisture, it is typically roasted with 1–3 wt.-% linseed oil which forms a protective barrier. $Mg/NaNO_3$ formulations yield the highest luminous efficiencies and hence are the most common payloads for illuminating and signaling flares. $NaNO_3$ is also applied as an oxidizer in military gas generators for air breathing propulsion (e.g. SA-6).

Formulation for LK-6TM (SA-6 motor)

– 65 wt.-% magnesium
– 25 wt.-% sodium nitrate
– 10 wt.-% naphthalene

Upon prolonged heating at T = 340 °C $NaNO_3$ dismutates to give an orthonitrate in accordance with

$$NaNO_3 + Na_2O \rightarrow Na_3NO_4$$

Rapid heating affords the peroxide as an intermediate:

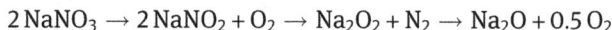

$$2\,NaNO_3 \rightarrow 2\,NaNO_2 + O_2 \rightarrow Na_2O_2 + N_2 \rightarrow Na_2O + 0.5\,O_2$$

– H. Singh, M. R. Somayajulu, R. B. Rao, A study on Combustion Behavior of magnesium-Sodium Nitrate Binary Mixtures, *Combust. Flame* **1989**, *76*, 57–61.
– J. R. Ward, L. J. Decker, A. W. Barrows, Burning Rates of Pressed Strands of a stoichiometric magnesium Sodium nitrate Mix, *Combust. Flame*, **1983**, *51*, 121–123.

Sodium perchlorate
Natriumperchlorat

Formula		NaClO$_4$
GHS		03
H-phrases		271
P-phrases		210-221-283-306+360-371+380+375-501a
UN		1502
EINECS		231-511-9
CAS		[7601-89-0]
m$_r$	g mol^{-1}	122.44
ρ	g cm^{-3}	2.499
Δ$_f$H	kJ mol^{-1}	−383.30
Mp	°C	468
Dp	°C	527
T$_{p(rh\leftrightarrow k)}$	°C	308
Λ	wt.-%	+52.27

Sodium perchlorate in moist air quickly forms a hydrate NaClO$_4$·H$_2$O [7791-07-3], (ρ = 2.02 g·cm^{-3}, −H$_2$O = 130 °C). Due to its high hygroscopicity (DRH = 43% RH), it seldom finds practical use. *Davies* has investigated its performance in flash compositions which had earlier been suggested by *Ellern:* NaClO$_4$/Ca (80/20). NaClO$_4$ decomposes exothermally (Δ$_{dec}$H = −28 kJ mol^{-1}).

- N. Davies, M. Bishop, The Luminous and Blast Performance of Flash Powders, *IPS*, Adelaide, **2001**, pp. 73–86.
- V. Klyucharev, A. Razumova, The Cooperative Processes of Magnesium Oxidation in Perchlorate Mixtures with Oxide, *ICT -Jata*, Karlsruhe, **1995**, 63.
- H. Ellern, *Military and Civilian Pyrotechnics*, Chemical Publishing Company, New York, **1968**, p.361 Formula #50.

Sodium peroxide
Natriumperoxid

Formula		Na$_2$O$_2$
GHS		03, 05
H-phrases		272-314
P-phrases		210-221-303+361+353-305+351+338-405-501a
UN		1504
EINECS		215-209-4
CAS		[1313-60-6]
m$_r$	g mol^{-1}	77.979

ρ	g cm^{-3}	2.805
Mp	°C	460
Dp	°C	675
$T_{p(??\to??)}$	°C	512
$\Delta_f H$	kJ mol^{-1}	−510.87
Λ	wt.-%	+20.52

Pure Na_2O_2 is colorless, whereas technical qualities are rather yellowish. With water, it forms hydrogen peroxide.

- J. E. Tanner Jr., Thermodynamics of Combustion of Various Pyrotechnic Compositions, *RDTR-277*, NAD Crane, **1974**, 25 pp.

Solid rocket propellants
Festtreibstoffe

Solid rocket propellants are energetic formulations with slightly negative oxygen balance that burn under pressure to yield products of high temperature and low molecular mass. The performance of a solid rocket motor (e.g. the specific impulse, I_{sp}) is influenced by
- chemical composition of the propellant which determines the heat of explosion and the mean molecular weight of the combustion products
- geometry and size of the propellant block
- operating pressure of the motor and nozzle design

The advantages of solid rocket propulsion are
- high thrust density
- simple handling
- readiness of use
- low technical complexity and hence enhanced availability (MTBF) when compared to liquid propulsion engines
- fewer production steps
- low cost

However, the disadvantages are
- short burn times
- no burn-stop and reignition
- very little modulation of thrust
- moderate specific impulse
- high sensitiveness (IM test)

Solid rocket propellants are categorized into three groups depending on their chemistry (Table S.8):

- composite propellants
- double base propellants
- composite-modified double base (CMDB)

Table S.8: Composition and properties of solid rocket propellants.

	Oxidizer	Fuel	ρ (g cm^{-3})	u (mm s^{-1})	I$_{sp}$(s)
Composite	NH$_4$ClO$_4$, AP	Polymers	1.50–1.90	1–380	170–260
	KClO$_4$, KP	HTPB			
	NH$_4$NO$_3$, AN	Metals			
	NH$_4$N$_3$O$_4$, ADN	Al, B			
Double base	Energetic: NC		1.50–1.64	5–20	170–220
	Plasticizer: NGl, DEGN, MTN, BTTN				
CMDB	Mixtures of the above compositions		1.70–1.90	7.5–25	240–270

For tactical reasons, military solid rocket propellants are categorized according to their electrooptical signature into nine classes (AA, AB, . . ., CB, CC) which is standardized in STANAG 6016 (Figure S.11). The first letter determines the primary aerosol formation. This can occur directly behind the nozzle due to expansion of the combustion products and is determined mainly by metal oxides, soot, and other substances condensed at ambient temperature and pressure. The second letter describes the amount secondary aerosol formation. The latter is determined by the relative humidity at given temperature and hence may vary appreciably. Secondary aerosol formation is caused by condensation of water vapor and deliquescence of reaction products such as HCl or HF. STANAG 6016 also details how the signature is determined based on thermochemical calculations. Double-base propellants yield very little secondary aerosol, just H$_2$O. However, composites based on ammonium perchlorate (AP), HTPB, and aluminum yield both condensed Al$_2$O$_3$ and hygroscopic HCl causing intense smoke.

Current research addresses the replacement of AP by ammonium dinitramide (ADN). As ADN is more energetic and also more sensitive than AP, typical hazard classifications obtained with ADN/GAP propellants are HD 1.1 while conventional AP/HTPB/Al yields 1.3 only. Hence, development of formulations that retain the energetic and signature advantages of ADN but with reduced sensitiveness and sufficient stability is one of the current challenges.

Other research focusses on new high oxygen density materials based on nitramide derivatives.

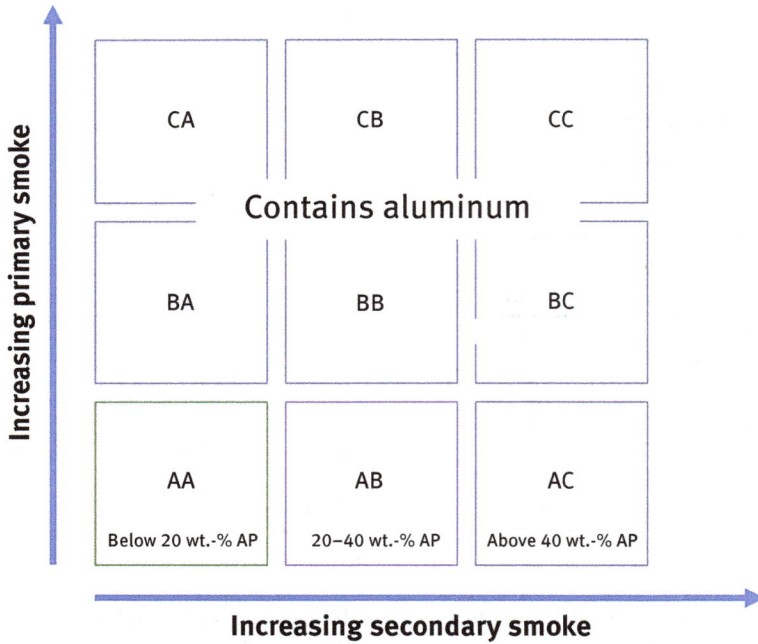

Figure S.11: NATO signature classes for solid rocket propellants.

- N. Wingborg, M. Skarstind, M. Sjöblom, A. Lindborg, M. Brantlind, J. Johansson, Ammonium dinitramide, *ICT-JATA*, **2019**.
- L. De Luca, T. Shimada, V. Sinditskii, M. Calabro (Eds.), *Chemical Rocket Propulsion*, Springer, **2017**, 1084 pp.
- N. Kubota, *Propellants and Explosives*, Wiley-VCH, 3. Aufl. **2015**, 534 pp.
- M. Rahm, T. Brinck, Green Propellants Based on Dinitramide Salts: Mastering Stability and Chemical Compatibility Issues in T. Brinck (Ed.), *Green Energetic Materials*, Wiley, New York, **2014**, 179–204.
- M. Rahm, *Green Propellants*, PhD Thesis, KTH, Stockholm, **2010**.
- *Solid Propellant Smoke Classification*, STANAG 6016, NATO Military Agency for Standardization, Brussels, **1994**.
- A. Dadieu, R. Damm, E. W. Schmidt, *Raketentreibstoffe*, Springer, **1968**, 118–176.

Soman, GD

3,3-Dimethylbutan-2-yl methylphosphonofluoridat

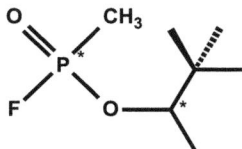

VR55.		
Formula		$C_7H_{16}FO_2P$
CAS		[96-64-0]
m_r	g mol^{-1}	182.175
ρ	g cm^{-3}	1.022
Mp	°C	−70 to −80
Bp	°C	167 to 200
Dp	°C	150
Λ	wt.-%	−193.22
LCt$_{50}$	ppm	10 (at high level physical stress)
	ppm	70 (at rest)
ICt$_{50}$	ppb min^{-1}	25–30
$\Delta_f H$	kJ mol^{-1}	−1070 (estimated)
$\Delta_c H$	kJ mol^{-1}	−3971.3
	kJ g^{-1}	−21.800
	kJ cm^{-3}	−22.279

Soman is a nerve agent that acts very fast through inhalation, dermal, or ocular resorption. GD contains two stereo centers (*) and hence occurs technically always as a blend of four different stereoisomers explaining the range of physical properties (volatility) and toxicity.

–   S. Franke, *Lehrbuch der Militärchemie, Band 1*, Militärverlag der Deutschen Demokratischen Republik, 2. Auflage, Leipzig, **1977**, pp. 398–422.

Sorguyl

Tetranitroglycoluril

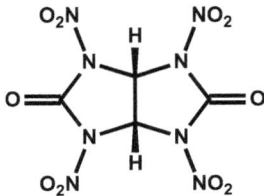

tngu		
Formula		$C_4H_2N_8O_{10}$
REACH		LPRS
EINECS		259-682-5
CAS		[55510-03-7]
m_r	g mol^{-1}	322.107
ρ	g cm^{-3}	1.98
Δ_fH	kJ mol^{-1}	41.48
$\Delta_{ex}H$	kJ mol^{-1}	−1878.6
	kJ g^{-1}	−5.832
	kJ cm^{-3}	−11.548
Δ_cH	kJ mol^{-1}	−1901.9
	kJ g^{-1}	−5.905
	kJ cm^{-3}	−11.691
Λ	wt.-%	+4.97
N	wt.-%	34.79
Dp	°C	250
Friction	N	70
Impact	J	2
V_D	m s^{-1}	9073 at ρ = 1.94 g cm^{-3} in silver tube (ø = 4 mm)
P_{CJ}	GPa	39.92 at ρ = 1.94 g cm^{-3}

Sorguyl is a high-performance explosive with a sensitiveness comparable to CL-20 named after the place of its invention at the SNPE- now EURENCO facility in Sorgues, France. Contrary to CL-20 it, however, suffers from being extremely sensitive to hydrolysis. Formal replacement of the carbonyl group by either methylene or ethylene yields the likewise hydrolysis-prone explosives K55 (BCHMX) and K56.

- J. Boileau, J.-M. L. Emeury, J.-P. Kehren, *Tetranitroglycuril, seine Herstellung und Verwendung als Explosivstoff*, DE-Patent 2435651, SNPE, France, **1980**.

Sparklers
Wunderkerzen

Sparklers are steel wires coated with a pyrotechnic composition by dipping. A typical sparkler formulation contains
- 54.0 wt.-% barium nitrate
- 9.0 wt.-% dextrin
- 28.0 wt.-% steel filings
- 7.0 wt.-% black aluminum
- 0.5 wt.-% charcoal
- 1.5 wt.-% calcium carbonate

The density of that composition is $\rho = 1.66$ g cm^{-3}. Its decomposition starts at T = 216 °C in DSC and the BAM impact energy is 15 J.

As the formulation is oxygen underbalanced, upon combustion only dextrin and the ultrafine aluminum burn in the condensed phase, while the majority of the steel filings is ejected hot and starts to burn in the air under the characteristic sparking. Upon combustion of sparklers nitrogen monoxide, NO, volatile iron nitrosyl compounds, $(Fe(NO)_x)$, and barium containing dust, $(BaCO_3, BaO)$, are formed. Hence, sparklers must not be used in closed rooms and should never be used to decorate food stuff (cakes and pies).

- C. Martin, T. deVries, Chemie der Wunderkerze – ein Thema nicht nur zur Weihnachtszeit, *Chemkon* **2004**, *11*, 13–20.

Specific heat and heat capacity
Spezifische Wärme, Wärmekapazität

The total differentials of the internal energy and the enthalpy read

$$dU = \left(\frac{\partial U}{\partial T}\right)_V dT + \left(\frac{\partial U}{\partial V}\right)_T dV$$

$$dH = \left(\frac{\partial H}{\partial T}\right)_p dT + \left(\frac{\partial H}{\partial V}\right)_T dp$$

For the cases dV = 0 and dp = 0 the meaning of the temperature coefficients

$$\left(\frac{\partial U}{\partial T}\right)_V$$

and

$$\left(\frac{\partial H}{\partial T}\right)_p$$

is the amount of heat required to heat 1 mol of a substance by 1 K under isochoric or isobaric conditions.

In this context,

$$\left(\frac{\partial U}{\partial T}\right)_V = c_v$$

is the specific heat at constant volume and

$$\left(\frac{\partial H}{\partial T}\right)_p$$

is the specific heat at constant pressure. For ideal gases we can write

$$c_p - c_v = R.$$

For real gases the difference significantly deviates from R. For condensed matter, the difference of $C_p - C_v$ is about R/10:

$$\frac{c_p}{c_v} = \gamma$$

is the isentropic exponent.

The temperature dependence of the specific heat of various gases and condensed materials occurring as reaction products of energetic materials is depicted in Figures S.12 and S.13.

- W. Schreiter, *Chemische Thermodynamik*, De Gruyter, **2014**, 546 pp.
- M. Binnewies, E. Milke, *Thermochemical Data of Elements and Compounds*, Wiley-VCH, **2002**, 928 pp.

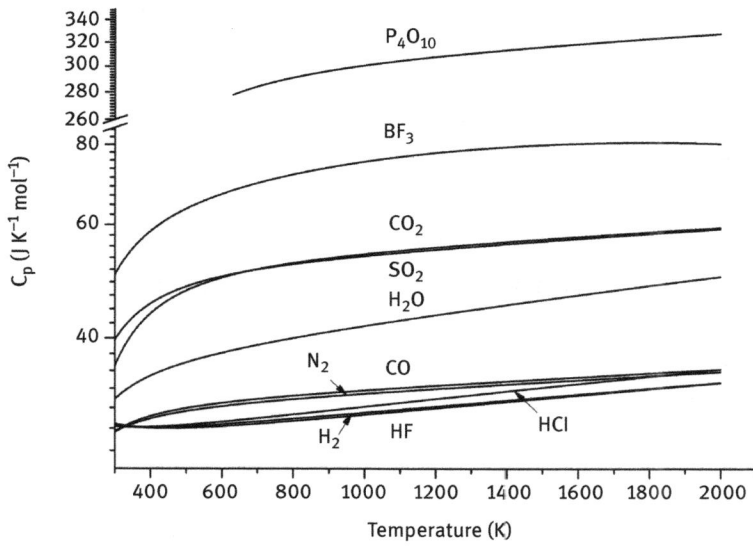

Figure S.12: Specific heat of some gaseous elements and compounds.

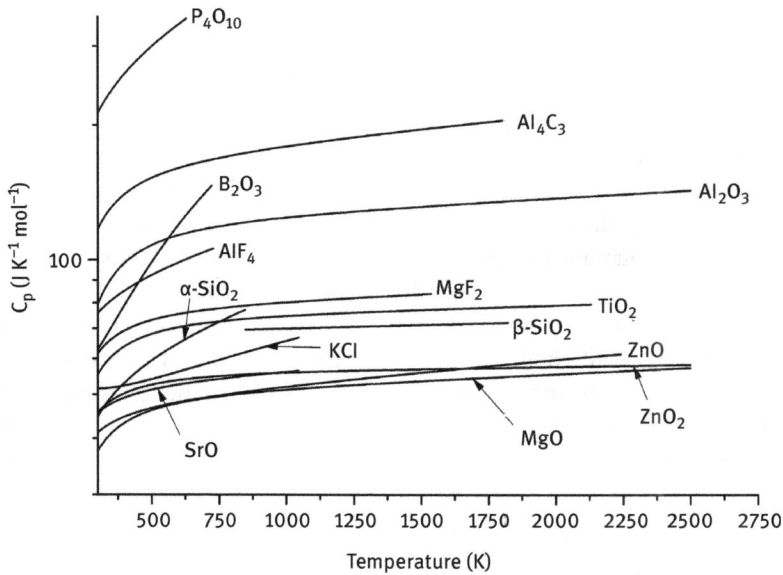

Figure S.13: Specific heat of some condensed compounds.

Specific impulse
Spezifischer Impuls

The specific impulse, I_{sp} (s), describes the thrust F (kg m s^{-1}) delivered by the unit mass of propellant $\acute{m}$ (kg s^{-1}) reacting per second:

$$I_{sp} = \frac{F}{\acute{m}g}$$

Its reciprocal is the specific consumption of propellant. With the exit velocity, w (m s^{-1}), there is the simple relationship,

$$I_{sp} = \frac{w}{g}$$

explaining the popular overall approximation of I_{sp} by the expression

$$I_{sp}\mu\sqrt{\frac{T_c}{\overline{M}}}$$

- A. Dadieu, R. Damm, E. W. Schmidt, *Raketentreibstoffe*, Springer, **1968**, 48–65.

Stabilizer
Stabilisatoren

The O–NO$_2$ bond in organic nitrates ($\Delta_{dis}H \sim$ 165–170 kJ mol^{-1}) is significantly weaker than the C–NO$_2$ bond in nitroaromatics ($\Delta_{dis}H \sim$ 295 kJ mol^{-1}). It is hence that organic nitrates and predominantly nitrocellulose decompose by splitting off NO$_2$, NO, and nitrous acid, HNO$_2$. The remaining glucose ring in NC then successively breaks down under extrusion of glyoxal and more NO$_2$. The NO$_2$ in turn can nitrate other parts of the NC strand releasing more energy (autocatalysis) furthering the decomposition. To counter the decomposition of NC, both organic and inorganic stabilizers are used.

The organic stabilizers are mostly aromatic amines (e.g. diphenylamine) and urea derivatives (acardite, centralite, etc.), which scavenge NO and NO$_2$ by forming stable nitroso and nitro compounds and thereby interrupt the reaction chain. Among the inorganic substances used as stabilizers, calcium carbonate and magnesium oxide play an important role to buffer acids.

- M. Bohn, NC-based energetic materials – stability, decomposition, and ageing, *Nitrocellulose – Supply, Ageing and Characterization*, April 24–25 **2007**, Aldermaston, AW, UK.
- T. Lindholm, Reactions in stabilizer and between stabilizer and nitrocellulose in propellants, *Propellants Explos. Pyrotech.* **2002**, *27*, 197–208.

Stemming
Besatz

Stemming is the tamper used to fill the supernatant void space above an explosive charge in a hole to reduce the loss of blast pressure. Typically, the stemming is an inert material like clay, sand, mortar, or water.

Strobe compositions
Blinksätze

Strobe compositions contain similar ingredients as illuminating or signaling compositions however with a significant lower fuel level, which and supported by the presence of combustion catalysts causes combustion irregularities. Hence after ignition the steady state consists of alternating phases of luminous combustion and a dark phase with a rather smoldering reaction. The duration of an interval (from start of luminous phase to end of dark phase) decreases with increasing oxygen balance. Strobe compositions have been suggested as distress signals, electrooptic decoys, and for use with stun grenades. While there is plenty of interest in such applications those compositions cannot deliver a stable frequency but burn rather erratically. It is hence that strobe compositions remain an oddity mainly exploited for firework applications. Typical compositions are given in Table S.9. Compositions yielding little light but spontaneous release of gas (rattle) are so-called "black strobes".

Table S.9: Pyrotechnic strobe compositions.

Ingredients (wt.-%)	Blue	Green	Yellow	Sliver	Red	Magenta	"Black"
Barium nitrate		67		66			
Strontium nitrate					66	5	
Tetramethylammonium nitrate	30					28	40
Mg/Al-alloy (100–200 mesh)		21	26	26	22		
Copper (100 mesh)	15					12	30
Ammonium chloride		7					
Ammonium perchlorate	55		58			55	30
Sodium sulfate			8				
Potassium perchlorate					8	7	
Barium perchlorate		5			5		
Graphite			4				
Frequency range (Hz)	?	10	?	5	7		

- J. Glück, T. M. Klapötke, J. J. Sabatini, Flare or strobe: a tunable chlorine-free pyrotechnic system based on lithium nitrate, *Chem. Comm.* **2018**, *54*, 821–824.
- C. J. White, Strobe Chemistry, *J. Pyrotech.* **2004**, *20*, 7–16.

- C. J. White, Blue Strobe Light Pyrotechnic Compositions, *Pyrotechnica* **1992**, *XIV*, 33–45.
- T. Shimizu, Studies on Strobe Light Pyrotechnic Compositions, *Pyrotechnica*, **1982**, *VIII*, 5–28.
- U. Krone, Strahlungsemission in Intervallen – Oscillierende Verbrennung pyrotechnischer Phrases, *ICT-Jahrestagung*, **1975**, 225–237.
- F.-W. Wasmann, Pulsierend abbrennende Pyrotechnische Systeme, *ICT-Jahrestagung*, **1975**, 239–250.

Strontium dichlorophthalate

Strontium 3,4-dichlorphthalat

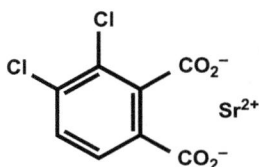

Formula		$C_8Cl_2H_2O_4Sr$
EINECS		304-292-3
CAS		[94248-20-1]
REACH		LPRS
m_r	g mol^{-1}	320.62
$\Delta_f H$	kJ mol^{-1}	−1200 (estimated)
Λ	wt.-%	−64.87

In analogy to the alkaline earth metal salts of tetrachlorophthalic acid (see *calcium tetrachlorophthalate)* strontium dichlorphthalate too is used as a color intensifier for red burning light signal and tracer formulations.

Strontium nitrate

Strontiumnitrat

Formula		$Sr(NO_3)_2$
GHS		03, 07
H-phrases		272-315-319
P-phrases		210-221-302+352-305+351+338-321-501a
UN		1507
EINECS		233-131-9
CAS		[10042-76-9]
m_r	g mol^{-1}	211.63
ρ	g cm^{-3}	2.986
Mp	°C	570

Bp	°C	645
$Tp_{(h\leftrightarrow k)}$	°C	160
Dp	°C	584 (600)
$\Delta_f H$	kJ mol^{-1}	−505.97
Λ	wt.-%	+37.80
N	wt.-%	13.24
Specification		MIL-S-20322B

Strontium nitrate in moist air forms the deliquescent tetrahydrate, $(Sr(NO_3)_2 \cdot 4\ H_2O)$ [13470-05-8] (ρ 2.249 g $\cdot$ cm^{-3}, −4H$_2$O at 36 °C, Dp: 1100 °C, $\Delta_f H$ −2154.8 kJ $\cdot$ mol^{-1}). Its thermal decomposition is very much alike that of calcium nitrate

$$Sr(NO_3)_2 \rightarrow Sr(NO_2)_2 + O_2$$

$$Sr(NO_2)_2 \rightarrow SrO + NO_2 + NO$$

At higher temperatures the decomposition occurs in accordance with

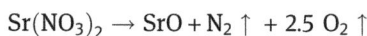

$$Sr(NO_3)_2 \rightarrow SrO + N_2 \uparrow + 2.5\ O_2 \uparrow$$

Between T = 635 and 715 °C the release of NO$_x$ is observed.

Strontium nitrate forms a coordination compound with hexamethylentetramine $(C_6H_{12}N_4)$ which burns with a smoke-free red flame. Though known for quite some time (*Scheele, Obhlidal*), it was only recently that *Kruszynski* could characterize the compound. Strontium nitrate is used in red signalling flare and tracer compositions. In the presence of chlorine compounds the important emitter SrCl is formed which is complemented by the other emitter SrOH which also forms in chlorine free flames. Blends from $Sr(NO_3)_2$ and barium nitrate with magnesium yield white (*Schladt*) or yellow light (*Tyroler*) due to subtractive color mixing. Due to its hygroscopicity, it is often protected with linseed oil before processing.

- J. J. Sabatini, E.-C. Koch, J. Poret, J. Moretti, S. M. Harbol, Chlorine-Free Red-Burning Pyrotechnics, *Angew. Chem. Int. Ed.* **2015**, *54*, 10968–10970.
- W. T. Scheele, *Pyrotechnic Composition*, USP 1.423.264, **1922**, USA.
- W. Obhlidal, E. Forster, *Verfahren zur Herstellung von farbig brennenden Illuminationskörpern*, DE873512, **1953**, Germany.
- T. Sierańksi, R. Kruszynski, On the governing of alkaline earth metal nitrate coordination spheres by hexamethylenetetramine, *J. Coord. Chem.* **2012**, *66*, 42–55.
- G. J. Schladt, *Pyrotechnic Composition*, USP 2.035.509, **1934**, USA.
- J. F. Tyroler, *Flare System*, USP 3.888.177, **1974**, USA.

Strontium oxalate

Strontiumoxalat

Formula		SrC_2O_4
GHS		07
H-phrases		302-312
P-phrases		280-302+352-322-301+312-312-501a
UN		3288
EINECS		212-415-6
CAS		[814-95-9]
m_r	g mol^{-1}	175.64
ρ	g·cm^{-3}	2.08
Dp	°C	477 (-CO)
$\Delta_f H$	kJ mol^{-1}	−1371
Λ	wt.-%	−9.11

Strontium oxalate in moist air readily forms the monohydrate, CAS-No [6160-36-7], which loses its crystal water between 90 and 170 °C. It serves as a flame color agent in red flare and tracer formulations.

Strontium perchlorate

Strontiumperchlorat

Formula		$Sr(ClO_4)_2$
GHS		03, 07
H-phrases		272-315-319-335
P-phrases		210-221-302+352-305+351+338-405-501a
UN		1508
EINECS		236-614-2
CAS		[13450-97-0]
m_r	g mol^{-1}	286.521
ρ	g·cm^{-3}	3.02
Dp	°C	477
$\Delta_f H$	kJ mol^{-1}	−762.79
Λ	wt.-%	+44.67

Douda reported the use of $Sr(ClO_4)_2$, as sole oxidizer and color agent in acrylate-polymer-based signalling formulations with magnesium as fuel. *Wasmann* formulated red flickering strobe flares based on strontiumperchlorate tetrahydrate, pentaerythrol dinitratediacrylate, and methacrylic acid methylester. *Douda* also developed a procedure to polymerize polycarylates containing $Sr(ClO_4)_2$ for both pyrotechnics and propellants. *Douda* also for the first time prepared the non-hygroscopic (sic!) coordination compound trisglycin-$\kappa N,N',N''$-strontium diperchlorate, $Sr[C_2H_5NO_2]_3(ClO_4)_2$ [12247-02-8], in very good yield. The compound burns with a highly saturated smokeless flame. This compound is the prototype of molecular pyrotechnics, as fuel, oxidizers and color agent are all present in one molecule.

- B. E. Douda, *Plastic Pyrotechnic compositions containing strontium perchlorate and acrylic polymer*, US Patent 3.258.373, **1964**, USA.
- B. E. Douda, *Process for Polymerizing acrylic monomers with strontium perchlorate for pyrotechnics and propellants*, US Patent 3.369.946, **1964**, USA.
- B. E. Douda, *Pyrotechnic compound tris (glycine) strontium (ii) perchlorate and method for making same*, US Patent 3296045, USA, **1968**.
- F. W. Wasmann, Festtreibstoffe mit pulsierendem Abbrand, *Explosivstoffe*, **1973**, *21*, 1–8.
- F. W. Wasmann, Pulsierend abbrennende pyrotechnische Systeme, *ICT -Jata*, Karlsruhe, **1975**, pp. 239–250.

Strontium peroxide
Strontiumperoxid

Formula		SrO_2
CAS		[1314-18-7]
GHS		07, 03, 05
H-phrases		272-315-318-335
P-phrases		220-261-280-305+351+338
UN		1509
EINECS		215-224-6
m_r	g mol^{-1}	119.619
ρ	g cm^{-3}	4.56
Dp	°C	488-512
$\Delta_f H$	kJ mol^{-1}	−633.46
Λ	wt.-%	+13.38
Specification		JAN-S-612, TL-1370-0001T002Bl1830

Colorless hygroscopic powder which easily forms the hydrate and is used as such in red and NIR tracers, first fires and delay formulations.

- N. E. Trickel; S. V. Strommen, *Tracer for ammunition*, US Patent 4.597.810, **1985**, USA.
- T. A. Doris Jr., K. D. Vest, K. D, *Igniter Composition*, US Patent 6.036.794, **1998**, USA.
- G. Henry III, IR Dim Tracer for Ammunition, *Annual Guns & Ammunition Symposium*, **2003**.

Stun grenade

Schockwurfkörper, Blendgranate

Stun grenades are pyrotechnic grenades that can be thrown by hand or delivered by other means. After a very short delay $t = 0.5–1.5$ s they yield either a single intense report or a series of irregular timed reports associated with highly intense flashes of light. The high light intensity in the range of $I_\lambda > 5$ Mcd and acoustic peak pressure of up to and beyond 170 dB(Ai) effect incapacitation of exposed persons – particularly if applied inside confined rooms. This is due to spontaneous stimulus *satiation* which makes reasonable and implied actions impossible at least for a distinct period. Typical compositions found in stun grenades are flash compositions. The application of stun grenades – though known as *non-lethal ammunition* – may still yield fatal injuries especially if used in close vicinity of a person.

Substitute explosives

Ersatzsprengstoffe

In the course of World War II, Germany experienced a severe shortage of standard HEs such as RDX, PETN, and TNT. To counter the unavailability of these materials a great variety of so-called Ersatzsprengstoffe was developed which were mixtures of standard explosives with greater fractions of either ammonium nitrate or all sorts of metal nitrates. Finally, by 1944, when even those nitrates were no longer available in sufficient quantity complete inert fillers such as sodium and potassium chloride and plaster of paris were used for blending with TNT.

The nitrate HE mixtures developed then are majorly non-ideal HEs with large critical diameters and reduced sensitiveness. It is hence that they have gained some renewed attention as insensitive HEs to meet with UN Test Series 7.

– B. T. Fedoroff, Dictionary of Explosives, Ammunition and Weapons (German Section), Picatinny Arsenal, *TR, 2510*, **1958**, pp. 43–44.

Sulfur (sulphur)

Schwefel

Aspect		Light yellow crystals
Formula		S_8
CAS		[7704-34-9]
GHS		02, 07
H-phrases		228-315
P-phrases		210-241-280-302+352-321-362
UN		1350
EINECS		231-722-6
m_r	g mol^{-1}	256.512
ρ	g cm^{-3}	2.07
Mp	°C	119
Bp	°C	445
$T_{p(orh \to mnkl)}$	°C	96-103
T_i	°C	~260
$\Delta_c H$	kJ mol^{-1}	−296.8
	kJ g^{-1}	−9.256
	kJ cm^{-3}	−19.161
Λ	wt.-%	−99.81

Apart from charcoal, sulfur is probably the oldest fuel in pyrotechnics. Though having a low heat of combustion, when compared with many other fuels, it easily ignites and hence is or has been a popular component in pyrotechnics to lower the ignition temperature. It is an advantage that sulfur undergoes fusion at 119°C which allows diffusion of sulfur into the lattice of an oxidizer thereby greatly enhancing reactivity. Another property only discovered recently is the so-called *spill-over* of crystalline sulfur when in contact with high porous activated charcoal. The sulfur migrates onto the surface of the charcoal thereby becoming amorphous and hence highly reactive – the underlying secret of black powder! The latter is the most prominent and also paramount application of sulfur with countless derivatives. The role of sulfur in black powder has been investigated in great detail by Shimizu and Seel. However, some details still remain unclear. In some systems, sulfur may take up a double role both as fuel and oxidizer. In other systems (e.g. Zn/S), sulfur is clearly an oxidizer. Sulfur for use in pyrotechnics must not be flowers of sulfur as these contain more than 0.1 wt.-% sulfuric acid which may trigger instability and spontaneous reactions with oxidizers and metallic fuels.

- R. Steudel, *Chemistry of the Non-Metals*, 2nd Edition, DeGruyter, Berlin, **2020**, 528–538.

– L. Medenbach, I. Escher, N. Kçwitsch, M. Armbrüster, L. Zedler, B. Dietzek, P. Adelhelm, Sulfur Spillover on Carbon Materials and Possible Impacts on Metal-Sulfur Batteries, *Angew. Chem. Int. Ed.* **2018**, *57*, 13666–13670.

Syntin, Synthin

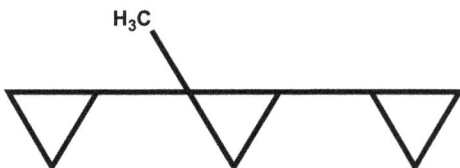

1-Methyl-1,2-dicyclopropyl-cyclopropane, MDCP, cyclin		
Formula		$C_{10}H_{16}$
CAS		[93223-46-2]
m_r	g mol^{-1}	136.234
ρ	g cm^{-3}	0.8504
$\Delta_f H$	kJ mol^{-1}	133
$\Delta_c H$	kJ mol^{-1}	−6354.8
	kJ g^{-1}	−46.646
	kJ cm^{-3}	−39.668
Λ	wt.-%	−328.83
Mp	°C	
Bp	°C	158
	°C	80 at 6.5 kPa
T_i in O_2	°C	323
$\Delta_{vap}H$	kJ mol^{-1}	47.1

Syntin is a hydrocarbon mixture consisting of four stereoisomers that has been developed in the USSR in 1960. It has been used as a fuel extensively between 1970 and 1990 in the upper stages of the Proton launch vehicle and was considered too for the Buran shuttle program. Its advantage over common fuels is its high positive heat of formation and density, both of which can be attributed to the cyclopropyl units (see *polynitrated polyhedra*). Syntin allegedly also possesses a lower viscosity than comparable hydrocarbon rocket fuels. Due to high production cost, its use was abandoned after 1996.

– T. Edwards, Liquid Fuels and Propellants for Aerospace Propulsion: 1903–2003, *J. Propul Power* **2003**, *19*, 1089–1107.

– Y.A. Lisochkin, V. I. Poznyak, Determination of Global Parameters of Gas-Phase Oxidation Reactions from Heat-Evolution Rates in a Plug-Flow Reactor, *Combust. Explos. Shock.* **1998**, *34*, 133–138.
– S.M. Pimenova, M.P. Kozina, V.P. Kolesov, The enthalpies of combustion and formation of *cis* and *trans*-l-methyl-1,2-dicyclopropylcyclopropane, *Thermochim. Acta* **1993**, *221*, 139–141.

T

Tabun, GA

(RS)-Dimethylphosphoramidocyanidsäureethylester, Trilon 83, T83

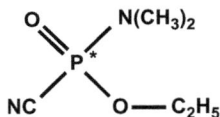

Formula		$C_5H_{11}N_2O_2P$
CAS		[77-81-6]
m_r	$g\ mol^{-1}$	162.128
ρ	$g\ cm^{-3}$	1.077
Mp	°C	−48
Bp	°C	237–240
Λ	wt.-%	−157.89
LCt_{50}	ppm	400
$\Delta_f H$	$kJ\ mol^{-1}$	−560 (estimated)
$\Delta_c H$	$kJ\ mol^{-1}$	−2979.7
	$kJ\ g^{-1}$	−18.379
	$kJ\ cm^{-3}$	−19.794

Pure tabun is a colorless limpid liquid with fruity odor. The technical product however due to hydrolysis (HCN splitt-off) smells like bitter almond.

- S. Franke, *Lehrbuch der Militärchemie, Band 1*, Militärverlag der Deutschen Demokratischen Republik, 2. Auflage, Leipzig, **1977**, pp. 464–468.

https://doi.org/10.1515/9783110660562-019

z-Tacot

Tetranitrobenzo-1,3a,4,6a-tetraazapentalen

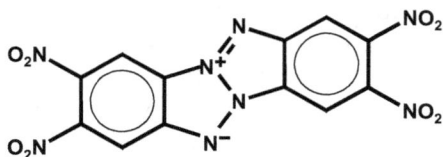

1,3,7,9-Tetranitro-6*H*-benzotriazolo[2,1-a] benzotriazol-5-ium hydroxide inner salt		
Aspect		Red orange crystals
Formula		$C_{12}H_4N_8O_8$
REACH		LPRS
EINECS		246-752-5
CAS		[25243-36-1]
m_r	g mol^{-1}	388.213
ρ	g cm^{-3}	1.85
$\Delta_f H$	kJ mol^{-1}	−461.2
$\Delta_{ex} H$	kJ mol^{-1}	−1883.8
	kJ g^{-1}	−4.852
	kJ cm^{-3}	−8.977
$\Delta_c H$	kJ mol^{-1}	−4832.9
	kJ g^{-1}	−12.449
	kJ cm^{-3}	−23.031
Λ	wt.-%	−74.18
N	wt.-%	28.86
Mp	°C	378 (dec)
Friction	N	>360 (BAM)
Impact	J	>50 (BAM)
	cm	>300 (tool 12)
V_D	m s^{-1}	7748 at ρ = 1.85 g cm^{-3}
P_{CJ}	GPa	27.2 at ρ = 1.85 g cm^{-3}

The designation *z*-TACOT derives from the z-alignment of nitrogen atoms in the tetraazapentalene unit. *z*-TACOT belongs to the very temperature insensitive high explosives (HEs). The calculated maximum safe temperature for processing z-Tacot is 263 °C.

–   T. Gołofit, Thermal behavior and safety of 1,3,7,9-tetranitrodibenzo-1,3a,4,6a-tetraazapentalen (z-TACOT), *Thermochim. Acta* **2018**, *667*, 59–64.

Taggant

Markierungsstoff

In accordance with the convention on the marking of plastic explosives signed March 1, 1991 military and non-military used brisant HEs based on PETN, HMX or RDX, and other comparatively powerful substances that are moldable at ambient temperature and whose components at ambient temperature have a vapor pressure lower than $P_{vap} = 10^{-4}$ Pa need a suitable taggant to enable detection. Taggants in accordance with the convention and their minimum concentrations in plastic explosives are given in Table T.1.

Table T.1: Taggants for plastic explosives.

Name of taggant	Sum formula	CAS-No.	P_{vap} (Pa)	Minimum content (wt.-%)
Ethylene glycol dinitrate (EGDN)	$C_2H_4N_2O_6$	628-96-6	10.17	0.2
2,3-Dimethyl-2,3-dinitrobutane (DMNB)	$C_6H_{12}N_2O_4$	3964-18-9		1
4-Nitrotoluene (*p*-MNT)	$C_7H_7NO_2$	99-99-0	6.52	0.5
2-Nitrotoluene (*o*-MNT)	$C_7H_7NO_2$	88-72-2	19.19	0.5

- Accessed on April 2, 2020 at https://treaties.un.org/doc/db/Terrorism/Conv10-english.pdf
- H. Östmark, S. Wallin, H. G. Ang, Vapor Pressure of Explosives: A Critical Review, *Propellants Explos. Pyrotech.* **2012**, *37*, 12–23.

Taliani test

In the Taliani test, gun propellants are stored at high temperatures two in a vial equipped with a manometer. The time to reach P = 100 mm Hg (13.33 kPa) after evacuation is determined. The test temperatures are NC = 135 °C and gun propellants = 110 °C. As solvents and moisture affect the test they have to be removed prior to testing. The Taliani test is the predecessor of today's vacuum stability test.

- C. Boyars, W. G. Gough, The Taliani Test as a Criterion of Propellant Stability, *US NAVORD-Report 3023*, US Naval Powder Factory, September **1953**, 53 pp.

Tapematch

A tapematch is a transparent adhesive tape (ca. 25 mm width), with a strip of fine-grained black powder (two or three-component) in the middle (ca. 10 mm width). Tapematch is used to transfer ignition and is a common material in both the fireworks field and military pyrotechnics (Figure B.12, p. 120).

Terephthalic acid, TPA

Terephthalsäure

COOH

COOH

Formula		$C_8H_6O_4$
EINECS		202-830-0
CAS		[100-21-0]
m_r	g mol^{-1}	166.13
$\rho_{25\,°C}$	g cm^{-3}	1.51
Mp	°C	402
Sub	°C	260
$\Delta_f H$	kJ mol^{-1}	−816
$\Delta_c H$	kJ mol^{-1}	−3189.7
	kJ g^{-1}	−19.200
	kJ cm^{-3}	−28.991
Λ	wt.-%	−144.46

TPA is used frequently in non-toxic training smokes. Anthony investigated the level of organic volatiles in TPA training smokes. Table T.2 depicts a formulation and performance data.

Table T.2: Composition and performance of a TPA obscurant.

Saccharose (wt.-%)	11
Potassium chlorate (wt.-%)	31
Terephthalic acid (wt.-%)	58
α_{VIS} at 20 °C, 50% RH (m^2 g^{-1})	5.3
Y_F at 20 °C, 50% RH (−)	0.58

- J. S. Anthony, W. T. Muse, S. A. Thomson, L. C. B. Crouse, C. L Crouse, Characterization of pyrotechnically disseminated terephthalic acid as released from light vehicle obscuration smoke system (LVOSS) canisters, *Smoke and Obscurants Symposium*, Aberdeen Proving Ground, **1998**.

Tetracene

Tetrazen

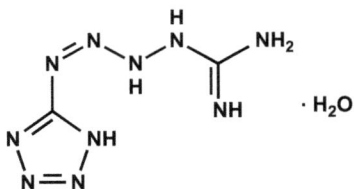

GNGT		
Formula		$C_2H_8N_{10}O$
REACH		LPRS
EINECS		608-603-6
CAS		[31330-63-9]
m_r	g mol^{-1}	188.152
ρ	g cm^{-3}	1.653
Δ_fH	kJ mol^{-1}	+189
$\Delta_{ex}H$	kJ mol^{-1}	−554.8
	kJ g^{-1}	−2.949
	kJ cm^{-3}	−4.874
Δ_cH	kJ mol^{-1}	−2119.4
	kJ g^{-1}	−11.264
	kJ cm^{-3}	−18.620
Λ	wt.-%	−59.52
N	wt.-%	74.42
Dp	°C	140
Friction	N	7
Impact	J	0.1–0.3
V_D	m s^{-1}	1500 at 1.60 g cm^{-3}
Trauzl	cm^3	155

Tetracene is a very feeble primary explosive, however, due to its pronounced mechanical and thermal sensitiveness, it is used to sensitize primary formulations. It is compatible with lead azide but not compatible with silver azide.

Tetramethylammonium nitrate

Tetramethylammoniumnitrat, Tetra-Salz

TMAN		
Formula		$N(CH_3)_4NO_3$
GHS		03, 07
H-phrases		272-315-319-335
P-phrases		210-221-302+352-305+351+338-405-501a
UN		1479
EINECS		217-723-4
CAS		[1941-24-8]
m_r	$g \cdot mol^{-1}$	136.151
ρ	$g \cdot cm^{-3}$	1.25
Dp	°C	~360
$\Delta_f H$	kJ mol^{-1}	−341.37
$\Delta_{ex}H$	kJ mol^{-1}	−505.4
	kJ g^{-1}	−3.712
	kJ cm^{-3}	−4.640
$\Delta_c H$	kJ mol^{-1}	−2948.1
	kJ g^{-1}	−21.653
	kJ cm^{-3}	−27.066
Λ	wt.-%	−129.26
N	wt.-%	20.58

TMAN is slightly hygroscopic. It was used in World War II in mixtures with nitrates as both gun propellant and substitute explosives. TMAN is obtained from reaction of triethylamine with methyl nitrate. TMAN is utilized in autoignition compositions and strobe flares.

– S. R. Jain, M. V. Rao, V. R. P. Verneker, Kinetics and mechanism of thermal decomposition of tetramethylammonium nitrate, *Proc. Ind. Acad. Sci. A, Chem. Sci*, **1978**, *87*, 31–36.

Tetramethylammonium perchlorate

Tetramethylammoniumperchlorat

Formula	$N(CH_3)_4ClO_4$
GHS	03, 07
H-phrases	272-315-319-335
P-phrases	210-221-302+352-305+351+338-405-501a
UN	1479
EINECS	219-805-5

CAS		[2537-36-2]
m_r	$g \cdot mol^{-1}$	173.595
ρ	$g \cdot cm^{-3}$	1.439
Dp	°C	~340
$\Delta_f H$	$kJ\ mol^{-1}$	−277
$\Delta_{ex}H$	$kJ\ mol^{-1}$	−809.6
	$kJ\ g^{-1}$	−4.664
	$kJ\ cm^{-3}$	−6.711
$\Delta_c H$	$kJ\ mol^{-1}$	−3012.1
	$kJ\ g^{-1}$	−17.351
	$kJ\ cm^{-3}$	−24.968
Λ	wt.-%	−87.56
N	wt.-%	8.07

TMAP is a colorless crystalline powder used mainly in pyrotechnic strobe compositions.

- E. Palacios, R. Burriel, P. Ferloni, The phases of [(CH$_3$)$_4$N](ClO$_4$) at low temperature, *Acta Cryst.*, **2003**, *B59*, 625–633.
- S. R. Jain, P. R. Nambiar, Effect of tetramethylammonium perchlorate on ammonium perchlorate and propellant decomposition, *Thermochim. Acta*, **1976**, *16*, 49–54.

Tetranitroadamantane

Tetranitroadamantan

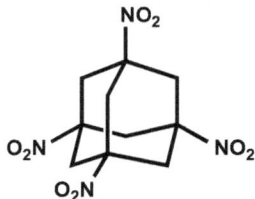

1,3,5,7-Tetranitrotricyclo[3.3.1.1^{3,7}]decane		
Formula		$C_{10}H_{12}N_4O_8$
CAS		[75476-36-7]
m_r	$g\ mol^{-1}$	316,224
ρ	$g\ cm^{-3}$	1.68
$\Delta_f H$	$kJ\ mol^{-1}$	−321.33
$\Delta_{ex}H$	$kJ\ mol^{-1}$	−1423.8
	$kJ\ g^{-1}$	−4.502
	$kJ\ cm^{-3}$	−7.564

$\Delta_c H$	kJ mol^{-1}	−5329.2
	kJ g^{-1}	−16.853
	kJ cm^{-3}	−28.312
Λ	wt.-%	−91.07
N	wt.-%	17.72
Mp	°C	361 (dec)
T_{5ex}	°C	400
Friction	N	>360 (BAM)
Impact	J	>50 (BAM)
V_D	m s^{-1}	6702 at 1.68 g cm^{-3}
P_{CJ}	GPa	18.17 at 1.68 g cm^{-3}

Tetranitroadamantane is an insensitive HE with high thermal stability and has performance comparable to TNT.

– G. P. Sollott, E. E. Gilbert, A Facile Route to 1,3,5,7-Tetraaminoadamantane. Synthesis of 1,3,5,7-Tetranitroadamantan, *J. Org. Chem.* **1980**, *45*, 5405–5408.

Tetranitrobenzopyrido-1,3A,6,6A-tetraazapentalene, BPTAP

Tetranitrobenzopyrido-1,3,A,6,6,A-tetraazapentalen

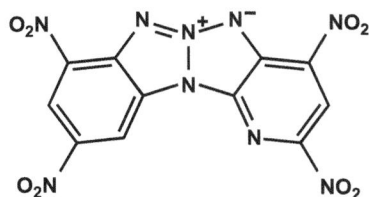

2,4,8,10-Tetranitro-,5H-pyrido[3′,2′:4,5][1,2,3]
triazolo[1,2-a]benzotriazol-6-ium, inner salt

Formula		$C_{11}H_3N_9O_8$
CAS		[86662-96-6]
m_r	g mol^{-1}	389.2
ρ	g cm^{-3}	1.84
$\Delta_f H$	kJ mol^{-1}	444
$\Delta_{ex} H$	kJ mol^{-1}	−1898.2
	kJ g^{-1}	−4.870
	kJ cm^{-3}	−8.960
$\Delta_c H$	kJ mol^{-1}	−5201.5
	kJ g^{-1}	−13.365
	kJ cm^{-3}	−24.965
Λ	wt.-%	−63.6

N	wt.-%	32.34
Dp	°C	375
Friction	N	>355
Impact	J	15 (BAM)
V_D	m s^{-1}	7430 at 1.78 g cm^{-3}
P_{CJ}	GPa	29.4 at 1.78 g cm^{-3}
$\varnothing_{cr}$	mm	<3

BPTAP is a temperature-resistant HE with a very small critical diameter and low sensitivity suitable for the use in advanced EFIs. *Hiskey et al.* developed a simple synthesis for it.

- H. V. Huynh, M. A. Hiskey, New Suitable Replacement for the High-Temperature Explosive HNS -4, *Nucl. Weap. J.* **2006**, 2–5.

Tetranitrocarbazole, TNC

Tetranitrocarbazol, Gelbmehl

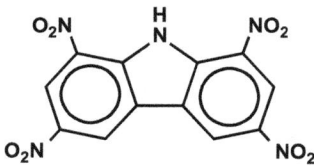

Aspect		Yellow crystal powder
Formula		$C_{12}H_5N_5O_8$
CAS		[28453-24-9]
m_r	g mol^{-1}	347.2
ρ	g cm^{-3}	1.893
$\Delta_f H$	kJ mol^{-1}	18
$\Delta_{ex}H$	kJ mol^{-1}	−1545.4
	kJ g^{-1}	−4.451
	kJ cm^{-3}	−8.426
$\Delta_c H$	kJ mol^{-1}	−5454.8
	kJ g^{-1}	−15.711
	kJ cm^{-3}	−29.741
Λ	wt.-%	−85.25
N	wt.-%	20.17
Mp	°C	296
T_{5ex}	°C	470
Impact	J	7.6 (Picatinny Arsenal machine)
Specification		MIL T-13723A

TNC was initially used in wartime Germany as a non-hygroscopic replacement for charcoal in black powder-type first fires and delay formulations. Today, TNC is used both in the USA and UK in gassy delay formulations (e.g. SR112).

- https://www.army.mil/article/121496/picatinny_scaling_up_in_house_chemicals_produc tion_to_shun_higher_costs

Tetranitroethylene, TNE

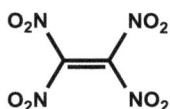

Tetranitroethylen

O$_2$N, NO$_2$

O$_2$N, NO$_2$

Aspect		Yellow green amorphous substance
Formula		$C_2N_4O_8$
CAS		[13223-78-4]
m_r	g mol^{-1}	208.04
ρ	g cm^{-3}	~2
Δ_fH	kJ mol^{-1}	85
$\Delta_{ex}H$	kJ mol^{-1}	−883.5
	kJ g^{-1}	−4.247
	kJ cm^{-3}	−8.494
Δ_cH	kJ mol^{-1}	−872.0
	kJ g^{-1}	−4.192
	kJ cm^{-3}	−8.379
Λ	wt.-%	+30.76
N	wt.-%	26.93

Baum obtained the green amorphous solid by careful pyrolysis of hexanitroethane (HNE). TNE is not stable and decomposes steadily. It is however a very strong dienophile (10× stronger than tetracyanoethylene) and hence suitable for the synthesis of polynitrocarbocycles. Therefore, TNE has not to be isolated in substance but HNE instead can be heated with the diene to yield TNE in situ.

- J. Yang, Y. Gong, H. Mei, T. Li, J. Zhang, M. Gozin, Design of Zero Oxygen Balance Energetic Materials on the Basis of Diels-Alder Chemistry, *J. Org. Chem.* **2018**, *83*, 14698–14702.
- K. Baum, D. Tzeng, Synthesis and reactions of tetranitroethylene, *J. Org. Chem.* **1985**, *50*, 2736–2739.
- T. S. Griffin, K. Baum, Tetranitroethylene. In situ formation and Diels-Alder reactions, *J. Org. Chem.* **1980**, *45*, 2880–2883.

Tetranitromethane, TNM

Tetranitromethan

Formula		CN_4O_8
GHS		03, 06, 08
H-phrases		271-301-315-319-330-335-351
P-phrases		220-260-281-284-301+310-305+351+338
REACH		LPRS
EINECS		208-094-7
CAS		[509-14-8]
m_r	$g\ mol^{-1}$	196.033
ρ	$g\ cm^{-3}$	1.638
$\Delta_f H$	$kJ\ mol^{-1}$	38.49
$\Delta_{ex}H$	$kJ\ mol^{-1}$	−443.9
	$kJ\ g^{-1}$	−2.264
	$kJ\ cm^{-3}$	−3.709
$\Delta_{ex}H$	$kJ\ mol^{-1}$	−431.5
	$kJ\ g^{-1}$	−2.201
	$kJ\ cm^{-3}$	−3.606
Λ	wt.-%	+48.97
N	wt.-%	28.58
Mp	°C	14.2
Bp	°C	126.2
$\Delta_v H$	$kJ\ mol^{-1}$	43.93
Dp	°C	192
Friction	N	60 (BAM)
Impact	J	19 (BAM)
	cm	>100 (tool 12)
V_D	$m\ s^{-1}$	6320 at ρ = 1.60 g cm^{-3}

TNM was used in suspensions with aluminum powder as a substitute explosive in World War II. Nowadays, TNM only serves as trinitromethylsynthon in the synthesis of other nitro compounds.

- Y. V. Vishnevskiy, D. S. Tikhonov, J. Schwabedissen, H.-G. Stammler, R. Moll, B. Krumm, T. M. Klapötke, N. W. Mitzel, Tetranitromethane: A Nightmare of Molecular Flexibility in the Gaseous and Solid State, *Angew. Chem. Int. Ed.*, **2017**, *56*, 9619–9623.

Tetranitrooxanilide, TNO

Tetranitrooxanilid

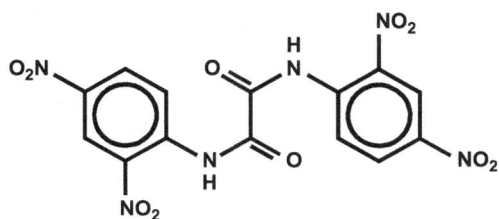

Aspect		Yellow crystal powder
Formula		$C_{14}H_8N_6O_{10}$
REACH		LPRS
EINECS		238-872-1
CAS		[14805-54-0]
m_r	g mol^{-1}	420.254
ρ	g cm^{-3}	1.82
$\Delta_f H$	kJ mol^{-1}	−418
$\Delta_{ex} H$	kJ mol^{-1}	−2224.8
	kJ g^{-1}	−5.294
	kJ cm^{-3}	−9.635
$\Delta_c H$	kJ mol^{-1}	−6234.6
	kJ g^{-1}	−14.836
	kJ cm^{-3}	−27.001
Λ	wt.-%	−83.76
N	wt.-%	20.00
Mp	°C	313
T_{5ex}	°C	392
Impact	J	15 (Picatinny Arsenal machine)

Tetranitrooxanilide is a popular fuel for gassy delays (SR 74) and spectral counter-measures in the UK.

Tetrasulfur tetranitride

Tetraschwefeltetranitrid

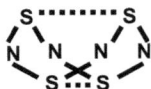

Aspect		Yellow-orange
Formula		S_4N_4
GHS		Not assigned
H-phrases		Not assigned
P-phrases		Not assigned
CAS		[28950-34-7]
m_r	g mol^{-1}	184.29
ρ	g cm^{-3}	2.23
Mp	°C	178
Bp	°C	185 (dec)
c_p	J mol^{-1} K^{-1}	
$\Delta_f H$	kJ mol^{-1}	+498
$\Delta_{ex}H$	kJ mol^{-1}	−491.32
	kJ g^{-1}	−2.666
	kJ cm^{-3}	−5.945
$\Delta_{comb}H$	kJ mol^{-1}	−1685.2
	kJ g^{-1}	−9.144
	kJ cm^{-3}	−20.398
Λ	wt.-%	−69.46
N	wt.-%	30.40
Impact	J	4
V_D	m s^{-1}	5600 at 1.70 g cm^{-3}

S_4N_4 is a thermochromic compound being pale crème at low temperature, orange yellow at ambient, and dark red at high temperature. It is reported to be sensitive to friction and impact but so far, those values have never been properly established. When carefully ignited it burns rapidly with a blue flame and explodes when hit hard.

- M. Becke-Goehring, E. Fluck, Developments in Inorganic Chemistry of Compounds containing the Sulphur Nitrogen Bond, in C. B. Colburn (Ed.) *Developments in Inorganic Nitrogen Chemistry*, Elsevier Publishing Company, Amsterdam, **1966**, 150–240.
- R. Steudel, *Chemistry of the Non-Metals*, 2nd. Ed. De Gruyter, Berlin, **2020**, 591–597.
- E.-C. Koch, M. Sućeska, Analysis of the Explosive Properties of Tetrasulfur Tetranitride, S_4N_4, *Z. Anorg. Allg. Chem.* https://onlinelibrary.wiley.com/doi/full/10.1002/zaac.202000406.

Tetryl

Tetryl

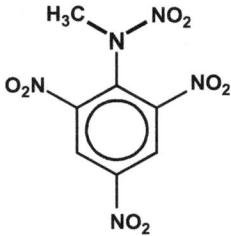

2,4,6-Trinitrophenylmethylnitramine		
Aspect		Light yellow crystal powder
Formula		$C_7H_5N_5O_8$
REACH		LPRS
EINECS		207-531-9
CAS		[479-45-8]
m_r	g mol^{-1}	287,145
ρ	g cm^{-3}	1.731
$\Delta_f H$	kJ mol^{-1}	20
$\Delta_{ex}H$	kJ mol^{-1}	−1477.6
	kJ g^{-1}	−5.146
	kJ cm^{-3}	−8.908
$\Delta_{ex}H$	kJ mol^{-1}	−3489.2
	kJ g^{-1}	−12.151
	kJ cm^{-3}	−21.034
Λ	wt.-%	−47.36
N	wt.-%	24.39
c_p	J K^{-1} mol^{-1}	62.45
Mp	°C	129.45
$\Delta_m H$	kJ mol^{-1}	26.65
Dp	°C	187
Friction	N	>360 (BAM)
Impact	J	3 (BAM)
V_D	m s^{-1}	7479 at ρ = 1.614 g cm^{-3}
P_{CJ}	GPa	22.64 at ρ = 1.614 g cm^{-3}
Trauzl	cm^3	410
$\varnothing_{cr}$	mm	<4 at ρ = 1.860 g cm^{-3}
Koenen	mm	6
Specification		STANAG 4021

For many decades, tetryl was "the" booster explosive of choice. However, its toxicity (it was used as finely powdered material which is prone to be inhaled by workers) led to its replacement by less toxic RDX in these applications. It is insoluble in water but soluble in acetone and in molten TNT to form a low melting eutectic at 65 °C with 20–30 wt.-% TNT which was used as tetrytol for grenades and torpedo warheads.

– T. R. Gibbs, A. Popolato, *LASL Explosive Property Data*, University of California Press, **1980**, pp. 163–171.

TEX, 4,10-dinitro-2,6,8,12-tetraoxa-4,10-diazaisowurtzitane

4,10-Dinitro-2,6,8,12-tetraoxa-4,10-diazaisowurtzitan

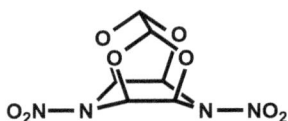

Formula		$C_6H_6N_4O_8$
CAS		[130919-56-1]
m_r	g mol^{-1}	262.136
ρ	g cm^{-3}	1.985
$\Delta_f H$	kJ mol^{-1}	−541
Λ	wt.-%	−42.72
N	wt.-%	21.37
$\Delta_{ex}H$	kJ mol^{-1}	−1137.3
	kJ g^{-1}	−4.339
	kJ cm^{-3}	−8.612
$\Delta_c H$	kJ mol^{-1}	−2677.6
	kJ g^{-1}	−10.215
	kJ cm^{-3}	−20.276
Dp	°C	282
Friction	N	>355
Impact	J	23-25 (BAM)
V_D	m s^{-1}	7075 at 1.87 g cm^{-3} (97 wt.-% TEX, 3 wt.-% wax)
P_{CJ}	GPa	29.2 at 1.87 g cm^{-3}
$\varnothing_{cr}$	mm	<20

TEX is a temperature-stable cheap HE with very high density, low sensitiveness, and large critical diameter suitable for use in large charges.

– E.-C. Koch, TEX – 4,10-Dinitro-2,6,8,12-tetraoxa-4,10-diazatetracyclo[5.5.0.0^{5,9}.0^{3,11}]-dodecane – Review of a Promising High Density Insensitive Energetic Material, *Propellants Explos. Pyrotech.* **2015** *40*, 374–387.

Thermal battery
Thermalbatterie

Thermally activated batteries work in a similar way as common alkali-manganese batteries. However, they do not contain an aqueous solution as electrolyte but mixtures of alkali halides and solid at service temperature and therefore do not conduct. Consequently, thermal batteries do not undergo any kind of slow discharge reactions and have service lifetimes of more than 25 years. For activation thermal batteries require heat to liquify the salts to give an ionic liquid melt having electrical conductivities 30 times higher than common aqueous electrolytes. It is hence that thermal batteries have much higher power densities than common batteries.

The long service lifetime, high power density, and high reliability make thermal batteries ideal sources of electrical energy in challenging applications like outer space and defence. Figure T.1 displays the schematic section view of a thermal battery.

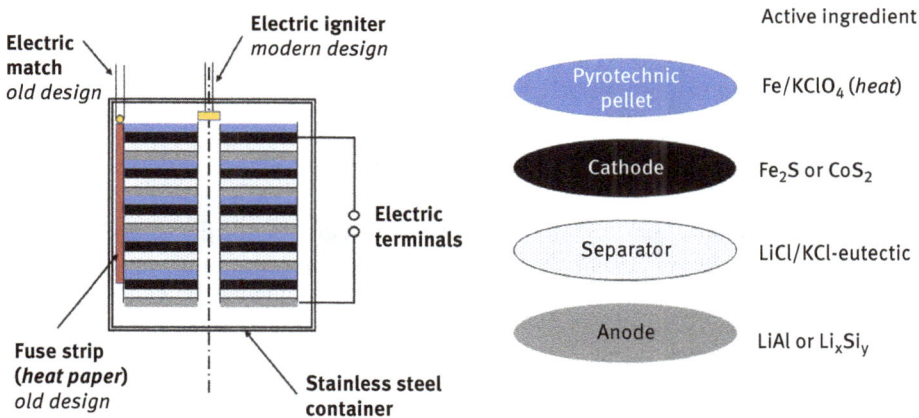

Figure T.1: Schematic of construction and components of a thermal battery.

The electric cell consists of the anode (temperature resistant alloy of lithium, e.g. LiAl or Li_xSi_y), the separator (electrolyte) (often a eutectic of alkali halides), the cathode (iron sulfide or thermally more robust cobalt sulfides), and the pyrotechnic heater ($Fe/KClO_4$, $\rightarrow$ *Heat*) which is conductive due to the large surplus of iron contained in the composition. Depending on the voltage, the thermal battery contains an integer number of cells. A fast burning lateral fuze strip provides simultaneous ignition of all cells. The cell stack is wrapped in a non-conductive thermally insulating material and is sealed airtight in a stainless steel cylinder with electrical terminals leading to outer side through glass-metal seals. Table T.3 shows typical cell configurations and their performance.

Table T.3: Typical materials and performance of thermal batteries.

System				
Anode		Calcium	LiX	LiSi
Electrolyte (separator)		LiCl–KCl	LiCl–KCl	LiCl–KCl-LiBr
Cathode		CaCrO$_4$	FeS$_2$	CoS$_2$
Voltage	V	2.60–2.20	X = Al: 1.8, S = Si: 1.95	1.7
Power density	mA cm^{-2}	100–800	100–1500	
Specific energy (theor.)	Wh kg^{-1}	540	1380	
Specific energy (pract.)	Wh kg^{-1}	5–15	50–80	110
Temperature range	°C	400–550	350–400	−650 °C
Operating time	min	Bis zu 5	Bis zu 60	Bis zu 120
Disadvantages		CaLi$_2$ formation	LiK$_6$Fe$_{24}$S$_{26}$Cl formation	CoS$_2$ is expensive

- R. A. Guidotti, P. Masset, Thermally Activated ("thermal") battery Technology Part I: An Overview, *Journal of Power Sources* **2006**, *161*, 1443–1449.
- E.-C. Koch, Special materials in Pyrotechnics VII: Pyrotechnics Used in Thermal Batteries, *Def. Technol.*, **2019**, *15*, 254–263.

Thermite
Thermit

Thermites are blends from metallic fuels (either elements or alloys) and a metal oxide reacting in an exothermic fashion after ignition in accordance to the following simplified equation:

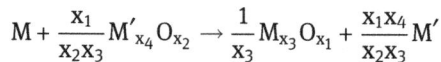

$$M + \frac{x_1}{x_2 x_3} M'_{x_4} O_{x_2} \rightarrow \frac{1}{x_3} M_{x_3} O_{x_1} + \frac{x_1 x_4}{x_2 x_3} M'$$

In particular a mixture consisting of coarse aluminum (*Grieß*) (ca. 25 wt.-%) and iron (II,III) oxide (FeO · Fe$_2$O$_3$) (*Hammerscale*) (ca. 75 wt.-%) is also synonymously called thermite. The latter burns with high temperature T ~ 2500 °C producing melt-liquid product iron used for welding tracks or other means of construction:

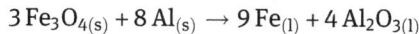

$$3\,Fe_3O_{4(s)} + 8\,Al_{(s)} \rightarrow 9\,Fe_{(l)} + 4\,Al_2O_{3(l)}$$

The high ignition temperature of thermite (T_i > 500 °C) makes this pretty safe to handle but also requires the use of ignition compositions reaching those temperatures (e.g. Mg/BaO$_2$). *Fischer* and *Grubelich* have collated tables with the heat of reaction of a great many number of thermite reactions. Thermites based on components of high volatility and/or extremely low particle size (in the one figure micrometric or even nanometric (*nanothermite*) can be extremely sensitive towards electrostatic discharge and mechanical stimuli. These thermites when present as porous materials may react with very high speed and show even supersonic reaction, that is, undergo LVD.

-   P. G. Shaw, *Thermitic Thermodynamics: A Computational Survey and Comprehensive Interpretation of Over 800 Combinations of Metals, Metalloids, and Oxides*, CRC Press, New York, **2020**, 1068 pp.
-   M. Comet, B. Siegert, F. Schnell, V. Pinchot, F. Cizsek, D. Spitzer, Phosphorus-Based Nanothermites: A New Generation of Pyrotechnics Illustrated by the Example of n-CuO/Red P Mixtures, *Propellants Explos. Pyrotech.* **2010**, *35*, 220–225.
-   S. H. Fischer, M. C. Grubelich, Theoretical Energy Release of Thermites, Intermetallics, and Combustible Metals, *IPS*, Monterey, **1998**, 231–286.
-   L. L. Wang, Z. A. Munir, Y. M. Maximov, Thermite reactions their utilization synthesis and processing of materials, *J. Mater. Sci.* **1993**, *28*, 3693–3708.

Thermobaric explosives, TBX
Thermobare Sprengstoffe

Thermobaric explosives are composite explosive charges/charge configurations containing surplus combustible metallic, non-metallic and/or organic fuels (having a high heat of combustion) resulting in an oxygen balance well below $\Lambda < 100$ wt.-%. The highly negative Λ of the TBX induces an intense afterburn reaction of the primary detonation products (CO, H_2, soot) and the dispersed extra fuel particles which must have reached at least ignition temperature. This afterburn yields a prolonged positive pressure phase in comparison to common HE charges (Figure T.2). For the same reason, TBX do not act as point sources but extend over a considerable volume of a couple of 10 or even 100 cubic meters. This behavior allows the blast effects of TBX to literally overcome and circumvent physical obstacles like walls, ramparts, trenches

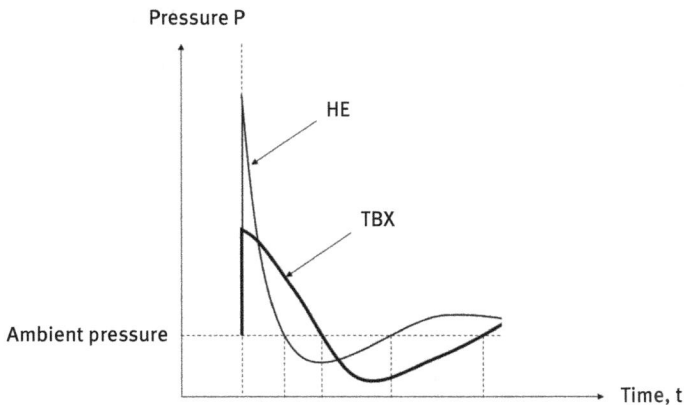

Figure T.2: Pressure–time curve for common (HE) and thermobaric high explosives (TBX).

and to have the pressure waves acting from different sides on a target structure su-
perseding the allowable load case for many structures. Hence, TBX are also often referred to as volumetric explosives.

A typical TBX is PBXN-133, which was initially designated PBXIH-135 before its
qualification. Table T.4 shows the composition and performance data of it. Figure T.3
depicts the highly resolved traces in two band pass regions characteristic for the
emission of vapor-phase species Al (λ = 396 nm) and AlO (λ = 484 nm). The initial
short excitation peak of aluminum follows a delayed but prolonged emission of AlO
indicative of the successive nature of occurrence that is formation of AlO from Al in
the post-detonation zone.

Table T.4: Composition and performance on PBXN-113.

Parameter	Value
Aluminum (wt.-%)	35
Octogen class 5 (wt.-%)	45
HTPB (wt.-%)	20
Λ (wt.-%)	−105
Density (g cm^{-3})	1.71
V_D (m s^{-1})	6980
$\varnothing_{cr}$ (mm)	<9.5
Impact (cm)	98
VST (mL g^{-1})	0.09
UN Test Series 7	Fulfills = EI(D)S*

*Extremely insensitive (detonating) substance.

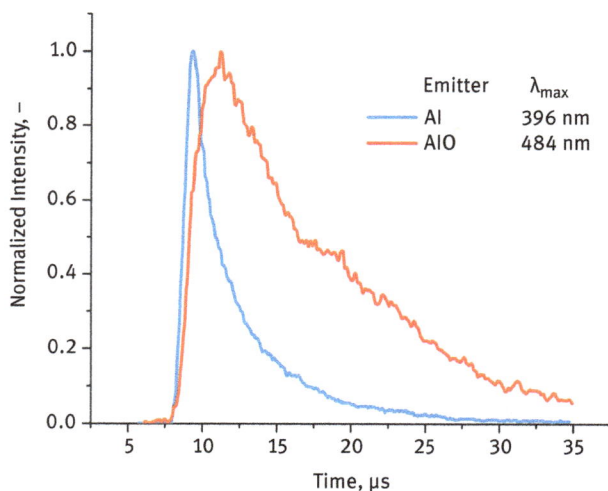

Figure T.3: Consecutive appearance of Al and AlO emission upon detonation of PBXN-113 (after *Carney*).

- L. Türker, Thermobaric and enhanced blast explosives (TBX and EBX), *Defence Tech.* **2016**, *12*, 423–445.
- E.-C. Koch, *L-165 Volumetric Explosives Part 1 Fuel/Air Explosives*, MSIAC, Brussels, **2010**, 25 pp.
- J. R. Carney, J. S. Miller, J. C. Gump, G. I. Pangilinan, Time-resolved optical measurements of the post-detonation combustion of aluminized explosives, *Rev. Sci. Instruments* **2006**, *77*, 1-063103–6-063103.

Thermochemical codes
Thermochemische Rechenprogramme

Thermochemical codes serve the determination of the equilibrium composition of reacted energetic materials. However, with few exceptions (CHEETAH, EXPLO), those codes do not take into account reaction dynamics and kinetics but determine the compositions exclusively by thermodynamic reasoning (e.g. minimization of the Gibbs free energy).

Systems containing the elements C, H, N, O, Cl, and F inherently yield high amounts of gases and are addressed by most codes with sufficient accuracy which can be verified by experimental methods (e.g. emission spectroscopy). However, systems producing a major amount of condensed products as this are the case with thermites or alloying reactions cannot be treated like ideal gases and only dedicated codes like CERV, EKVI, or Factsage yield reasonable output.

Freeware (*f*) commercial (*c*) accessible as well as restricted governmental (*g*) codes are CEA(*f*), CERV(*g*), CHEETAH(*g*), COPPELIA(*g*), EKVI(*c*), EXPLO(*c*), FactSage(*c*), ICT(*c*), Real(*f*), and AISTJAN(*c*). A round robin comparison of codes has been conducted and was published by *Koch et al.* for CEA, CERV, Cheetah, EXPLO, ICT, REAL, and Tanaka (AISTJAN).

- E.-C. Koch, R. Webb, V. Weiser, Review on Thermochemical Codes, O-138, MSIAC, Brussels, **2010**, 35 pp.

Thulium

Aspect	Silvery
Formula	Tm
GHS	02
H-phrases	260-228
P-phrases	210-231+232-240-241-280-501a
UN	3178
EINECS	231-140-2

CAS		[7440-30-4]
m_r	g mol^{-1}	168.93421
ρ	g cm^{-3}	9.321
Mp	°C	1545
Bp	°C	1944
c_p	J mol^{-1} K^{-1}	27.02
$\Delta_m H$	kJ mol^{-1}	16.84
$\Delta_v H$	kJ g^{-1}	190.7
$\Delta_c H$	kJ mol^{-1}	−944.35
	kJ g^{-1}	−5.59
	kJ cm^{-3}	−52.11
Λ	wt.-%	−14.21

Thulium filings and powder burn with a green corona.

- E.-C. Koch, V. Weiser, E. Roth, S. Kelzenberg, Consideration of some 4 f-Metals as New Flare Fuels Europium, Samarium, Thulium and Ytterbium, *ICT -Jata*, **2011**, Karlsruhe, V1.
- E. Roth, S. Knapp, A. Raab, V. Weiser, E.-C. Koch, Emission Spectroscopy of Some 4 f-Metal Flames, *ICT -Jata*, **2013**, Karlsruhe, P68.

Titanium
Titan

Aspect		Silvery grey
Formula		Ti
GHS		02
H-phrases		250-251
P-phrases		210-222-235+410-280-420-422a
UN		2546
EINECS		231-142-3
CAS		[7440-32-6]
m_r	g mol^{-1}	47.88
ρ	g cm^{-3}	4.540
Mp	°C	1666
Bp	°C	3358
c_p	J mol^{-1} K^{-1}	25.05
$\Delta_m H$	kJ mol^{-1}	14.15
$\Delta_v H$	kJ g^{-1}	410.0
c_L	m s^{-1}	6260
T_i	°C	250–400
$\Delta_c H$	kJ mol^{-1}	−944.7
	kJ g^{-1}	−19.73
	kJ cm^{-3}	−89.58
Λ	wt.-%	−66.83

Titanium burns in air with a dazzling white flame and spectacular sparking. Ti is frequently used in fireworks (cone formulations, flying squibs, etc.) to create sparks. The reason for the sparking is that the melt liquid droplets of burning titanium at first dissolve both oxygen and nitrogen and with increasing temperature the solubility product of N_2 is exceeded causing N_2 to volatilize again thereby bursting the droplet:

$$TiO_xN_{y(l)} \longrightarrow \gamma\text{-}TiO + \beta - Ti_2O_3 + N_2$$

In technical pyrotechnics, titanium is frequently used in igniter formulations, first fires and tracer formulations. Table T.5 shows typical titanium qualities of rockwood lithium (previously Degussa, before that Chemetall).

Table T.5: Ti powder and properties.

Name	H-content (wt.-%)	ESD (µJ)	Ti (°C)	Rim value (50 cm s^{-1})	Particle diameter (µm)	∅-Particle diameter according to Blaine (µm)
Ti-E	n.a.	0.32	>240	35 ± 10	<45 µm, min 99.9%	3 ± 1
Ti-S 9.5	<0.1	1.0	>400	35 ± 10	<45 µm, min 99.9%	9.5 ± 1.5

Titanium hydride

Titanhydrid

Aspect		Dark gray
Formula		$TiH_{x\ (x \leq 2)}$
GHS		02
H-phrases		228
P-phrases		210-240-241-280-370+378a
UN		1871
EINECS		231-726-8
CAS		[7704-98-5]
m_r	g mol^{-1}	49.92
ρ	g cm^{-3}	3.91
Mp	°C	1000 (dec)
c_p	J mol^{-1} K^{-1}	30.3
$\Delta_f H$	kJ mol^{-1}	−144.3
$\Delta_c H$	kJ mol^{-1}	−1097.7
	kJ g^{-1}	−22.000
	kJ cm^{-3}	−86.02
Λ	wt.-%	−96.15
T_i	°C	250−400

TiH_x decomposes with water. It is used mainly with $KClO_4$ in igniter formulations (Table T.6).

Table T.6: TiH_x and typical properties.

Name	H content (wt.-%)	ESD (µJ)	T_i (°C)	Particle diameter (µm)	$\varnothing$-Particle size after Blaine (µm)
TiH_2-N	>3.8	56	>400	<63 µm, min 99.9%	5 ± 1

– M. A. Cooper, M. S. Oliver, Titanium Subhydride Potassium Perchlorate ($TiH_{1.65}/KClO_4$) Burn Rates from Hybrid Closed Bomb-Strand Burner Experiments, *SAND 2012-7381*, Sandia National Laboratory, **2012**, 42 pp.

Titanium(IV) chloride

Titantetrachlorid, FM

Formula		$TiCl_4$
GHS		06, 08, 05
H-phrases		330-314-370-372
P-phrases		280-310-304+340-301+330+331-303+361+353-305+351+338
REACH		LRS
EINECS		231-441-9
CAS		[7550-45-0]
m_r	g mol^{-1}	189.692
ρ	g cm^{-3}	1.726
$\Delta_f H$	kJ mol^{-1}	−804.2
$\Delta_v H$	kJ mol^{-1}	−35.8
$\Delta_m H$	kJ mol^{-1}	−10
Mp	°C	−24.3
Bp	°C	136.5
Λ	wt.-%	0
c_p	J K^{-1} mol^{-1}	145.2

$TiCl_4$ is a limpid liquid, which due to spontaneous hydrolysis fumes very intensely. In the course of reaction various hydroxychlorides and hydrochloric acid form which successively yield $Ti(OH)_4$ and more HCl:

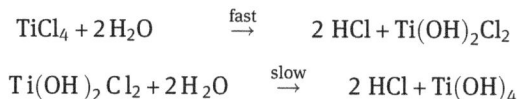

$$TiCl_4 + 2\,H_2O \quad \xrightarrow{fast} \quad 2\,HCl + Ti(OH)_2Cl_2$$

$$Ti(OH)_2Cl_2 + 2\,H_2O \quad \xrightarrow{slow} \quad 2\,HCl + Ti(OH)_4$$

$TiCl_4$ is very corrosive for non-protected metals and also very dangerous when brought into contact with the skin or inhaled. $TiCl_4$ is often used as non-flammable marking charge for ammunition.

- M. Rigo, P. Canu, L. Angelin, G. Della Valle, Kinetics of $TiCl_4$ Hydrolysis in a Moist Atmosphere, *Ind. Eng. Chem. Res.* **1998**, *37*, 1189–1195.
- E. Murray, F. Llados, *Toxicological Profile for Titanium Tetrachloride*, U.S. Department of Health and Human Services, Public Health Service Agency for Toxic Substances and Disease Registry September **1997**, 145 pp; accessible at: https://www.atsdr.cdc.gov/ToxProfiles/tp101.pdf

TKX-50

Hydroxylammonium 5,5'-bistetrazolat 1,1'-dioxid

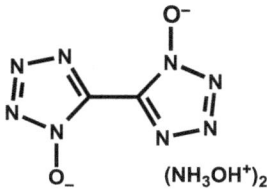

(NH₃OH⁺)₂

Hydroxylammonium-5,5′-bistetrazolate-1,1′-dioxide, ⊢ATO		
Formula		$C_2H_8N_{10}O_4$
CAS		[1403467-86-6]
m_r	g mol⁻¹	236.15
ρ	g cm⁻³	1.877
$\Delta_f H$	kJ mol⁻¹	193
$\Delta_{ex}H$	kJ mol⁻¹	−1172.2
	kJ g⁻¹	−4.964
	kJ cm⁻³	−9.317
$\Delta_c H$	kJ mol⁻¹	−2123.4
	kJ g⁻¹	−8.992
	kJ cm⁻³	−16.877
Λ	wt.-%	−27.10
N	wt.-%	59.31
Dp	°C	187−221
Friction	N	80−144 (BAM)
Impact	J	17.5−20 (BAM)
V_D	m s⁻¹	6596 at 1.15 g cm⁻³
	m s⁻¹	8233 at 1.725 g cm⁻³ (T./paraffin/graphite: 94.4/4.5/1; ∅21 mm)
P_{CJ}	GPa	13.3 at 1.15 g cm⁻³
Koenen	mm	10, type H

TKX-50 is an insensitive HE with non-ideal character as indicated by a pronounced diameter dependence of its detonation velocity. The violence of its reaction in the Koenen test however indicates difficulties to meet IM requirements in fast cook-off scenarios. Experimental calorimetry by *Konkova et al.* shows that the heat of formation of TKX-50 is significantly lower than initially predicted with CBS-4 M (Δ_fH: 444 kJ mol^{-1}, *Klapötke, 2012*) which appears to be a general issue when using CBS-4 M for ionic compounds.

- N. Fischer, D. Fischer, T. M. Klapötke, D. G. Piercey, J. Stierstorfer, Pushing the limits of energetic materials – the synthesis and characterization of dihydroxylammonium 5,5'-bistetrazole-1,1'-diolate, *J. Mater. Chem.*, **2012**, *22*, 20418–20422.
- T. M. Klapötke, T. G. Witkowski, Z. Wilk, J. Hadzik, Determination of the Initiating Capability of Detonators Containing TKX-50, MAD-X1, PETNC, DAAF, RDX, HMX or PETN as a Base Charge, by Underwater Explosion Test, *Propellants Explos. Pyrotech.* **2016**, *41*, 92–97.
- T. S. Konkova, J. N. Matjushin, E. A. Miroshnichenko, A. F. Asacehnko, P. B. Dzhevakov, Thermochemical Properties of TKX-50, *ICT-JATA*, **2016**, P-90.
- P. Gerber, Properties of Charges based on TKX-50, *16th International Detonation Symposium*, Cambridge, ML, 15–20.7. **2018**.

TMA, thermal mechanical analysis
Thermomechanische Analyse

The TMA serves the measurement of the elongation of a sample body as function of temperature. From this the thermal coefficient of expansion α (K^{-1}) is obtained. Table T.7 depicts typical values for pure HE and formulations:

Table T.7: Linear thermal coefficient of expansion of selected high explosives.

Substance	α (K^{-1})	T range
Baratol	$3.4 \times 10^{-5} + 2.8 \times 10^{-7}$ T	−40 to 60 °C
Comp. B	5.46×10^{-5}	6 to 25 °C
HMX	$2.69 \times 10^{-5} + 6.06 \times 10^{-7}$ T	0 to 90 °C
PBX 9011	2.22×10^{-5}	25 to 74 °C
PBX 9404	4.70×10^{-5}	25 to 70 °C
PBX 9501	4.91×10^{-5}	54 to 74 °C
PETN	$8.55 \times 10^{-5} + 1.82 \times 10^{-7}$ T $+ 6.3 \times 10^{-10}$ T$^2 + 2.17 \times 10^{-12}$ T^3	−160 to 100
RDX	$1.83 \times 10^{-4} + 3.625 \times 10^{-7}$ T $+ 5.48 \times 10^{-10}$ T^2	−100 to 135
TATB	2.36×10^{-4}	−60 to 10
TNT	$5.00 \times 10^{-5} + 7.8 \times 10^{-8}$ T	−40 to 60

- *STANAG 4525*, ed. 1, October **2001** Explosives, Physical/Mechanical Properties, Thermomechanical Analysis for Determining Coefficient of Linear Thermal Expansion (TMA)
- Gibbs Popolato
- R. K. Weese, A. K. Burnham, Coefficient of Thermal Expansion of the Beta and Delta Polymorphs of HMX, *Propellants Explos. Pyrotech.* **2005**, *30*, 344–350.

TNT, trinitrotoluene

Trinitrotoluol

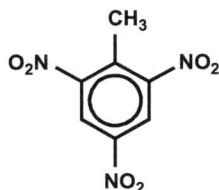

Trotyl		
Aspect		Crème to light yellow crystals
Formula		$C_7H_5N_3O_6$
REACH		LRS
EINECS		204-289-6
CAS		[118-96-7]
m_r	g mol^{-1}	227.133
ρ	g cm^{-3}	1.654
$\rho_{lq@T}$	g cm^{-3}	$1.545 - 1.016 \times 10^{-3}$ T (°C)
$\Delta_f H$	kJ mol^{-1}	−67.07
$\Delta_{ex} H$	kJ mol^{-1}	−1056.8
	kJ g^{-1}	−4.653
	kJ cm^{-3}	−7.696
$\Delta_c H$	kJ mol^{-1}	−3402.2
	kJ g^{-1}	−14.979
	kJ cm^{-3}	−24.775
Λ	wt.-%	−73.96
N	wt.-%	18.50
c_p	J K^{-1} mol^{-1}	314.55
κ	W K^{-1} m^{-1}	0.260
Mp	°C	80.8
Bp	°C	240 (dec)
Friction	N	>360 (BAM)
Impact	J	15 (BAM)
V_D	m s^{-1}	7290 at ρ = 1.65 g cm^{-3}
P_{CJ}	GPa	22 at = 1.65 g cm^{-3}

Trauzl	cm^3	300
$\varnothing_{cr}$	mm	13 at ρ = 1.62 g cm^{-3}
Koenen	mm	5
Specification		STANAG 4025

TNT is the most widely used military HE of the twentieth century. The ease of handling and processing favored by relatively low mechanical sensitiveness and a convenient melting point well below 100 °C facilitates melt casting with machinery heated by hot water. Its broad international use led to it becoming a standard reference HE. Since the 1940s, the detonation energy of both conventional and nuclear warheads gives *TNT equivalents*. TNT is also used as an energetic binder for non-meltable HEs such as GUDN (Guntol), HMX (Octol), nitroguanidine (Nigutol), PETN (Pentolite), and RDX (Comp B, Cyclotol) and eventually also pyrotechnics. Molten TNT is more sensitive to impact and shock than solid TNT.

- T. R. Gibbs, A. Popolato, *LASL Explosive Property Data*, University of California Press, **1980**, pp. 172–187.
- E.-C. Koch, V. Weiser, E. Roth, 2,4,6-Trinitrotoluene: A Surprisingly Insensitive Energetic Fuel and Binder in Melt-Cast Decoy Flare Compositions, *Angew. Chem. Int Ed.* **2012**, *51*, 10038–10040.

TNT equivalent

TNT-Äquivalent

The TNT equivalent serves the approximate performance comparison of HEs with TNT as the reference HE. As different methods exist to evaluate the performance consequently there are also different TNT equivalents which do not compare directly with one another. Important TNT equivalents are based on

- blast pressure (MPa)
- lead block volume aka Trauzl test (cm^3)
- heat of explosion (kJ g^{-1})
- Dent depth in plate-dent test (mm)

And are depicted for selected HEs at given density in Table T.8.

The TNT equivalent, M_{TNT}, (kg), based on any performance parameter of an explosive M_{Ex}(kg) can be determined with the following equation:

$$M_{TNT} = \left(\frac{Parameter(Ex)}{Parameter(TNT)} \right) \cdot M_{Ex}$$

Table T.8: TNT equivalents of selected high explosives.

	RDX	HMX	PETN	Comp B	C4	Octol*
Reference density (g cm^{-3})	1.81	1.906	1.778	1.65		1.81
Blast pressure	1.14–1.18	1.56	1.27	1.11	1.37	1.37
Heat of explosion	1.24	1.26	1.27	1.09	1.20	1.22
Plate dent	6.12	6.04	6.07	5.14		
Trauzl	1.60–1.84	1.42	1.73	1.3		

*90/10 (HMX/TNT).

- M. Held, TNT-Equivalent, *Propellants Explos. Pyrotech.* **1983**, *8*, 158–167.
- M. M. Swisdak, Simplified Kingery Airblast Calculations, *Proceedings of the 26th DoD Explosives Safety Seminar,* Miami, 16–18 August **1994**.
- J. L. Maienschein, Estimating Equivalency Of Explosives Through A Thermochemical Approach, *Preprint UCRL- JC-147683*, **2002**, 14 pp.
- Z. Bajić, J. Bogdanov, R. Jeremić, Blast Effects Evaluation Using TNT Equivalent, *Sci. Tech. Rev.* **2009**, *59*, 50–53.
- V. Karlos, G. Solomos, Calculation of Blast Loads for Application to Structural Components, *Administrative Arrangement No JRC 32253-2011 with DG-HOME Activity A5 – Blast Simulation Technology Development,* Ispra, **2013**, 58 pp.

Tracer

Leuchtspur

To allow optical tracking of a projectile a tracer composition is usually pressed into a cavity in the back of the projectile (small arms and direct fire large caliber (e.g. tank guns).

A typical red tracer composition contains
- 36 wt.-% magnesium
- 44 wt.-% strontium nitrate, $Sr(NO_3)_2$
- 10 wt.-% PVC
- 10 wt.-% chlorinated paraffin

With increasing spin of the projectile both burn rate and light intensity increase (Figure T.4). The burning surface of a composition burning at high spin attains a convex shape. To avoid both blinding the shooter and disclosing the position of fire, a dark burning intermediate composition is pressed on top of the tracer composition (dim tracer/dark tracer). The latter is finally topped with tracer ignition composition. The challenges in the development of a tracer are to achieve reliable ignition and stable burning ignition composition while being exposed to enormous

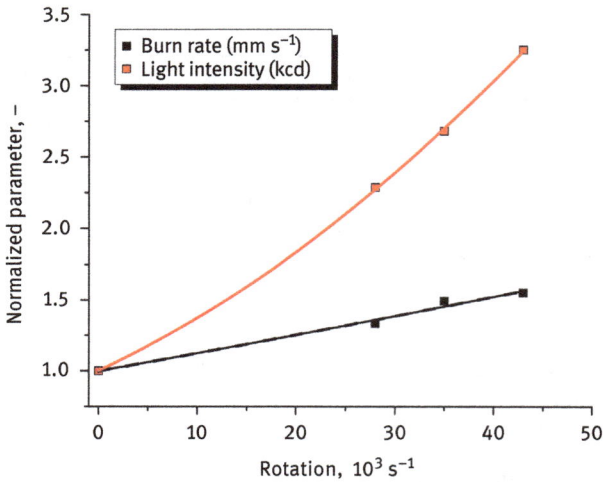

Figure T.4: Burn rate and light intensity as a function of rotational speed with a tracer.

spin and a severe pressure drop from ~400 MPa within the gun barrel down to <0.1 MPa after leaving the barrel in the wake of the projectile.

A typical dark tracer is the following:

- 72 wt.-% barium peroxide, BaO_2
- 10 wt.-% barium sulfate, $BaSO_4$
- 6 wt.-% strontium oxalate, SrC_2O_4
- 5 wt.-% silicon, amorphous
- 3 wt.-% limetree charcoal
- 4 wt.-% calcium resinate

Similar as with *base bleed* charges in indirect fire projectiles, tracers reduce the bottom drag of a conical projectile and hence yield higher terminal velocity and precision.

Triacetone peroxide, TATP

Triacetonperoxid

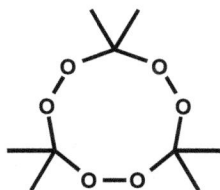

Formula		$C_9H_{18}O_6$
CAS		[17088-37-8]
m_r	g mol^{-1}	222.238
ρ	g cm^{-3}	1.272
$\Delta_f H$	kJ mol^{-1}	−664.5
$\Delta_{ex}H$	kJ mol^{-1}	−902.93
	kJ g^{-1}	−4.063
	kJ cm^{-3}	−5.168
$\Delta_c H$	kJ mol^{-1}	−5449.7
	kJ g^{-1}	−24.522
	kJ cm^{-3}	−31.192
Λ	wt.-%	−151.18
Sub	°C	93.6
$\Delta_{subl}H$	kJ mol^{-1}	80.6
Dp	°C	>150
Friction	N	0.1 (BAM)
Impact	J	2-3 (BAM)
V_D	m s^{-1}	5300 at 1.18 g cm^{-3}
Trauzl	cm^3	250

TATP is a colorless crystalline compound with a distinct spicy smell. Though TATP is a primary explosive, it is not technically used due to its high ambient temperature vapor pressure ($p_{25\,°C} \sim 8$ Pa) effecting migration and recrystallization in thin needles which are extremely sensitive to fracture and detonation. TATP has been misused frequently in terroristic attacks both as primary and main charge explosive. TATP does not dissolve in water but is soluble in many organic solvents like acetone and chloroform. Upon synthesis of TATP other peroxides (e.g. DADP) are usually formed too.

- V. P. Sinditskii, V. I. Kolesov, V. Y. Egorshev, D. I. Patrikeev, O. V. Dorofeeva, Thermochemistry of cyclic acetone peroxides, *Thermochim. Acta* **2014**, *585*, 10–15.

Triamino-2,4,6-trinitrobenzene, TATB

1,3,5-Triamino-2,4,6-trinitrobenzol

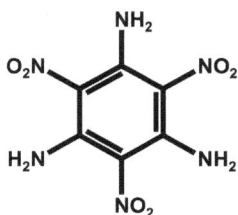

Aspect		Canary-yellow crystals
Formula		$C_6H_6N_6O_6$
REACH		LPRS
EINECS		221-297-5
CAS		[3058-38-6]
m_r	g mol^{-1}	258.15
ρ	g cm^{-3}	1.937
$\Delta_f H$	kJ mol^{-1}	−154
$\Delta_{ex}H$	kJ mol^{-1}	−1136.4
	kJ g^{-1}	−4.402
	kJ cm^{-3}	−8.527
$\Delta_c H$	kJ mol^{-1}	−3064.6
	kJ g^{-1}	−11.872
	kJ cm^{-3}	−22.995
Λ	wt.-%	−55.78
N	wt.-%	32.55
c_p	J K^{-1} mol^{-1}	64.05
Mp	°C	448–449 (accompanied by rapid decomposition)
$\Delta_{sub}H$	kJ mol^{-1}	168
Friction	N	>360 (BAM)
Impact	J	>50 (BAM)
	cm	>320 (tool 12)
V_D	m s^{-1}	7748 at ρ = 1.847 g cm^{-3}
P_{CJ}	GPa	25,9 at ρ = 1.847 g cm^{-3}
Trauzl	cm^3	175
$\varnothing_{cr}$	mm	<4 at ρ = 1.860 g cm^{-3}

TATB is an extremely insensitive high explosive. Due to the strong intermolecular hydrogen bonding in the crystals, TATB is poorly soluble in any common solvent. Even in DMSO at 21 °C, the solubility is only 0.23 g L^{-1} (6.6 g L^{-1} at 148 °C). The solubility increases appreciably in the presence of ionic liquids through the formation of *Meisenheimer-type* complexes (Figure T.5). Mixture of DMSO and ethylmethylimidazolium acetate dissolve 25 g L^{-1} at 21 °C.

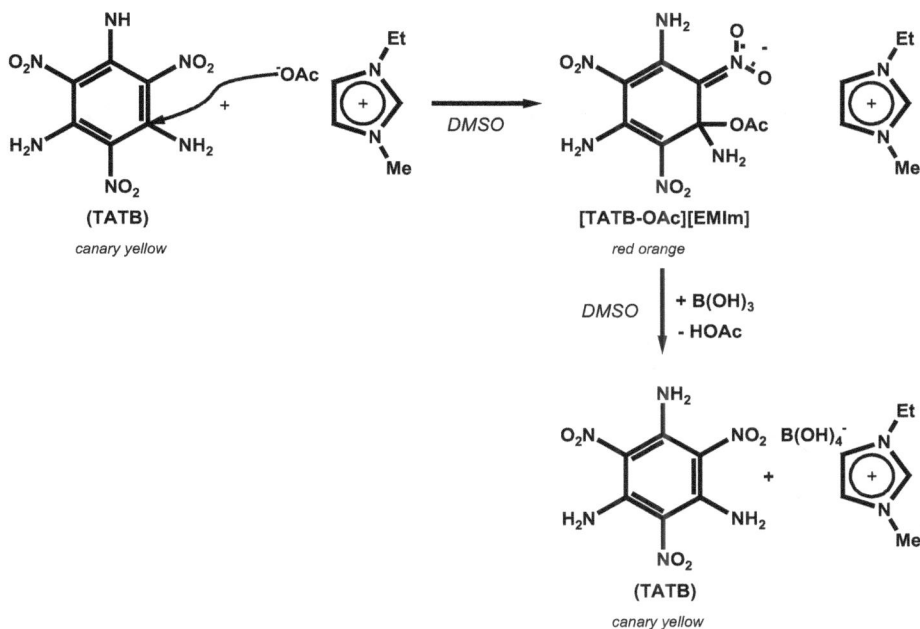

Figure T.5: Dissolution of TATB in DMSO/EMImOAc through formation of Meisenheimer complex [TATB-OAc][EMIm] and subsequent precipitation by addition of boric acid.

Upon impact and shock TATB undergoes endothermal reaction to form various furazans and furoxanes among which CL-14 itself is a substance made deliberately (Fig. T.6). It is hence that TATB is extremely insensitive towards shock waves and shows no reaction up to shock pressure of 6.8 MPa in the NOL-LSGT. TATB prepared sonochemically obtains pores which assist in the adiabatic shock heating (pore collapse). Hence preparation parameters (frequency and power of ultrasonic waves) can be used to adjust the porosity and consequently shock sensitivity of TATB.

- P. Zhao, S. Lee, T. Sewell, H. S. Udaykumar, Tandem Molecular Dynamics and Continuum Studies of Shock-Induced Pore Collapse in TATB, *Propellants Explos. Pyrotech.* **2020**, *45*, 196–222.
- T. Yong-Jin Han, P. F. Pagoria, A. E. Gash, A. Maiti, C. A. Orme, A. R. Mitchell, L. E. Fried, The solubility and recrystallization of 1,3,5-triamino-2,4,6-trinitrobenzene in a 3-ethyl-1-methylimidazolium acetate–DMSO co-solvent system, *New J. Chem.* **2009**, *33*, 50–56.
- B. M. Dobratz, The Insensitive High Explosive Triaminotrinitrobenzene (TATB): Development and Characterization, 1884–1994, *LA-13014-H*, Los Alamos National Laboratory, **1995**, 151 pp.
- J. Sharma, J. W. Forbes, C. S. Coffey, T. P. Liddiard, The Physical and Chemical Nature of Sensitization Centers Left from Hot Spots Caused in Triaminotrinitrobenzene by Shock or Impact, *J. Phys. Chem.* **1987**, *91*, 5139–5144.
- J. Sharma, J. C. Hoffsommer, D. Glover, M. Gibson, C. S. Coffey, J. W. Forbes, T. P. Lippard, W. L. Elban, F. Santiago, Sub-Ignition Reactions at Molecular Levels in Explosives subjected to impact and underwater shock, *8th International Detonation Symposium*, Albuquerque, NM, 15–19 July, **1985**, 725–733.

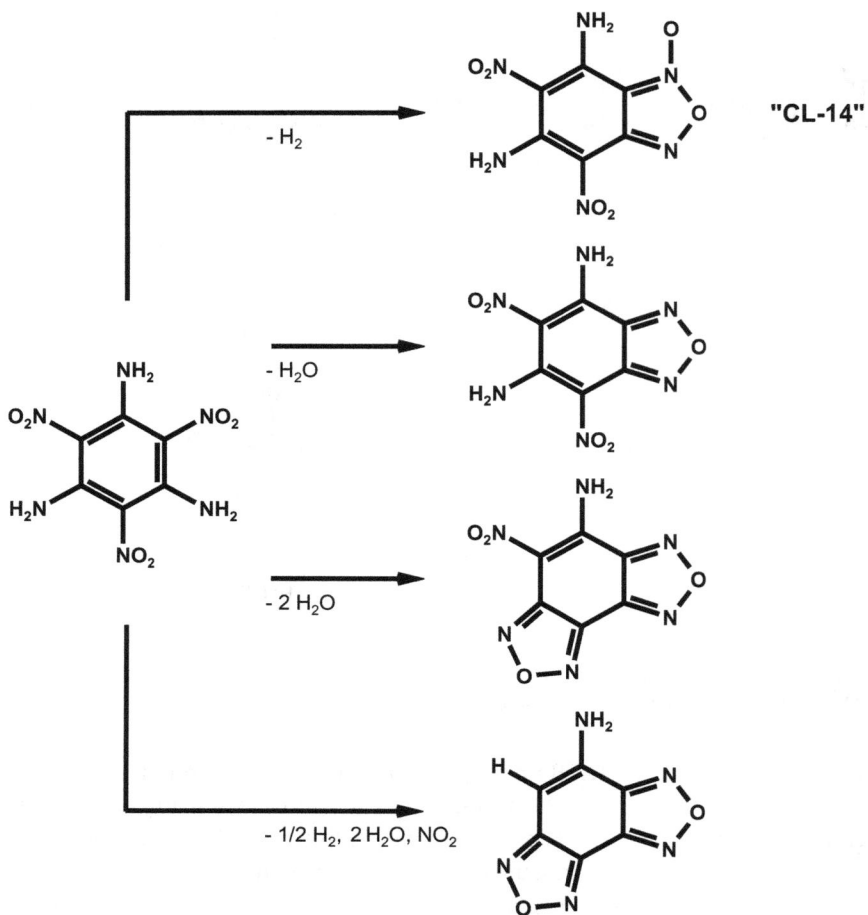

Figure T.6: Reaction products of TATB upon impact and shock after *Sharma et al.*

Triamino-3,5-dinitropyridine, TADNP

2,4,6-Triamino-3,5-dinitropyridin

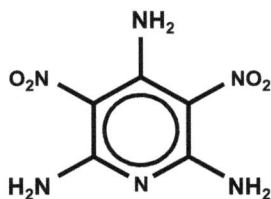

Aspect		Yellow crystals
Formula		$C_5H_6N_6O_4$
CAS		[39771-28-3]
m_r	g mol^{-1}	214.14
ρ	g cm^{-3}	1.819
Δ_fH	kJ mol^{-1}	−110
$\Delta_{ex}H$	kJ mol^{-1}	−813.2
	kJ g^{-1}	−3.798
	kJ cm^{-3}	−6.908
Δ_cH	kJ mol^{-1}	−2715.1
	kJ g^{-1}	−12.679
	kJ cm^{-3}	−23.063
Λ	wt.-%	−67.24
N	wt.-%	39.25
Mp	°C	342 (dec)
P_{CJ}	GPa	26.9 at
		ρ = 1.81 g cm^{-3}

- R. A. Hollins, L. H. Merwin, R. A. Nissan, W. S. Wilson, R. Gilardi, Aminonitropyridines and their N-oxides, *J. Heterocycl. Chem.* **1996**, *33*, 895–904.

Triaminoguanidinium 5,5'-azotetrazolate, TAGzT

Triaminoguanidinium-5,5'-azotetrazolat

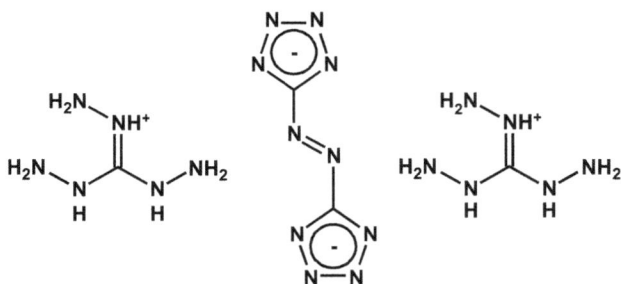

Aspect		Yellow crystals
Formula		$C_4H_{14}N_{22}$
CAS		[2165-23-3]
m_r	g mol^{-1}	374.334
ρ	g cm^{-3}	1.608
$\Delta_f H$	kJ mol^{-1}	1075.3
$\Delta_{ex} H$	kJ mol^{-1}	−1374.4
	kJ g^{-1}	−3.672
	kJ cm^{-3}	−5.904
$\Delta_c H$	kJ mol^{-1}	−5221.9
	kJ g^{-1}	−13.950
	kJ cm^{-3}	−22.431
Λ	wt.-%	−72.66
N	wt.-%	82.32
Mp	°C	195 (dec)
Friction	N	84
Impact	J	4

TAGZT is used as fuel in low-smoke indoor pyrotechnics and due to its high nitrogen content also on gas-generating compositions for airbags.

- N. R. Kumbhakarna, M. Khichar, K. J. Shah, A. Chwodhury, L. Patidar, S. T. Thynell, Liquid-phase decomposition mechanism for bis(triaminoguanidinium) azotetrazolate (TAGzT), *Phys. Chem. Chem. Phys.* **2020**, *22*, 7314–7328.
- N. N. De, N. R. Cummock, I. E. Gunduz, B. C. Tappan, S. F. Son, Photoflash and laser ignition of select high-nirogen materials, *Combust. Flame* **2016**, *167*, 207–217.
- B. Tappan, A. N. Ali, S. F. Son, T. Brill, Decomposition and Ignition of the High-Nitrogen Compound Triaminoguanidinium Azotetrazolate (TAGzT). *Propellants Explos. Pyrotech.*, **2006**, *31*, 163–168.

Triaminoguanidinium dinitramide, TAGN

Triaminoguandiniumdinitramid

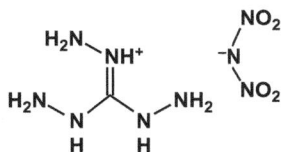

Formula		$CH_9N_9O_4$
CAS		[252062-65-0]
m_r	g mol^{-1}	211.17
ρ	g cm^{-3}	1.581
$\Delta_f H$	kJ mol^{-1}	183
$\Delta_{ex}H$	kJ mol^{-1}	−1149.7
	kJ g^{-1}	−5.445
	kJ cm^{-3}	−8.608
$\Delta_c H$	kJ mol^{-1}	−1862.8
	kJ g^{-1}	−8.822
	kJ cm^{-3}	−13.948
Λ	wt.-%	−18.94
N	wt.-%	59.70
Mp	°C	85
$\Delta_{sm}H$	J g^{-1}	93
Dp	°C	150
Friction	N	12-24 (BAM)
Impact	J	2-4 (BAM)
Koenen	mm	10, type H
V_D	m s^{-1}	5300 at ρ = 0.95 g cm^{-3}

TAGDN is extremely sensitive and even in prilled quality does not meet the requirements for UN Test Series 3 (transport through public). The results of the Koenen test (type H at 10 mm orifice) clearly show TAGDN also to be an extremely hazardous material.

- T. S. Kon'kova, Y. N. Matyushin, E. A. Miroshnichenko, A. B. Vorob'ev, Thermochemical properties of dinitramidic salts, *Russ. Chem. Bull.* **2010**, *58*, 2020–2027.
- T. M. Klapötke, J. Stierstorfer, Triaminoguanidinium dinitramide – calculations, synthesis and characterization of a promising energetic compound, *Phys. Chem. Chem. Phys.* **2008**, *10*, 4340–4346.
- N. Wingborg, N. V. Latypov, Triaminoguanidine Dinitramide, TAGDN Synthesis and Characterization, *Propellants Explos. Pyrotech.* **2003**, *28*, 314–318.

Triaminoguanidinium nitrate, TAGN

Triaminoguanidiniumnitrat

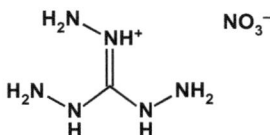

Formula		$CH_9N_7O_3$
REACH		LPRS
EINECS		223-647-2
CAS		[4000-16-2]
m_r	g mol^{-1}	167.128
ρ	g cm^{-3}	1.594
$\Delta_f H$	kJ mol^{-1}	−48.12
$\Delta_{ex} H$	kJ mol^{-1}	−748.4
	kJ g^{-1}	−4.477
	kJ cm^{-3}	−7.136
$\Delta_c H$	kJ mol^{-1}	−1631.8
	kJ g^{-1}	−9.764
	kJ cm^{-3}	−15.563
Λ	wt.-%	−33.51
N	wt.-%	58.67
Dp	°C	215
Friction	N	120 (BAM)
Impact	J	4 (BAM)
V_D	m s^{-1}	5300 at ρ = 0.95 g cm^{-3}
Trauzl	cm^3	350

TAGN due to its low explosion temperature (T_{ex} ~ 2500 K) and the low molecular mass of its gaseous products ($\bar{M}$ = 18.6 g mol^{-1}) was investigated for some time as an ingredient in low-erosive gun propellants.

- N. Kubota, N. Hirata, S. Sakamoto, Decomposition of TAGN, *Propellants Explos. Pyrotech.* **1988**, *13*, 65–68.
- S. Eisele, F. Volk, K. Menke, Gas Generator Materials Consisting of TAGN and Polymeric Binders, *Propellants Explos. Pyrotech.* **1992**, *17*, 155–160.
- F. Volk, H. Bathelt, Influence of Energetic Materials on the Energy-Output of Gun Propellants, *Propellants Explos. Pyrotech.* **1997**, *22*, 120–214.
- U. Bley, A. Hoschenko, P. S. Lechner, *Thermal Pre-Ignition Agent*, US Patent Application 20180127328A1, Germany, **2018**.

Tribromo-2,4,6-trinitrobenzene, TBTNB

1,3,5-Tribrom-2,4,6-trinitrobenzol

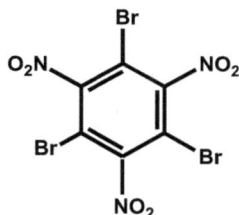

Formula		$C_6Br_3N_3O_6$
CAS		[83430-12-0]
m_r	g mol^{-1}	449.79
ρ	g cm^{-3}	2.39
$\Delta_f H$	kJ mol^{-1}	+209
Ω	wt.-%	−21.34
N	wt.-%	9.34
Mp	°C	297
Dp	°C	337
Impact	J	21.4
V_D	m s^{-1}	6600 at 2.21 g cm^{-3}

TBTNB is relevant only as an intermediate compound in the synthesis (e.g. leading to TATB). While it has both high positive heat of formation and high density it has only marginal detonation properties stressing the importance of low molecular mass of products to achieve high detonation performance which heavy bromine (79.904 u) cannot tender.

- J. C. Bennion, L. Vogt, M. E. Tuckerman, A. J. Matzger, Isostructural Cocrystals of 1,3,5-Trinitrobenzene Assembled by Halogen Bonding, *Cryst. Growth Des.* **2016**, *16*, 4688–4693.

Trichloro-2,4,6-trinitrobenzene, TCTNB

1,3,5-Trichlor-2,4,6-trinitrobenzol

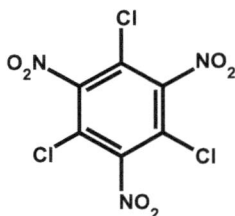

Formula		$C_6Cl_3N_3O_6$
REACH		LPRS
EINECS		220-115-1
CAS		[2631-68-7]
m_r	$g\,mol^{-1}$	316.441
ρ	$g\,cm^{-3}$	1.92
$\Delta_f H$	$kJ\,mol^{-1}$	−150
$\Delta_{ex}H$	$kJ\,mol^{-1}$	−1035.5
	$kJ\,g^{-1}$	−3.272
	$kJ\,cm^{-3}$	−6.283
$\Delta_c H$	$kJ\,mol^{-1}$	−2211.1
	$kJ\,g^{-1}$	−6.987
	$kJ\,cm^{-3}$	−13.416
Λ	wt.-%	−22.75
N	wt.-%	13.28
Mp	°C	193
Bp	°C	318
Impact	J	21.5

TCTNB is an important intermediate in the synthesis of TATB.

- L. Philip, M. Bange, Biphasic production of 2,4,6-triamino-1,3,5-trinitrobenzene (tatab), US-Patent 10,421,709, USA, **2019**.

Tricyanotriazine, TCT

Tricyanotriazin

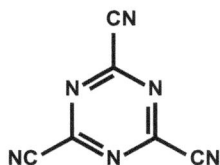

Formula		C_6N_6
CAS		[7615-57-8]
m_r	g mol^{-1}	156.10
ρ	g cm^{-3}	1.391
$\Delta_f H$	kJ mol^{-1}	+657
$\Delta_c H$	kJ mol^{-1}	−3018.1
	kJ g^{-1}	−19.334
	kJ cm^{-3}	−26.894
Λ	wt.-%	−122.99
N	wt.-%	53.837
Mp	°C	119
Bp	°C	215 (dec)

TCT, the formal trimer of pseudohalogen dicyan, is a highly energetic compound that has been tested both as fuel for spectral flares and as precursor for carbon nitride materials.

- E. C. Koch, Experimental Advanced Infrared Flare Compositions, *33rd International Pyrotechnics Seminar*, 16–21 July **2006**, Fort Collins, CO, USA, 71–79.
- S. Fanetti, M. N. Nobrega, K. Dziubek, M. Citroni, A. Sella, P. F. McMillan, M. Hanfland, R. Bini, Structure and reactivity of 2,4,6-tricyano-1,3,5-triazine under high-pressure conditions, *CrystEngComm*, **2019**, *21*, 4493–4500.

Triethylaluminum, TEA

Triethylaluminium

Formula		$Al(C_2H_5)_3$
GHS		02, 05
H-phrases		250-260-314
P-phrases		210-231+232-280-302+334-303+361+
		353-304+340+310-305+351+338-370+378-422
REACH		LRS
EINECS		202-619-3
CAS		[97-93-8]
m_r	$g\ mol^{-1}$	114.167
ρ	$g\ cm^{-3}$	0.837
$\Delta_f H$	$kJ\ mol^{-1}$	−192.05
$\Delta_c H$	$kJ\ mol^{-1}$	−5150.6
	$kJ\ g^{-1}$	−45.115
	$kJ\ cm^{-3}$	−37.762
Λ	wt.-%	−294.30
c_p	$J\ K^{-1}\ mol^{-1}$	239
Mp	°C	−46
Bp	°C	192
$\Delta_m H$	$kJ\ mol^{-1}$	10.6
$\Delta_v H$	$kJ\ mol^{-1}$	73.2

TEA is a colorless limpid pyrophoric liquid. TEA is used as an incendiary (M202 Flame Assault Shoulder) and serves in small quantities as igniter for fuel air explosives. Since the 1970s, TEA was also used as payload in spectrally adapted flares. The spectral energy at an altitude of h = 9000 m, $E_{3.6-4.5\ \mu m}$ = 311 J g^{-1} sr^{-1} and is twice the energy of any other state-of-the-art spectral payload (2020).

- B. Gelin, Använding av pyrofora material i motmedelsfacklor, FOA Rapport C 20471-D1, **1982**, 30 pp.
- B. Gelin, *Pyrofora alkylaluminum föreningar i motmedelsfacklor*, FOA Rapport C 20562-E4, **1984**, 38 pp.

Triethylene glycoldinitrate, TEGDN

Triethylenglykoldinitrat

Formula		$C_6H_{12}N_2O_8$
REACH		LPRS
EINECS		203-847-6
CAS		[111-22-8]
m_r	g mol^{-1}	240.17
ρ	g cm^{-3}	1.327
$\Delta_f H$	kJ mol^{-1}	−629
Λ	wt.-%	−66.62
N	wt.-%	11.66
$\Delta_{ex}H$	kJ mol^{-1}	−1066.0
	kJ g^{-1}	−4.439
	kJ cm^{-3}	−5.890
$\Delta_c H$	kJ mol^{-1}	−3447.1
	kJ g^{-1}	−14.353
	kJ cm^{-3}	−19.046
Mp	°C	−19
Bp	°C	160
p_{vap}	Pa	0.48
Dp	°C	195
Friction	N	n.a.
Impact	J	12
Trauzl	cm^3	320
Specification		AOP-4719

TEDGDN was first suggested by *Gallwitz* as an energetic plasticizer in gun propellants. Due to the lower sensitiveness when compared with DEGDN, TEGDN is used for insensitive gun propellants, for example, *HUX* (Table T.9).

Table T.9: Composition and performance of HUX gun propellant.

Component	(wt.-%)	Performance	
NC	52	ρ (g cm^{-3})	1.59
RDX	11	Heat of explosion (J g^{-1})	4688
NGu	9	Force (J g^{-1})	1065
TEGDN	26	Flame temperature (K)	2820
Ethylcentralite	1	Covolume (cm^3 kg^{-1})	0.1339
Additives	1		

Trifluoro-2,4,6-trinitrobenzene

1,3,5-Trifluor-2,4,6-trinitrobenzol

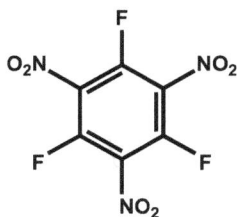

Formula		$C_6F_3N_3O_6$
CAS		[1423-11-6]
m_r	g mol^{-1}	267.08
ρ	g cm^{-3}	1.92
$\Delta_f H$	kJ mol^{-1}	−536
$\Delta_{ex} H$	kJ mol^{-1}	−1131.2
	kJ g^{-1}	−4.235
	kJ cm^{-3}	−8.132
$\Delta_c H$	kJ mol^{-1}	−1822.1
	kJ g^{-1}	−6.823
	kJ cm^{-3}	−13.099
Λ	wt.-%	−35.94
N	wt.-%	15.73
Mp	°C	87
V_D	m s^{-1}	7000 at 1.80 g cm^{-3}

TFTNB is very prone to hydrolysis.

- W. M. Koppes, H. G. Adolph, M. E. Sitzmann, *Process for the Preparation of 1,3,5-Trifluoro-2,4,6-trinitrobenzene*, US-Patent 4,173,591, USA, **1979**.

Triiodo-2,4-6-trinitrobenzene

1,3,5-Triiod-2,4,6-trinitrobenzol

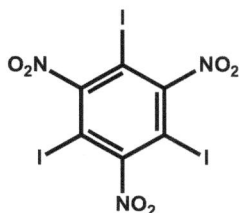

Formula		$C_6I_3N_3O_6$
CAS		[1698044-20-0]
m_r	g mol^{-1}	590.79
ρ	g cm^{-3}	3.057
$\Delta_f H$	kJ mol^{-1}	+1054
Λ	wt.-%	−16.24
N	wt.-%	7.11
I	wt.-%	64.44
Mp	°C	274
Dp	°C	367
Impact	J	7.8
V_D	m s^{-1}	5632 at 2.96 g cm^{-3} (Cheetah 7.0)
P_{CJ}	GPa	24.5 at 2.96 g cm^{-3} (Cheetah 7.0)

TITNB is a colorless crystalline compound with a very high positive heat of formation. It has been considered for ADW applications for detonative release of iodine.
In comparison to the other 1,3,5-trihalogeno (Cl or Br)-trinitrobenzenes it is more
impact sensitive.

- G. Zhao, C. He, D. Kumar, J. P. Hooper, G. H. Imler, D. A. Parrish, J. M. Shreeve, 1,3,5-Triodo
 -2,4,6-trinitrobenzene (TITNB) from benzene: Balancing performance and high thermal stability
 of functional energetic materials, *Chem. Eng.* **2019**, *378*, 122119.
- K. B. Landenberger, O. Bolton, A. J. Matzger, Energetic-Energetic Cocrystals of Diacetone
 Diperoxide (DADP): Dramatic and Divergent Sensitivity Modifications via Cocrystallization,
 J. Am. Chem. Soc. **2015**, *137*, 5074–5079.

Trinitroazetidine, TNAZ

1,3,3-Trinitroazetidin

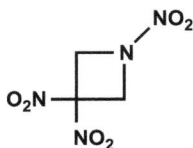

Formula		$C_3H_4N_4O_6$
CAS		[97645-24-4]
m_r	g mol^{-1}	192.088
ρ	g cm^{-3}	1.84
$ρ_{105 °C}$		1.554
$ρ_{120 °C}$		1.522
$Δ_fH$	kJ mol^{-1}	11.72
$Δ_{ex}H$	kJ mol^{-1}	−1172.8
	kJ g^{-1}	−6.105
	kJ cm^{-3}	−11.234
$Δ_cH$	kJ mol^{-1}	−1788
	kJ g^{-1}	−9.309
	kJ cm^{-3}	−17.129
Ω	wt.-%	−16.66
N	wt.-%	29.17
Mp	°C	101
$Δ_mH$	kJ mol^{-1}	30
$Δ_vH$	kJ mol^{-1}	66 at 122 °C
p_{vap}	Pa	50 at 122 °C
DSC-onset	°C	270
Friction	N	160 (BAM)
Impact	J	7.4 (BAM)
V_D	m s^{-1}	7300 at ρ = 1.64 g cm^{-3}
Trauzl	cm^3	325

TNAZ is a melt-castable HE which is more powerful than TNT (Table T.10). Its large-scale use is impeded by high production cost and a high vapor pressure at the melting point (50 Pa at 122 °C). Hence, eutectic components are explored to facilitate casting at lower temperatures.

Table T.10: Comparison of comp B versus ARX 4007.

	Comp B	ARX-4007
RDX (wt.-%)	60	60
TNAZ (wt.-%)	–	40
TNT (wt.-%)	40	–
Wax (wt.-%)	+1	–
ρ (g cm^{-3})	1.68	1.76
V_D (m s^{-1})	7920	8890
P_{CJ} (GPa)	29.5	34.8
Friction (N)	112	72

- D. S. Watt, M. D. Cliff, TNAZ Based Melt-Cast Explosives Technology Review and AMRL Research Directions, *DSTO-TR-0702*, July **1998**.
- N. Liu, S. Zeman, Y. Shu, Z. Wu, B. Wang, S. Yin, Comparative study on melting points of 3,4-bis(3-nitrofurazan-4- yl)furoxan(DNTF)/1,3,3-trinitroazetidine (TNAZ) eutectic compositions with molecular dynamic simulations, *RSC Adv.*, **2016**, *6*, 59141–59149.

Trinitrobenzene, TNB

1,3,5-Trinitrobenzol

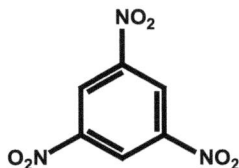

Aspect		Pale yellow green crystals
Formula		$C_6H_3N_3O_6$
REACH		LPRS
EINECS		202-752-7
CAS		[99-35-4]
m_r	g mol^{-1}	213.106
ρ	g cm^{-3}	1.76
Δ_fH	kJ mol^{-1}	−37.24
$\Delta_{ex}H$	kJ mol^{-1}	−1118.9
	kJ g^{-1}	−5.250
	kJ cm^{-3}	−9.241
Δ_cH	kJ mol^{-1}	−2752.9
	kJ g^{-1}	−12.918
	kJ cm^{-3}	−22.736
Λ	wt.-%	−56.31

N	wt.-%	19.72
Mp	°C	123
$\Delta_m H$	kJ mol^{-1}	14.32
Dp	°C	232
Friction	N	>355 (BAM)
Impact	J	7.4 (BAM)
V_D	m s^{-1}	7300 at ρ = 1.71 g cm^{-3}
Trauzl	cm^3	325

Though more powerful than TNT and more stable than latter, the costly synthesis prevents TNB from becoming a substitute for TNT.

4,4,4-Trinitrobutyric acid methylester

γ-Trinitrobuttersäuremethylester

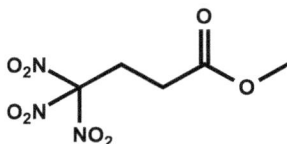

Formula		$C_5H_7N_3O_8$
CAS		[5857-63-6]
m_r	g mol^{-1}	237.124
ρ	g cm^{-3}	1.66 (estimated in accordance with Ammon)
$\Delta_f H$	kJ mol^{-1}	−373.21
$\Delta_{ex} H$	kJ mol^{-1}	−1194.0
	kJ g^{-1}	−5.035
	kJ cm^{-3}	−8.358
$\Delta_c H$	kJ mol^{-1}	−2594.8
	kJ g^{-1}	−10.943
	kJ cm^{-3}	−18.165
Λ	wt.-%	−37.11
N	wt.-%	17.72
Mp	°C	29.25
Bp	°C	90 at 266 Pa
V_D	m s^{-1}	*7573 at TMD*
P_{CJ}	GPa	*23.55 at TMD*

4,4,4-Trinitrobutyric acid methylester is an alleged explosive ingredient in the PFM-1 anti-personnel mine.

- K. Schimmelschmidt, *Verfahren zur Herstellung trinitromethylsusbtituierter Verbindungen*, DE-Patent 852684, Germany, **1952(1944)**.

2,2,2-Trinitroethyl acetate

Trinitroethylacetat

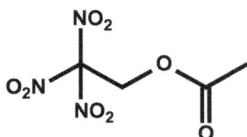

Formula		$C_4H_5N_3O_8$
CAS		[4998-90-7]
m_r	g mol^{-1}	223.10
ρ	g cm^{-3}	1.474
$\Delta_f H$	kJ mol^{-1}	−306
$\Delta_{ex}H$	kJ mol^{-1}	−1196.9
	kJ g^{-1}	−5.365
	kJ cm^{-3}	−7.908
$\Delta_c H$	kJ mol^{-1}	−1982.7
	kJ g^{-1}	−8.887
	kJ cm^{-3}	−13.099
Ω	wt.-%	−17.93
N	wt.-%	18.84
Mp	°C	26
Bp	°C	118 at 0.5 kPa
	°C	64.5 at 0.1 kPa
V_D	m s^{-1}	*7401 at TMD*
P_{CJ}	GPa	*20.6 at TMD*

Trinitroethylacetate is an alleged explosive ingredient in the PFM-1 anti-personnel mine.

- N. S. Marans, R. P. Zelinski, 2,2,2-Trinitroethanol: Preparation and Properties, *J. Am. Chem. Soc.* **1950**, *72*, 5329–5330.
- N. D. Lebedeva, V. L. Ryadnenko, I. N. Kuznetosova, heats of combustion and formation of nitroacetone, ethyl nitroacetate, trinitroacetonitrile, trinitroethanol and trinitroethyl acetate, *Zhur. Fiz. Khim.* **1968**, *42*, 1827–1830.

Triphenylamine

Triphenylamin

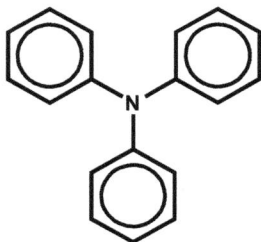

Formula		$C_{18}H_{15}N$
REACH		LPRS
EINECS		210-035-5
CAS		[603-34-9]
m_r	g mol^{-1}	245.324
ρ	g cm^{-3}	0.774
Mp	°C	127
Bp	°C	365
$\Delta_f H$	kJ mol^{-1}	234.72
$\Delta_c H$	kJ mol^{-1}	−9461.8
	kJ g^{-1}	−38.569
	kJ cm^{-3}	−29.853
Λ	wt.-%	−283.7
N	wt.-%	5.71

Triphenylamine is a non-toxic effective stabilizer for NC and compatible with the latter. It reacts more slowly with NGl but more rapidly than Akardite or 2-NDPA. It does not form any carcinogenic products in gun propellants. For double- and triple-base propellants it is probable that requirements in accordance with AOP48 are not met when TPA is used.

- S. Wilker, G. Heeb, B. Vogelsanger, J. Petrzilek, J. Skladal, Triphenylamine – a 'New' Stabilizer for Nitrocellulose Based Propellants – Part I Chemical Stability Studies, *Propellants Explos. Pyrotech.* **2007**, *32*, 135 – 148.

Tris(2-chloroethyl)amine, HN-3

2-Chlor-N,N-bis(2-chlorethyl)ethan-1-amin, Stickstofflost

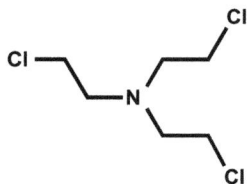

T9, TBA		
Formula		$C_6Cl_3H_{12}N$
CAS		[555-77-1]
m_r	$g\ mol^{-1}$	204.527
ρ	$g\ cm^{-3}$	1.2348
Mp	°C	−4
Bp	°C	230 (dec)
$\Delta_f H$	$kJ\ mol^{-1}$	−256 (estimated)
$\Delta_c H$	$kJ\ mol^{-1}$	−3668.3
	$kJ\ g^{-1}$	−17.936
	$kJ\ cm^{-3}$	−22.147
Λ	wt.-%	−129.07
LCt_{50}	ppm	1500
ICt_{50}	$ppm\ min^{-1}$	200

In the pure state HN-3 is a colorless and odorless oily liquid which has been developed and utilized mainly as blister agent. HN-3 quickly decomposes near its boiling point and also decomposes at prolonged exposure to heat and sunlight.

– S. Franke, *Lehrbuch der Militärchemie, Band 1*, Militärverlag der Deutschen Demokratischen Republik, 2. Auflage, Leipzig, **1977**, pp. 292–304.

Tungsten
Wolfram

Aspect	Silvery
Formula	W
GHS	02
H-phrases	228
P-phrases	210-240-241-280-370+378a
UN	3089
EINECS	231-143-9

CAS		[7440-33-7]
m_r	g mol^{-1}	183.84
ρ	g cm^{-3}	19.300
Mp	°C	3407
Bp	°C	5658
c_p	J mol^{-1} K^{-1}	24.30
$\Delta_m H$	kJ mol^{-1}	35.40
$\Delta_v H$	kJ g^{-1}	806.8
c_L	m s^{-1}	5320
Λ	wt.-%	−17.40
$\Delta_c H(O_2)$	kJ mol^{-1}	−843
$\Delta_c H(F_2)$	kJ mol^{-1}	−1721

Tungsten is a non-toxic heavy metal. It is used as fuel in delay formulations. In alloys with cobalt it serves for DIME charges. Tungsten carbide, WC CAS-No. [12070-12-1], ρ = 15.70 g cm^{-3} is used to produce KE penetrators as armor-piercing projectiles.

Tungsten trioxide
Wolframtrioxid

Aspect		Citron powder
Formula		WO_3
GHS		07
H-phrases		302-315-319-335
P-phrases		261-302+352-305+351+338-321-405-501a
EINECS		215-231-4
CAS		[1314-35-8]
m_r	g mol^{-1}	231.85
ρ	g cm^{-3}	7.16
Mp	°C	1473
Bp	°C	1800
c_p	J mol^{-1} K^{-1}	72.8
$\Delta_m H$	kJ mol^{-1}	73.4
Λ	wt.-%	6.90
$\Delta_f H$	kJ mol^{-1}	−842.9

WO_3 is used as an oxidizer in thermites, nanothermites, and enhanced blast explosives.

– M. Comet, C. Martin, F. Schell, D. Spitzer, Nanothermite foams: From nanopowder to object, *Chem Eng.* **2017**, *316*, 807–812.
– M. Comet, C. Martin, M. Klaumunzer, F. Schell, D. Spitzer, Energetic nanocomposites for detonation initiation in high explosives without primary explosives, *Appl. Phys. Lett.* **2015**, *107*, 243108–1–4.

TZZ

TZZ stands for tungsten–zinc–zirconium and describes a pressed metal powder mixture used for reactive fragments.

– D. P. Chonowski, MS-Thesis, *Small Scaled Reactive Materials Combustion Test Facility*, Urbana, **2011**, 107 pp.

U

Underwater detonation

Unterwasserdetonation

A detonation of a high-explosive charge submerged in water at first yields a shock wave very similar to that in open air. Behind the shock front, the pressure decays exponentially down to a value of P_{max}/e, which is the decay constant. Beyond this point, the pressure decreases significantly slower. The peak pressure P_{max} at any location R, and the decay constant P_{max}/e are dependent on the type of explosive and the mass of the charge, M. Following the initial shock front from the center of detonation with some delay successive shocks of lower peak pressure occur. Those shocks result from the oscillation of the gas bubble of the detonation products. The gas bubble contains the primary detonation products as well as the products from the rather slow reaction of the initial detonation products with the surrounding water. It is hence that for an increase in energy of gas bubble shock waves (short *bubble energy*) those underwater charges often contain fuels reactive with water to give large amounts of hot hydrogen.

The peak pressure, P_{max} (MPa), at the location R (m) correlates with the charge mass M (kg) as follows:

$$P_{max} = K \cdot \left(\frac{\sqrt[3]{M}}{R} \right) \alpha$$

Typical values for underwater explosives (Table U.1).

Table U.1: Detonative properties of various high explosives under water.

High explosive	ρ (g cm^{-3})	K	J	α	Range of validity (MPa)
H-6	1.76	59.2	4.08	1.19	10–138
HBX-1	1.72	56.1–56.7	3.95	1.15–1.37	3–60
HBX-3	1.84	50.3	4.27	1.14	3–60
Pentolite	1.71	56.5	3.52	1.14	3–138
TNT	1.60	52.4	3.50	1.13	3–138

The radius of the gas bubble, A_{max} (m), in a depth, H (m), can be calculated with

$$A = J \left(\frac{\sqrt[3]{M}}{H} \right)_{max}$$

https://doi.org/10.1515/9783110660562-020

The general influence of the ratio of Al/oxygen content of various CHNO high explosives on bubble energy and shock energy is depicted in Figure U.1. To evaluate the shock pressure and gas bubble energy of various high explosives, *Swisdak* has introduced performance figures, equivalent weight shock wave energy (*SWE*) and relative bubble energy (*RBE*), which can be compared with a reference high explosive (pentolite).

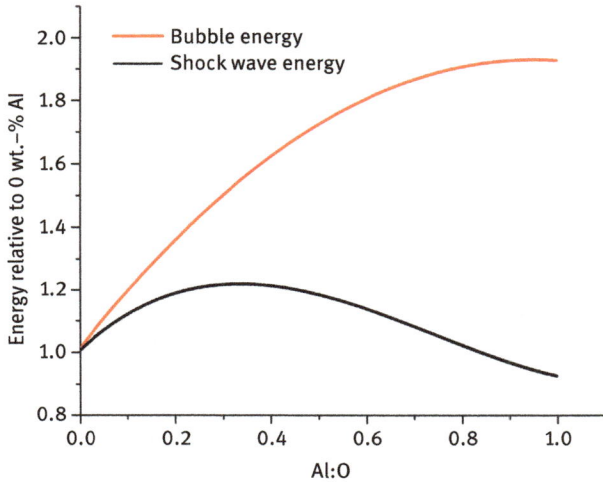

Figure U.1: Influence of Al:O ratio on the bubble and shock energy in underwater detonations after Swisdak.

Doherty showed that both SWE and RBE values can be estimated from the *Kamlet & Jacobs* figures *N*, *M* and the heat of detonation of aluminized high explosives with Al/O ≤ 1.

- R. H. Cole, *Underwater Explosions*, Princeton Press, **1948**, 437 pp.
- M.M. Swisdak, *Explosion Effects and Properties Part II -Explosion Effects in Water, NSWC/NOL TR 76-116*, NSWC, Silver Spring, **1978**, 112 pp.
- D.A. Chichra, R. M. Doherty, Estimation of Performance of Underwater Explosives, *Detonation Symposium*, Portland, **1989**, pp. 633–639.
- P.V. Satyaratan, R. Vedam, Some Aspects of Underwater Testing Method, *Propellants Explos.* **1980**, *5*, 62–66.

Urea nitrate

Harnstoffnitrat, UN

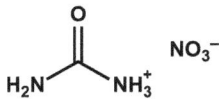

Formula		$CH_5N_3O_3$
REACH		LPRS
EINECS		204–703–5
CAS		[124-47-0]
m_r	g mol^{-1}	123.068
ρ	g cm^{-3}	1.69
$\Delta_f H$	kJ mol^{-1}	−562.75
$\Delta_{ex}H$	kJ mol^{-1}	−366.2
	kJ g^{-1}	−2.975
	kJ cm^{-3}	−5.028
$\Delta_c H$	kJ mol^{-1}	−545.3
	kJ g^{-1}	−4.431
	kJ cm^{-3}	−7.489
Ω	wt.-%	−6.5
N	wt.-%	34.15
Mp	°C	152
Dp	°C	186
Friction	N	>355
Impact	J	50 (250–400 μm)
V_D	m s^{-1}	4700 at 1.20 g cm^{-3} at 30 mm $\varnothing$
P_{CJ}	GPa	26.11 at 1.666 g cm^{-3} at 60 mm $\varnothing$
Trauzl	cm^3	270
Koenen	mm	1, type A

Due to its severe lack of thermal stability, urea nitrate does not find any technical use but has been suggested as an improvised explosive. The reaction of UN with concentrated sulfuric acid yields *nitrourea*.

– Improvised Munitions Handbook, Department of the Army Technical Manual, *TM 31-210*, **1969**, I–13.
– J. C. Oxley, J. L. Smith, S. Vadlamannati, A. C. Brown, G. Zhang, D. S. Swanson, J.Canino, Synthesis and Characterization of Urea Nitrate and Nitrourea, *Propellants Explos. Pyrotech.* **2013**, *38*, 335–344.

UXO

UXO is the acronym for *unexploded ordnance* and summarizes all kinds and conditions of unreacted ammunition found.

- J. Byrne (Ed.), Unexploded Ordnance Detection and Mitigation, Springer, **2009**, 286 pp.

V

Vacuum stability test, VST
Vakuumstabilitätstest

The VST in accordance with STANAG 4556 serves the general determination of the thermal stability of an explosive material or stability (reactivity) of its combinations with other energetic or structural materials (1:1 mixture by weight). The test determines the permanent gas volume developed by 5 g of an explosive (just 0.25 mg in case of a primary explosive!) under heating. Therefore, high explosives, single-base propellants, and pyrotechnics are heated for 40 h at 100 °C, and double-base propellants and nitrate-polyester-based propellants are kept for 40 h at 90 °C. Less than 3 mL gas are uncritical, 3–5 mL are critical, and more than 5 mL are considered to be incompatible.

– *STANAG 4556 PPS (Ed. 1) - Explosives Vaccum Stability Test*, NATO Military Agency for Standardization, Brussels, November **1999**, 20 pp.

Vaporific effect

Upon collision of a projectile, fragment, or a jet from shaped charge with a target structure the vaporization of one or both materials may occur, if the impact velocity exceeds the sonic velocity of at least one material. If the impact occurs in air, then the sheared metal fragments and vapors will ignite to yield a deflagration, the blast and thermal effects of which are responsible for the major damage on the target.

It is hence that the liner materials for shaped charges may contain easily oxidized metals like aluminum, magnesium, zinc, or any alloys of these materials. The vaporific effect is also accountable for the devastative effects of fragments on aluminum structures.

– Warren W. Hillstrom, Impact thresholds for the Initiation of Metal Sparking, *ARBRL-MR-02820*, US Army Armament Research and Development Command Ballistic Research Laboratory, Aberdeen, **1978**, 36 pp.
– M. A. Cook, *The Science of High Explosives*, ACS Publishing, **1958**, pp. 259–263.

Vieille test

The Vieille test requires heating a powder sample to 110 °C in a vial with a litmus paper in the vapor phase above it. The heating is maintained until the litmus turns red. The sample is removed from the vial for 10 h and the test is repeated with a new

https://doi.org/10.1515/9783110660562-021

litmus paper. The test is completed when the heating duration is less than 1 h. The accumulated heating times is a measure of stability.

Vieille, Paul (1854–1934)

Paul Vieille was a French engineer who, together with M. Berthelot, discovered that the shock wave is an essential characteristic of a detonation. He developed the first shock tube to generate shock waves for measurement purposes. He derived a combustion law for gun powder named after him (Vieille's law) and a stability test for nitrocellulose and powders based thereon. On the occasion of the 100th anniversary of Vieille's development of the first French smokeless gun powder "Poudre B" in 1984, a scientific meeting was held in his honor "Journee Paul Vieille" and has since been conducted nine times.

Figure V.1: Paul Vieille, reprinted from Bollé.

- E. Bollé, "Paul Vieille ", *Z. Sch. Spreng.* **1934**, *29*, 323–326 and 368–371.
- Accessed at http://www.af3p.org/fr/Evenements/Congres-et-colloques/Journees-Paul-Vieille.html

Vieille's law
Vieillesches Gesetz

Paul Vieille found that the combustion front progressing along the surface normal of an energetic material (then gun propellant) is a function of temperature and pressure:

$$\frac{dx}{dt} = a \cdot p^n$$

The temperature dependence is described with a coefficient a, while the pressure sensitivity of combustion is described with the exponent n. Both parameters are a function of the composition of the energetic material and cannot be predicted ab initio.

Viton® A

1,1,2,3,3,3-Hexafluoro-1-propen-1,1-difluoroethen-copolymer		
Aspect		Opaque, rubbery
Formula		$[C_3F_6]_n - [C_2H_2F_2]_m$, n = 1 m ~ 3.5
REACH		LPRS
EINECS		618-470-6
CAS		[9011-17-0]
m_r	g mol^{-1}	374.145
ρ	g cm^{-3}	1.75–1.80
$\Delta_f H$	kJ mol^{-1}	−2784
Λ	wt.-%	72.7
LOI	vol.-% O_2	31.5
κ	W K^{-1} m^{-1}	0.226
Mp	°C	n.a.
Dp	°C	>400

Viton® A is the trade name for a copolymer from hexafluoropropene, C_3F_6, and vinylidenefluoride, $C_2H_2F_2$, in a typical molar ratio of the monomers of 13:5. Viton® A is freely soluble in ketones, supercritical CO_2, mixtures of ionic liquids and ketones, but insoluble in non-polar solvents like hydrocarbons. Viton® A has a glass transition temperature of T_g = −27 °C, and starts to decompose at T>400 °C, without having undergone fusion before. Viton® A is used as a binder in high explosives (e.g. PBXN-5) and pyrotechnics (e.g. MTV).

- E.-C. Koch, *Metal-Fluorocarbon Based Energetic Materials*, Wiley-VCH, **2012**, pp. 27–28.
- H. G. Ang, S. Pisharath, *Energetic Polymers*, Wiley-VCH, **2012**, pp. 156–159.

Volk, Fred (1930–2005)

Fred Volk was born in Karlsruhe. He studied chemistry at the Technical University of Karlsruhe and completed his dissertation in 1960 with a thesis on combustion mechanisms. Right then he entered the ICT which had been founded at the same time and started to built-up the analytical laboratory. In 1975, he was promoted to deputy director of the institute. Volk gained much international attention and developed a high reputation as scientist with his extensive research into the thermodynamics and thermochemistry of detonating and deflagrating explosives which built the foundation for the ICT-Thermodynamic Code which he conjointly developed with his collaborator Helmut Bathelt (1928*). In numerous publications and reports, he addressed the chemistry of detonation and in an international collaboration was able to demonstrate the first successful detonative synthesis of nanodiamonds. He retired from ICT in 1999 but remained active in the field thereafter.

- F. Volk, Detonation Products as a Function of Initiation Strength, ambient gas and binder systems of explosives charges, *Propellants Explos. Pyrotech.* **1996**, *21*, 155–159.
- N. R. Greiner, D. S. Phillips, J. D. Johnson, F. Volk, Diamonds in detonation soot, *Nature* **1988**, *333*, 440–442.
- F. Volk, H. Bathelt, Application of the Virial Equation of State in Calculating Interior Ballistic Quantities, *Propellants Explos.* **1976**, *1*, 7–14.
- Entry in the German National Library: http://d-nb.info/gnd/105874256

VX

Ethyl({2-[bis(propan-2-yl)amino]ethyl}sulfanyl)(methyl)phosphinat, EA 1701

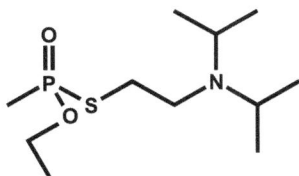

*(RS)-O-*Ethyl-*S*-2-diisopropylamino-ethylmethylphosphonothiolate, TX 60		
Formula		$C_{11}H_{26}NO_2PS$
CAS		[50782-69-9]
m_r	g mol^{-1}	267.367
ρ	g cm^{-3}	1.008
Mp	°C	−51
Bp	°C	298
$\Delta_f H$	kJ mol^{-1}	−740 (estimated)
Λ	wt.-%	−224.40
LCt$_{50}$	ppm	36
ICt$_{50}$	ppm min^{-1}	5

VX is a colorless and odorless oily liquid. VX possess the highest dermal toxicity when compared with any other common nerve agents.

– S. Franke, *Lehrbuch der Militärchemie, Band 1*, Militärverlag der Deutschen Demokratischen Republik, 2. Auflage, Leipzig, **1977**, pp. 438–445.

W

Welding and cladding
Sprengschweißen

In explosive welding or bonding applications, two or more plates are driven together at high velocities, and surface jetting occurs. This results in a bond between the two driven metals and can result in a wavy interface of high strength between the welded materials. If the plate velocities are too high (relative to the material properties), the interface material will jet as a shaped charge, and the bonding will not occur at the interface. The velocity must be such that the plastic deformation occurs at the interface and only local surface jetting occurs. The requirements for welding to occur are:
1. The existence of a jet at the surface.
2. An increase in pressure, associated with the rapid dissipation of kinetic energy, to a sufficient level for a sufficient time to achieve stable interatomic bonds. The pressure is determined by the impact velocity, and the time available for bonding is determined by the velocity of the collision point.

The final geometry of the explosive-metal system will vary depending on the geometry and materials of the parts to be welded. The nature of the interface between the welded parts will vary depending on the materials welded and the nature of the explosive-metal interaction system.

Figure W.1 shows a surface-cladding arrangement. The top plate, or flyer plate, is driven by the explosive force in order to weld a thin sheet of material onto a heavy sheet. The buffer sheet is optional made out of low-density materials, typically plastics. Its task is to dampen the steep and hard shock front generated by the detonation wave. Furthermore, detonation waves usually do have rather rough surfaces, thus also roughen the surface of the clad sheet, which should be avoided. This technique is

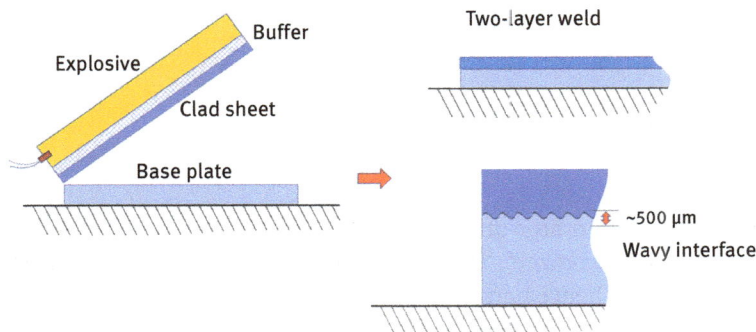

Figure W.1: Angled stand-off cladding.

https://doi.org/10.1515/9783110660562-022

useful in plating a good structural material with a thin cladding to protect it from a corrosive or hazardous environment.

Explosive welding is similar to explosive cladding, but two flyer plates are used to form an explosively welded final product. Figure W.2 depicts an explosively welded sandwich or a three-layer weld. Explosive welding is a solid-phase welding process in which high explosives are used to join the weld surfaces in a high-velocity collision. This collision produces severe, localized plastic flow at the surface between the two surfaces. Explosive welding is used to weld metal combinations, many of which cannot be welded by conventional means.

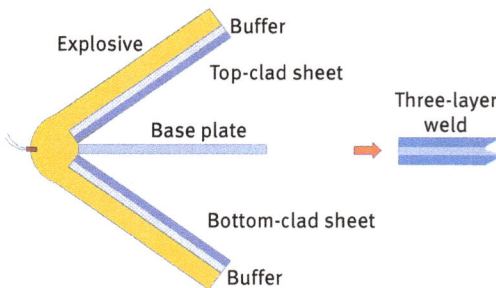

Figure W.2: Three-layer welding.

The weld is usually formed by oblique impact of the plate surface with the weld, progressing from the apex along the collision interface by a process similar to the collapse of a shaped charge liner. The parameters involved in explosive welding include the physical and mechanical properties of the metals to be welded, the type and amount of explosive used, the mode of initiation, the initial geometry of the weld operation, and the type and geometry of buffer sheets (if used) between the metal and the explosive. These parameters influence the collision angle, the impact velocity, and the collision- point velocity.

– W. P. Walters. J. A. Zukas, *Fundamentals of Shaped Charges*. A Wiley-Interscience Publication, USA, **1989**, 111–115.
– P. W. Cooper, *Explosives Engineering*, Wiley-VCH, New York, **1996**, pp. 444–451.

Whistling compositions
Pfeifsätze

Whistling compositions contain aromatic carboxylic acid derivatives as fuels and chlorates or perchlorates as oxidizers. They are near oxygen balanced (similar to the strobe compositions) and hence burn unstable with alternating phases of strong and poor gas evolution. The formulations are extremely sensitive and can yield disastrous explosions when present in larger quantities.

The free-standing consolidated compositions upon combustion yield a hissing sound (see for an exception **B** in Table W.1). Upon combustion under partial confinement in a tube after a phase of strong gas evolution, a rarefaction at the orifice of the tube travels back and impacts on the burning composition thereby triggering the combustion of the next layer. Typical formulations are given in Table W.1. Formulations B yields a whistling sound without any confinement at all. Whistling compositions such as SR136 have been suggested as payloads in spectrally adapted flares.

Table W.1: Selected whistling compositions (in wt.-%).

Components	A	B	C	D*	E	F
Potassium nitrate					30	50
Potassium chlorate	73	73				
Potassium perchlorate			75.2	72		
Gallic acid	24					
Potassium picrate						50
Potassium benzoate				28		
Potassium dinitrophenolate					70	
Sodium salicylate		20	19.8			
Gum Arabic	3					
Paraffin			3			
Vaseline		6				
Iron (III) oxide		1	2			

*SR-136.

–   M. L. Davies, A Review of the Chemistry and Dynamics of Pyrotechnic Whistles, J. Pyrotech. **2005**, *21*, 1–12.

WS-6D, *kyrillic*: ВС-6Д

WS-6D is a liquid explosive used in the PFM-1 anti-personal mine (Figure W.3). Its composition is given in several documents as 1,5-dichloro-3,3-dimethoxy-2,2,4,4-tetranitropentane (no CAS assigned). Other sources give the below information indicating that it is a blend of known explosive substances such as

–   9 wt.-% *Bis(2,2,2-trinitroethyl) formal*
–   49 wt.-% *Bis(2-chloro-2,2-dinitroethyl) formal*
–   21 wt.-% *Trinitromobutyric acid methyl ester*
–   21wt.-% *Trinitroethyl acetate*
–   DOS – 2.5 wt.-% additive.

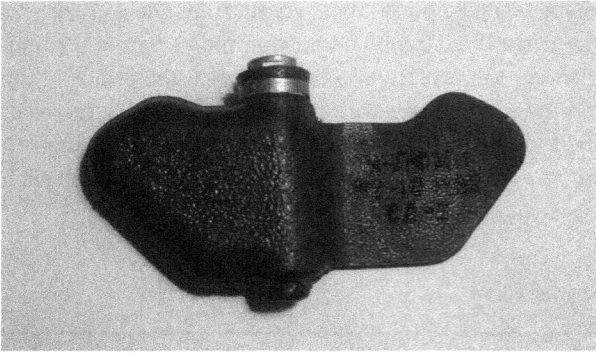

Figure W.3: PFM-1 anti–personal mine containing 37 g WS-6D liquid explosive. https://commons.wikimedia.org/wiki/File:PFM1_mine.jpg.

WTD 91

The WTD 91, *Wehrtechnische Dienststelle für Waffen und Munition,* Engl: Defence technology establishment for weapons and ammunition is the center of competence for weapons and ammunition of the German Armed Forces. With its instrumentation capabilities stretching over 200 km^2, it is by far the largest testing ground in western Europe. Historically, it originates from the Krupp AG shooting range established in 1876. The farthest shooting distance today is 28 km. The staff of the WTD 91 comprises 700 persons. The WTD 91 runs departments in charge of measuring equipment, explosives, weapon–ammunition–missiles and protection, as well as reconnaissance simulation and sensors. The director of WTD 91 is Gerhard Wallrich.

- Accessed at https://www.bundeswehr.de/de/organisation/ausruestung-baainbw/organisation/wtd-91

X

Xenon difluoride

Xenondifluorid

Formula		XeF_2
GHS		03, 05, 06
H-phrases		272–301–330–314
P-phrases		221–301+310–303–361+353–304
		+340–305+351+338–320–330–405–591a
EINECS		237–251–2
CAS		[13709–36–9]
m_r	g mol^{-1}	169.287
ρ	g cm^{-3}	4.32
Mp	°C	127–129
Sub	°C	114
c_p	J mol^{-1} K^{-1}	
$\Delta_{sub}H$	kJ mol^{-1}	51.46
∅	wt.-%	22.45
	g cm^{-3}	0.97
$\Delta_f H$	kJ mol^{-1}	−128 to −173

The colorless, heavily refracting xenon difluoride (XeF_2) is the only commercially available noble gas compound. XeF_2 is stable at ambient temperature and dissolves in water (25 g L^{-1}) and only slowly undergoes decomposition to give HF, Xe, and O_2. While it is mainly used as fluorinating agent in organic synthesis it was already discussed as a potential oxidizer in rocket propulsion soon after its discovery in 1968. It was investigated 2008 for the first time as an oxidizer in pyrotechnics with magnesium as fuel.

- E.-C. Koch, V. Weiser, E. Roth, S. Kelzenberg, Magnesium / Xenon difluoride (MAX) - A New High Energy Density Material, *39th International Annual Conference of ICT*, June 24–27, **2008**, Karlsruhe, pp. 127.1–127.4

Xylokoll®

Xylokoll is the former trade name of Nitrochemie, Aschau for a powdered free-flowing, stabilized nitrocellulose with low nitrogen content. Xylokoll burns slowly and is used in stage pyrotechnics and as a flame expander in flare compositions (VIS/IR).

- A. Hahma, *Wirkmasse für ein beim Abbrand im Wesentlichen spektral strahlendes Infrarotscheinziel mit Raumwirkung*, DE 10 2011 120 454, Germany, **2014**.
- J. Wraige, *Energetic Compositions*, WO9730954, **1997**, UK.
- E. Lohmann, *Pyrotechnischer Satz zur Erzeugung von Lichtblitzen*, DE3402546A1, **1985**, Germany.

https://doi.org/10.1515/9783110660562-023

Y

Ytterbium

Aspect		Silvery
Formula		Yb
GHS		02
H-phrases		261-228
P-phrases		210-231+232-370+378b-402+404
UN		3208
EINECS		231-173-2
CAS		[7440-64-4]
m_r	g mol^{-1}	173.04
ρ	g cm^{-3}	6.965
Mp	°C	824
Bp	°C	1192
c_p	J mol^{-1} K^{-1}	26.74
$\Delta_m H$	kJ mol^{-1}	7.66
$\Delta_v H$	kJ g^{-1}	128.83
c_L	m s^{-1}	1820
$\Delta_c H$	kJ mol^{-1}	−907.25
	kJ g^{-1}	−5.243
	kJ cm^{-3}	−36.518
Λ	wt.-%	−13.87

Atomized ytterbium (<40 μm) oxidizes less quickly than atomized magnesium of comparable size. It hence distinguishes itself from many other rare earth metals which oxidize within days of exposure to air and moisture. Ytterbium satisfies the Glassman criterion for vapor-phase combustion and yields in air and in the presence of F or Cl a distinct green chemiluminescence due to YbO, YbF, and YbCl. Yb-containing flames exhibit a greater spectral emissivity in the mid-IR than comparable Mg flames.

- E.-C. Koch, V. Weiser, E. Roth, S. Kelzenberg, Consideration of some 4*f*-Metals as New Flare Fuels: Europium, Samarium, Thulium and Ytterbium, *ICT -Jata*, **2011**, Karlsruhe, V1.
- E.-C. Koch, V. Weiser, E. Roth, S. Knapp, S. Kelzenberg, Combustion of Ytterbium Metal, *Propellants Explos. Pyrotech.* **2012**, *37*, 9–11.
- E.-C. Koch, A. Hahma, Metal-Fluorocarbon-Pyrolants. XIV High Density-High Performance Decoy Flare Compositions Based on Ytterbium /Polytetrafluorethylene/Viton ®, *Z. Anorg. Allg. Chem.* **2012**, *638*, 721–724.
- E.-C. Koch, V. Weiser, E. Roth, S. Knapp, J. vanLingen, J. Moorhoff, Metal-Fluorocarbon Pyrolants XV Combustion of Two Ytterbium -Halocarbon Formulations, *J. Pyrotech.* **2012**, *31*, 3–9.

https://doi.org/10.1515/9783110660562-024

Z

Zinc
Zink

Aspect		Grey green (as fine powder)
Formula		Zn
GHS		02, 09
H-phrases		250–251–261–400–410
P-phrases		210–222–231+232–280–422a–501a
UN		1436
EINECS		231–175–3
CAS		[7440–66–6]
m_r	g mol^{-1}	65.37
ρ	g cm^{-3}	7.140
Mp	°C	419.6
Bp	°C	907
c_p	J mol^{-1} K^{-1}	26.74
$\Delta_m H$	kJ mol^{-1}	6.67
$\Delta_v H$	kJ g^{-1}	114.2
c_L	m s^{-1}	3700
$\Delta_c H$	kJ mol^{-1}	−350.5
	kJ g^{-1}	−5.362
	kJ cm^{-3}	−38.283
Λ	wt.-%	−24.47

Zinc burns in air with a bluish-white flame accompanied by the formation of copious amounts of dense white smoke of ZnO. Though zinc satisfies the Glassman criterion for vapor-phase combustion, its flame is very dim and hence of no use in illuminating flares. Due to its high reactivity, metallic zinc does not find any use in commercial or military pyrotechnics but is occasionally used in hand-crafted fireworks. *Brown* studied the combustion behavior of various binary zinc/oxidizer mixtures (PbO_2, Pb_3O_4, PbO, SrO_2, BaO_2, and $KMnO_4$).

- M. J. Tribelhorn, D. S. Venables, M. G. Blenkinsop, M. E. Brown, Comparison of Iron and Zinc as pyrotechnic Fuels, *NIXT'94*, Pretoria, **1994**, 180- 190,
- R. Cardwell, Zinc Stars, *Pyrotechnica*, **1977**, *1*, 9–11.
- I. Schmied, Licht- und Farbintensitäten pyrotechnischer Phrases durch Variation der Metallkomponente, *14. Arbeitstagung Wehrtechnik- Chemie und Physik der Explosivstoffe*, **1968**, Mannheim.

https://doi.org/10.1515/9783110660562-025

Zinc peroxide

Zinkperoxid

Formula		ZnO$_2$
GHS		03, 07
H-phrases		272-332-315-319-335
P-phrases		210-221-302+352-305+351+338-405-501a
UN		1516
EINECS		215-226-7
CAS		[1314-22-3]
m$_r$	g mol^{-1}	97.39
$\rho_{25°C}$	g cm^{-3}	5.5
Mp	°C	246 (dec)
$\Delta_f H$	kJ mol^{-1}	−300
Λ	wt.-%	+12.3

Zinc peroxide is used as an alternative oxidizer in primer and igniter formulations, where it has replaced toxic barium, lead, and mercury compounds.

- R. Hagel, K. Redecker, *Verwendung von Zinkperoxid in sprengstoffhaltigen oder pyrotechnischen Gemischen*, DE2952069C2, **1980**, Germany.
- R. Hagel, K. Redecker, *Blei- und Bariumfreie Anzündsätze*, WO 97/1639, **1996** Germany.
- L. Guindon, C. Jalbert, D. Lepage, *Non-toxic, heavy metal-free zinc peroxide-containing IR tracer compositions and IR tracer projectiles containing same generating a dim visability IR trace*, EP2360134A2, **2011**, USA.

Zirconium

Aspect		Silvery grey
Formula		Zr
GHS		02
H-phrases		250-260-252
P-phrases		210-222-231+232-280-422a-501a
UN		2008
EINECS		231-176-9
CAS		[7440-67-7]
m$_r$	g mol^{-1}	91.22
ρ	g cm^{-3}	6.506
Mp	°C	1857
Bp	°C	4200
c$_p$	J mol^{-1} K^{-1}	25.20
$\Delta_m H$	kJ mol^{-1}	20.92

$\Delta_v H$	kJ g^{-1}	561.3
c_L	m s^{-1}	4360
$\Delta_c H$	kJ mol^{-1}	−1097.5
	kJ g^{-1}	−12.031
	kJ cm^{-3}	−78.276
Λ	wt.-%	−35.08

Zirconium (Zr) is a grey powder with Blaine particle sizes in the one-figure μm range. Zr powder is extremely sensitive towards electrostatic discharge with the sensitiveness increasing with decreasing particle size and decreasing hydrogen content. To phlegmatize Zr powder, it is often blended with much less sensitive ZrH$_2$. Likewise, as the sensitiveness the rim burn rate also decreases with increasing particle size and increasing hydrogen content.

Table Z.1 displays the properties of commercial zirconium powder of the manufacturer Rockwood Lithium (previously Chemetall before that Degussa). Zr ignites in air at around 180 °C and burns with a dazzling white flame.

Table Z.1: Zr powder and their properties.

Name	H content (wt.-%)	ESD (µJ)	T_i (°C)	Rim value (50 cm s^{-1})	Particle size (µm)	∅ particle size after Blaine (µm)
Zr CA	<0.2	1.8	180 ± 20	13 ± 5	<45 µm, >99.9%	2.0 ± 0.3
Zr FA	0.3 ± 0.05	3.2	200 ± 30	65 ± 20	<45 µm, >99.9%	2.3 ± 0.5
Zr GA	<0.25	18	240 ± 25	70 ± 20	<45 µm, >99.9%	5.5 ± 1.0
Zr GH	0.8 ± 0.2	56	265 ± 50	460 ± 75	<45 µm, >99.9%	5.5 ± 1.0

http//www.albemarle-lithium.com/fileadmin/media/Global/Documents/PDF-documents/Rockwood-Lithium-Overview-Zirconium-Products-January-2013.pdf

Zr was first introduced to military pyrotechnics by *Feistel* in 1930 who recognized its superior caloric properties and good corrosion resistance.

Due to its high price Zr is not used in consumer fireworks but it is a frequent ingredient in high performance igniters such as in vehicle restraint systems or in military pyrotechnics.

The alloys of Zr with nickel are cheaper than the pure metal and due to greater brittleness also process more easily to give powder. In addition, the presence of another metal with only 1/3 of the heat of combustion helps to adjust the energy content of ZrNi alloys (→).

Phase diagrams of Zr important for pyrotechnics can be found in *Ondiks* monograph. *Kubota* investigated the combustion behavior of Zr Zr/KNO$_3$ while *Rao* studied Zr/NaNO$_3$ mixtures. Zr/BaCrO$_4$ first fires were also investigated by *Kuwahara*. Zr is an advantageous fuel in RP-based obscurants. The ZrP forming in a side reaction is neither prone to hydrolysis by water or even acids and hence does not form obnoxious PH_3.

- T. Kuwahara, C. Tohara, Ignition Characteristics of Zr/BaCrO$_4$ Pyrolant, *Propellants Explos. Pyrotech.* **2002**, *27*, 284–289.
- E.-C. Koch, A. Dochnahl, *Pyrotechnische Wirkmasse zur Erzeugung eines im Infraroten stark emissiven und im Visuellen undurchdringlichen Aerosols*, DE 19914097, **2000**, Germany.
- H. M. Ondik, H. F. McMurdie, *Phase Diagrams for Zirconium and Zirconia Systems*, NIST, **1998**, USA.
- N. Kubota, K. Miyata, Combustion of Ti and Zr Particles with KNO$_3$, *Propellants Explos. Pyrotech.* **1996**, *21*, 29–35.
- R. B. Rao, P. N. Rao, H. Singh, H. A Combustion Study of Metal Powders in Contact with Sodium Nitrate, *Combust. Sci. and Tech.* **1995**, *110–111*, 185–195.

Zirconium hydride

Zirconiumhydrid, Zirkonwasserstoff

Formula		ZrH_2
GHS		02, 07
H-phrases		228-315-319-335
P-phrases		210-241-302+352-305+351+338-405-501a
UN		1437
EINECS		231-727-3
CAS		[7704-99-6]
m_r	g mol^{-1}	93.236
$\rho_{25°C}$	g cm^{-3}	5.61
Mp	°C	180 (dec)
$\Delta_f H$	kJ mol^{-1}	−169
$\Delta_c H$	kJ mol^{-1}	−1217
	kJ g^{-1}	−13.053
	kJ cm^{-3}	−73.227
Λ	wt.-%	−51.48

ZrH_2 is used in igniter formulations that need to be insensitive towards accidental ignition. Typical ZrH_2 qualities are displayed in Table Z.2.

- R. N. Broad, Replacement of First Fire Composition in M127A1 Ground Illumination Signal, Technical Report ARWEC-TR-97002, Picatinny Arsenal, New Jersey, December 1997, 23 pp.

Table Z.2: ZrH$_2$ powder and their properties.

Name	H content (wt.-%)	ESD (µJ)	T$_i$ (°C)	Rim value (50 cm s^{-1})	Particle size (µm)	Ø-particle size after Blaine (µm)
ZrH$_2$-F	>1.4	56		350 ± 60	<45 µm, >99.9%	2.3 ± 0.5
ZrH$_2$-S	>1.9	3200	250±50	600 ± 150	<45 µm, >99.9%	2.6 ± 0.6
ZrH$_2$-G	>1.9	5600	255±55	1300 ± 600	<45 µm, >99.9%	5.5 ± 1.0

http://www.albemarle-lithium.com/fileadmin/media/Global/Documents/PDF-documents/Rockwood-Lithium-Overview-Zirconium-Products-January-2013.pdf

Zirconium–nickel alloy
Zirconium Nickel Legierungen

Zirconium forms eight phases with nickel having the compositions NiZr$_2$–Ni$_5$Zr (Fig. Z.1). Phases with a content of 30% nickel (NiZr$_2$), 50% nickel (NiZr), and 70% nickel (Ni$_7$Zr$_2$) are produced on a large scale for use in pyrotechnics (Table Z.3). As ZrNi is more brittle than pure Zr, it facilitates production of fine powders.

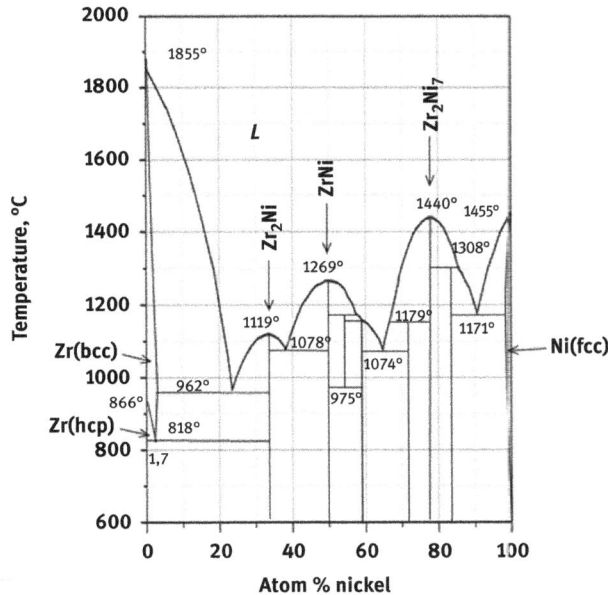

Figure Z.1: Phase diagram of Zr–Ni after *Zaitseva et al.*

Table Z.3: Zr–Ni powder and properties.

Name	H content (wt.-%)	ESD (µJ)	T_I (°C)	Rim value (50 cm s⁻¹)	Particle size (µm)	∅-particle size after Blaine (µm)
70/30 A		0.1	>225	200 ± 75	<45 µm, >99.9%	4 ± 2
70/30 B	<1.5	18	>225	1400 ± 600	<45 µm, >99.9%	4 ± 2
70/30 C		–		1400 ± 500	<45 µm, >99.9%	4 ± 2
30/70 A		3.3	>240	575 ± 200	<45 µm, >99.9%	5 ± 2
30/70 C				800 ± 400	<45 µm, >99.9%	5 ± 2

http//www.albemarle-lithium.com/fileadmin/media/Global/Documents/PDF-documents/
Rockwood-Lithium-Overview-Zirconium-Products-January-2013.pdf

– I. Zaitseva, N. E. Zaitsevaa, E. Kh. Shakhpazova, A. A. Kodentsov, Thermodynamic properties and phase equilibria in the nickel–zirconium system. The liquid to amorphous state transition, *Phys. Chem. Chem. Phys.* **2002**, *4*, 6047–6058.
– *Ellern*, p. 384.

Tables

Plastic-bonded explosives (PBX), pressable

Explosive	CL-20 (wt.-%)	HMX (wt.-%)	RDX (wt.-%)	TATB (wt.-%)	ADNBF (wt.-%)	DATB (wt.-%)	Al (wt.-%)	Binder (wt.-%)	P_{CJ} (GPa)	V_D (m s^{-1})	$\varnothing_{cr}$ (mm)
PBXN-1			68				20	12 Nylon			
PBXN-2		95						5 Nylon			
PBXN-3			86					14 Nylon			
PBXN-4						94		6 Nylon		7200 @ 1.70	
PBXN-5		95						5 Viton A	27 @ 1.86	8820 @ 1.86	
PBXN-6			95					5 Viton		8440 @ 1.77	
PBXN-7			35	60				5 Viton		7690	
PBXN-8			98*					1 Cellulose 1 Stearin			
PBXN-9		92						2 Hytemp 6 DOA	31.0 @ 1.73	8490 @ 1.73	
PBXN-10			94.00					1.5 Hytemp 4.5 DOA		8250 @ 1.69	
PBXN-11		96						1 Hytemp 3 DOA	35.4 @ 1.80	8820 @ 1.80	
PBXN-12 (PBXIH-18)		64.4					30	1.4 Hytemp 4.2 DOA			
PBXC-18		35			55			10 Viton	28.8		

(continued)

https://doi.org/10.1515/9783110660562-026

(continued)

Explosive	CL-20 (wt.-%)	HMX (wt.-%)	RDX (wt.-%)	TATB (wt.-%)	ADNBF (wt.-%)	DATB (wt.-%)	Al (wt.-%)	Binder (wt.-%)	P_{CJ} (GPa)	V_D (m s^{-1})	$\varnothing_{cr}$ (mm)
PBXC-19	95							5 EVA	34.5 @ 1.896	9083 @ 1.896	
PBXW-7			35	60				5 PTFE		7600 @ 1.747	<6.4 @1.84
PBXW-14		50$		45				5 Viton		7540 @ 1.7 98	<3.175
PBXW-16		x						x DOS/Hytemp			

*50% RDX Class 5, 50% RDX Class 7; $37.5 Class 1, 12.5 Class 5.

Plastic-bonded explosives (PBX), cure castable

Explosive	CL-20 (wt.-%)	HMX (wt.-%)	RDX (wt.-%)	NTO (wt.-%)	Al (wt.-%)	AP (wt.-%)	Binder (wt.-%)	P_{CJ} (GPa)	V_D (m s^{-1})	$\varnothing_{cr}$ (mm)
PBXN-101		82					Polystyrene			
PBXN-102		59			23					
PBXN-103					27	40	33 NC		5400	27–29 @ 1.89
PBXN-104		70					30			
PBXN-105			7		25.8	49.8			5180	60–90 @ 1.90
PBXN-106			75				25 PEG/BDNPA/F		7840	3 @ 1.65
PBXN-107			86						8120 @ 1.65	
PBXN-108			82–85				HTPB		8004 @ 1.54	
PBXN-109			64		20		16 HTPB		7500 @ 1.65	5–7 @ 1.66
PBXN-110		88					HTPB, IDP		8330 @ 1.672	5–7 @ 1.67
PBXN-111			20		25	43	HTPB		5597 @ 1.79	80 @ 1.79
PBXN-112		89					11 Lauryl methacrylate			
PBXN-113		45			35		HTPB		6980 @ 1.71	9.5 @ 1.71
PBXC-117			71		17		4.7 EHA, 3.4 DOM		7923 @ 1.752	

(continued)

(continued)

Explosive	CL-20 (wt.-%)	HMX (wt.-%)	RDX (wt.-%)	NTO (wt.-%)	Al (wt.-%)	AP (wt.-%)	Binder (wt.-%)	P_{CJ} (GPa)	V_D (m s^{-1})	$\emptyset_{cr}$ (mm)
PBXC-119		82					18 ?	27.4 @ 1.635	8075 @ 1.635	
PBXC-121		83					12.5 Lauryl methacrylate, 4 DOA		8230 @ 1.62	
PBXC-125			82				16.6 Formrez YA 23-4,1,4LDIM-100		7900 @ 1.6	
PBXC-126		79					8.4 GAP, 8.0 TMETN, 2.5 TEGDN	32.1 @ 1.74	8360 @ 1.74	
PBXIH-134		X		X			Binder			<12.7
PBXIH-136			X		X	X	TMETN, polycaprolactone			
PBXIH-137			82				Binder		7780	<0.925
PBXW-107			64		20		18.55 BDNPA/F, 4.5 PEG, 1.95 others			
PBXW-114		78			10		5.37 HTPB, 5.37 IDP, others		8230 @ 1.72	6.1 @ 1.74
PBXW-119		80					13 FEFO, % others	33.9 @ 1.82	8300 @ 1.82	
PBXW-120		80					15 FEFO, 5 FPF	34 @ 1.82	8500 @ 1.82	
PBXW-121				63	15		12 HTPB			90
PBXW-122			5	47	15	20	7 IDP, HTPB			>127
PBXW-123				44.8	30.2		18.8 TMETN, 6.2 PCL		5500 @ 1.92	180

PBXW-124	20	27	20	20	7 IDP, HTPB			
PBXW-125	20	22	18	20	6.5 IDP, HTPB			
PBXW-126	20	22	26	20	6.5 IDP, HTPB	16	6470 @ 1.80	
PBXW-127		33	50		BTTN, PEG			
PBXW-128	77				11 IDP, 11 HTPB		7973 @ 1.48	<4.7
PBXW-129		38	49		6.5 TMETN, 6.5 PCL			
PBXW-131	88				HTPB		8652	
PBXW-132			?					

*50% RDX Class 5, 50% RDX Class 7.

Plastic-bonded explosives (PBX), extrudable

Explosive	PETN (wt.-%)	RDX (wt.-%)	Binder (wt.-%)	P_{CJ} (GPa)	V_D (m s^{-1})	$\varnothing_{cr}$ (mm)
PBXN-201		83	12 Viton/5 PTFE			
PBXN-202		91	9 EVA			
PBXN-301	80		20 Silicone			0.36

- R. L. Beauregard, Navy Explosives, 219–240.
- G. Antic, V. Dzingalasevic, Characteristics of cast PBX wit aluminium, *Sci. Tech. Rev.* **2006**, *56*, 52–58.
- R. Weinheimer, Properties of Selected Explosives, **IPS, 2000**, Grand Junction, 649–661.
- H. S. Kim, B. S: Park, Characteristics of the Insensitive Pressed Plastic Bonded Explosive, DXD-59, *Propellants Explos. Pyrotech.* **1999**, *24*, 217–220.

German plastic-bonded explosives

Explosive	RDX (wt.-%)	HMX (wt.-%)	HNS (wt.-%)	Al (wt.-%)	AP	Binder (wt.-%)	E_G (m s^{-1})	V_D (m s^{-1})	E-modulus (N mm^{-2})	Application
DXP-1380		92				Hytemp/DOA		8400 @ 1.75		
DXP-1340		96				Hytemp/DOA		8600 @ 1.78		
DXP-1460		94				Hytemp/DOA		1.80		
DXP-2340	96					Hytemp/DOA		8350 @ 1.70		
DXP-2380	92					Hytemp/DOA		8200 @ 1.68		
KS 11	85					HTPB				
KS 13	88					HTPB				
KS 22	67			18		HTPB		7350 @ 1.64	5	Underwater
KS 32		85				HTPB			16	
KS 33		90				HTPB		1.71	30	
KS 51	X			X	X	HTPB				
KS 54c	28			18	40	HTPB				Underwater
KS 56	32			22	32	HTPB				
KS 57/5	24			24	40	HTPB		5620 @ 1.84	2	Underwater

(continued)

(continued)

Explosive	RDX (wt.-%)	HMX (wt.-%)	HNS (wt.-%)	Al (wt.-%)	AP	Binder (wt.-%)	E_G (m s^{-1})	V_D (m s^{-1})	E-modulus (N mm^{-2})	Application
KS 58	X			X	X	HTPB				Underwater
KS 59	X			X	X	HTPB				Underwater
KS 63			96			PU				
Rh 26	90 (i-RDX)					HTPB, IDP	2650	8150 @ 1.66	10.6	155 mm
Rh 29	70			20		HTPB		7900	13.5	155 mm

- P. Wanninger, E. Rottenkolber, E. Kleinschmidt, Detonative Properties of Charges Containing Ammonium Perchlorate, *4e Congres International de Pyrotechnie*, 5–9 June, **1989**, La Grande Motte, France, pp. 55–59
- M. Held, Steady Detonation Velocity D$_\infty$ of Infinite Radius Derived from Small Samples, *Propellants Explos. Pyrotech.* **1992**, *17*, 275–277.
- K. P. Rudolf, Improved Insensitive Hytemp/DOA Bonded HMX and RDX Mixtures by Paste Process, *Insensitive Munitions Energetic Materials Technology Symposium*, March 10–13, **2003** Orlando, FL, USA.
- P. Wanninger, Rh26 – An Insensitive Charge for Gun Ammunition, *36th International Annual Conference of ICT*, 28 June- 1 July **2005**, Karlsruhe, Germany, V-7.
- W. Arnold, Significant Parameters Influencing the Shock Sensitvity Part I and II, *36th International Annual Conference of ICT*, 28 June- 1 July **2005**, Karlsruhe, Germany, P-188.

Fraunhofer-ICT: plastic-bonded explosives (GAP)

Explosive	RDX (wt.-%)	CL-20 (wt.-%)	Al (wt.-%)	AP	Binder (wt.-%)	V_D (m s^{-1})	Gurney-E (m s^{-1})	Bubble-E (10^6 J kg^{-1})	Impact (J)	Friction (N)
GHX 78	67		15		GAP					
GHX 82	27		30 (5 µm)	25	GAP	6766 @ 1.91			2	20
GHX 83	62		20		GAP					
GHX 84	57		25		GAP					
GHX 85	52		30 (5 µm)		GAP	7653 @ 1.87		3.49	5	80
GHX 86	82				GAP			2.06		
GHX 87	42		40		GAP					
GHX 89	27		50		GAP					
GHX 99	47		30 (150 µm)		GAP			3.56		
GHX 100	47		30 (50 µm)		GAP			3.66		
GHX 101	47		30 (5 µm)		GAP			3.65		
GHX 106		27	30	25	GAP	6864 @ 1.95	2420		5	32
GHX 107		22	35	25	GAP	6560 @ 1.96	2215		7.5	36
GHX 116	27		25 (5 µm)	30 (200 µm)	GAP	6732 @ 1.88			3	24
GHX 117	27		25 (5 µm)	30 (15:200 µm, 15:5 µm)	GAP	7072 @ 1.87			4	30

GHX 118*	29	25 (5 μm)	35 (15:200 μm, 20:2.2 μm)	GAP	5	24
GHX 119*	29	25 (50 nm)	35 (15:200 μm, 20:2.2 μm)	GAP	2	16 (explosion)
GHX 145						
GHX 147	75 (I-RDX)			GAP	20	324
GHX 148	75			GAP	15	324

*+1 wt.-% Graphit.

- T. Keicher, G. Langer, T. Rohe, A. Kretschmer, W. Ehrhardt, S. Kölle, A. Happ, Herstellung und Charakterisierung von Underwatersprengstofffen, *30th International Annual Conference of ICT*, 29 June-2 July **1999**, Karlsruhe, P-121.
- G. Langer, T. Keicher, W. Ehrhardt, A. happ, A. Keßler, A. Kretschmer, The influence of particle size of AP and AL on the performance of underwater explosives, *34th International Annual Conference of ICT*, 24–27 June **2003**, Karlsruhe, V-12.
- M. A. Bohn, M. Herrmann, P. Gerber, L. Borne, Investigation oft he change in thermal and shock sensitivity by ageing of RDX charges bonded by HTPB-IPDI and GAP-N100, *IMEMTS*, October **2013**, San Diego, USA.

Fraunhofer-ICT: plastic-bonded explosives (HTPB and others)

Explosive	RDX (wt.-%)	NTO (wt.-%)	HMX (wt.-%)	NGu (wt.-%)	Al (wt.-%)	AP	Binder (wt.-%)	V_D (m s^{-1})	Gurney_E (m s^{-1})	Bubble-E (10^6 J kg^{-1})	Impact (J)	Friction (N)
HX-72	80 (10 µm)						HTPB	7750 @ 1.48				
HX-76	30			55			HTPB	7420 @ 1.55				
HXA-123	70				15		HTPB	7350 @ 1.62				
HXA-171	52				30		HTPB	7278 @ 1.67	2040		15	360
HXA-172	42				40		HTPB	6860 @ 1.72	1840		15	360
HXA-173	32				50		HTPB	6552 @ 1.77	1380		15	360
HXA-174	27				30	25	HTPB	5852 @ 1.70			10	192
HXA-177	67				15		HTPB	7580 @ 1.60	2760		10	360
HXA-178	42				15	25	HTPB	6617 @ 1.63	2230		20	120
HXA-179	27				30	25	Poly-NIMMO	6325 @ 1.87	2230		7.5	36
HXA-180	52				30		Poly-NIMMO	7487 @ 1.80	2550		10	120
HXA-181	42				15	25	Poly-NIMMO	6916 @ 1.78	2970		7.5	40
HXA-182	67				15		Poly-NIMMO	7815 @ 1.72	3030		7.5	144
HXA-192	64				20		HTPB	7100 @				
HXA-193	64				20		HTPB	6980 @				

(continued)

(continued)

Explosive	RDX (wt.-%)	NTO (wt.-%)	HMX (wt.-%)	NGu (wt.-%)	Al (wt.-%)	AP	Binder (wt.-%)	V_D (m s^{-1})	Gurney_E (m s^{-1})	Bubble-E (10^6 J kg^{-1})	Impact (J)	Friction (N)
HXA-194	64				20		HTPB					
HXA-195	64				20		HTPB					
HXA-196	64				20		HTPB					
HXA-197	64				20		HTPB					
HXA-201			64				HTPB					
HXA-202			64				HTPB					
HXA-213			64				HTPB	6800 @ 1.713				
HXA-224			64				HTPB					
HXA-244			X	X	X							
HX 310		25	42	10			HTPB	7750 @ 1.57				
PHX 31	85						Cariflex 1107	7960 @ 1.57				
PHXA-81			65		25		PIB					
PHXA-82			60		30		PIB					

- F. Volk, F. Schedlbauer, Detonation Products of Less Sensitive High Explosives formed under Different Pressures of Argon and in Vacuum, *9th International Detonation Symposium*, **1989**, pp. 962–971.
- S. Cumming, R. W. Torry, D. F. Debenham, B. J. Garaty, Insensitive High Explosives and Propellants – The United Kingdom Approach, *IMEMTS-1994*, **1994**, pp. 348–356.
- P. Lamy, C. O. Leiber, A. S: Cumming, M. Zimmer, Air Senior National Representative Long Term Technology Project on Insensitive High Explosives (IHEs) Studies of High Energy Insensitive High Explosives, *27th International Annual Conference of ICT*, 25 – 28 June, **1996**, Karlsruhe, V-1.

Lawrence Livermore National Laboratory – qualified formulations

LX	$\rho_{20\,°C}$ (g cm^{-3})	HMX	PETN	DATB	HNS	TATB	CL-20	Binder	P_{CJ} (GPa)	V_D (m s^{-1})	$\varnothing_{cr}$ (mm)
01								*			
02			73.5					17.6 BR; 6.9 ATEC, 2 carbosil			
03		70		20				10 Viton			
04	1.865	85						15 Viton	34.5	8470	
07		90						10 Viton			
08	1.42		63.7					34.3 PDMS, 2 carbosil	36.2	8640	
09	1.84	93						7 DNPA/FEFO	37.7	8840	
10	1.864	94.5						5.5 Viton	38.6	8820	
11		95.5						4.5 Estane			
13			80					20 PDMS			
14	1.823	95.5						4.5 Estane	37.4	8800	
15					95			5 Kel-F			
16			96.5					3.5 FPC-461			
17	1.915					92.5		7.5 Kel-F		7600	6.6
18					99.5			0.5 Epoxy			
19	1.927						95.8	4.2 Estane			
20		74						20 PU/6 TMETN			

- ID. Tran, C.M. Tarver, J. Maienschein, P. Lewis, R. Pastrone, R.S. Lee, F. Roeske, Characterization of Detonation Wave Propagation in LX-17 Near the Critical Diameter, *Det. Symp.*, San Diego, **2002**, pp. 684–692.

Literature

Monographies – Book chapters– Reviews

General

M. H. Keshavarz, T. M. Klapötke, *Energetic Compounds/Methods for Prediction of their Performance*, DeGruyter, Berlin, **2020**, 144.

S. Bhattacharya, A. K. Agarwal, T. Rajagopalan, V. K. Patel, (Eds.), *Nano-Energetic Materials*, Springer, Singapore, **2019**, 290.

T. M. Klapötke, *Chemistry of High Energy Materials*, 5. Ed. DeGruyter, Berlin, **2019**, 429.

E.-C. Koch, *Sprengstoffe Treibmittel Pyrotechnika*, 2. Aufl., De Gruyter, Berlin, **2019**, 627.

A. S. Cumming, M. S. Johnson, *Energetic Materials and Munitions: Life Cycle Management, Environmental Impact and Demilitarization*, Wiley-VCH, Weinheim, **2019**, 264.

T. M. Klapötke, *Energetic Materials Encyclopedia*, DeGruyter, Berlin, **2018**, 505.

E.-C. Koch, *Sprengstoffe Treibmittel Pyrotechnika*, 1. Aufl. Lutradyn, Kaiserslautern, **2018**, 476.

M. H. Keshavarz, T. M. Klapötke, *The Properties of Energetic Materials: Sensitivity, Physical and Thermodynamic Properties*, De Gruyter, Berlin, **2018**, 199.

M. Shukla, V. M. Boddu, J. A. Steevens, R. Damavarapu, J. Leszczynski (Eds.), *Energetic Materials – From Cradle to Grave*, Springer, New York, **2017**, 482.

D. Dilhan, *Dictionnaire de Pyrotechnie*, 7. Aufl., AF3P, **2016**, 358.

J. Akhavan, *The Chemistry of Explosives*, Edition 3, RSC, London, **2011**, 193.

J. P. Agrawal, *High Energy Materials*, Wiley-VCH, **2010**, 464.

J. Köhler, R. Meyer, A. Homburg, *Explosives*, 6th Ed, Wiley-VCH, Weinheim, **2007**, 421.

U. Teipel (Eds.), *Energetic Materials*, Wiley-VCH, **2005**, 621.

A. Bailey, S. G. Murray, *Explosives, Propellants and Pyrotechnics*, Brassey's, London, **1989**, 187.

T. R. Gibbs, A. Popolato (Eds.), *LASL Explosive Property Data*, University of California Press, Berkeley, **1980**, 471.

G. Gorst, *Pulver und Sprengstoffe*, Militärverlag der Deutschen Demokratischen Republik, Berlin, **1977**, 227.

B. Fedoroff, S. Kaye (Eds.), *Encyclopedia of Explosives and Related Items, Volume 1 – 10*, Picatinny Arsenal, Dover, USA, **1960–1983**.

T. Urbanski, Chemie und Technologie der Explosivstoffe, 3 Volumes, VEB Deutscher Verlag für Grundstoffindustrie, Leipzig, **1961–1964**.

B. T. Fedoroff, H. A. Aaronson, G. D. Clift, E. F. Reese, *Dictionary of Explosives Ammunition and Weapons (German Section)*, Dover, **1958**, 345.

H. Kast, L. Metz (Eds.), *Chemische Untersuchung der Spreng- und Zündstoffe*, Friedrich Vieweg, Braunschweig, **1931**, 583 pp.

A. Stettbacher, *Schiess- und Sprengstoffe*, 2. Auflage, Verlag von Johann Ambrosius Barth, Leipzig, **1933**, 459 pp.

Ballistics

B. Kneubühl, *Ballistik*, Springer Verlag, Heidelberg, **2018**, 437.

Z. Rosenberg, E. Dekel, *Terminal Ballistics*, Springer, **2012**, 323 pp.

I. G. Assovskiy, *Physics of Combustion and Interior Ballistics*, Nauka, Moscow, **2005**, 357 pp. (in Russian)

https://doi.org/10.1515/9783110660562-027

G. Weihrauch, *Ballistische Forschung im ISL*, ISL, **1994**, 393 pp.

L. Stiefel (Eds.) Gun Propulsion Technology, Volume 109, Progress in Astronautics and Aeronautics, AIAA, **1979**, 563 pp.

H. Krier, M. Summerfeld (Eds.), *Interior Ballistics of Guns*, Volume 66, Progress in Astronautics and Aeronautics, AIAA, **1979**, 384 pp.

E. Schneider, *Beiträge zur Ballistik und Technischen Physik – Gedenkschrift für Hubert Schardin*, Mittler Verlag, **1967**, 327 pp.

W. Wolff, *Raketen und Raketenballistik*, Deutscher Militärverlag, **1964**, 342 pp.

R. E. Kutterer, *Ballistik*, 3. Aufl., Vieweg, **1959**, 304 pp.

W. C. Nelson (Eds.), *Selected Topics on Ballistics, Cranz Centenary Colloquium*, Pergamon Press, **1959**, 280 pp.

H. Athen, *Ballistik*, 2. Ed. Quelle & Meyer, **1958**, 258 pp.

N. N. *Internal Ballistics*, His Majesty's Stationary Office, London, **1951**, 311 pp.

P. Curti, *Äussere Ballistik*, Verlag Huber, Frauenfeld, **1945**, 392.

U. Gallwitz, *Die Geschützladung*, J. Neumann-Neudamm, Berlin, **1944**, 179 pp.

T. Vahlen, *Ballistik*, 2. Ed. deGruyter, Berlin, **1942**, 267 pp.

E. Bollé, G. Seitz, *Einführung in die innere Ballistik*, Freidrich Vieweg, Braunschweig, **1941**, 139 pp.

H. Schardin, *Beiträge zur Ballistik und Technischen Physik*, Verlag von Johann Ambrosius Barth, Leipzig, **1938**, 216 pp.

L. Hänert, *Geschütz und Schuß*, 2. Verb. Auflage, Springer, Berlin, **1935**, 370 pp.

C. Cranz, K. Becker, *Ballistik Bde 1-3* + Ergänzungsband, Springer Verlag, Berlin, **1926–1935**.

Detonations

C. O. Leiber, *Assessment of Safety and Risk with a Microscopic Model of Detonation*, Elsevier, **2004**, 594.

C. L. Mader, *Numerical Modeling of Explosives and Propellants*, 2. Ed., CRC Press, Boca Raton, **1998**, 439.

P. W. Cooper, *Explosives Engineering*, Wiley-VCH, **1996**, 460.

R. Chéret, *Detonation of Condensed Explosives*, Springer, **1993**, 427.

W. Fickett, W. C. Davis, *Detonation*, University of California Press, Berkeley, **1979**, 386.

H. D. Gruschka, F. Wecken, *Gasdynamic Theory of Detonation*, Gordon and Breach, Ney York, **1971**, 198.

C. H. Johansson, P. A. Persson, *Detonics of High Explosives*, Academic Press, London, **1970**, 330.

N.N. *Les Ondes de Détonation*, CNRS, Paris, **1962**, 486.

S. S. Penner, F. A. Williams (Eds.), *Detonation and Two Phase Flow*, Volume 6, Progress in Astronautics and Aeronautics, AIAA, **1962**, 368.

M. A. Cook, *The Science of High Explosives*, American Chemical Society, **1958**, 440.

J. Taylor, *Detonation in Condensed Explosives*, Clarendon Press, Oxford, **1952**, 192.

R. H. Cole, *Underwater Explosions*, Princeton University Press, **1948**, 437.

Pyrotechnics and pyrotechnic ammunition

A. P. G. Shaw, *Thermitic Thermodynamics*, CRC Press, Boca Raton, **2020**, 1067.

J. A. Conkling, C. J. Mocella, *Chemistry of Pyrotechnics*, 3rd Ed., Taylor & Francis, **2019**, 297.

E. Lafontaine, M. Comet, *Nanothermite*, ISTE-Wiley, New York, **2016**, 327.

V. E. Zarko, A. A. Gromov, *Energetic Nanomaterials*, Elsevier, Amsterdam, **2016**, 374.

C. Rossi, *Al-based Energetic Nanomaterials*, ISTE-Wiley, New York, **2015**, 154.

E.-C. Koch, *Metal-Fluorocarbon Based Energetic Materials*, Wiley-VCH, Weinheim, **2012**, 342.

J. A. Conkling, C. J. Mocella, *Chemistry of Pyrotechnics*. 2. Ed., CRC Press, New York, **2011**, 225.

L. Scheit, *German Flare Pistols and Signal Ammunition*, Brad Simpson Publishing, Koblenz, **2011**, 703.

T. Shimizu, *Fireworks*, 4. Ed., Pyrotechnica Publications, Midland, **2010**, 390.

G. Steinhauser, T. M. Klapötke, "Green" Pyrotechncs: A Chemists' Challenge, *Angew. Chem. Int. Ed.* **2008**, 47, 3330–3347.

R. Lancaster, *Fireworks*, 4. Ed., Chemical Publishing, New York, **2006**, 497.

D. Brunel, *Le grand livre des feux d'artifice*, CNRS, Paris, **2004**, 312.

K. Kosanke, B. Kosanke, *Pyrotechnic Chemistry*, Journal of Pyrotechnics, Whitewater, CO, **2004**.

A. P. Hardt, *Pyrotechnics*, Pyrotechnica Publications, Post Falls, **2001**, 429.

M. S. Russell, *The Chemistry of Fireworks*, 2. Ed. RSC, London, **2009**, 169.

R. T. Barbour, *Pyrotechnics in Industry*, McGraw Hill, New York, **1981**, 190.

J. H. McLain, *Pyrotechnics*, The Franklin Institute Press, **1980**, 243.

T. Shimizu, *Feuerwerk vom physikalischen Standpunkt aus*, Hower Verlag, Hamburg, **1976**, 252.

K. O. Brauer, *Handbook of Pyrotechnics*, Chemical Publishing Company, New York, **1974**, 402.

H. Ellern, *Military and Civilian Pyrotechnics*, Chemical Publishing Company, **1968**, New York, 464.

E. W. Lawless, I. C. Smith, *Inorganic High-Energy Oxidizers*, Marcel Dekker Inc, New York, **1968**, 304.

T. F. Watkins, J. C. Cackett, R. G. Hall, *Chemical Warfare, Pyrotechnics and the Fireworks Industry*, Pergamon Press, Oxford, **1968**, 114.

A. A. Shidlovski, *Basic Pyrotechnics*, **1965**.

J. C. Cakett, *Monograph on Pyrotechnic Compositions*, Ministry of Defence, **1965**, 131.

G. W. Weingart, *Pyrotechnics*, 2nd Ed., Chemical Publishing Company, New York, **1947**, 244.

H. B. Faber, *Military Pyrotechnics*, 3 Volumes, Government Printing Office, Washington, **1919**.

Missile technology and -propulsion

R. H. Schmucker, M. Schiller, *Raketenbedrohung 2.0*, Mittler, Hamburg, **2015**, 407.

N. Kubota, *Propellants and Explosives*, 3. Ed. Wiley-VCH, Weinheim, **2015**, 534.

G. D. Roy, *Advances in Chemical Propulsion*, CRC Press, Boca Raton, **2002**, 528.

F. S. Simmons, *Rocket Exhaust Plume Phenomenology*, The Aerospace Press, El Segundo, **2000**, 286.

G. P. Sutton, *Rocket Propulsion Elements*, Wiley, New York, **1992**, 636.

R. T. Holzmann, *Chemical Rockets and Flame and Explosives Technology*, Marcel Dekker, New York, **1969**, 449.

A. Dadieu, R. Damm, E. W. Schmidt, *Raketentreibstoffe*, Springer, Vienna, **1968**, 805.

R. T. Holzmann, *Advanced Propellant Chemistry*, American Chemical Society, **1966**, 290.

J. Taylor, *Solid Propellent and Exothermic Compositions*, Interscience, New York, **1959**, 153 pp.

A. J. Zaehringer, *Solid Propellant Rockets*, 2. Ed., American Rocket Corp. Wyandotte, USA, **1958**, 306.

H. G. Mebus, *Berechnung von Raketentriebwerken*, C. F. Wintersche Verlagsbuchhandung, Füssen, **1957**, 120.

Chemistry, Synthesis and Theory of explosives

D. S. Viswanath, T. K. Ghosh, V. M. Boddu, *Emerging Energetic Materials: Synthesis, Physicochemical, and Detonation Properties*, Springer, New York, **2018**, 478 pp.

S. Venugopalan, *Demystifying Explosives: Concepts in High Energy Materials*, R. Sivabalan (Eds.), Elsevier, Amsterdam, **2015**, 224 pp.

T. Brinck (Eds.), *Green Energetic Materials*, Wiley, New York, **2014**, 290.

J. R. Sabin (Eds.), *Advances in Quantum Chemistry – Energetic Materials*, Elsevier, Amsterdam, **2014**, 344.

R. Matyas, J. Pachman, *Primary Explosives*, Springer, Berlin, **2013**, 338.

H. G. Ang, S. Pisharath, *Energetic Polymers*, Wiley-VCH, Weinheim, **2012**, 218.

M. R. Manaa, C.-S. Yoo, E. J. Reed, M. S. Strano, *Advances in Energetic Materials Research*, Volume 1405, MRS, Pittsburgh, **2011**, 165.

J. P. Agrawal, R. D. Hodgson, *Organic Chemistry of Explosives*, Wiley, **2007**, 384.

T. M. Klapötke (Eds.), *High Energy Density Materials*, Volume 125 of *Structure and Bonding*, Springer, Berlin, **2007**, 286.

N. N., *Multifunctional Energetic Materials*, Volume 896, MRS, Pittsburgh, **2005**, 242.

P. Politzer, J. S. Murray (Eds.) Energetic Materials Part 2. Detonation, Combustion, Elsevier, Amsterdam, **2003**, 453.

P. Politzer, J. S. Murray (Eds.) Energetic Materials Part 1. Decomposition, Crystal and Molecular Properties, Elsevier, Amsterdam, **2003**, 465.

N. N., *Synthesis, Characterization and Properties of Energetic/Reactive Nanomaterials*, Volume 800, MRS, Pittsburgh, **2003**, 366.

T. B. Brill, T. P. Russell, W. C. Tao, R. B. Wardle (Eds.) *Decomposition, Combustion, and Detonation Chemistry of Energetic Materials*, Volume 418, MRS, Pittsburgh, **1996**, 454.

A. T. Nielsen, *Nitrocarbons*, Wiley-VCH, Weinheim, **1995**, 190.

D. H. Liebenberg, R. W. Armstrong, J. J. Gilman (Eds.), *Structure and Properties of Energetic Materials*, Volume 296, MRS, **1992**, Pittsburgh, 390.

G. Olah, D. R Squire, *Chemistry of Energetic Materials*, Academic Press, San Diego, **1991**, 212.

S. N. Bulusu (Eds.), *Chemistry and Physics of Energetic Materials*, Kluwer Academic Publishers, **1990**, 764.

H. Feuer, A. T. Nielsen (Eds.), Nitrocompounds, Wiley-VCH, Weinheim, **1990**, 636.

G. A. Olah, R. Malhotra, S. C. Narang, *Nitration*, Wiley-VCH, Weinheim, **1989**, 330.

K. B. G. Torssell, *Nitrile Oxides, Nitrones, and Nitronates in Organic Synthesis*, Wiley-VCH, Weinheim, **1988**, 332.

J. H., Boyer, *Nitroazoles*, Wiley-VCH, Weinheim, **1986**, 368.

K. Schofield, *Aromatic Nitration*, Cambridge University Press, **1980**, 376.

Testing, reactivity, stability, thermochemistry

A. Koleczko, N. Eisenreich, A. B. Vorozhtsov (Eds.), *Phenomena in Combustion of Propellants and Explosives*, Fraunhofer Verlag, Stuttgart, **2017**, 194.

S. R. Ahmad, M. Cartwright, *Laser Ignition of Energetic Materials*, Wiley, New York, **2015**, 283.

A. S. Shteinberg, *Fast Reactions in Energetic Materials*, Springer Verlag, Berlin, **2008**, 201.

M. Sućeska, *Test Methods for Explosives*, Springer, New York, **1995**, 225.

K. K. Andrejev, *Thermische Zersetzung und Verbrennungsvorgänge bei Explosivstoffen*, Erwain Barth Verlag, Mannheim, **1964**, 150.

3D-Printing (*Additive Manufacturing*) of pyrotechnics, propellants- & explosives

M. H. Straathof, C. A. van Driel, J. N. J. van Lingen, B. L. J. Ingenhut, A. T. ten Cate, H. H. Maalderink, Development of Propellant Compositions for Vat Photopolymerization Additive Manufacturing, *Propellants Explos. Pyrotech.* **2020**, *45*, 36–52.

L. J. Groven, M. J. Mezger, Printed Energetics: The Path toward Additive Manufacturing of Munitions. In *Energetic Materials – Advanced Processing Technologies for Next-Generation Materials* CRC Press, **2017**, pp. 115–128.

A. K. Murray, T. Isik, V. Ortalan, I. E. Gunduz, S. F. Son, G. T. C. Chiu, J. F. Rhoads, Two-component additive manufacturing of nanothermite structures via reactive inkjet printing. *J. Appl. Phys.* **2017**, *122*, 184901-1–184901-5.

M. Sweeney, L. L. Campbell, J. Hanson, M. L. Pantoya, G. F. Christopher, Characterizing the feasibility of processing wet granular materials to improve rheology for 3D printing. *J. Mater. Sci.*, **2017**, *52*, 13040–13053.

T. J. Fleck, A. K. Murray, I. E. Gunduz, S. F. Son, G. T.-C Chiu, J. F. Rhoads, Additive manufacturing of multifunctional reactive materials, *Additive Manufacturing*, **2017**, *17*, 176–182.

A. K. Murray, W. A. Novotny, T. J. Fleck, I. E. Gunduz, S. F. Son, G. T. C. Chiu, J. F. Rhoads, Selectively-deposited energetic materials: A feasibility study of the piezoelectric inkjet printing of nanothermites. *Additive Manufacturing*, **2018**, *22*, 69–74.

J. van Lingen, M. Straathof, C. van Driel, A. den Otter, 3D printing of Gun Propellants, *43rd International Pyrotechnics Seminar*, 8–13. July **2018**, Fort Collins, CO, USA, 129–141.

Chemical Warfare Agents & Fuels

M. H. Keshavarz, *Liquid Fuels as Jet Fuels and Propellants: A Review of their Productions and Applications*, Nova Publishers, **2018**, 175.

M. H. Keshavarz, *Combustible Organic materials*, DeGruyter, Berlin, **2018**, 220.

J. A. Romano Jr., B. J. Lukey, H. Salem, (Eds.) *Chemical Warfare Agents: Chemistry, Pharmacology, Toxicology and Therapeutics*, CRC Press, **2007**, 732.

T. C. Marrs, R. L. Maynard, F. R. Sidell (Eds.), *Chemical Warfare Agents Toxicology and Treatment*, Wiley, New York, **2007**, 738.

J. F. Bunnett, M. Mikolajczyk (Eds.), *Arsenic and Old Mustard: Chemical Problems in the Destruction of Old Arsenical and Mustard Munitions*, Kluwer Academic Publishers, **1998**, 200.

K. Lohs, W. Spyra, *Chemische Kampfstoffe als Rüstungsaltlasten mit einem Anhang von W. Bretschneider*, EF-Verlag, Munich, **1992**, 314.

S. M. Somani, *Chemical Warfare Agents*, Academic Press, San Diego, **1992**, 443.

R. Stöhr, *Chemische Kampfstoffe und Schutz vor chemischen Kampfstoffen*, Militärverlag der Deutschen Demokratischen Republik, Berlin, **1977**, 446.

S. Franke (Eds.) *Lehrbuch der Militärchemie, Band 1 und 2*, Militärverlag der Deutschen Demokratischen Republik, 2. Auflage, Berlin, **1977**, 512+615.

Forensics & Detection of Explosives

K. Evans-Nguyen, H. Hutches (Eds.), *Forensic Analysis of Fire Debris and Explosives*, Springer, New York, **2020**, 364.

J. Yinon (Ed.), *Counterterrorist Detection Techniques of Explosives*, Elsevier, Amsterdam, **2007**, 440.

R. L. Woodfin (ed.), *Trace Sensing of Explosives*, Wiley, New York, **2007**, 363.

H. Schubert, A. Rimski-Korsakov (Eds.), *Stand-off Detection of Suicide Bombers and Mobile Subjects*, Springer, Dordrecht, **2006**, 165.

H. Schubert, A. Kuznetsov, *Detection and Disposal of Imprivised Explosives*, Springer, Dordrecht, **2006**, 239.

J. Yinon, S. Zitrin, *Modern Methods an Applications in Analysis of Explosives*, Wiley, New York, **1996**, 305.

Conferences

Airbag Symposium
AIAA Propulsion and Energy Forum and Exposition (AIAA)
Ballistics Symposium (Ball Symp)
Combustion Symposium (Comb Symp)
Dinitramide & FOX-Meeting, last held in **2011**.
Gordon Research Conference on Energetic Materials (GRC)
ICT-Annual Conference (ICT-JaTa)
International Autumn Seminar on Pyrotechnics Explosive Propellants (IASPEP)
International Detonation Symposium (IDS)
International High Energy Materials Conference & Exhibits (HEMCE)
International Heat Flow Calorimetry Symposium on Energetic Materials (HFCEM)
International Fireworks-Symposium (IFWS)
Insensitive Munitions and Energetic Materials Symposium (IMEMTS)
International Pyrotechnics Seminar (IPS)
Korean International Symposium on High Energy Materials (KISHEM)
New Trends in Research of Energetic Materials (NTREM) (http://www.ntrem.com/)
Nitrocellulose Symposium
Ordnance, Munitions and Explosives Symposium (OME)
Workshop on Pyrotechnic Combustion Mechanism (WPC) (https://www.lutradyn.com/home/wpc/)

Periodics with ISSN

Journal of Pyrotechnics, 1995–2016, not published since, 1082–3999
Pyrotechnica, 1977–1994, publication suspended, 0272–6251
Explosivstoffe, 1952–1974, publication suspended, 0014–505X
Propellants Explosives Pyrotechnics, 0721–3115
Central European Journal of Energetic Materials, 1733–7178
Journal of Energetic Materials, 0737–0652
Combustion Explosion and Shock Wave, 0010–5082
Combustion and Flame, 0010–2180
Thermochimica Acta, 0040–6031
Journal of Hazardous Materials, 0304–3894
Journal of Propulsion and Power, 0748–4658
Sprenginfo, 0941–4584
Nobel-Hefte, 0029–0858
Defence Technology, 2214–9147
Combustion Science and Technology, 0010-2202
Zeitschrift für das gesamte Schieß- und Sprengstoffwesen, 1906–1944, publication suspended, 0372–8935

Index

CAS-No Index

https://doi.org/10.1515/9783110660562-029

Sum Formula Index

https://doi.org/10.1515/9783110660562-030